Robotic Exploration of the Solar System
Part 1: The Golden Age 1957–1982

Paolo Ulivi with David M. Harland

Robotic Exploration of the Solar System

Part 1: The Golden Age 1957–1982

 Springer

Published in association with
Praxis Publishing
Chichester, UK

Dr Paolo Ulivi
Cernusco Sul Naviglio
Italy

Dr David M. Harland
Space Historian
Kelvinbridge
Glasgow
UK

SPRINGER–PRAXIS BOOKS IN SPACE EXPLORATION
SUBJECT *ADVISORY EDITOR*: John Mason B.Sc., M.Sc., Ph.D.

ISBN 978-0-387-49326-8 Springer Berlin Heidelberg New York

Springer is a part of Springer Science + Business Media (*springer.com*)

Library of Congress Control Number: 2007927751

Cover design: Jim Wilkie
Copy editing: Alex Whyte
Typesetting: BookEns Ltd, Royston, Herts., UK

Printed in Germany on acid-free paper

To Elena Sofia
who will see the next return of Halley

Table of contents

List of illustrations. ix
List of tables xvii
Foreword. xix
Author's preface. xxi
Acknowledgments. xxiii

Introduction xxv
Mercury: extremes of heat and cold xxv
Venus: a swamp or a greenhouse?. xxvii
Mars, life and the 'canali' xxxii
Jupiter: a ball of hydrogen xxxix
Saturn, its rings and moons xliii
Uranus and Neptune: outer giants xlv
Pluto: the incredible shrinking planet l
Asteroids: those fantastic points of light lii
Comets: flying sandbanks or dirty snowballs?. liii
Phantoms: Vulcan, trans-Plutonian planets and the like lv

1. **The beginning** 1
Space race 1
Humans or robots?. 2
The first 'artificial planets' 5
The first interplanetary probe. 5
The first JPL projects. 9
The first Soviet probes 12
The first success 18
Product 2MV. 26
The 'Zond' probes 31
Farewell to the 'little green men'. 33
Korolyov's last probes 45

Solar probes. 48
Together to Venus 52
A Voyager without sails 65
A repeat mission. 70
Mars again. 73
Other players 88

2.   Of landers and orbiters 97
A new decade. 97
To the surface! 97
Into the storm 99
A first look beyond the asteroids 125
The taste of Venus 156
The curse of the transistor 160
Soviet soil from the Red Planet 167
The planet of contradictions. 171
Hot and hotter. 196
Snowballs will wait. 206
Postcards from Hell 209
Landing in Utopia 216
Pigeons, rovers, sniffers 256
The Venusian fleet 262
The color of Venus. 284
'Purple Pigeons' from the cold 289

3.   The grandest tour 301
The journey of three lifetimes. 301
Grand Tour reborn 309
The spacecraft that could 311
Launch and teething troubles 318
Jupiter: ring, new moons and volcanoes!. 323
The return to Jupiter: life, perhaps? 346
Saturn and mysterious Titan. 363
The final one–two punch 382
Dull planet, incredible moon 398
To a blue planet. 422
The larger perspective. 441

Glossary 457
Appendices. 465
Chapter references. 477
Further reading 523
Index. 525

List of illustrations

A 1958 artist's impression of Mercury. xxviii
An artist's impression of Mars as seen from Deimos. xxxvi
An artist's impression of the south polar cap of Mars. xxxviii
An artist's impression of Jupiter and its Great Red Spot. xlii
A ring around Saturn as inferred by Christiaan Huygens xliv
An artist's impression of Uranus as seen from Miranda xlvii
How Neptune was inferred from the motions of Uranus. xlix
Pioneer 4. 3
Pioneer 5. 6
Pioneer 5's instrument platform. 8
A 1959 drawing of NASA's planned standard launchers 10
Mariner A. 11
Mariner B. 12
The Molniya rocket. 13
A model of Venera 1. 17
A line drawing of Mariner R folded for launch. 20
Preparing Mariner R. 21
Mariner 2's microwave radiometer 23
Mariner 2's plasma spectrometer. 24
Radiometer scans of Venus by Mariner 2 25
The Soviet 2MV-1 spacecraft. 27
Mars 1: the 2MV-4 configuration 29
Zond 1: the 3MV-1 configuration 32
Preparing a Mariner C probe for launch 36
A line drawing of Mariner 4 37
Launching the ill-fated Mariner 3 39
The Mariner 4 imaging track on a telescopic map of Mars 41
Some Mariner 4 images. 43
A model of Venera 3. 47
A Pioneer solar wind spacecraft 50
An 'exploded' view of a Pioneer solar wind spacecraft. 51

The NASA–MIT Sunblazer. 53
The standard second-generation Venera bus 54
A cutaway showing the Venera 4 capsule 55
A Venera 4 instrument 56
A Mariner design for a cometary flyby 57
The Marshall Space Flight Center's proposal for a Venus entry probe 59
A line drawing of Mariner 5 60
Mariner 5 61
Launching Mariner 5 62
The trajectories of Venera 4 and Mariner 5 at Venus 64
An early idea for the Voyager Mars–Venus orbiter and lander 66
A cutaway of the Philco–Ford Voyager lander 67
The definitive Voyager Mars orbiter and lander 69
Testing a Martian lander. 71
The temperatures and pressures measured by Veneras 4, 5 and 6. 73
The M-69 Mars spacecraft 75
A Mariner releasing a Mars entry probe 77
Preparing Mariner 6 79
One of the Mariner 69 spacecraft folded for launch. 80
A map of Mars showing the imaging tracks of Mariners 6 and 7 82
Some Mariner 6 and 7 far-encounter images 84
Some near-encounter images by Mariner 6 85
Some near-encounter images by Mariner 7 86
Accuracy of astronautics guidance. 88
An ESRO cometary probe. 89
The European MESO Mercury flyby spacecraft 90
The 1961 Solar Monitor 91
Propulsion systems for deep-space probes 92
The Venera 7 capsule 99
A mockup of the Soviet Mars-71P spacecraft 101
Detail of the Mars-71P mockup 102
Another view of the Mars-71P mockup. 103
The M-71S configuration 104
A mockup of the M-71 lander. 105
The M-71 lander in its stowed configuration 106
The M-71 landing sequence. 107
The tiny PrOP-M walking robot 109
The 1971 Mariner spacecraft. 111
The ill-fated Mariner 8 spacecraft 113
The narrow-angle camera of the 1971 Mariner spacecraft 114
An Atlas–Centaur lifts off with Mariner 9. 115
Some Mariner 9 images. 119
Seasonal changes in the south polar cap of Mars 121
Images by Mariner 9 of Phobos and Deimos. 121
A photometric map of Mars by Mars 3. 123

A line drawing of the Pioneer Jupiter spacecraft 127
Preparing Pioneer 10. 129
How the Pioneer imaging photopolarimeter operated 130
The Pioneer plaque 132
Mounting the plaque. 133
A Pioneer 10 image of Jupiter. 135
The small red spot on Jupiter imaged by Pioneer 10 136
An image by Pioneer 10 of Jupiter's satellite Europa. 137
The geometry of Pioneer 10's flyby of Jupiter 139
Pioneer 10 looks back at Jupiter 140
The Jovian magnetosphere 141
A Pioneer 11 image of the Great Red Spot 142
The trajectories of Pioneers 10 and 11. 144
A Pioneer 11 view of Jupiter's northern hemisphere 145
Images by Pioneer 11 during its Saturn encounter. 148
Saturn's infrared radiation measured by Pioneer 11. 149
The trajectories of Pioneers 10 and 11 in the solar system 152
The proposed Pioneer Outer Planets Orbiter 155
A drawing of Venera 8 on the surface of Venus 158
Venera 8's entry trajectory. 159
The M-73S Mars configuration 161
The VPM-73 photopolarimeter carried by Mars 5 163
An image of Mars taken by Mars 5 166
The 5NM Mars sample-return probe. 169
The 4GM prototype for a tracked rover 170
A 3M bus to carry two heavy Mars landers 171
A Soviet model of an advanced spacecraft. 173
The Mariner 10 spacecraft. 175
A line drawing of Mariner 10 176
Mariner 10 is prepared for launch. 177
The optics of the Mariner 10 camera. 178
A picture of Earth taken by Mariner 10 179
The geometry of Mariner 10's flyby of Venus 180
The first image of Venus ever returned by a spacecraft 181
An ultraviolet view of Venus by Mariner 10 183
Atmospheric circulation on Venus. 184
The heliocentric trajectory of Mariner 10 185
The geometry of Mariner 10's first encounter with Mercury 187
A long-range view of Mercury by Mariner 10 189
The Caloris basin of Mercury 190
A part of the floor of the Caloris basin. 191
The region antipodal to the Caloris basin 192
A closer look at the antipodal region 193
The intercrater plains of Mercury 194
Scarps on Mercury 195

Preparing Helios 1. 197
The trajectory of Helios 1 199
The trajectory of Helios 2 201
Temperatures endured by Helios 1 203
The Solar Orbiting Relativity Experiment 205
Trajectories to reach comet Encke. 207
Using a Mariner spacecraft to investigate comet Encke 208
Using Helios C to investigate comet Encke 209
The configuration of the Venera spacecraft for 1975 210
A model of the Venera 9 lander 212
The surface of Venus as seen by Venera 9. 213
The surface of Venus as seen by Venera 10 215
An image of Venusian clouds by the Venera 9 orbiter. 216
A bistatic radar map provided by Veneras 9 and 10 217
An artist's impression of a Viking spacecraft approaching Mars 218
The configuration of the Viking Orbiter 219
The configuration of the Viking Lander 221
The Viking biology package 224
Preparing the Viking 2 Lander 227
The Viking 1 Lander's first pictures. 232
Some Viking 1 Lander pictures 233
The Viking 2 Lander's first pictures. 236
The site sampled by the Viking 2 Lander's arm. 237
Pictures by the Viking 2 Lander 238
The Viking 2 Lander on the surface 240
A teardrop-shaped 'island'. 244
An image of Cydonia 245
The summit of Olympus Mons 246
The Argyre basin 247
Seasonal changes in the northern polar cap of Mars 248
A dust storm sweeping over the Viking 1 landing site 249
Valles Marineris 250
Phobos 252
Deimos 253
'Big Joe' at the Viking 1 site 255
The 'Purple Pigeon' Ganymede lander. 257
Scenarios for post-Viking Mars missions. 261
An ESRO Venus Orbiter. 263
How Venus interacts with the solar wind. 264
Preparing the Pioneer Venus Orbiter 265
Preparing the Pioneer Venus Multiprobe. 267
Testing the Pioneer Venus parachute. 269
The Pioneer Venus large probe 270
One of the Pioneer Venus small probes 271
An instrument on the Venera 11 and 12 landers 273

The entry sites of the Pioneer Venus probes 274
Pioneer Venus Orbiter imagery of Venus. 278
Infrared measurements of the north pole of Venus 280
An altimetry radar map of Venus by the Pioneer Venus Orbiter 282
An image of comet Halley by the Pioneer Venus Orbiter. 283
The surface of Venus as seen by Venera 13 286
The surface of Venus as seen by Venera 14 287
The French Eole satellite. 290
A rover prototype for use on Venus 292
Gary Flandro 302
The J–S–P trajectory 303
The J–U–N trajectory 304
The J–S–U–N trajectory 305
The Marshall Space Flight Center's proposal for the Grand Tour. 306
The TOPS spacecraft. 308
A Mariner–Jupiter–Uranus spacecraft 310
The configuration of the Voyager spacecraft 312
The scan platform of one of the Voyager spacecraft 314
The optical arrangement of Voyager's cameras 316
Preparing Voyager 2 317
A Titan IIIE–Centaur lifts off with Voyager 2. 319
A picture of Earth and the Moon by Voyager 1 321
The trajectory of Voyager 1 through the Jovian system. 324
Jupiter and the Great Red Spot by Voyager 1. 325
Jupiter, Io and Europa 327
Detail of the Great Red Spot 328
A large brown oval on Jupiter. 329
Jupiter's ring. 331
How the Jovian ring was discovered by Voyager 1 332
Amalthea by Voyager 1. 332
Io by Voyager 1 333
Europa by Voyager 1 333
The large heart-shaped feature on Io. 334
Temperature and pressure profiles of the Jovian atmosphere. 335
The night-side of Jupiter imaged by Voyager 1 336
Ganymede by Voyager 1. 337
A closer look at the grooved terrain on Ganymede 338
The Valhalla impact structure on Callisto by Voyager 1 339
How active volcanoes were discovered on Io. 341
The lava lake in the caldera of Loki by Voyager 1 342
An infrared radiometer scan of the Loki caldera 343
Mesas and tilted massifs in the south polar region of Io by Voyager 1 344
Linda Morabito 345
Jupiter imaged by Voyager 2. 348
How the wind speed varies with latitude on Jupiter. 349

An image taken during Voyager 2's first Jovian ring-plane crossing. 351
Callisto viewed by Voyager 2 352
The cratered surface of Callisto. 353
Voyager 2's passage through the Jovian system 354
Galileo Regio on Ganymede by Voyager 2 355
Grooved terrain on Ganymede by Voyager 2 356
Europa revealed by Voyager 2. 357
The evening terminator of Europa. 358
Amalthea in transit of Jupiter 359
Volcanoes on Io by Voyager 2 360
The Voyager 2 occultations in the Jovian system. 361
The two ansae of the Jovian ring by Voyager 2. 362
A closer look at part of Jupiter's ring 363
Voyager 1's trajectory through the Saturnian system 365
A distant Voyager 1 image of Saturn 367
Haze layers over the north polar region of Titan. 369
Two views of Mimas. 371
Viewed from down-Sun the ring spokes appear bright to Voyager 1 373
Voyager 1 occultations in the Saturnian system. 374
The leading hemisphere of Dione 375
Janus by Voyager 1. 376
Epimetheus by Voyager 1 376
Braids, kinks and knots in the F ring by Voyager 1. 377
The heavily cratered north polar region of Rhea 378
Voyager 1 views the 'ying yang moon' Iapetus 379
Saturn as a crescent phase by Voyager 1. 381
Voyager 2's trajectory through the Saturnian system 384
Iapetus by Voyager 2 385
The irregular 'hamburger moon' Hyperion 386
Titan's north polar hood. 386
Some of Saturn's minor satellites. 387
Tethys. 388
Voyager 2 views dark spokes on the B ring. 390
The structure of the F ring 391
Voyager 2's close flyby of Enceladus. 392
A foreshortened view of the ring system by Voyager 2 393
The unilluminated face of the rings by Voyager 2 394
Phoebe by Voyager 2 395
Structures in the atmosphere of Saturn by Voyager 2 396
The geometry of Voyager 2's encounter with Uranus 402
An early view of the Uranian rings 403
Voyager 2 views Uranus's bland disk 405
Moons associated with Uranus's rings. 406
Puck by Voyager 2 408
Oberon by Voyager 2 412

Titania by Voyager 2 413
Umbriel by Voyager 2 414
Ariel by Voyager 2 415
Miranda by Voyager 2 416
The rings of Uranus are silhouetted against the planet 417
The rings viewed against space 418
The night-side of the rings 419
Voyager 2 leaves Uranus behind 421
Voyager 2's flybys of Neptune and Triton 425
Occultations in the Neptunian system by Voyager 2 427
Images of Neptune by Voyager 2 429
Nereid, Larissa and Proteus 431
Neptune's high-level clouds casting shadows 432
The south polar region of Triton 433
Ruach Planitia on Triton 435
The 'cantaloupe' terrain of Triton's southern hemisphere 436
The Liberté, Egalité and Fraternité ring arcs 438
The night-side of the Neptunian rings 439
Voyager 2 leaves Neptune and Triton behind 440
The trajectories of the two Voyagers as they leave the solar system 442
The plan to take a 'portrait' of the solar system 444
The Voyager 'portrait' of the solar system 445
The structure of the heliopause 448
Voyager 2's line of sight of the impact of comet Shoemaker–Levy 9 449
The rates at which spacecraft are leaving the solar system 452
The trajectories of the spacecraft leaving the solar system 453
A tangential Hohmann transfer orbit 466
Type 1 and Type 2 transfer orbits 466
How gravity-assist slingshots work 468

List of tables

The solar system at the beginning of the 'space age' lvii
The performance of the Saturn V as a launcher for planetary missions 70
Star encounters by the Pioneer spacecraft 155
Star encounters by the Voyager spacecraft. 452
Hohmann transfer ellipses from Earth. 467
Soviet planetary probe designations. 469
Reported spacecraft discoveries of planetary satellites 471
Chronology of solar system exploration 1952–1982 473
Planetary launches 1960–1981 475

Foreword

In commemoration of 50 years of space exploration, Paolo Ulivi, with the assistance of David Harland, has undertaken to provide a detailed description of all the robotic missions to have ventured out of Earth's gravity well. They refer to the first 25 years as the "golden age", when scientists, engineers and technicians developed innovative means of achieving their mission goals under tight schedules and small budgets.

In the '50s, engineers leveraged the rapid developments in rocketry during World War II to create a new class of launch vehicle capable of placing a payload into orbit. It was a time of master draftsmen and engineers who utilized slide rules to calculate parameters. But the accelerating rate of technological progress throughout this period enabled mission designs to become ever more sophisticated, leading ultimately to the Voyager missions that were dispatched beyond the edge of the solar system, never to return. As a testimony to the expertise of the space pioneers, these craft, launched in the 1970s, are still actively exploring in their dotage.

On receiving my copy of the manuscript, I turned immediately to the sections on Pioneer Venus (1978) and Pioneer Jupiter (1975)/Saturn (1979), which are of special significance to me for the reason that I began my career at the University of Arizona supporting these projects. In 1975, I did a thesis on the radii of the Galilean satellites using the newly returned data from the Imaging PhotoPolarimeter. Using the masses measured for these four satellites, I calculated a more accurate density for each. After graduating, I joined the Solar Flux Radiometer team led by Dr. Martin Tomasko, and we participated in the first American landing on Venus by measuring the solar flux as the probe descended through the atmosphere. By balancing the solar input with the infrared output, we provided insight into the thermal balance that keeps the surface of the planet at a temperature at about 900°F – the trickle of solar energy that reaches the surface drives the massive greenhouse affect. When Pioneer 11 provided our first look at the Saturnian system in September 1979, I was in the control center at Ames, helping to transmit commands to the camera. Since the one-way-light-time was about 1.5 hours, the shortest time from issuing commands to receiving images was 3 hours. Nevertheless, we were able to target most of the satellites, and sweep by Saturn and its beautiful rings. I was there for the discovery of the F ring, and the moon 1979S1. Other data were not easily decoded; indeed, it would take me 10 years to figure out why sunlight reflected from Titan's atmosphere was highly polarized.

The descriptions of these missions in *Robotic Exploration of the Solar System* took me back to that primitive time when textbooks lacked any description of major features of the neighboring planets. Small teams of scientists supported by dedicated engineers – by now using hand calculators and mainframe computers – unraveled the fundamental truths of our most significant solar system neighbours over the course of two decades; it truly was the golden age.

Much of the information on other missions is completely new to me. The Russian attempts to land on Mars without any knowledge of the density of the atmosphere reminds me of the spirit of the American pioneers who set out into the unknown at tremendous risk of life and limb. Later, the Viking missions to Mars survived despite numerous issues that I have been blissfully ignorant of, including swapping Viking A with B on the launch pad after the batteries had become depleted. When missions are successful, these technical glitches are soon forgotten, but there are many and varied lessons to be learned from these early experiences.

The rivalry between the USSR and the US motivated the first 25 years of space exploration: a peaceful, but intense competition during the cold war. The scientists and engineers who experienced these missions went on to play pivotal roles in the second 25 years of the space age, on missions that will be related in further volumes of this series. But few of these explorers are actively engaged in missions today, and the task of exploration will become the responsibility of a new generation, one that has been brought up using high-speed personal computers and the Internet. This book serves as a valuable reference for these newcomers to inform them of the pioneering work that has preceded them.

Peter H. Smith
Lunar and Planetary Laboratory
University of Arizona
June 2007

Author's preface

It was the late summer of 1981 and I was a 10-year old spending the final weeks of my school vacation with relatives on the Tuscany coast. There, as we watched the daily news one evening, we stared in awe at pictures of the distant planet Saturn that had been returned in vivid colors and amazing detail by a robotic spacecraft named Voyager 2. The next day, I spent most of my time sifting through the newspapers that my grandfather had brought me to read more about this amazing robot. I still have the cuttings from that day! Over the ensuing days, I read the reports that stated that the motor that steered the camera had jammed, but was then recovered, thereby enabling the robot to resume sending amazing views of the planet, its rings and its satellites. Upon returning home, I drew from our shelf one of the astronautics books, published in the early 1960s, to learn more about the intrepid emissary, only to read that the goal of the Voyager mission was to land a capsule on Mars. There must be something wrong, I thought! That was my introduction to solar system exploration and its complex history.

The book you now hold is the result of a 25-year fascination with the robots sent to explore the solar system. To borrow what a reviewer wrote of my first book on *Lunar Exploration*, this is a veritable "spotter's guide" to the robots that explored the solar system. Unlike most of the popular books on the topic, I have used as much 'first generation' material as possible; ranging from conference papers and reports, to the mission results as published in the scientific press. It contains accounts of triumphs and failures, and a number of hitherto untold stories. I hope the book will not only make interesting reading for space enthusiasts, but also serve as a reliable starting point for graduate students, space engineers and planetary scientists setting out to conduct in-depth studies. If you use this book for this purpose, then I would be delighted to hear from you.

Paolo Ulivi
Milan, Italy
March 2007

Acknowledgments

As usual, there are many people that I must thank. First of all, my family; and in particular my mother, my father (who made most of the translations from Russian) and my brother for his constant help with that black-box people call a "computer". I must also acknowledge the assistance of the staffs of the library of the aerospace engineering department of Milan Politecnico, the Italian national library located in Florence, and the Historical Archives of the European Union, as well as members of many Internet forums in which I participate, including the Friends and Partners in Space, the History of Astronomy Discussion Group, Unmannedspaceflight.com, the Interplanetary Communication forum, the Italian forumastronautico.it, the French Forum de la Conquête Spatiale and Histoire de la Conquête Spatiale, and the Russian forum of the magazine Novosti Kosmonavtiki. Thanks also to Sergei V. Andreyev, Charles A. Barth, Jacques Crovisier, Keay Davidson, Dwayne Day, Audouin Dollfus, Gérard Epstein, Ben Evans, James Garry, Jon Giorgini, Brian Harvey, Joseph V. Hollweg, David W. Hughes, Stefano Innocenti, Ivan Ivanov, Viktor Karfidov, Gunther Krebs, Alan J. Lazarus, Jean-François Leduc, David Lozier, Franco Mariani, Ed B. Massey, Sergei Matrossov, David J. McComas, Don P. Mitchell, Dominique Moniez, Dmitry Payson, Basil Pivovarov, Michel Poquérusse, David Portree, Joel Powell, Patrick Roger-Ravily, Mario Ruggieri, Olivier Sanguy, Henning Scheel, Jean-Jacques Serra, Bradford A. Smith, Philip J. Stooke, G. Leonard Tyler, Jan van Casteren, Ronald J. Vervack Jr., Victor Vorontsov, Paul Wiegert, Anatoly Zak, Gary P. Zank, and I sincerely apologize if I have left out anyone. A hearty 'thank you' goes to all my friends and colleagues for their encouragement over the last two years; in particular to Aldo, Alessia, Andrea, Antonella, Antonio, Aurora, Ciro, Cristina, Elena, Elisa, Emanuele, Federico, Filippo, Flavia, Francesco, Giorgio (thanks again for the copy of *Aelita* and for the interesting discussions of it), Giulio, Giuseppe, Luigi, Renato, Roberto, Silvia, Stefania, Ugo, Virginia. Of course, a particular thank you goes to David Harland for his invaluable assistance in the preparation of both this book and of the earlier *Lunar Exploration*, to Peter Smith for contributing the Foreword, and to Clive Horwood for his patience and kindness.

Regarding the illustrations: I have endeavored to use the 'raw' mission imagery, some of which included 'reseau' markings. Although not particularly beautiful, they have the benefit of conveying an idea of the kind of data that the scientists actually used. With a few exceptions, all the US planetary images were downloaded from the Planetary Data System Imaging node at JPL (pdsimg.jpl.nasa.gov). One picture is a special tribute to my thesis advisor Professor Amalia Ercoli Finzi. Although I have managed to identify the copyright holders of most of the drawings and photographs, in those cases where this has not been possible and I deemed an image to be important in illustrating the story, I have used it and attributed as full a credit as possible; I apologize for any inconvenience this may create.

Introduction

Early in human history, as a result of peering at the night sky using the unaided eye, it was realized that in addition to the stars that remained in fixed patterns, there were points of light that moved, and these were referred to as planets – the name meaning 'wanderers'. Initially, it seemed self-evident that the celestial realm revolved around Earth once per day, but a few bold philosophers, most notably Nicolaus Copernicus, realized that this was not the case, and that, with the exception of the Moon, which really did travel around the Earth – although on a monthly rather than a daily basis – the planets, including Earth, orbited the Sun. This 'heliocentric theory' was the first recognition that there was such an entity as the 'solar system'. In order of increasing heliocentric range, the planets were Mercury, Venus, Earth, Mars, Jupiter and Saturn. At the time of the invention of the telescope all that was known about the planets was how they traveled across the sky. By turning his crude telescope to the heavens Galileo Galilei revolutionized astronomy. The following is what we knew, or thought we knew, of the solar system at the start of the 'space age'.

MERCURY: EXTREMES OF HEAT AND COLD

As it never strays far from the Sun in the sky as viewed from Earth, Mercury is the most difficult to observe of the five planets that are able to be seen by the naked eye. Its rapid motion and brief apparitions led the ancient Greeks (following the trend of the preceding cultures) to name it Hermes, after the messenger of the gods and the god of thieves. Accordingly, the Romans named it Mercurius. At the time of the invention of the telescope early in the seventeenth century, all that was known of Mercury was the time it took to resume a given configuration in the sky: its synodic period of 115 days. Although it was Galileo's report in 1609 that Venus showed phases that demonstrated the veracity of the Copernican hypothesis, Mercury was a much more difficult target, in part because its disk is smaller, but mainly because his telescopes were of low optical quality, and it was not until Giovanni Battista Zupi used a better instrument 30 years later that the phases were observed. After several generations of astronomers had accumulated drawings of an almost featureless disk, it was decided that the planet must have a permanently cloudy atmosphere.

Mercury is an interior planet, which means that it orbits nearer the Sun than does Earth. When the planet is at inferior conjunction, it can pass across the solar disk as viewed by a terrestrial observer. However, since the plane of the planet's orbit is slightly inclined to that in which the Earth travels around the Sun – known as the ecliptic – the planet will more often than not pass either just above or below the solar disk. A transit, as it is called, can only occur if the conjunction falls in May or November. The first such event was observed on 7 November 1631 in France by Pierre Gassendi.

In the late nineteenth century a new and more powerful type of telescope enabled astronomers to challenge the belief that the surface of Mercury must be obscured by clouds. First, a number of people reported seeing mountains, bright spots and lines. Then two astronomers, one an amateur, William Denning in England, and the other a professional, Giovanni Virginio Schiaparelli in Italy, independently took a serious interest in the planet. In 1882 Denning began to observe it in twilight, and noted that once it was above the disturbed air near the horizon he could see dark markings and bright spots. By the time that Schiaparelli in Milan heard of Denning's observations, he had already been observing Mercury for more than a year, but unlike Denning, he did this in daylight. Schiaparelli had also observed dark areas and bright spots, some of which he thought were surface features and others, which seemed to vary, he took to be cloud patterns. He was determined to measure the rotational period of the planet. The fact that the markings seemed to move only very slowly from one day to the next implied that the rotation was rather slow. Armed with a mathematical study by George Darwin, the physicist son of the famous naturalist, who had determined that tides are responsible for the Moon's rotation being synchronized with its orbital motion around Earth, Schiaparelli attempted to fit the features that he had identified to the 88-day period of the planet's solar orbit, his reasoning being that as a result of being so close to the Sun it might have become 'tidally locked'. After several years, he announced that its rotational period was indeed 88 days. Although he had had some difficulty in reconciling observations in which some of the most prominent markings could not be recognized, he concluded that these must have been masked by clouds at the time. Evidently, one side of the planet was permanently scorched and the other was in eternal night. In fact, since the planet's orbit was elliptical, the Sun would rise by several degrees in the region near the boundary dividing day from night – known as the terminator – and promptly set again, creating periods of twilight. Meanwhile, other observers had monitored the brightness of the planet's disk at different phases and found that this was similar to how the Moon varied under different illumination, which prompted them to suggest that Mercury was airless. However, despite Schiaparelli's caveats, his conclusions were accepted by most of the scientific world.

Over the next decade or so the foremost astronomers observed Mercury, but their observations received very little attention, and in 1897 Percival Lowell drew a map that included a network of black criss-crossing lines. During 3 years of observations at the Meudon Observatory near Paris, Eugène Antoniadi noticed that some of the markings, which were often quite obvious, seemed at times to be fixed for hours, yet showed a pronounced movement from one day to the next, which led him to believe that Schiaparelli's figure for the rotational period was correct. In 1934 he published a chart of the Sun-facing side that was much more detailed than that of Schiaparelli in 1889, despite showing most of the same markings. As a Greek-born astronomer, Antoniadi annotated the map with names drawn from Hellenic and Egyptian legends about Hermes. As there was no logic in calling the dark areas 'maria' (seas), as had been done on the Moon and Mars, Antoniadi used the more apt 'solitudo' (desert). He named one large gray patch Solitudo Hermae Trismegisti after the mythological inventor of all sciences, including astronomy, and a whitish spot Liguria as a tribute to Schiaparelli, who was born in Savigliano in the ancient Roman region of Liguria. Like Schiaparelli, Antoniadi explained away the fact that some markings sometimes disappeared by supposing that they had been hidden by clouds, but was mystified by the type of clouds that could form on a body so close to the Sun. Shortly after Antoniadi's observations, American astronomers published measurements of the temperature of the planet's surface at different phase angles, again noting that it was similar in this respect to the Moon.

When astronomers at the high altitude of the Pic du Midi in southern France began to observe Mercury in the 1930s, they made significant advances. Some of the finest drawings and the first good photographs were produced in the ensuing decades, and after the true rotational period was identified in the 1960s it was possible to compile these into maps of the albedo features visible to the naked eye. At the time, however, no one questioned that Mercury was not tidally locked. Also starting in the 1930s, day-time observations of the polarization of sunlight reflected by Mercury's surface were made at Meudon and Pic du Midi. The polarization of the planet was found to be very similar to that of the Moon; so much so, in fact, that in 1950 the astronomers at Pic du Midi predicted that it, too, must be heavily cratered. The polarimetry also ruled out an atmosphere at a surface pressure greater than 10^{-5} of that at sea level on Earth.

Until the 1950s, the uncertainty of such a basic datum as the diameter of Mercury was relatively large, which meant that its bulk density was so poorly known that it was impractical to propose a hypothesis of its internal composition. In this respect, a particularly important event was Mercury's transit of the Sun on 7 November 1960, during which a variety of techniques – including one that compared the flux of light from the partially covered portion of the Sun to that of a clean portion – enabled the diameter of the planet to be measured to better than 1 per cent. As the planet's mass had already been determined from how it perturbed the orbit of the asteroid Eros, it was then possible to calculate its density, but the very high value indicated the presence of a metallic core that was at odds with contemporary theories of how the solar system formed.

VENUS: A SWAMP OR A GREENHOUSE?

The next planet from the Sun was of particular significance to ancient civilizations. Mediterranean and 'fertile crescent' cultures associated its brightness and beauty

A 1958 drawing of Mercury as seen from a distance of 9,500 km, showing the Solitudines of Martis, Jovis and Lycaonis. Mountains and craters are visible at the terminator. In fact, it was widely believed even before the space age that Mercury would have craters and a surface resembling that of the Moon. (Artist: the Italian amateur astronomer Guido Ruggieri; reprinted with kind permission)

with love and its goddesses: Ishtar to the Babylonians, Aphrodite to the Greeks, and Venus to the Romans. In contrast, on noting that 8 terrestrial years precisely equated to 5 times the planet's synodic period of 584 days, the Mayans of Central America made this interval their most important unit of time. Venus was the subject of some of Galileo's earliest telescopic observations. Although the only thing his instruments were capable of showing was the changing phase of the planet as it orbited the Sun, that finding was very significant since it proved the heliocentric model proposed by Copernicus. Meanwhile, Johannes Kepler had realized that because Venus was nearer the Sun it should be able to be observed crossing the solar disk. In fact, transits by Venus are much rarer than those of Mercury, occurring in pairs at 8-year intervals, spaced more than a century apart. Kepler computed that a transit was due in 1631, but because the Sun would be below the horizon at the time for viewing in Europe, where most astronomers were then located, it was not observed. When English amateur Jeremiah Horrock repeated Kepler's calculations and realized that a second transit was due on 4 December 1639, and that it would occur during daylight, he and a friend became the first – and for more than a century the only – people to observe Venus transit the disk of the Sun. The most recent transit occurred on 8 June 2004 and was well timed for Europe, but on 6 June 2012 the second of the current pair will not favor European observers.

Giovanni Riccioli in 1643 was the first to notice the unusual but well-documented phenomenon of 'ashen light' by which when Venus is in the crescent phase its dark part is sometimes faintly visible. A century and a half previously, Leonardo da Vinci had reasoned that the condition at crescent phase known as 'the old moon in the young moon's arms' was due to the dark part of the Moon being illuminated by sunlight reflecting from a 'full Earth'. It was initially suspected that Earth might be doing the same for Venus, but calculations showed that this was not feasible. Various theories have been advanced over the centuries to explain the glow over the dark hemisphere, ranging from oceanic bioluminescence to celebratory firework displays, but it is now thought to be an electrical phenomenon akin to an aurora or airglow. Although the disk of Venus was of greater angular diameter than that of Mercury, it was frustratingly featureless. In the seventeenth and eighteenth centuries efforts to find markings that would enable the rotational period to be measured gave the false impression that this was 24 hours. Meanwhile, Edmund Halley had realized that if the planet's crossing of the solar disk during a transit was precisely timed at two widely separated sites, it would be possible to triangulate the distance between Earth and Venus and apply the laws of orbital motion formulated by Kepler to calculate the mean distance from the Sun to the Earth – a distance known as the Astronomical Unit (AU). In anticipation of the transits of 1761 and 1769, many European governments sent out expeditions. In effect, it was a kind of eighteenth-century 'space race' in which English astronomers traveled to India and French astronomers went as far afield as Siberia. However, the best-known of these expeditions was that of James Cook to Tahiti in 1769. Although the 1769 transit was seen from 80 sites across the world, and 150 timings were taken, the accuracy of the calculation was limited by the quality of the telescopes of that time. Nevertheless, in 1761 the Russian astronomer Mikhail V. Lomonosov made the first significant discovery regarding the nature of the planet: looking at the appearance of the edge of the black disk of the planet against the Sun he realized that it had to be surrounded by an atmosphere "similar (or even possibly larger) than that [which] is poured over our Earth". Unfortunately, this discovery is usually assigned to Johann Schröter who, in 1790, was one of the first to notice that the phase when the planet's disk is exactly half illuminated and half in darkness – known as dichotomy – occurs a little earlier than it should when Venus is in the evening sky and a little later when it is in the morning sky, from which he reasoned that it must possess a dense and extended atmosphere. However, Schröter seems to have been the first to detect faint markings and correctly interpret them as being of atmospheric origin. In fact, at times he reported that there were bright spots on the dark side of the terminator, and by analogy with such a phenomenon on the Moon he thought these were the summits of high mountains catching the rays of the Sun.

Strange as it may now seem, there were even reports in the early telescopic era of a Venusian moon; indeed, it had even been named Neith. It was first sighted in 1645 by Francesco Fontana, but it was not until Giovanni Domenico Cassini announced in 1686 that he had seen an object close by Venus that mimicked its phase that people began to search for it. Sightings continued for over a century, but ceased upon the development of better telescopes, which suggests that all such sightings were merely artifacts of the unwieldy telescopes of that time – some of which used a lens with a diameter of only a few centimeters and a focal length measured in tens of meters.

By the nineteenth century the consensus was that the rotational periods of Earth, Venus and Mars were all about 24 hours, but in 1877 Schiaparelli saw two diffuse bright spots near the cusps that seemed not to move from day to day, leading him to the view that Venus took the same time to rotate as it did to travel around the Sun: 225 days. Despite criticism that synchronous rotation would have caused the atmosphere on the permanently dark side to freeze and create an ice cap, and studies that suggested a variety of other rotational periods, Schiaparelli's assertion that Venus, like Mercury, was tidally locked, remained the presumption to the dawn of the space age. In 1896 Percival Lowell saw a number of markings that radiated out from a point near the equator in a pattern reminiscent of the spokes of a wheel; furthermore, the pattern remained in a fixed position on the disk from day to day. Unable to see these markings, other astronomers dismissed them as illusory. After a nervous breakdown, during which Lowell's assistant cast doubt on the markings, Lowell – for the first and last time in his career as an amateur astronomer – issued a retraction. Nevertheless, he commissioned the building of a spectrograph to measure the Doppler shift of the planet's atmosphere to prove that it was indeed immobile, but the results were inconclusive. The mystery of Lowell's spokes has only recently been clarified: by 'stopping down' his telescope to a fraction of its aperture – as he did to reduce chromatic aberration when observing a very bright object – he would, on applying a high magnification, perceive the shadow of the blood vessels in his eye, whose structure is a good likeness of his sketches. In retrospect, significant observations were made in 1897 and 1898 by Leo Brenner in Croatia. Although he inferred an incorrect rotational period of just under 24 hours – which he specified to an accuracy of 10^{-5} second! – his sketches may have recorded what we now know to be C- and Y-shaped atmospheric patterns. However, he was regarded by most of his contemporaries as an unreliable witness. (Indeed, he published by a false name; his real name was Spiridon Gopcevic.) During the early part of the twentieth century various visual studies suggested a number of rotational periods ranging from 2.8 to 8 days, but in 1927 Frank Ross at the Mount Wilson Observatory exploited advances in photography to take pictures of Venus at infrared and ultraviolet wavelengths just beyond the visible spectrum. In the infrared Venus was as bland as in the visible, but in the ultraviolet it showed streaks and bands paralleling the equator, evidently due to unidentified ultraviolet-absorbing compounds in the clouds. Similar observations were made in the 1950s by G.P. Kuiper and R.S. Richardson in America and also by N.A. Kozyrev in the Soviet Union. Richardson suspected that the planet might rotate in the opposite sense to its motion around the Sun – that is, in a retrograde manner.

Meanwhile, scientists such as Svante Arrhenius, who had won the Nobel Prize for chemistry in 1903, had portrayed Venus as a swamp-covered planet that was teeming with life, somewhat similar to the Carboniferous period on Earth 300 million years ago. Since the only clouds known on Earth were those containing droplets of water vapor, Arrhenius concluded that cloud-enshrouded Venus must be "dripping wet". In 1929 Bernard Lyot undertook a polarization study and concluded that the clouds were only brilliant because sunlight was reflecting from a dense suspension of tiny droplets of liquid. There was no spectroscopic evidence for water in the atmosphere of Venus, nor for formaldehyde, which could form in a mixture of carbon, hydrogen and oxygen irradiated by solar ultraviolet. However, a spectrogram taken in 1932 by W.S Adams and T. Dunham included an infrared absorption feature that they could not identify, but was later found to be due to carbon dioxide. This transformed the view of Venus into an arid desert and, because carbon dioxide cannot form clouds of droplets and therefore could not be responsible for the uniform cloud cover, it was inferred that the clouds must be laden with dust particles. Nevertheless, in 1955, in what would turn out to be a bumper year for theorizing about Venus, Fred Hoyle suggested that the carbon dioxide was derived from the dissociation by radiation of hydrocarbons, and there must be a global ocean of oil! That same year, F.L. Whipple and D.H. Menzel accepted both the polarimetry indicating that the clouds were made of droplets and the spectroscopy indicating that the atmosphere was rich in carbon dioxide, and they proposed that there must be a vigorous hydrological cycle in which the carbonic acid that forms when water absorbs carbon dioxide must have so eroded the land as to transform the planet into a vast ocean of carbonated water. In fact, in the late 1950s, American and French astronomers finally found spectroscopic evidence for water, but only in trace amounts. The French results were obtained in a most unusual and adventurous way. Audouin Dollfus ascended to 14,000 meters in a pressurized nacelle suspended from a cluster of helium balloons to take his spectra from above most of the vapor in the Earth's atmosphere. Meanwhile, Kozyrev reported auroral displays on the night-side, and said he had identified emissions of molecular and ionized nitrogen, but this was disputed as it could not be replicated.

When the microwave brightness of Venus was measured for the first time at the inferior conjunction of 1956, a dramatic and unexpected result was obtained. With certain assumptions, this could provide the surface temperature. At the wavelength of the initial observations, the planet appeared to be as bright as a 'black body' at 330°C. There were two possible explanations: either the high temperatures was due to electronic effects in the planet's ionosphere (this result being evidence that there was one) or the surface was at a very high temperature (in which case there could be no ocean of any form). Measurements at other wavelengths gave lower temperatures that were difficult to explain in terms of ionospheric effects, but could be explained by a cloud layer opaque at these wavelengths that was masking the surface. Infrared data showed the cloud tops to be less than 0°C, and also showed this temperature to be more or less uniform over both the day-side and night-side, which in turn cast doubts on the popular presumption that the rotation was synchronous. In July 1959 astronomers across the world witnessed the extremely rare event of Venus passing in front of (occulting) a bright star, in this case Regulus in the constellation of Leo, and the manner in which the starlight diminished provided some firm data; in particular the opacity of the planet's atmosphere, from which it was possible to infer density, temperature and pressure profiles.

One final discovery that was made just as the space age began, but which passed almost unnoticed at the time, was the true rotational period of Venus's atmosphere. In 1957 French amateur astronomer Charles Boyer, who was in Brazzaville in the Congo, began to photograph the planet in the ultraviolet to assist in the research of Henri Camichel, an astronomer at the Pic du Midi Observatory in the Pyrenees. On scrutinizing his pictures, Boyer noticed that some atmospheric markings appeared to repeat at 4-day intervals. Three years later Boyer and Camichel announced this fact, initially in a popular French astronomy magazine and later in a more official French publication. When they submitted a paper to an American journal that published planetary research, it was rejected. It would seem that the scientists and engineers planning the early Soviet and American planetary missions were unaware of this work. In the 1960s, Boyer was able to see other markings including the Y-shaped pattern that would later be 'discovered' by spacecraft.

MARS, LIFE AND THE 'CANALI'

The red hue of Mars and the brightness that it attained at times prompted the ancient Mediterranean cultures to associate it with the blood-thirsty god of war: Ares for the Greeks and Mars for the Romans. It played a key role in the scientific revolution of the European Renaissance. Tycho Brahe used an accurate quadrant and the naked eye to compile a record of the planet's positions. Although insightful, Copernicus had retained the ancient belief that the planets moved in circles, but following Brahe's death his assistant, Johannes Kepler, analyzed the data and realized that the planet actually moved in an elliptical orbit. The three laws of orbital motion that Kepler developed empirically would later be given a firm mathematical basis by Isaac Newton. Galileo in 1610 and Fontana in 1636 were the first to observe Mars telescopically, but at best they were able to see only its illumination phase, which, since it lies beyond the Earth's orbit, is never less than 85 per cent. The

Neapolitan Jesuit Father Bartoli in 1644, and Riccioli and his assistant Francesco Grimaldi of the Collegio Romano, may have been the first people to discern surface features, but it was Christiaan Huygens in November 1655 who left the first reliable drawing, showing a dark triangular feature that would later be named Syrtis Major. Huygens was also able to establish the period of rotation to be slightly over 24 hours – a measurement that was confirmed by Cassini a decade later. At one point Cassini observed Mars pass in front of a star, and upon seeing the star dim prior to being physically occulted he concluded that the planet had an extended atmosphere. Although later observers with better telescopes saw no such dimmings, the existence of an atmosphere was never in doubt since clouds of various extent and color had been seen. In 1672, Huygens appears to have been the first to notice that Mars had a bright south polar cap. Cassini later noted that there was a cap at each pole. In 1719 Giacomo Maraldi noted that the caps waxed and waned in size. After this early start, the planet was more or less neglected until 1783, when William Herschel began a detailed study, paying particular attention to how the caps varied with the seasons. He noted that the south polar cap was more extensive in winter than its counterpart, and on discovering that it was not centered on the geographic pole, determined the obliquity of the spin axis relative to the plane of the planet's orbit.

By about 1830 the art of telescope-making had advanced sufficiently to prompt an age of 'Areography'. The first map would appear to be that of two German amateur astronomers William Beer and J.H. Mädler, published in 1840. Among other things, they selected a small roundish spot as the origin of the longitudes (a sort of 'Martian Greenwich'). They correctly inferred that the northern spring was colder and longer than the southern one owing to the marked eccentricity of the planet's orbit. William Rutter Dawes made drawings at the oppositions of 1862 and 1864, and Richard Anthony Proctor compiled these into a map that was published in 1867 and remained the standard for a decade. Beer and Mädler had annotated their map by letters that were explained in an accompanying legend, but Proctor decided to name features after astronomers who had observed the planet, and he developed a nomenclature based on terms such as 'Continent', 'Ocean', 'Sea', 'Bay' and 'Strait'. By the 1870s Mars appeared to be remarkably similar to Earth: its day was just a little longer; it had a substantial atmosphere; it had similar seasons, although owing to the size of the planet's orbit they lasted for twice as long; it had polar caps that were most likely made of water ice; and, in addition to continents, it might also have large bodies of open water – all of which led naturally to the conclusion that it must host some form of life.

The most dramatic year for observing Mars prior to the space age was the 'great' perihelic opposition of 1877. Asaph Hall at the US Naval Observatory searched for satellites, and found two tiny ones: the first on 10 August and the second, orbiting even closer in, the following week. He named them respectively Deimos and Phobos (Terror and Panic) after the attendants of Ares, as mentioned by Homer and Hesiod. On turning his attention to Mars, Schiaparelli found it difficult to reconcile his observations with the existing maps, so he decided to make his own map. But in contrast to his predecessors, who had simply sketched the appearance of the planet, he decided to exploit the fact that his telescope had been fitted with a micrometer for measuring the angular separation of double stars, and took accurate measurements of the latitude and longitude of the features. Consequently, his map was not only much more precise and contained more detail than its predecessors, but it was drawn in a very different style. Schiaparelli derived his nomenclature from classical literature and the Bible, and many of the names that he assigned are still in use today. At times of exceptionally good 'seeing' during his observations in September and October he saw thin straight lines on the bright areas, and, using a term that Angelo Secchi had introduced several decades earlier in his drawings of Mars, Schiaparelli referred to them as 'canali'. At this time, he also saw the first Martian dust storm to be historically documented: it lasted until early 1878. Remarkably, however, observations made at the same time by Nathaniel E. Green under the much better and far steadier sky of the Atlantic island of Madeira showed no such lines. Astronomers world-wide endeavored to see the canali, with only a few succeeding. Although no two observers agreed on the appearance, size or location of the canali, the fact of their existence was an accepted truth. In the meantime, Schiaparelli reported more and more amazing aspects of the planet, including that the shape of some of the dark areas seemed to change from one opposition to the next, which he attributed to their being seas that invaded dry lands in some places and retreated in others. However, he soon realized that if there were bodies of water on Mars they should act like mirrors and reflect the Sun back to Earth – and such a glint had never been observed. Schiaparelli thought the canali to be of natural origin, but the term could be translated as artificial canals, and in 1892 Camille Flammarion, a French astronomer and advocate of life on other worlds, speculated in *Mars and its Conditions of Habitability* that the canali might be a global irrigation system built by an advanced civilization.

Observing Mars from the Peruvian Andes at the near-perihelic opposition of 1892, William Henry Pickering was surprised to see a faint dark line on one of the dark areas. This meant that the dark areas could not be seas. In 1860 Emmanuel Liais had suggested that bodies of water were unlikely to undergo the observed seasonal cycle of darkening, and argued that they were the beds of dried seas on which vegetation bloomed in response to the retreating polar caps infusing vapor into the atmosphere. Adopting Liais's hypothesis, Pickering suggested that the canali were simply swaths of vegetation living off volcanic gases that leaked from vast cracks in the crust of an otherwise inhospitable desert. Pickering accepted an invitation from Percival Lowell to observe the opposition of 1894. In writing up his conclusions the following year in a book entitled simply *Mars*, Lowell accepted that the dark areas were vegetation, but concluded that the canali were an irrigation system – as far as he was concerned, there was no other plausible explanation. Initially skeptical, Schiaparelli succumbed to Lowell's argument, as advanced in the form of a series of personal letters, and wrote popular accounts of how an irrigation system might be built and operated – to the point of even discussing the political organization that the 'Martials' would need to ensure the most equitable distribution of this precious resource. "Mars," he wrote, "must be the paradise of socialists as well [as hydraulic engineers]". The debate over the nature of the canali and the possibility of a Martian civilization raged for the remainder of the nineteenth century and the first decade of the new century, with Lowell publishing a succession of books. Meanwhile, although Lowell's work received popular acclaim, his credibility as a telescopic observer was undermined in professional circles by his reports of networks of fine lines on both Mercury and Venus. E.E. Barnard and C.A. Young in the United States, E.W. Maunder in England, and Vincenzo Cerulli in Italy all reported being able to resolve the canali into a plethora of detail right at the limit of visibility which, in lesser 'seeing', the human eye and brain readily interpreted as straight lines. But the canali enthusiasts and Lowell's public remained unconvinced.

It was also during this time that Martians made their debut in the popular culture. In 1896 Herbert George Wells, inspired by Lowell's first book, wrote his novel *The War of the Worlds*, and in 1911 Edgar Rice Burroughs issued the first of his series of 'pulp fiction' about Mars. *Aelita*, a novel by Alexei Tolstoy that would have a great impact on the Soviet space and Mars programs, was published in 1923. This, and the amazing silent movie that director Iakov Protazanov based upon it, were just part of the Soviet fascination with the Red Planet; Schiaparelli's "socialist paradise". The Russian rocket pioneer Fridrikh Tsander even made "On to Mars!" his motto.

In 1867 William Huggins and P.J.C. Janssen each reported detecting the presence of 'aqueous vapor' in the Martian atmosphere using spectroscopy, as did H.C. Vogel in 1872 and E.W. Maunder in 1875, but the visual observing method that they used was not rigorous. In 1894 W.W. Campbell found no evidence of vapor. However, in 1895 Vogel obtained photographic spectra that he insisted confirmed the presence of vapor. In 1908 V.M. Slipher took a spectrogram of Mars using a film sensitive to the near-infrared and reported absorption by vapor. At the perihelic opposition of 1909 Campbell was able to demonstrate definitively that the Martian atmosphere was arid. His spectra would remain the last word on the issue until the beginning of the space age. That same year, Antoniadi, a long-time observer of Mars and one of the few professional astronomers to have seen the canali, was invited to use the 83-cm-diameter 'Grand Lunette' refractor at Meudon in Paris. On his first session on 20 September, which was an exceptionally clear night, he was astonished to see that the continuous lines that he had previously observed were revealed as a multitude of knots, dark streaks, spots, bands and half tones – as claimed by Barnard, Young, Maunder and Cerulli. Although, for most astronomers, 1909 signaled the end of the Lowellian vision of an ancient civilization struggling to survive as their planet dried up, the belief that the dark areas were patches of vegetation persisted until the beginning of the space age. Nevertheless, the map adopted by NASA for use in planning its first Mars missions had been compiled by E.C. Slipher and was liberally adorned with canali.

After the 1909 opposition, studies of Mars concentrated not so much on mapping but on estimating the surface pressure and temperatures, obtaining detailed spectra, and identifying the composition of the polar caps. By reasoning backward from the presumption that water must be stable in liquid form on the surface, Lowell had inferred that, as he expressed it, the temperature compared favorably with that of a summer's day in England. The first thermocouple measurements of Mars's temperature were made in 1924 by Edison Pettit and S.B. Nicholson, and gave a global temperature of about –30°C. The fact that the dark areas were up to 20°C warmer was interpreted as the result of insulation by a layer of vegetation. If Lowell's estimate of 87 hPa was correct, then the polar caps must be water ice. The first spectroscopic evidence of any particular gas in the atmosphere came in 1947,

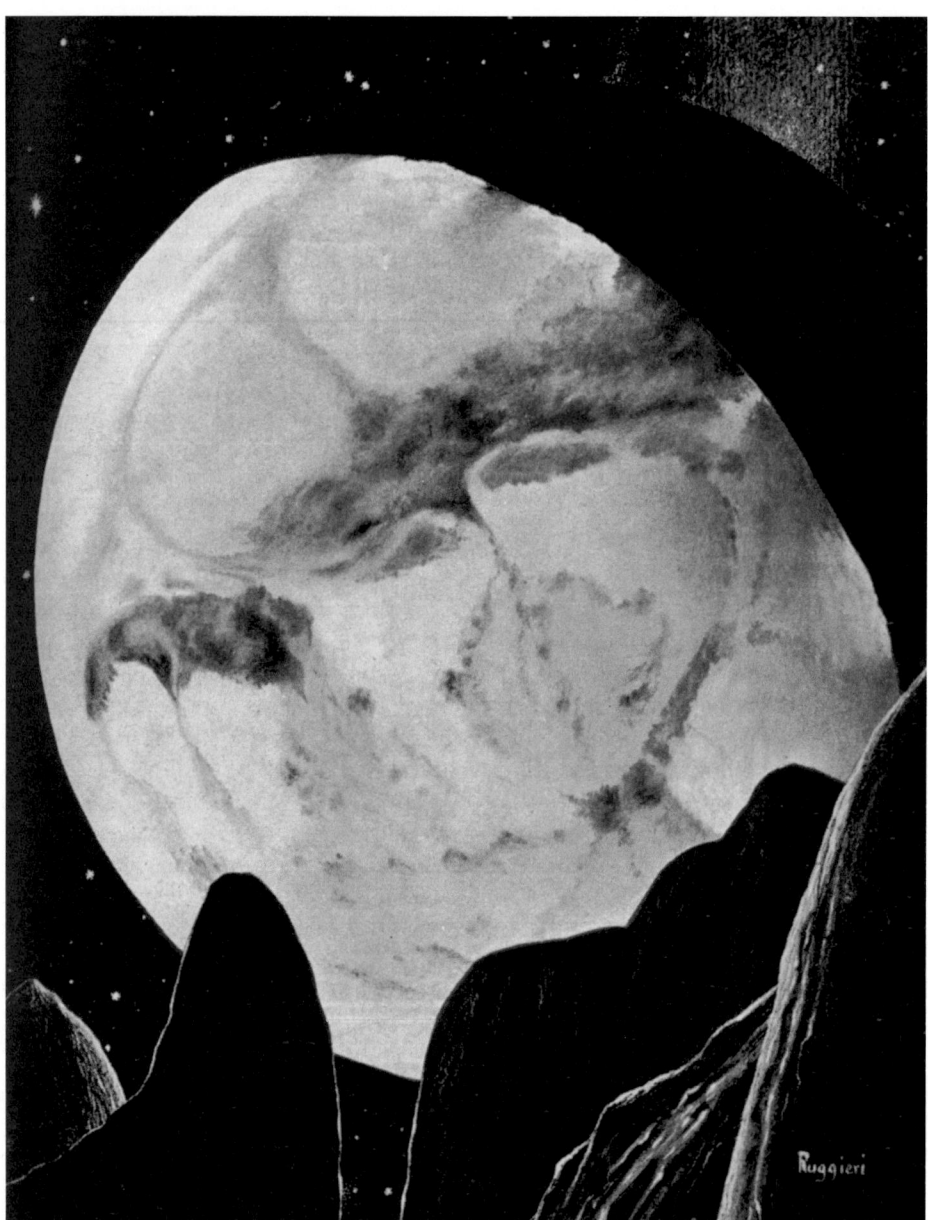

Mars seen from Deimos. The dark areas are Mare Sirenum and Mare Cimmerium (by coincidence, this is the area that would be imaged by Mariner 4 in 1965) and the dark spot below is Trivium Charontis. The three branches of the Avernus, Laestrygon and Antaeus 'canali' are also visible. (Artist: Guido Ruggieri)

when G.P. Kuiper used a new lead-sulfide infrared detector developed by the military and he detected carbon dioxide. Since this is a minor constituent of the Earth's atmosphere because it is removed by water and stored in carbonate rock, it seemed reasonable to presume that it must be a minor part of the Martian atmosphere. A variety of technical advancements provided estimates of the surface pressure ranging from a few tens to a hundred hectopascals. By further analogy with Earth, the principal gas was believed to be nitrogen, but this was not amenable to detection by spectroscopy from the Earth's surface because its absorption lines occur in the ultraviolet, and the Earth's atmosphere is opaque in this region of the spectrum. Although spectroscopy found no evidence for near-infrared reflection by chlorophyll on the dark areas, A.P. Kutyreva proposed that since Mars receives less sunlight, its vegetation might absorb energy across a broader portion of the spectrum, and Y.L. Krynov found that some terrestrial plants do indeed absorb in the near-infrared. In 1954 the biologist Hubertus Strughold argued that the Martian vegetation was a lichen, which is a symbiotic association of a fungus with an alga in which the fungus provides an isolated environment for the alga and lives off the wastes of the alga's photosynthesis – together they are able to thrive in an environment in which neither could survive alone. Gavril A. Tikhov postulated a surface pressure of 80–120 hPa, an atmosphere of predominantly nitrogen with a lesser quantity of carbon dioxide, and an environment not unlike the Siberian tundra – frozen and squalid, but still full of life. As director of the Alma Ata Observatory, he would influence the choice of payloads and the design of the first Soviet missions. In 1957 the American astronomer William M. Sinton detected near-infrared spectral absorption features that he suggested were due to the carbon–hydrogen bond of organic molecules. The 'Sinton bands' appeared to confirm that vegetation was present on Mars. Unfortunately, George Pimentel would establish in 1965 that these features in Sinton's spectra were caused by the presence of deuterated 'heavy' water in our own atmosphere.

With improved instrumentation, the search for water vapor resumed. At the near-perihelic opposition of 1954 the French flew a telescope on a high-altitude balloon in order to observe from above most of the vapor in our own atmosphere, but were unsuccessful. However, in 1963 Audouin Dollfus set up a special spectroscope high in the Swiss Alps and managed to utilize the Doppler effect arising from the relative motions of the two planets to isolate the spectral features of the Martian atmosphere from those of our own atmosphere. At about the same time, Hyron Spinrad at Mount Wilson obtained a spectrogram with a new infrared-sensitive film. Their results were confirmed later in the year by an American team using a telescope on a stratospheric balloon. Although the vapor was present only in tiny amounts, it was calculated that liquid water would be able to exist on the surface in the widely presumed 87 hPa as long as the temperature did not exceed 35°C. Harold Urey had made the radical suggestion in 1961 that no nitrogen was present, and if this was true, it would make Mars so hostile that not even hardy lichens could survive. In 1964 Spinrad completed his spectroscopic analysis by estimating the partial pressure of carbon dioxide at about 4.2 hPa and imposing an upper limit on the surface pressure of 25 hPa, which suggested that Mars was a very hostile environment.

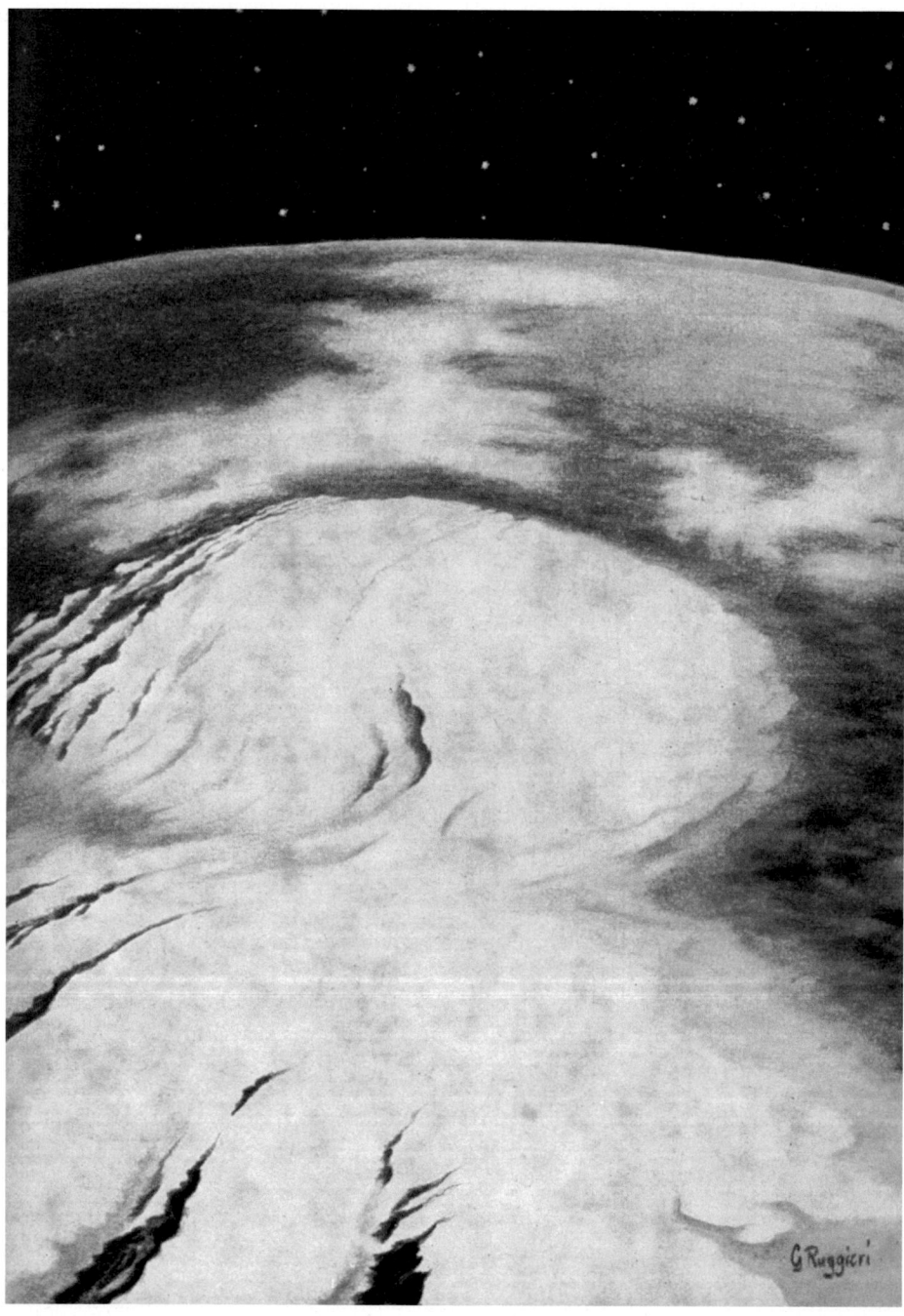

The south polar cap of Mars. Note how remarkably similar this drawing is to space probe views! (Artist: Guido Ruggieri)

Starting in the late 1950s, Dollfus had also taken polarization measurements of the disk of Mars, and concluded that the soil was similar to pulverized rock that was rich in iron oxides, which confirmed the impression of the planet as an arid desert. The polarization data also showed that the polar caps were made of ice of some kind, and that the white clouds that were observed from time to time were made of icy crystals and, therefore, were similar to terrestrial cirrus clouds. During the 1950s, there were suggestions that volcanoes might be active on Mars. When in 1954 an area the size of Texas rapidly turned dark, D.B. McLaughlin in America suggested that the dark areas were blankets of volcanic ash. On hearing of this, Tsuneo Saheki in Japan said that an unusually bright "flare" that he had seen at the terminator in 1951 might have been an eruption caught in the act. A number of observers reported anomalous flares at later oppositions, but these are now thought to have been due to sunlight glinting off clouds, hazes and fogs in lowlands.

On 9 March 1965, Mars reached opposition for the 167th time since the invention of the telescope; it was not a particularly favorable one for telescopic study, but just 4 months later the planet would be visited for the first time by a robotic emissary.

JUPITER: A BALL OF HYDROGEN

The planet Jupiter was the subject of one of the first observations of the modern era of astronomy, and quite possibly the single most important observation in the history of science. On 7 January 1610, Galileo turned one of his first crude telescopes to the planet and, in addition to the fact that it displayed a disk, he saw four bright star-like points close alongside it that appeared to change their mutual positions from night to night. He immediately appreciated that these were satellites revolving around Jupiter in orbits that were viewed from an almost edge-on perspective. The revolutionary significance of this discovery was that it proved that Earth, as Copernicus had said, was not alone as a center of celestial motions. Although Galileo called the satellites the 'Medicean Stars' after his protector, Cosimo de Medici, the Grand Duke of Tuscany, they were popularized by Kepler as the Galilean satellites, and this collective noun is still used today. In 1614 the German astronomer Simon Meyer, also known as Simon Marius, who may have seen them before Galileo, was inspired by four mythological lovers of Jupiter and assigned the satellites proper names in order of increasing distance from the planet: Io, Europa, Ganymede and Callisto. Although much less well known than his discovery of the satellites, Galileo also noticed that the planet's disk was appreciably flattened at the poles. The markings on Jupiter were so striking that they were able to be discerned by the telescopes of the first half of the seventeenth century. Dark equatorial belts were reported in 1630, and in 1664, Robert Hooke and G.D. Cassini independently noticed a dark spot located about one-third of a planetary radius south of the equator that spanned one-tenth of its diameter. Named the Great Red Spot, it often faded and reappeared – for example, it was not visible in the second half of the eighteenth century. By tracking atmospheric features, Cassini was able to determine that the planet rotates in just less than 10 hours, and this rapid rotation explained why the disk was flattened at the poles. Cassini was also the first to notice that spots near the equator moved at a greater rate than those more distant from it. The axis of rotation was inclined at 10 degrees to the orbital plane. This meant that during the 12-year-long circuit of the Sun there was little seasonal variation. Given the periods of the orbits of the satellites, it was possible to calculate the mass of the planet, which proved to be more than 300 times heavier than Earth; in fact, as would later be realized, Jupiter is not only the most massive planet in the solar system, but it is more massive than all the other planets combined. A measurement of its volume provided a figure for the bulk density, which was surprisingly only 35 per cent greater than that of water, which in turn implied that the atmosphere must be very deep and comprise lightweight gases. The planet was, in fact, a 'gas giant'.

In 1675 the Jovian satellites facilitated another key discovery of modern physics, namely that light travels at a finite speed. On noting that the times for the transits, occultations and eclipses of the satellites differed from the times predicted by their orbital motions, the Danish astronomer Olaus Roemer realized that the magnitude of the discrepancy correlated with the relative positions and distances of Earth and Jupiter, such that the light had farther to travel when Jupiter was far away, giving the impression of events being some minutes late, and vice versa. This enabled Roemer to calculate a figure for the speed of light very close to its actual value.

In 1860 G.P. Bond estimated that Jupiter radiated almost twice as much energy to space as it received from the Sun, from which he reasoned that the planet must be in the process of contracting and transforming gravitational potential into heat. It was concluded that the interior was a hot gas. A decade later R.A. Proctor wrote: "Jupiter is still a glowing mass, fluid probably throughout, still bubbling and seething with the intensity of the primeval fires, sending up continuous enormous masses of cloud, to be gathered into bands under the influence of the swift rotation of the giant planet." He was thinking of Jupiter as a 'failed' star. In the 1920s Harold Jeffreys proved that the visible surface could not be hot. He argued instead for a rocky core englobed by a mantle of ice and solid carbon dioxide, surrounded by a very deep but tenuous gaseous envelope. Logic suggested that the atmosphere should be composed primarily of hydrogen and helium, but since neither had lines in the region of the spectrum to which the Earth's atmosphere was transparent, this could not be verified. The most prominent lines in Jupiter's spectrum remained a mystery until the 1930s when Rupert Wildt realized that they corresponded to methane and ammonia, which were simple hydrogenated compounds of carbon and nitrogen, respectively. Other constituents were present only in trace amounts. The fact that the atmosphere was hydrogen-rich meant that, in terms of chemistry, it was a 'reducing' environment. The presence of hydrogen and helium suggested to Wildt that Jupiter could have a 'cosmic composition' similar to that of the Sun – that is, the planet's intense gravity had enabled it to retain these lightweight gases, whereas they would have had sufficient thermal agitation speeds to leak to space from the Earth's upper atmosphere. Wildt refined Jeffreys' idea by proposing that the rocky core was surrounded first by a thick layer of water ice and, in turn, by an ocean of condensed gases. The next real advance was the independent suggestion in 1951 by W.R. Ramsey and W. DeMarcus that the core was not rock but metallic hydrogen, and that this was surrounded first by an ocean of liquid hydrogen and then by the hydrogen-rich gaseous envelope. Such a metallic core would readily conduct electrical currents that would, in turn, generate a magnetic field, although at the time it was not apparent how this prediction could be put to the test. In 1955 B.F. Burke and K.L. Franklin found that Jupiter emitted radio waves. It was a chance discovery, since they were not studying the planet; in tracking down a 'noise' they found that it came from a celestial source. All planets should emit radio radiation of thermal origin, but at such a low intensity as to be almost undetectable. In the case of Jupiter, not only was the emission much stronger, but its characteristics implied high-energy processes. Soviet scientist Iosif S. Shklovsky identified the process as synchrotron radiation, which electrons emit as they spiral along the lines of force of a powerful magnetic field. With further observations, it became evident that there was a periodicity of 9 hours 55 minutes which, assuming the hypothesis that, as on Earth, the magnetic field was more or less rigidly tied to the planet's rotation, gave a measurement of the rate at which the core rotated. Remarkably, this differed by several minutes from the period obtained by tracking the atmospheric features.

In 1892 E.E. Barnard found a fifth satellite. Named Amalthea, it orbited closer in than Io, and was much fainter than the Galileans. Although this was the last satellite to be discovered visually at the telescope, the retinue was further expanded between 1904 and 1951 by seven photographic discoveries – all of which occupied distant irregular orbits inclined to the equatorial plane, in some cases in a retrograde sense, with periods ranging from 8 months to almost 2 years, and were at most several hundred kilometers in size. In 1975 the International Astronomical Union assumed responsibility for naming planetary satellites. It promptly decided to name the objects in distant irregular orbits after Jupiter's lovers, of which there was no shortage in mythology, with the proviso that those in prograde orbits must have names that end with 'a' and those in retrograde orbits must have names that end with 'e'. The names by which astronomers had referred to their satellite discoveries were dismissed; thus, for example, Hades became Sinope.

When W.H. Pickering was observing from Peru in 1892, he gained the impression that the disk of Io varied between circular and elliptical, suggesting that the satellite was egg shaped. Efforts to derive a period of rotation from these observations were inconclusive. On perceiving similar changes in the shapes of the other satellites, he even speculated that the moons had formed by the agglomeration of either meteoric dust or the dust of a ring system that existed early in the history of the system. When scrutinized with larger and better telescopes, Io appeared elongated during transits of Jupiter, at which times its dark polar regions and bright equator were viewed against the backdrop of Jovian clouds; otherwise it failed to reveal any deviation from circularity, and it was concluded that the distortions reported by Pickering must have been due to flaws in his telescope. In 1925 Paul Guthnick published 'light curves' assembled from photometric observations over an extended time which showed that, just like the Earth's Moon, the Galilean satellites were tidally locked. In 1849 W.R. Dawes had observed that Ganymede seemed to have a bright patch at its north pole. Further observations were made by E.E. Barnard and E.M. Antoniadi, and in 1951 E.J. Reese made a map that integrated their best observations. A few

Jupiter and its Great Red Spot as they would appear from the surface of Amalthea, based on 1954 observations of the planet. (Artist: Guido Ruggieri)

years later B.F. Lyot published maps of all four Galileans based on observations made by observers at Pic du Midi. In 1961 Audouin Dollfus published another set of maps. The general impression was that Io was yellowish with dark polar regions; Europa was uniformly white and of such a high albedo as to suggest that it had a frosty surface; Ganymede, in addition to its bright polar spot, had some very dark regions; and Callisto was not unlike the Earth's Moon in having a dull gray disk. In the 1940s, G.P. Kuiper found no spectroscopic evidence of atmospheres, but an infrared absorption band implied that Europa was covered by a substantial layer of water snow and that, on Ganymede, the snow was confined to the north polar region; neither Io nor Callisto showed any such feature. Upon realizing that Ganymede would occult a bright star on 13 August 1911, Friedrich Ristenpart had been able to measure its diameter, and, by assuming that the other Galileans had similar albedos, inferred their diameters. The subsequent studies of the actual albedos enabled these sizes to be refined. Ganymede proved to be one of the largest moons in the entire solar system; in fact, it is larger than the planet Mercury. Whereas the bulk density of Io was found to be comparable to that of the Earth's Moon, suggesting that Io was primarily rock, the lower densities of the other Galileans suggested that they contained significant fractions of ice, with this fraction increasing with distance from the planet.

If the features observed on Jupiter represented only the outermost few kilometers of an immensely deep atmosphere, then what was the Great Red Spot? An early idea was that it was a static cloud above a tall mountain. Although the varying altitude of the ammonia clouds could explain the observed changes in the size and color of the spot, it sometimes being submerged, the fact the spot was known to move at varying speed made this untenable. The spot was evidently some sort of a stable structure in the upper atmosphere. Interestingly, in the 1930s several 'white ovals' erupted at the edge of the southern equatorial belt. In mutual encounters over the ensuing decades they have progressively merged, and now form a single feature which, as a result of a newly acquired hue, has been named 'Red Junior'.

SATURN, ITS RINGS AND MOONS

When Galileo observed Saturn through a telescope in July 1610 he was astounded to see what appeared to be a small disk in apparent alignment at each side of the planet, in a configuration that did not change from one night to the next. His puzzlement was increased when he pointed an improved telescope at the planet two years later and found the smaller disks to be absent, yet present again in 1613. The mystery of the 'companions' of Saturn was resolved in 1655. On 25 March of that year Christiaan Huygens in the Hague pointed a telescope of his own making at Saturn, and although he was disappointed to find that the companions were barely visible, he saw a thin dark line spanning the planet just north of the equator. Over the ensuing months the companions completely disappeared, and by early 1656 so too had the line. But then the line reappeared south of the equator, and the companions once again began to be visible. In a moment of epiphany, Huygens realized that the companions were views of parts of a continuous ring that was centered on, but detached from, the planet. He also realized that since the ring was inclined to the plane of the planet's orbit around the Sun, the line of sight from Earth would pass through the ring plane twice in each of the planet's 29.5-year orbits, at which time the ring would be rendered invisible – as it had been in 1612 for Galileo. Furthermore, during his observations of Saturn, Huygens also discovered that it was accompanied by a bright satellite, later named Titan, with an orbital period of 16 days.

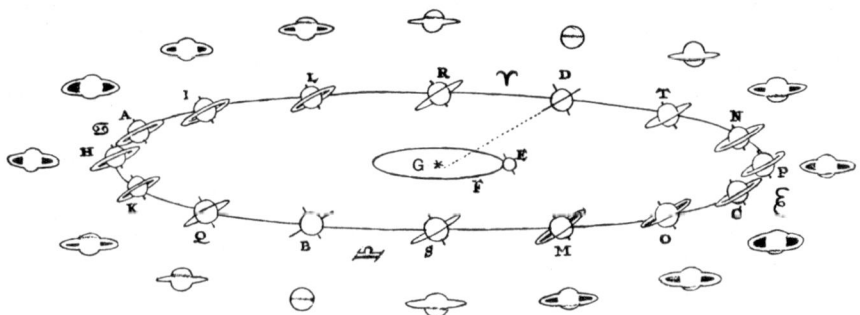

In his *Systema Saturnium* in 1659, Christiaan Huygens showed how an inclined ring explained the cyclical manner in which Saturn's appearance varies. As Earth passes through the ring plane, the ring becomes invisible.

Giovanni Domenico Cassini made several discoveries about Saturn, its ring and moons. In 1671 he discovered a second satellite, Iapetus, and the next year he found Rhea much closer to the planet. Iapetus was surprising, because it was easy to see on one side of its orbit but much fainter on the opposite side. He correctly inferred that the moon must rotate synchronously with its orbit, and that its 'leading' hemisphere must be unusually dark. In 1675 he identified a dark gap about two-thirds of the way out on the ring. This became known as the Cassini Division, with the outer ring labeled A and the inner ring labeled B. In 1684 Cassini spotted the fainter satellites Dione and Tethys. Until the 1780s the two rings were believed to be solid disks, but Pierre Simon de Laplace showed that such objects would not be stable against gravitational tides and argued that they might be made of a multitude of narrower 'ringlets' that were so closely packed as to be beyond the resolution of telescopes. In 1855 James Clerk Maxwell proved that such a configuration would be unstable, but in 1857 he pointed out that the rings *could* comprise millions of smaller bodies that traced more or less orderly circular orbits, bestowing the appearance of solidity. This hypothesis was verified by spectroscopic measurements by J.E. Keeler in 1895, which showed that the velocity of rotation of the rings varied from the inner to the outer edge in accordance with Kepler's laws. Meanwhile, as William Herschel was testing his new 1.2-meter-diameter reflector in 1789, he discovered Mimas and Enceladus orbiting just outside the ring system. From the fact that even with such a powerful telescope the rings were invisible when viewed edge-on, Herschel calculated that they could be no more than 500 km in thickness. Despite the evanescent nature of the features in the planet's atmosphere, he was able to derive an accurate measurement of its 10.6-hour rotational period. In 1848 Hyperion was independently spotted by astronomers in England and America, orbiting just beyond Titan. In 1837 J.F. Encke saw a thin gap near the outer edge of the A ring, and this became known as the Encke Division. In 1850 W.C. Bond and W.R. Dawes independently discovered a faint ring inward of the B ring that became known as the C ring. In 1871 Daniel Kirkwood found that there were statistically significant 'zones of avoidance' due to orbits at certain distances from the planet being in resonance with the orbits of some of the moons that orbit just beyond the ring system, and he showed that these were responsible for clearing the observed divisions. At almost 80 per cent, the albedo of the rings indicated that their particles must be mostly ice.

The presence of the satellites enabled the density of Saturn to be calculated, and it proved to be even more lightweight than Jupiter, its bulk density being just 70 per cent that of water. If an ocean of sufficient size were available, then the planet would float! The study of the physical conditions of Saturn began in 1905, as V.M. Slipher obtained the first detailed spectra of its atmosphere. He noted that the spectrum was reminiscent of that of Jupiter, but could not identify the strongest absorption bands. When Rupert Wildt reanalyzed Slipher's spectra in 1931 he suggested that methane and ammonia were responsible, and this was confirmed in 1933 by laboratory tests by Theodore Dunham. Wildt had also suggested that Saturn must have a rocky and metallic core surrounded by a layer of water, ammonia and methane ice, which, in turn, was enshrouded by a dense atmosphere of primarily hydrogen. However, because Saturn was colder and of weaker gravity than Jupiter, he inferred that its cloud layers would be located at a greater depth and be masked by haze – which was why the planet's appearance was generally so bland.

When W.H. Pickering discovered Phoebe in 1898 this was the first photographic discovery of a planetary satellite. In addition to its orbit being elliptical with a period of 546 days, it was tilted 30 degrees to the equatorial plane of the Saturnian system. However, the fact that the equator of Saturn was tilted 26.75 degrees to the plane in which the planet orbits the Sun meant that the moon's orbit almost matched the ecliptic, which suggested that it was a 'captured' object. This was reinforced by the fact that the orbit was retrograde. Phoebe would remain the last satellite of Saturn to be discovered prior to the space age.

Perhaps the most remarkable discovery of the first half of the twentieth century in respect of a member of Saturn's retinue was made in 1908 when José Comas Solá observed a limb-darkening effect on the tiny disk of Titan, which he took to indicate that the moon must have a significant atmosphere. This remained speculative until a spectroscopic analysis in 1944 by G.P. Kuiper established the presence of gaseous methane, indicating a reducing environment. In fact, of all the planetary satellites in the solar system, only Titan has a significant atmosphere.

URANUS AND NEPTUNE: OUTER GIANTS

On 13 March 1781, while the German-born English amateur astronomer William Herschel was sweeping the sky with a telescope of his own construction, he found a green blob that was more akin to a comet or nebula. Four days later he noticed that it had moved, which suggested that it was a comet. On scrutinizing it more closely, he noted that the object had no coma or tail, such as would be expected of a comet, and, in fact, far from appearing fuzzy, it displayed a small disk not unlike that of a planet. He reported the discovery to Nevil Maskelyne, the Astronomer Royal, who ventured that it might actually be a planet. Further observations during that summer enabled A.J. Lexell in St Petersburg and P.S. de Laplace in Paris to independently determine its orbit, which proved to be almost circular, was about twice as far from the Sun as Saturn, and had a period of 84 years. Herschel wanted to name the new planet 'Georgium Sidus' after King George III, but this was rejected by his colleagues on the continent. J.J. Lalande in Paris suggested 'Herschel's Planet', but this did not find favor. J.E. Bode in Berlin proposed Uranus, because in mythology Uranus was the father of Saturn, just as Saturn fathered Jupiter, who in turn fathered Mars, and this was adopted.

In 1787 Herschel found that Uranus had two satellites. He suspected it of having four others, in addition to a faint ring, but was not certain and later decided that he had been mistaken. Over the next 10 years or so, he found that the plane in which they orbited the planet was tilted 98 degrees from the ecliptic! In 1851 the English amateur astronomer William Lassell discovered two additional satellites and, inspired by the literature of William Shakespeare, he and John Herschel decided to name the first pair of moons Titania and Oberon and the second pair Ariel and Umbriel. In 1948 G.P. Kuiper discovered a fifth satellite orbiting much nearer the planet, and in keeping with the scheme this was named Miranda.

The fact that there were satellites enabled the mass of the planet to be determined. It was presumed that the axis of the planet's rotation was perpendicular to the plane in which the satellites orbited it – that is, that for some mysterious reason the entire system was tipped over – and this was verified by G.V. Schiaparelli who, on seeing that the planet's disk was oblate, also noted that the 'equatorial bulge' matched the satellite orbits. There were occasional reports of faint banding, but no features that would enable the rotational period to be determined. By virtue of having its spin axis oriented within 10 degrees of the plane in which it orbits the Sun, sometimes Uranus faces its northern hemisphere toward the Sun and at other times it faces its southern hemisphere toward the Sun, giving it a unique seasonal cycle. E.M. Antoniadi made a study of Uranus when it was at equinox in the 1920s and evenly illuminated as it rotated: under the best 'seeing' he observed a faint band on each side of the equator and indications of some even fainter bands, but no individual atmospheric features. Angelo Secchi was the first to examine the spectrum of the planet, and saw several broad absorption bands, but had no idea of their cause. In 1932 Rupert Wildt identified these as methane. The planet's green hue arose because the red end of the spectrum was being absorbed by methane. As the space age began, therefore, the remarkable Uranian system was an almost total mystery.

As soon as Uranus was realized to be a planet, astronomers searched the archives for previous sightings, and it transpired that its position had been noted 22 times – the fact that its true nature had not been recognized was excusable, as the telescope made by Herschel was far superior. In fact, because the earliest sighting was by John

Uranus at spring equinox as seen from Miranda, 130,000 km away. No one expected Miranda to turn out to be a world far more exciting than Uranus! (Artist: Guido Ruggieri)

Flamsteed in 1690, it was possible to determine the orbit using data from a complete circuit of the Sun. However, as observations continued, it was soon realized that the planet was drawing ahead of its predicted position. Either Newton's laws of gravity were not working as they should, or something was perturbing the planet. The initial reaction was to reject the prediscovery sightings, but their veracity was self-evident, and even when this was done and the orbit recomputed, the planet continued to draw ahead. Intriguingly, after years of progressively accelerating, the planet settled down around 1822, but when it then started to lag behind, astronomers realized that it was being perturbed by an undetected planet. In principle, Newton's laws would enable the position of the perturber to be determined from its observed effects, and two young mathematicians, John Couch Adams in England and Urbain Jean Joseph Le Verrier in France, independently took up the challenge.

In 1766, while pondering mathematical relationships, J.D. Titus had noted that the mean heliocentric distances of the planets obeyed a simple numerical progression – although for some reason there was no planet for the member of the series between those for Mars and Jupiter. In 1772 J.E. Bode suggested that there might be a planet at that position, and (as related below) a search revealed rather more than had been expected. It was not known whether this numerical progression was purely empirical or related to a physical law that determined the spacing between planetary orbits, but in order to begin their process of computation both Adams and Le Verrier decided to exploit the relationship to gain the mean heliocentric distance of the perturber. After two years of work, in October 1845 Adams had a solution. He sent a note outlining his calculation to G.B. Airy, the Astronomer Royal. In fact, in 1834 T.J. Hussey had urged Airy, a noted mathematician in his own right, to undertake such a computation himself, and Airy, after giving the matter some thought, had concluded that the tools available to mathematics were inadequate to the task. He was therefore unwilling to believe that a recent graduate with no prior record of research could have succeeded. Airy was further put off by the fact that Adams had neglected to account for both of the observed anomalies (in longitude and in distance) of the motion of Uranus, and had not given coordinates for where the new planet should be found. In fact, Adams had actually pinned down the planet's position to within 2 degrees, and if a search had been made that autumn the planet would very likely have been located. On 1 June 1846 Le Verrier told the French Academy of Sciences that he had shown mathematically that there *had* to be a planet perturbing Uranus, but he was not taken seriously. By 31 August he had a solution and decided to contact an observer with whom he had had personal contact to explain his results. On 18 September he wrote to J.G. Galle at the Berlin Observatory, ending his letter by saying that if Galle were to study a particular patch of sky he should find the new planet. Galle received the letter on 23 September and that evening he and H.L. d'Arrest began to compare their star chart with the view through the telescope, and within an hour had spotted the planet as a tiny blue disk – it was within one apparent diameter of the Moon of the predicted spot. Its planetary nature was proved by the fact that the next day it had moved. Although Galle suggested the name Janus, Le Verrier had already chosen the name Neptune. In fact, on hearing of Le Verrier's calculation, Airy had asked J.C. Challis at the Cambridge

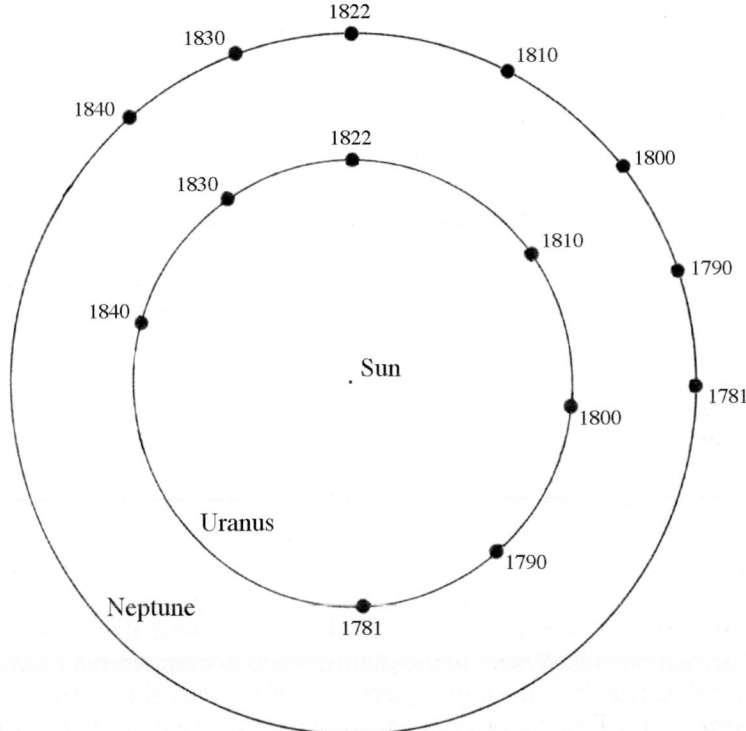

This plot shows the relative positions of Neptune in relation to Uranus during the years after the latter's discovery in 1781, when the planet was being mysteriously accelerated ahead of its calculated position. After 1822, astronomers were perplexed to find Uranus begin to lag behind. A mathematical *tour de force* that calculated the location of the perturbing object led directly to the discovery of Neptune.

Observatory to conduct a search for the planet, but Challis set out to make a detailed map of the specified patch of sky as a preliminary, and plotted the planet on 4 August and again on 12 August – only on switching to a higher magnification on 29 September did he perceive the disk that indicated its true character. Although there was a dispute about priority, the credit is usually assigned to Le Verrier.

As in the case of Uranus, it was found that Neptune had been repeatedly included on old star charts. In fact, the first observation proved to have been made in 1610 by none other than Galileo, who witnessed a rare close conjunction between Jupiter and Neptune, during which the latter appeared to pass through the Medicean satellites. Tantalizingly, Galileo even seems to have suspected the 'background star' of having moved from one night to the next, but he chose not to pursue the matter. As a result of the prior sightings, Neptune's orbit was readily computed. Its mean distance from the Sun proved to be less than expected on the basis of the Titus–Bode relationship. On 10 October, less than a month after Neptune's discovery, Lassell saw a relatively bright star nearby which he presumed to be a satellite, and this proved to be the case.

The moon, later named Triton, enabled the mass of the planet to be calculated. The calculations by Le Verrier and Adams had suggested that it would be twice the mass of Uranus, but they proved to be similar. Nevertheless, the fact that Neptune's orbit was closer than expected to Uranus meant that it was able to account for most of the observed perturbations.

Triton proved to be an unusual satellite: its orbit of Neptune was retrograde and significantly inclined to the planet's equator. Some satellites of other planets are known to orbit in this manner, the most celebrated case being Phoebe at Saturn, but they are all small bodies. Despite its large size, Triton's motion suggested that it had been captured – but until recently the manner in which this might have been done faced severe mathematical problems. The reason that Triton was discovered so soon after Neptune, was that it was very bright, which in turn suggested that it must be of considerable size and, being so far from the Sun, would be so chilly that it could even retain an atmosphere of some sort. However, it was not possible to confirm this from its star-like appearance.

The few astronomers with telescopes capable of showing detail on the minuscule disk of Neptune reported bright equatorial belts and occasional dark spots. Some of the best drawings were made in 1948 at Pic du Midi, and while they showed no trace of bands or belts they exhibited diffuse and irregular spots; however, it was not possible to determine a reliable rotational period. As with the other giant planets, the spectra of Neptune taken early in the twentieth century remained a puzzle until the 1930s. Although hydrogen was believed to be the principal constituent of all the giant planets, the methane bands were stronger on Uranus and Neptune, almost certainly because they were sufficiently cold to freeze most of the ammonia out of the atmospheres, while a thick hydrogen haze would obliterate most details from their disks. In fact, the temperature at Neptune's cloud tops might allow methane to condense. Although direct observations had not been able to measure the planet's rotational period, spectroscopic studies imposed some constraints. In 1949 G.P. Kuiper discovered a second satellite. Named Nereid, it was extremely faint, and in such an eccentric orbit that at its furthest distance it was over 9 million km from the planet. With an estimated diameter of 300 km Nereid was a fairly large object, but as the space age began this was all that was known about it.

PLUTO: THE INCREDIBLE SHRINKING PLANET

Although the presence of Neptune accounted for most of the observed perturbations of Uranus, the fact that there remained a discrepancy raised the prospect that another planet had yet to be discovered. Using a development of the technique used by Le Verrier, Percival Lowell computed an orbit for it. The initial search found nothing, but upon his death in 1916 he bequeathed funds to continue the search, and he had himself buried in the grounds of his observatory to spur on his successors. In 1929 Clyde Tombaugh, a young amateur astronomer, was hired to take and analyze photographic plates of the sky where the putative planet was expected to be. On 18 February 1930, when comparing plates taken near the star delta Geminorum that had been taken on 19 and 23 January, he noted a faint star that appeared to be moving in the manner of a distant planet. After further plates confirmed this, the discovery was announced on 13 March in order to celebrate the 75th anniversary of Lowell's birth. The observatory received many suggestions for names for the planet, and decided in favor of the suggestion by Venetia Burney, an 11-year-old schoolgirl in Oxford in England. On 1 May it was announced that the planet had been named Pluto, after the Roman god of the underworld – which seemed fitting for an object that spent its time in the perennial darkness of the outer solar system – and that its official symbol would be 'PL', signifying the initials of Percival Lowell.

The first scientific task was to identify Pluto's orbit, and here the surprises started. Whereas the other planets followed nearly circular orbits, the eccentricity of Pluto's orbit was 0.25, meaning that its perihelion distance was 40 per cent less than that of its aphelion. Owing to this eccentricity, the perihelion of Pluto's 248-year orbit was inside the orbit of Neptune, but the fact that the plane of Pluto's orbit was inclined at 17 degrees meant that the paths of the two planets did not intersect. Even before Pluto's discovery was announced, the astronomers at Lowell's observatory had tried to resolve its disk without success, and it was so faint that the usual spectroscopic, photometric and polarimetric techniques could not be used; the only data available was its orange-yellowish color. When its diameter was measured by G.P. Kuiper in 1950 by fitting the largest telescope in the world with a device to project a calibrated disk into the field of view alongside the image of Pluto, he measured just 5,900 km. The fact that Pluto was so small posed a problem: to account for the observed perturbations its mass would have to be comparable to Earth, but with a diameter of only 46 per cent that of the Earth its density would have to be implausibly high – in fact, it would have to be fully metallic – and there was no reason to expect that such an object would have formed so far from the Sun.

In 1936 A.C.D. Crommellin suggested that Pluto's surface was so smooth that what we saw was a specular reflection from just the central portion of its disk, and that in reality its diameter was much greater. If so, if the planet were to be observed to pass in front of a bright nebula or to occult a star then its size would become measurable. When Pluto 'failed' to occult a star in 1965, this established that its diameter did not exceed 6,800 km. The presence of a satellite would have enabled the issue of Pluto's mass to be resolved, but despite searches, first by Tombaugh immediately after the discovery and in the 1950s by Milton Humason, none was found until 1973. As we now know, Pluto is considerably smaller than Kuiper's estimate, making it even less likely to have caused the perturbations that were used to predict its existence, and raising the possibility that its discovery was a lucky happenstance.

In 1955 photometric observations established that Pluto's period of rotation was 6 days 9 hours, and the presence of marked variations in brightness within a single cycle that did not change over time suggested that the surface was made up of bright and dark areas, and that there was no noticeable atmosphere. One proposal for how Pluto came to occupy such a strange orbit posited that it was originally a satellite of Neptune, and was ejected during the same event that placed Triton into a retrograde orbit, but this has since been rejected as virtually impossible. At the start of the space age, therefore, we knew very little about Pluto.

ASTEROIDS: THOSE FANTASTIC POINTS OF LIGHT

At the close of the eighteenth century, European astronomers formed a 'sky police' to coordinate efforts to find the 'missing planet' between Mars and Jupiter suggested by the Titus–Bode series. On 1 January 1801, while compiling a star catalog for the Palermo Observatory, Giuseppe Piazzi discovered an object that he described as "something better than a comet". He tracked its motion for several weeks before losing it to daylight. The data allowed young Karl Friedrich Gauss, probably the greatest mathematician of all time, to compute its orbit, proving it to reside between Mars and Jupiter. Armed with Gauss's calculation, astronomers were able to recover it on 31 December. Following tradition, Piazzi decided to name it Ceres Ferdinandea in homage to Ferdinand, the Borbone king of Naples and Sicily, but it is now known simply as Ceres. Just as astronomers were celebrating the fact that the missing planet had been found, on 28 March 1802 H.W.M. Olbers spotted Pallas, on 2 September 1804 K.L. Harding discovered Juno, and on 29 March 1807 Olbers found Vesta – all traveling in similar orbits. In 1830 K.L. Hencke began a systematic search and found others, and the discoveries continued. After attempting in vain to discern the disk of Ceres, William Herschel reasoned that since such objects were too small to be resolved they shouldn't be referred to as planets but as 'asteroids' – objects that look like stars. This term is still widely used today, although they are now officially known as 'minor planets'. When it became evident that the early discoveries were simply the largest members of a large group that shared similar orbits, the term 'asteroid belt' was coined. As the number approached 100, statistical analyses discovered unsuspected characteristics. In 1866 Daniel Kirkwood found that the distribution of objects was not continuous, because there were 'zones of avoidance' at several mean heliocentric distances. He further found that the periods of such orbits were in resonance with Jupiter, such that if an object were to have an orbit at such a mean heliocentric distance it would be in conjunction with Jupiter always at the same position around the Sun, and would soon be perturbed into another orbit. As a result, Jupiter has cleared gaps in the belt. Kirkwood and Japanese astronomer Kiyotsugu Hirayama then discovered that some asteroids have such similar orbits that they may well be relics of larger asteroids that became fragmented.

In fact, for a long time it was thought that the asteroids were the remains of a planet that had broken up – but because the combined mass of the asteroids was no greater than the mass of our Moon, it was evidently only a small planet. G.P. Kuiper then proposed a scenario in which, as the planets formed, a dozen 'primitive' objects were left just sunward of Jupiter, and that by a succession of mutual impacts and fragmentations they gave rise to the belt. It was to be expected therefore that the asteroids would be irregularly shaped and heavily cratered. In fact, long-term studies established that the varying 'cross section' as an asteroid rotated produced complex changes in brightness, and a large number of rotational periods were determined this way. However, Kuiper's research marked a high point in professional interest in the asteroids, and there was then a lull until there was the prospect of sending spacecraft to inspect them.

Nevertheless, a few asteroids received particular attention. The 433rd asteroid was discovered by Gustav Witt on 13 August 1898, and named Eros. It was noteworthy because the eccentricity of its orbit enabled it to cross the orbit of Mars and almost reach the orbit of Earth. In fact, it proved to be the prototype of the Amor group of asteroids. Asteroid Apollo, discovered by K.W. Reinmuth on 24 April 1932, was the first member of the Near-Earth Asteroids. Unlike the Amors, the Apollos can cross the orbit of the Earth and therefore pose a risk of collision. The case of Hermes is particularly interesting. At the time of its discovery by Reinmuth on 28 October 1937, it was in the process of making a close encounter with Earth, passing by at a range not quite twice that of the Moon's orbit. It was several kilometers in size, and if it had struck Earth it would have done so with an energy many times greater than even the most powerful nuclear explosion. In fact, on 30 June 1908 either a small asteroid or a comet disintegrated several kilometers above Siberia and leveled a vast area of the Tunguska pine forest. Other groups of asteroids have also fascinated astronomers. Discovered by Max Wolf on 22 February 1906, Achilles shared Jupiter's orbit, some 60 degrees ahead of the planet. In fact, in 1772 Joseph-Louis de Lagrange had predicted such a configuration as a stable solution of a gravitational system with two large bodies (in this case the Sun and Jupiter) and a third of negligible mass (the asteroid). When an asteroid was found later that year in the corresponding stable point in the trail of Jupiter, it was suspected that there were probably many more, and this proved to be the case. As Achilles was the hero of Homer's epic poem *The Iliad*, which relates the Trojan War, it was decided to name the leading group after Greek heroes and the trailing group after the heroes of Troy – although before this scheme was introduced one of the leading group had been named after a Trojan and a member of the trailing group was a Greek! On 30 October 1920 Walter Baade discovered Hidalgo, which was traveling in an orbit ranging from the main belt to the orbit of Saturn and was highly inclined to the ecliptic – an orbit that was typical of a comet, suggesting that perhaps it was a 'dead' cometary nucleus. Hidalgo would remain the only asteroid known beyond Jupiter until Chiron was discovered by C.T. Kowal on 18 October 1977 in an orbit between Saturn and Uranus that made it the prototype of the Centaur group. At the time of its discovery it was near aphelion and asteroidal in character, but as it approached perihelion in the late 1980s it developed a cometary coma. Nevertheless, at no time had any of the main belt objects shown any signs of a gaseous envelope. At the start of the space age, in excess of 1,600 asteroids had been listed and it was evident that there were many more. Indeed, to astronomers exposing photographic plates they represented 'vermin of the sky'.

COMETS: FLYING SANDBANKS OR DIRTY SNOWBALLS?

Although ancient philosophers knew of the existence of comets, they were uncertain whether comets were of celestial or meteorological character. Aristotle believed that comets were a phenomenon not too dissimilar to clouds, but Egyptian and Chaldean astronomers believed them to be celestial. The Roman philosopher Lucius Anneus Seneca, one of the former preceptors of emperor Nero, devoted an entire chapter of his 'scientific' treatise *Naturales Quaestiones* of 62–63 AD to comets, in which he supported the celestial interpretation and gave proofs in favor of it. The apparitions of hundreds of bright comets were documented by observers in China, Korea, Japan, Europe, the Middle East and possibly also in pre-Columbus America, but it was not until the sixteenth century that the first fundamental discoveries were made. After observing three comets in 1531 and 1532, P. Apianus and G. Fracastorius noted that as the objects traveled across the sky their tails were always pointing away from the Sun, which argued against the meteorological interpretation. During the appearance of a bright comet in 1577, Tycho Brahe noted that, when seen by observers at widely separated sites, the comet appeared in more or less the same part of the sky, meaning that it was at least four times further out than the Moon. But establishing the celestial nature of comets merely raised the issue of the means by which they traveled across the sky.

A bright comet was discovered in 1682 and seen by many observers, including Edmund Halley, a 26-year-old astronomer and member of the Royal Society. In analyzing reports of this and a number of other comets employing a technique for computing a heliocentric parabolic orbit from observations, he realized in 1695 that the orbit of the 1682 comet resembled that of comets that had been seen in 1607 and 1531, and he predicted that it would return again in 1758. Halley did not live to see it because he died in 1742, but after it returned as predicted it became known as Halley's comet. It proved conclusively that comets were astronomical objects in closed orbits around the Sun, and fulfilled the 'prophecy' by Seneca: "Someone will discover one day in which region of the sky comets travel." Spotting comets and computing their orbits became a something of a sport for astronomers. Then, during the second half of the nineteenth century two major advances were made. First, in 1867, in the work that first made him known in the astronomical community, G.V. Schiaparelli provided a link between comets and the meteors that penetrate and are destroyed in the Earth's atmosphere, by pointing out that the Perseid shower, which is most active in early August, occurs when Earth crosses the orbit of periodic comet Swift–Tuttle. He was also able to establish a link between the Leonid shower in November and Tempel–Tuttle. In addition to a coma and a yellowish tail, the brightest comets also possessed a bluish tail. Spectroscopy revealed that the light of a coma and yellowish tail derived primarily from sunlight scattered by dust, whereas the bluish tails glowed by emissions from ionized compounds of hydrogen, carbon, oxygen and nitrogen. One of these compounds was cyanogen, a molecule composed of carbon and nitrogen. The fact that this is poisonous caused a small scare when it was realized that Earth would pass through the tail of Halley's comet in 1910. In the 1940s it was established that all the chemical species observed in comets were derived from the dissociation of stable molecules like water, methane, etc., although these could not be observed directly.

For centuries after Halley's epiphany, astronomers were at odds to explain what a comet actually was. Until the 1940s it was widely thought that a comet was a 'flying sandbank', whereby dust and debris traveled similar but independent orbits. As the comet neared perihelion, the individual particles would jostle each other and generate a cloud of dust. But this model could not explain the presence of gas; the tendency of cometary nuclei to split around perihelion; the outbursts that would make comets brighten in the course of a few hours; their 'rocket like' perturbation effects; and the star-like appearance of the only nucleus that had been seen at aphelion – because the duration of comet Encke's orbit was just 3.3 years, its aphelion was at only 4.1 AU, and in 1913 it had been photographed near the time of its aphelion. In 1950, after studying comet Encke, F.L. Whipple proposed an alternative that would become the standard model. Whipple said that a cometary nucleus was an asteroid-like object made mostly of ice mixed with dust, in effect a 'dirty snowball'. On approaching perihelion, the ice would sublime to gas, in the process releasing some trapped dust that would form the dense spherical coma around the nucleus. Gases would then be ionized to form the bluish tail that pointed away from the Sun, while the dust formed the yellowish tail. The orientations of the tails suggested that comets were immersed not only in an interplanetary magnetic field but also in a 'wind' of energetic charged particles streaming out from the Sun. While spewing out matter, the nucleus would be subject to a rocket-like recoil (or 'non-gravitational effects') that slightly changed its trajectory – just as had been observed.

As with asteroids, as soon as a good number of comets were known astronomers were able make statistical inferences; in particular that there were families of short-period comets that were gravitationally associated with the giant planets, usually by having their aphelia near the planetary orbits. However, because their orbits vary over time as a result of perturbations not all members of a particular family have similar orbits. The largest family was associated with Jupiter, as its enormous mass could readily influence the path of an object as small as a comet. In some cases they could also be captured by the planet, as was demonstrated by comet Shoemaker–Levy 9, discovered in 1993. By the start of the space age, a census of comets had been compiled, listing all their characteristics and relationships. It detailed how they had been observed to brighten and fade without any apparent reason, to split, to disintegrate, to disappear, to change orbit, to graze the Sun to within less than its radius or to approach the Sun no closer than Jupiter, to graze Earth, etc., or, indeed, to abandon the solar system for good.

PHANTOMS: VULCAN, TRANS-PLUTONIAN PLANETS AND THE LIKE

By the first half of the nineteenth century, Uranus was no longer the only planet to 'misbehave' and trace an orbit slightly different from that predicted. In fact, after all perturbations by the other planets were allowed for, it was found that the semimajor axis of Mercury's orbit was rotating by an unaccountable 43 arcseconds per century. Armed with the success of finding Neptune by theory alone, Le Verrier studied this problem by positing that Mercury was being perturbed by a small planet (or perhaps a number of smaller bodies) orbiting closer to the Sun. Although this perturber was named Vulcan in the expectation that it would soon be found, and observers reported mysterious dark spots transiting the Sun and bright planet-like objects near the Sun in the sky during eclipses, the search petered out after several decades. And then the anomalous advancement of Mercury's perihelion was explained in 1915 by

Albert Einstein as a proof that his general theory of relativity was mathematically superior to the account of gravity given by Newton. In fact, the precession of the planet's orbit derives from the fact that it is so close to the Sun. Although the existence of an intra-Mercurian planet had been so definitively disproved, there were, and remain today, suggestions that there might a belt of 'Vulcanoids' within Mercury's orbit, or perhaps even at its Lagrangian points.

An issue that astronomers faced at the start of the space age was whether the solar system ended at Pluto and, if so, why? Studying the orbits of periodic comets, C.H. Schuette noted the existence of a family of eight comets whose orbits suggested that they might be associated with a hypothetical object at a heliocentric distance of 77 AU. As his researches had not revealed any other families of this type, he suggested that this trans-Plutonian planet would be the final planet. But decades of searching by a few dedicated astronomers proved fruitless. To his credit, after finding Pluto, Tombaugh had continued to scan the sky for other trans-Neptunian planets, and the fact that he found nothing was itself strong evidence against the existence of such a 'Planet-X'. The issue of why the solar system appeared to end abruptly at Pluto was addressed independently by Kenneth E. Edgeworth in Britain in 1949 and G.P. Kuiper in 1951, both of whom suggested that there might be a second 'asteroid belt' beyond Neptune and Pluto whose population density diminished with increasing heliocentric distance. In contrast to the asteroids of the inner solar system, which were primarily made of rock, the objects in this Edgeworth–Kuiper Belt (or just Kuiper Belt as it is usually referred to) would be icy bodies; in effect, 'pristine' relics of the nebula out of which the Sun and planets formed. Decades later, it was realized that the objects in this belt (if indeed it existed, because there was no observational evidence) might be perturbed by the giant planets onto trajectories that caused them to penetrate the inner solar system, where they would appear as comets. Another 'comet reservoir' was theorized by Dutch astronomer J.H. Oort. By performing a statistical analysis of long-period comets with reliably computed orbits, he noticed that many had aphelia near 150,000 AU, and this prompted him to posit that this must be the heliocentric distance of a cloud of comet nuclei, for which he assumed a model similar to the 'dirty snowball' proposed by Fred Whipple. Oort supposed that these objects would remain in the cloud until perturbations from a passing star either expelled them into interstellar space or caused them to 'fall' toward the Sun. In contrast to the Kuiper Belt, which should be flattened into a thick disk in the plane of the ecliptic, the Oort Cloud should be spherical, which explained why comet orbits display a wide range of orientations. Should one of the comets then happen to cross the path of a planet, it could be deflected into a tighter orbit and become a short-period comet. Oort also addressed the issue of the origin of this cloud, recognizing that nuclei could not have had sufficient time to coalesce at such a distance from the Sun, where a single orbit might last millions of years. He reasoned that the nuclei had actually formed within the realm of the giant planets, and at an early stage in the history of the solar system had suffered close encounters that had deflected them far from the Sun. Remarkably, Oort's calculations implied that up to 97 per cent of the original comets should have been expelled to interstellar space, and that if other stars had done likewise then we should see an interstellar comet once per century on average, this being recognizable from its orbital characteristics. However, despite more than three centuries of observations, we have yet to see a single bona fide interstellar comet.

The solar system at the beginning of the 'space age'

Planet	Satellite	Satellite discovery
Mercury	–	–
Venus	–	–
Earth	Moon	–
Mars	Phobos	1877
	Deimos	1877
Jupiter	Amalthea	1892
	Io	1610
	Europa	1610
	Ganymede	1610
	Callisto	1610
	Himalia	1904
	Lysithea	1938
	Elara	1905
	Ananke	1951
	Carme	1938
	Pasiphae	1908
	Sinope	1914
Saturn	Mimas	1789
	Enceladus	1789
	Tethys	1684
	Dione	1684
	Rhea	1672
	Titan	1655
	Hyperion	1848
	Iapetus	1671
	Phoebe	1898
Uranus	Miranda	1948
	Ariel	1851
	Umbriel	1851
	Titania	1787
	Oberon	1787
Neptune	Triton	1846
	Nereid	1949
Pluto	–	–

Plus 1,616 numbered asteroids and 48 short-period comets that had been seen on at least two apparitions.

SOURCES

Antoniadi, E., "The Markings and Rotation of Mercury", Journal of the Royal Astronomical Society of Canada, 27, 1933, 403–410

Arpigny, C., "Propriétés Physiques et Chimiques des Comètes: Modèles et Problèmes Pendants" (Physical and Chemical Properties of Comets: Models and Standing Problems). In: "Le Comete nell'Astronomia Moderna: Il Prossimo Incontro con la Cometa di Halley", Naples, Guida, 1985, 229–250 (in French)

Baum, R., "An Observation of Mercury and its History", Journal of the British Astronomical Association, 107, 1997, 38

Baum, R., Sheehan, W., "In Search on Planet Vulcan: The Ghost in Newton's Clockwork Universe", Cambridge, Basic Books, 1997

Boyer, C., Camichel, H., "Observations photographiques de la planète Venus", Annales d'Astrophysique, 24, 1961, 531–535

Cattermore, P., Moore, P., "Atlas of Venus", Cambridge University Press, 1997, 1–29

Cunningham, C.J., "Introduction to Asteroids", Richmond, Willmann-Bell, 1988

Danielson, R.E., et al., "Mars Observations from Stratoscope II", The Astronomical Journal, 69, 1964, 344–352

Dobbins, T.A., Sheehan, W., "The Martian-Flares Mystery", Sky & Telescope, May 2001, 115–119

Dobbins, T., Sheehan, W., "The Story of Jupiter's Egg Moons", Sky & Telescope, January 2004, 114–120

Dollfus, A., "History of Planetary Science. The Pic di Midi Planetary Observation Project: 1941–1971", Planetary and Space Science, 46, 1998, 1037–1073

Edgeworth, K.E., "The Origin and Evolution of the Solar System", Monthly Notices of the Royal Astronomical Society, 109, 1949, 600–609

Goody, R.M., "The Atmosphere of Mars", Journal of the British Interplanetary Society, 16, 1957, 69–83

Hess, S.L., "Atmospheres of Other Planets", Science, 128, 1958, 809–814

Maffei, P., "La Cometa di Halley" (Halley's Comet), Milan, Mondadori, 1987 (in Italian)

Marov, M. Ya., "Mikhail Lomonosov and the Discovery of the Atmosphere of Venus during the 1761 Transit". In: Transits of Venus, Proceedings IAU Colloquium No. 196, 2004

Morrison, D., Samz, J., "Voyage to Jupiter", Washington, NASA, 1980, 1–9

Morrison, D., "Voyages to Saturn", Washington, NASA, 1982, 1–7

Müller, G., "Über die Lichtstärke des Planeten Mercur" (On the Brightness of Planet Mercury), Astronomische Nachrichten, 133, 1893, 47–52 (in German)

Odrway, F.I., "The Legacy of Schiaparelli and Lowell", Space Chronicle, 39, 1986, 19–27

Owen, T., "Titan", Scientific American, February 1982, 98–109

Pettit, E., Nicholson, S.B., "Radiation from the Planet Mercury", Astrophysical Journal, 83, 1936, 84–102

Ross, F.E., "Photographs of Venus", Astrophysical Journal, 68, 1928, 57–92

Ruggieri, G., "La Macchia Rossa di Giove" (Jupiter' Red Spot), Coelum, 21, 1953, 1–6 and 41–46 and 22, 1953, 8–13. (in Italian)

Ruggieri, G., "Mondi nello Spazio" (Worlds in Space), Rome, A.S.A., 1958 (in Italian)

Russell, H.N., "The Atmospheres of the Planets", Science, 81, 1935, 1–9

Sagan, C., "The Planet Venus", Science, 133, 1961, 849–858

Sandage, T., "The Neptune File", New York, Walker & Company, 2000

Schiaparelli, G.V., "Sur la Relation qui Existe entre les Comètes et les Etoiles Filantes" (On the Relationship Between Comets and Shooting Stars) Astronomische Nachrichten, 68, 1867, 331–332 (in French)

Schiaparelli, G.V., "Sulla Rotazione di Mercurio" (On the Rotation of Mercury), Astronomische Nachrichten, 123, 1890, 241–250 (in Italian)

Schiaparelli, G.V., "La Vita sul Pianeta Marte" (Life on Planet Mars), Natura ed Arte, 4 No. 11, 1985, 81–89 (in Italian)

Schuette, C.H., "Two New Families of Comets", Popular Astronomy, 57, 1949, 176–182

Sheehan, W., "The Planet Mars: A History of Observation and Discovery", Tucson, The University of Arizona Press, 1996

Sheehan, W., Dobbins, T.A., "Charles Boyer and the Clouds of Venus", Sky & Telescope, June 1999, 56–60

Sheehan, W., Dobbins, T., "Mesmerized by Mercury", Sky & Telescope, June 2000, 109–114

Sheehan, W., Dobbins, T.A., "Lowell and the Spokes of Venus", Sky & Telescope, July 2002, 99–103

Sheehan, W., Dobbins, T.A., "Lowell's Spokes on Venus Explained", Sky & Telescope, October 2002, 12–14

Sheehan, W., Kollerstrom N., Waff, C.B., "The Case of the Pilfered Planet", Scientific American, December 2004, page unknown

Sinton, William M., "Further Evidence of Vegetation on Mars", Science, 130, 1959, 1234–1237

Slipher, V.M., "A Photographic Study of the Spectrum of Saturn", Astrophysical Journal, 26, 1907, 59–62

Smith, A.G., "Radio Spectrum of Jupiter", Science, 134, 1961, 587–595

Stangl, M., "The Forgotten Legacy of Leo Brenner", Sky & Telescope, August 1995, 100–102

Stern, A., Mitton, J., "Pluto and Charon", New York, John Wiley & Sons, 1998, 7–40 and 138–143

Strom, R.G., "Mercury the Elusive Planet", Washington, Smithsonian Institution Press, 1987, 4–14

Stroobant, P., "Etude sur le Satellite Enigmatique de Vénus" (Study on the Enigmatic Satellite of Venus), Astronomische Nachrichten, 118, 1888, 5–10 (in French)

Struve, O., "The Origin of Comets", Sky & Telescope, February 1950, 82 (Reprinted in: Page, T, Page, L.W. (ed.), "The Origin of the Solar System", New York, Macmillan, 1966, 252–259)

Taylor, R.L.S., "Life on Mars – An Historical Perspective". In: Hiscox, J.H. (ed.), "The Search for life on Mars", London, British Interplanetary society, 1999, 3–17

Tombaugh, C., "Reminiscences of the Discovery of Pluto", Sky & Telescope, March 1960, 264 (Reprinted in: Page, T, Page, L.W. (ed.), "Wanderers in the Sky", New York, Macmillan, 1965, 65–73)

Whipple, F.L., "A Comet Model. I. The Acceleration of Comet Encke", The Astrophysical Journal, 111, 1950, 375–394

Wildt, R., Meyer, E.J, "Das Spektrum des Planeten Jupiter" (The Spectrum of Jupiter), Veroeffentlichungen der Universitaets-Sternwarte zu Goettingen, 2, 1931, 142–156. (in German)

Wildt, R., "Über das Ultrarote Spektrum des Planeten Saturn" (On the Infrared Spectrum of Saturn), Veroeffentlichungen der Universitaets-Sternwarte zu Goettingen, 2, 1932, 216–220

1

The beginning

SPACE RACE

The ballistic missile, one of the legacies of the Second World War, has changed the world both for good and bad. While it enabled a nation to attack the population and facilities of another state with impunity, it also provided a means of launching satellites that have not only revolutionized science and technology but also changed the everyday lives of people on the planet.

This process began in the 1950s with the Soviet Union and the United States, the two rival superpowers, developing missiles capable of delivering nuclear warheads over ranges of many thousands of kilometers. The Soviets designed a rocket that had four large boosters to assist with lift off. It was designated the 8K71, but was widely referred to as the 'Semyorka' – meaning 'number seven', after the military identifier: R-7. In America, competition between the Army, the Air Force and the Navy gave rise to different medium-, intermediate- and long-range missiles. Each superpower then announced that it was to launch a satellite for the International Geophysical Year, which was to run from mid-1957 through to the end of 1958. The potential to refine the military missiles to deliver a payload into space was now clear, but while the Soviets opted to introduce a straightforward modification to the 8K71, the White House ordered the US Navy to develop a new missile, based on the Viking rocket that had been used by scientists to probe the upper atmosphere. This was a decision that would later have grave repercussions. Despite having announced the intention to do so, when the Soviets put Sputnik – meaning 'satellite' or 'fellow traveler' – into orbit on 4 October 1957 the Americans were shocked, and their state of concern was compounded on 3 November by the launch of a considerably larger satellite carrying the dog named Laika. After the first US Vanguard rocket spectacularly failed on 6 December, the White House ordered the Army to restore American pride by using a modified Redstone missile to launch a satellite, and Explorer 1 was put into orbit on 1 February 1958.[1]

During 1958 the superpowers also started work on versions of their missiles that were capable of dispatching small probes to the Moon or, with just a little more power, to our nearest planetary neighbors. The Soviets added an upper stage to the 8K71 to produce the 8K72, which was capable of sending a 400-kg payload to the Moon. The Americans created the more modest Thor–Able which, as with many other American projects of the day, was born in the immediate aftermath of Sputnik. US engineers calculated that by adding the second stage of the Vanguard launch vehicle and a solid-fuel third stage to the Thor intermediate-range ballistic missile they could produce a vehicle that was capable of dispatching small payloads not only to the Moon, but also to Mars or Venus. Unfortunately, between the summer and fall of 1958 four American and three Soviet lunar launches failed due to the low reliability of their respective technologies, but in 1959 attempts fared rather better.[2] On 2 January 1959 the Soviets launched Luna 1, a 170-kg package designed to collect scientific data in transit to the Moon prior to impacting on its surface. The rocket thrust of the 8K72 fell short of that planned, however, with the result that after 34 hours the probe flew past the Moon at an altitude of 6,000 km. Nevertheless, because it had achieved 'escape velocity', it became the first artificial object to orbit the Sun. In celebration, the Soviets named the probe 'Mechta' – meaning 'Dream'. Only two months later, the American Pioneer 4 flew a similar trajectory, relaying 82 hours of data on the deep-space environment. In September, the Soviets dispatched another probe, Luna 2, which succeeded in impacting on the Moon, and in October they sent a new type of probe, Luna 3, on a trajectory that enabled it to photograph the far side of the Moon, which is never visible from Earth. One significant result from these early missions was data on the predicted phenomenon of the 'solar wind' – a flux of plasma that originates from our star and permeates the entire solar system.[3,4]

By the end of the 1950s, therefore, the two superpowers were about to embark on a 'space race' to explore the nearby planets.

HUMANS OR ROBOTS?

While the idea of navigating to the Moon and the planets has been a factor in human cultures for centuries, the first study to demonstrate its practicality using existing (or near-term future) technologies seems to have been by the American rocketry pioneer Robert H. Goddard who, in March 1920, elaborating on his earlier suggestion that a rocket "without operator" could be sent to the Moon, outlined a concept by which the planets could be explored. In particular, Goddard noted that astronomy could be greatly advanced by flying a telescope near Mars to map its surface in much greater detail than was possible using even the most powerful terrestrial telescopes, and he further suggested that it would be possible to communicate with any residents using "artificial meteors" that contained "metal sheets stamped with geometrical shapes, with the constellations, emphasizing the Earth and the Moon".[5]

Despite this early work, solar system exploration is generally regarded as having started in 1952 – five years before the launch of Sputnik. This was the year in which the German astronautics pioneer and father of the Nazi V2 missiles, Wernher von Braun, published his book *Das Marsprojekt* ('The Mars Project') presenting a study

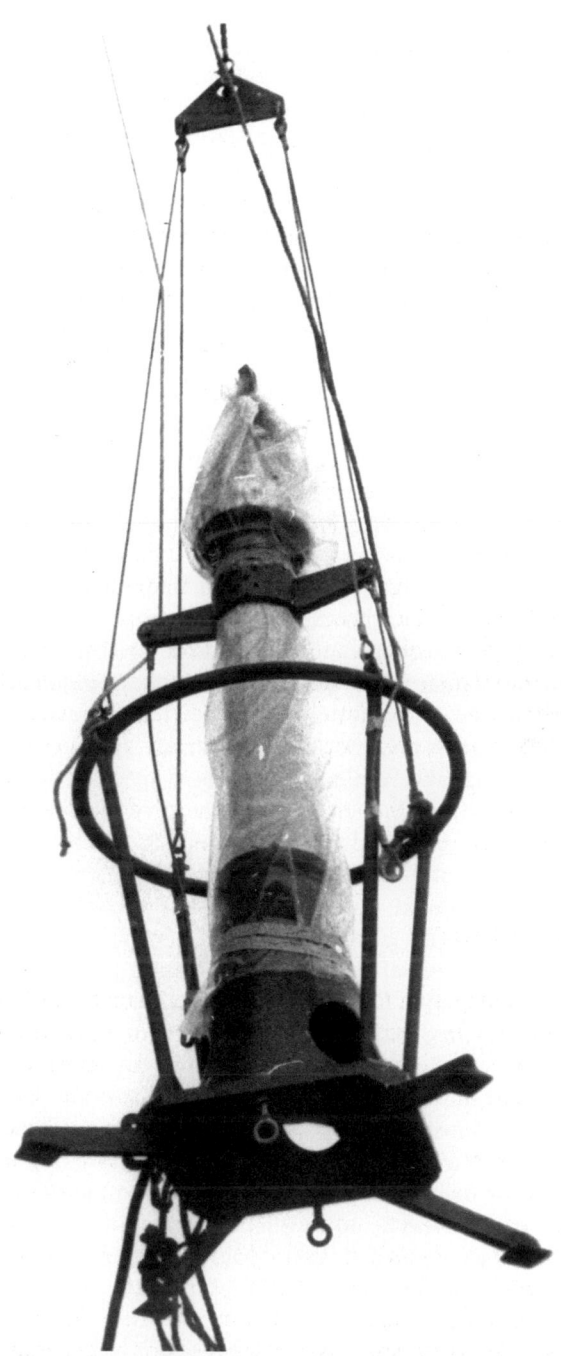

Pioneer 4 and the upper stage of its launch vehicle being hoisted to the Juno rocket. This spacecraft would become the second to venture in solar orbit, and the first US-made.

of the requirements for an expedition using 10 huge 4,000-tonne spaceships. Once in the vicinity of Mars, this fleet would release large gliders that would set down on the polar ice cap to enable the crew to explore for about 400 days. As it was not evident that advances in automation and control technology would soon render this scenario obsolete, von Braun envisaged a crew of 70 specialists.[6,7] Such a project would have cost the equivalent of hundreds of billions of dollars today, and was, in fact, beyond the technology of that era. An early start in the exploration of the solar system would have been unlikely but for a presentation by Eric Burgess and Charles A. Cross at the British Interplanetary Society on 20 September 1952. Influenced by a historical article published the previous year showing that it would be possible to build and launch a spacecraft weighing several kilograms using existing technology, the two authors proceeded to discuss the possibility of a Mars exploration mission using automatic systems. The article, called 'The Martian Probe' (the first use of the word 'probe' for an automatic craft for deep space) identified two optimal launch windows to the Red Planet in 1973 and 1988. On entering orbit around Mars, the probe would transmit television images of its surface to Earth, and analyze its topography, temperature and reflectance spectrum. Moreover, the article considered the problems of power generation and of returning data to Earth. The main merit of the short article was its proof that the exploration of the solar system would be feasible using technologies that were only slightly more advanced than those needed for the launch of an Earth satellite, and hence rather more accessible than those required to mount a mission such as von Braun was promoting. Over the following years, the calculation of trajectories to the planets and the study of (usually manned) interplanetary missions became almost 'fashionable' among specialists in space technology. The 'state of the art' had therefore matured considerably by the time Sputnik was launched.[8]

In parallel, the first spaceflight mechanics papers detailing the possibility of using the gravity fields of planets to change a craft's trajectory were published. The priority of this 'invention' is disputed, but it was a well-known effect in solar system dynamics, stemming from the study of the orbits of comets, with many well-documented examples dating back several centuries. To mention just two: periodic comet Lexell was never recovered after its 1770 apparition because it later passed so close to Jupiter that it was moved to an orbit with a period exceeding 200 years (and may even have been ejected from the solar system); in contrast, comet Brooks 2 was in a long-period orbit beyond Jupiter until 1886, and when it approached the planet and flew inside the orbit of Io, the innermost of the Galilean moons, it was deflected into a short-period orbit, in which it was identified in 1889.[9,10,11] Notable among the early gravity-assist essays was a paper delivered in 1956 by the Italian astronautics pioneer Gaetano Arturo Crocco in which he demonstrated the feasibility of using the gravity of both Mars and Venus to accomplish an Earth–Mars reconnaissance round trip in as little as one year. This is one of the first 'multiplanet' mission profiles ever proposed.[12]*

---

* In his 1951 novel *The Sands of Mars*, Arthur C. Clarke has one of his characters refer to using a Jovian flyby to achieve a gravity-assist in transit to Saturn.

THE FIRST 'ARTIFICIAL PLANETS'

The first human artifacts to be fired into interplanetary space may have been the side effects of an experiment that failed. At the end of the Second World War, the US Army shipped German V2 rockets to the United States to be used to test rocketry techniques and carry scientific instruments to investigate the upper atmosphere, the ionosphere, the Sun and cosmic rays. During one early meeting of the V2 scientific planning group, the astronomer Fritz Zwicky proposed releasing small grenades at very high altitude which, on re-entering the atmosphere, would simulate the behavior of meteors. This would not only serve to calibrate natural meteor observations, but would also yield data on hypersonic aerodynamics and conditions in the upper atmosphere. The project was approved, and a three-grenade release system was prepared. The experiment was attempted on 17 December 1946, but unfortunately no artificial meteor was seen before the rocket accidentally blew up 440 seconds after launch.

A related experiment was made on 16 October 1957, just 12 days after the launch of Sputnik. By now the last of the remaining V2 rockets had been donated to museums, and a variety of modern 'sounding' rockets developed. In this test an unguided liquid-propellant Aerobee rocket was used, and instead of grenades it was to carry three solid aluminum spheres, several centimeters in diameter, embedded in a 'shaped charge' designed to eject them at a rate of about 15 km/s. The rocket was launched by the US Air Force from White Sands in New Mexico, and when it reached a height of 87 km the charge was detonated, producing a bright flash that was recorded as far as Mount Palomar, 1,000 km distant. Although the pictures showed a meteor-like trail as bright as the planet Venus due to the re-entry of one of the spheres, no trace was found of the others, and it is possible that the angle at which they were ejected enabled them to escape from Earth and enter orbit around the Sun, thereby becoming the first (if inadvertent) 'artificial planets' some 14 months before Luna 1.[13,14]

THE FIRST INTERPLANETARY PROBE

As early as 1958, the United States set out to investigate the feasibility of launching interplanetary probes. The issues faced were unprecedented, since no spacecraft had been required to be exposed to the solar heat for months at a time, or to return data from distances of tens of millions of kilometers. The first American interplanetary exploration program was managed by Space Technology Laboratories on behalf of NASA's newly created Goddard Space Flight Center. This called for four probes to reach and orbit the Moon and Venus. The planet Venus could be reached by 150 days of flight after launch in June 1959, compared to 250 days for Mars, for which a launch opportunity opened 16 months later. The plan also called for a smaller probe to be launched one day after the Venus orbiter, to make a flyby of the planet. After the apparently successful flight of Luna 1 (the Soviets never announced that it was to have impacted on the Moon, and gave the impression that the objective had been to

The body of the Pioneer 5 deep-space probe was a fiberglass sphere painted with a black and white pattern for passive thermal control.

enter solar orbit), NASA revised its plan: the Venus flyby would proceed, but the focus of the orbital missions would be switched to the Moon. When the development of the flyby craft and its scientific payload fell so far behind schedule that it became evident that it would miss its launch window, it was decided to continue and inject the probe into a 'transfer orbit' that would intersect the orbit of the planet (even though by then the planet would not be nearby) in order to collect data on the deep-space environment and to test long-range transmission techniques. Although the 'minimum success criterion' called for the craft to operate for just one month, NASA engineers hoped that it would transmit through all of the first solar orbit.[15,16,17]

The design of the Venus flyby probe was simple, being derived from the 'paddle wheel' Explorer 6 satellite. A 66-cm-diameter spin-stabilized fiberglass sphere held all the scientific payload and electronics on a central deck, plus two transmitters – one of 5 W, the second of no less than 150 W, selectable from the ground, and designed to send data at three different speeds depending on the distance from Earth and the size of the receiving antenna. Data and telemetry transmission used a simple omnidirectional antenna that was mounted on 'top' of the craft and required the more powerful transmitter. Estimates suggested that it would be able to transmit data over a range of 80 million km. Four solar panel 'paddles' were installed at the ends of short arms, whose fixed orientations were optimized to ensure that at least a given minimum percentage of their surface would be exposed to the Sun regardless of the attitude of the spacecraft. The scientific payload accounted for a record 18.1 kg of the 43.2-kg mass of the spacecraft. It had a proportional radiation counter to detect high-energy particles, an ionization chamber and Geiger–Mueller counter with which to measure the total radiation flux, a coil magnetometer to measure the interplanetary magnetic field component perpendicular to the spin axis, and a micrometeoroid detector. To assist the interpretation of the data, a sensor would record the orientation of the probe by detecting when it was pointed toward the Sun. A number of thermometers on the interior of the sphere and on the paddles would record temperatures. Between transmission sessions, data would be stored on an on-board tape recorder.[18,19,20]

The Thor–Able launch vehicle was stacked on pad 17A at Cape Canaveral on the Atlantic coast of Florida in October 1959, but its launch had to be delayed, first from December to January and again later, and it did not finally lift off until 11 March 1960. Although the third and final stage of the rocket boosted Pioneer 5 to a speed of 11.115 km/s and placed it into an orbit ranging between 148.4 and 120.6 million km from the Sun with a period of 312 days, the departure velocity had fallen slightly short of the 11.26 km/s required to enable the trajectory to intersect with the orbit of Venus. As a result, the probe approached the planet no closer than 11 million km.[21] Some 27 minutes into the flight, the 250-foot-diameter Jodrell Bank radio telescope in England issued the command to jettison the third stage. Despite malfunctions in the transmission system, the spacecraft was able to be tracked for the first 2 days by short-range helical antennas at Cape Canaveral and Singapore, then the larger, more sensitive, radio telescopes at Jodrell Bank and Hilo in Hawaii were required. An early finding was that the Earth's magnetosphere was considerably larger than had been believed, extending out to 14 instead of 6 Earth radii. The most interesting data for future missions was that of the micrometeoroid detector, which recorded 87 impacts during one week, in which the spacecraft had traveled some 1.6 million km. At the end of March the Sun suffered several 'flares', whose effects were noted by the probe after an interval of 7 to 8 days, and at Earth some 20 minutes later, thereby giving an indirect measure of the mean speed of the plasma in the solar wind. When, during one such storm, the flux of high-energy charged particles known as galactic cosmic rays markedly decreased, it was realized that this modulation is a solar effect that is localized in interplanetary space, and is independent of the presence of either Earth or its magnetic field. When the Sun was less active, a weak interplanetary magnetic field was detected whose intensity correlated with the number of spots on the Sun's 'surface' – or photosphere. Unfortunately, data from the magnetometer was garbled and was difficult to interpret due to the instrument's position within the spacecraft and the fact that it was close to metallic masses.[22,23,24,25,26]

Pioneer 5 operated the 5-W transmitter up to 30 April 1960. When Jodrell Bank had the only antenna in the world still able to communicate with the probe, and was at times having difficulty, the probe was instructed to activate its 150-W transmitter, whereupon it was realized that the batteries to power it had been venting electrolytes into space. The solar panels could produce just one-tenth of the power to operate it, but with their output augmenting the deteriorating batteries, the 150-W transmitter was able to be operated on several occasions by ground command between 8 and 21 May. The 5-W system enabled the spacecraft to continue to be tracked, but no more

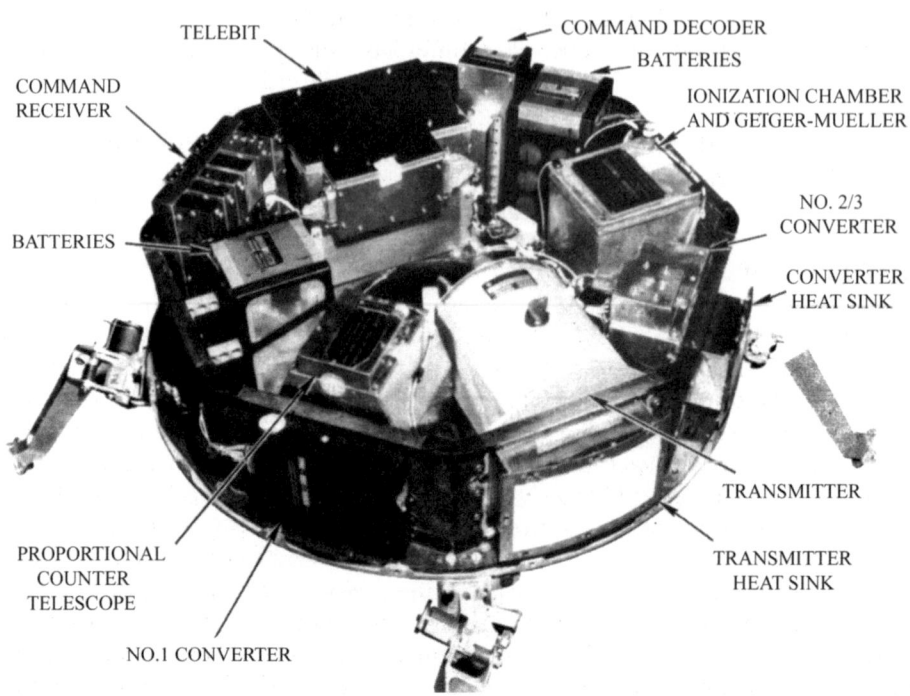

COMMAND DECODER

TELEBIT

BATTERIES

COMMAND
RECEIVER

IONIZATION CHAMBER
AND GEIGER-MUELLER

NO. 2/3
CONVERTER

BATTERIES

CONVERTER
HEAT SINK

TRANSMITTER

PROPORTIONAL
COUNTER
TELESCOPE

TRANSMITTER
HEAT SINK

NO.1 CONVERTER

The Pioneer 5 instrument platform showing the scientific payload and electronics.

usable scientific data was returned. Tracking became more and more sporadic, and the last session was at 11:31 UTC on 26 June from a distance of 36.4 million km. Pioneer 5 had operated for 107 days during which it had traveled almost 275 million km and traced out one-third of a solar orbit, and had thereby greatly exceeded the 'minimum success criterion'. On 10 August it reached perihelion. When it returned to the vicinity of the Earth in 1963, consideration was given to resuming tracking, but by then the batteries had become useless and the probe had to be written off. According to contemporary predictions, Pioneer 5 will remain in solar orbit for at least 100,000 years before its trajectory causes it to enter Earth's atmosphere and burn up as a meteor.[27,28,29,30,31] Overall, the probe returned 3 megabits of data during a total of 138 hours 54 minutes of transmissions. Precise measurements of the line-of-sight velocity of the probe relative to Earth were obtained. A carrier signal sent to the probe was returned to Earth at exactly 16/17th of the received frequency, and by correcting the observed frequency shift for the daily motion of the antenna as the Earth rotated on its axis – and also for the motion of the Earth around the center of gravity of the Earth–Moon system – it was possible to calculate the Doppler shift of the signal due to the relative motions of the Earth and the probe around the Sun, taking into account perturbations by the other planets. This analysis provided the first *direct* measurement of the Astronomical Unit (AU), which is defined to a first approximation as the mean distance between the Earth and the Sun. The best previous measure was 149,527,000 ± 10,000 km, and Pioneer 5 provided a value of 149,544,360 ± 13,700 km. These values differ from the true value (now known to a high degree of accuracy after more than 40 years of tracking Venus by radar) by 70,000 and 50,000 km respectively. Further analysis of the Pioneer 5 tracking revealed the perturbations of solar radiation pressure on the probe's motion which had influenced the measurement of the AU.[32,33,34]

THE FIRST JPL PROJECTS

The Jet Propulsion Laboratory (JPL) of Caltech (California Institute of Technology) had built the first Explorer satellites and the Pioneer 4 lunar probe under contract to NASA. At the end of the 1950s JPL initiated a study of interplanetary exploration projects. The first project, called 'Vega', would need the Atlas–Vega launcher. This would mate an Atlas intercontinental-range ballistic missile with a second stage derived from the Vanguard launcher and a JPL-designed stage capable of sending up to 360 kg to the Moon or the nearer planets. The plan called for at least three flights: one each to the Moon and Mars in 1960, and one to Venus in 1961. But the project faced three obstacles: (1) the financial status of NASA, which had just begun the expensive Project Mercury to build a craft capable of orbiting the Earth carrying an astronaut; (2) the shift of priorities to lunar exploration; and (3) the fact that the military was developing the Atlas–Agena, which offered similar performance using only two stages. At the end of 1959 NASA decided to adopt the Atlas–Agena and canceled Vega. JPL promptly switched its attention to a new family of interplanetary probes called 'Mariner' which, being heavier, would require the Centaur stage that was to be powered by liquid hydrogen.[35,36,37] Two different Mariner classes were envisaged: Mariner A and Mariner B.

Mariner A was to weigh up to 686 kg, be launched in 1962, and be equipped for a 27,000-km flyby of Venus. It would be built around a central hexagonal framework that would house the electronic packages and support the gimbal of a parabolic high-gain antenna, an omnidirectional antenna, two 1.8-m^2 solar panels (together rated for 300 W at Earth's distance from the Sun; twice as much at Venus) and four booms to position sensors. The scientific payload was to include a radiometer to measure the temperature field over the face of the planet, an ultraviolet spectrometer to study the composition of its upper atmosphere, a magnetometer to study the interplanetary magnetic field in detail and determine whether the planet had its own field, a plasma detector, a radiation counter, and a micrometeoroid detector. While Pioneer 5 had been spin-stabilized, Mariner A was to be a 3-axis-stabilized platform able to maintain a fixed orientation in space. To enable the ultraviolet spectrometer and radiometer to be accurately pointed at Venus during the flyby, they were to be set on a gimbaled platform, while the other instruments would be placed in fixed positions on the structure. Whereas Pioneer 5 had been incapable of correcting its trajectory, Mariner A would include a monopropellant hydrazine engine to refine its course and eliminate errors inherited from the launch vehicle. It would be fired once several weeks after launch, and again at the end of the mission – but this time to

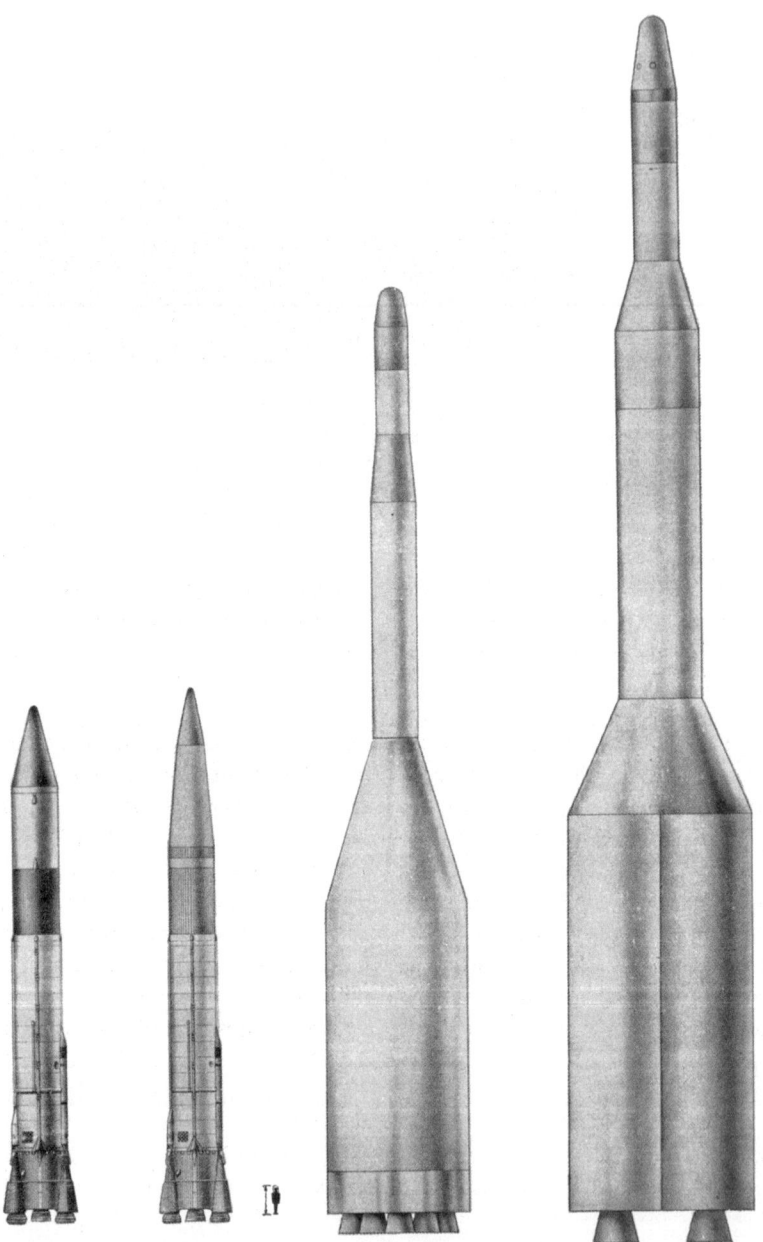

A 1959 drawing of NASA's planned standard launchers (left to right): Atlas–Centaur, Atlas–Vega, Saturn and Nova. A human figure provides a sense of scale. The Atlas–Vega would soon be replaced by the USAF-designed Atlas–Agena. The Saturn evolved into the Saturn I rocket that launched the first Apollo tests. The Nova was to be a launcher for a manned lunar flight using the direct-ascent strategy.

Mariner A: the spacecraft with which JPL intended to make the first planetary flyby missions. The solar panels and high-gain antenna are deployed, while the instrument booms are in their stowed position.

depletion as an engineering test. If the first flight was a success, further missions to Venus could carry either a television system or a ground-mapping radar. A variant of Mariner A was to make a Mars reconnaissance flyby carrying a vidicon camera capable of taking and transmitting a minimum cache of images.[38,39]

Mariner B would be a larger and more ambitious probe designed to explore either Venus or Mars. As the spacecraft approached the planet on the Mars mission it was to release a small capsule that would enter the atmosphere and land by parachute. It would study the atmosphere using a barometer, a thermometer, a mass spectrometer and a gas chromatograph, and after landing would use a panoramic camera to reveal the site. The spacecraft was to pass by the planet at a range of 15,000 km and would have a scientific payload similar to that of Mariner A, including a camera with a resolution of 1 km. No fewer than four such missions were scheduled: two to go to Mars in 1964 and two to go to Venus in 1965.[40,41]

JPL also started to study more ambitious 'Voyager' probes with which to explore Venus and Mars later in the decade. These would weigh more than 1 tonne and be launched by a version of the extremely powerful Saturn rocket that Wernher von Braun was developing at NASA's Marshall Space Flight Center as part of the effort to send humans to the Moon. Instead of making a brief flyby, a Voyager would enter orbit around its target planet and release a very large lander equipped with a comprehensive scientific payload. Even more ambitious was a series of spacecraft called 'Navigator', which were to explore Mercury, cross the asteroid belt to Jupiter, rendezvous with comets and fly close to the Sun.[42]

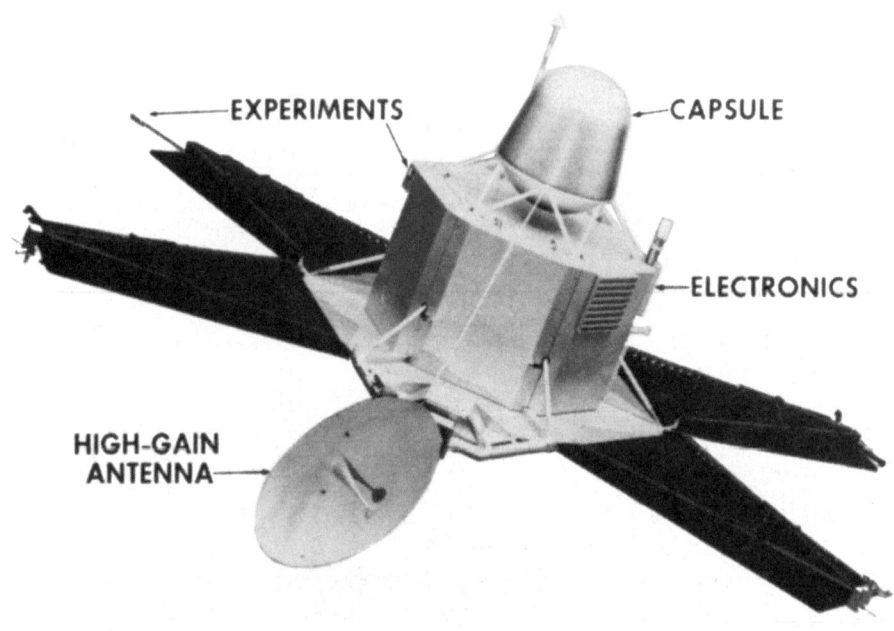

The 'heavy' Mariner B flyby spacecraft and atmospheric entry probe.

THE FIRST SOVIET PROBES

The first Soviet plans for interplanetary probes were just as ambitious. A conference reviewing the status of knowledge of the Moon and planets, in anticipation of their future exploration, was first convened at the State University of Leningrad in 1956. Soon after Sputnik was launched, the team (led by Sergei Pavlovich Korolyov) that had built the Semyorka missile, the Sputniks and the early lunar probes, set to work on a new project to use the 8K72 to launch a spacecraft to Mars in August 1958 (designated the 1M probe) and to Venus in June 1959 (the 1V probe). The Academy of Sciences assisted in computing the requisite interplanetary trajectories. However, the technical difficulties suffered by the early lunar probes pushed the launch of the Mars mission to the more realistic October 1960 launch window, and the launch of the Venus mission to 1961. Also, it was decided to switch to the 8K78, which was a variant of the 8K71 with two upper stages – stage I and stage L, the latter of which would achieve escape velocity. This launch vehicle would later acquire the name of 'Molniya' (Lightning) when, in 1964, it began to be used to launch communication satellites of that name.[43,44]

The 1M and 1V spacecraft were to be sophisticated platforms stabilized on three axes using Sun and star sensors in addition to gyroscopes, and bipropellant engines burning dimethyl hydrazine and nitric acid.[45] Korolyov initially planned to perform

The Molniya rocket, derived from the R-7 'Semyorka' ICBM, was the mainstay of the Soviet planetary program in the 1960s and early 1970s. This picture shows the launch of Venera 7 in 1970.

the complex mission of dropping a landing module into the atmosphere of a target planet and braking it by a cluster of parachutes – three at Mars and two at Venus. However, he faced some major problems. No *direct* data was available on the fundamental parameters of the atmospheres of these planets, particularly their compositions and profiles of pressure, density and temperature, with the result that his engineers had to rely on the dubious accuracy of data collected in three centuries of telescopic study. In designing the parachute for use on Mars, they relied on a study by Gavril Tikhov, who had inferred that the atmosphere was relatively dense. Also, the delivery of the atmospheric probe was complicated by the low accuracy of the ephemeris for the planet – the uncertainty in the position of the planet exceeded its diameter! It was therefore decided to send several spacecraft to perform scientific observations as they flew by the planet at ranges between 5,000 and 30,000 km, with a single lander trailing behind. However, the launch of the lander was canceled – according to some sources as late as the early autumn of 1960, several weeks before its scheduled departure but probably much sooner – since it would not be completed in time. For example, it was realized during development that the characteristics of the lander's penetration into the Martian atmosphere could not be replicated in wind tunnels or by air-drops. The R-11A sounding rocket (based on the R-11, SS-1b 'Scud' short-range ballistic missile) was therefore modified, and five of the resulting R-11A-MV were launched from the Soviet Union's oldest cosmodrome at Kapustin Yar to carry development models of the lander to a height of 50 km to test the parachutes. Although the 1M lander had been canceled by the time these tests were made in 1962, the data collected undoubtedly assisted in the development of the next generation of Mars and Venus probes.[46,47,48,49]

The instruments of the 1M flyby probe comprised a magnetometer, an infrared radiometer, a charged-particle sensor, a micrometeoroid detector, a camera, and an infrared spectrometer that was to investigate whether the seasonal variations of the dark areas of the planet's surface were due to vegetation – as inferred from the near-infrared 'Sinton bands'. All of the instruments were mounted on the outside of the probe with the exception of the camera, which was in a pressurized compartment that had an optical port. The imaging sequence would be initiated when a sensor noted the presence of Mars in the field of view of the camera, and the special photographic film would be developed on board, scanned and transmitted to Earth.[50]

To facilitate the planetary missions, the Soviets built the 'Pluton' communication center at Yevpatoria in the Crimea, which included three radio-telescopes capable of communicating with spacecraft over a distance of up to 300 million km.[51] This center, however, did not become available until January 1961, when its antennas had been calibrated against celestial radio sources.[52] Meanwhile, the 8K78 launch vehicle had been successfully demonstrated twice in January 1960 – although on both occasions with a mockup stage L. The assembly of the engineering prototype of the flyby probe was completed on 21 August 1960, just one month before the launch window opened. With time running out, the incomplete 'flight model' probes were sent to the Baikonur cosmodrome in Kazakhstan. (The facility was actually situated several hundred kilometers southwest of the town of that name, at a rail junction on the steppe at Tyuratam.) The optimal date of launch to maximize the payload was between 20 and 25 September, with a Mars flyby around 28 April of the following year.[53] As the departure trajectory would oblige the stage L to fire over the Atlantic Ocean, beyond the range of the network of tracking stations that spanned the Soviet Union, three ships were outfitted and stationed in open ocean to track and record the telemetry from the probes.[54] Unfortunately, it was not possible to start the final tests on the probes until 27 September, at which time a number of faults were found.

In fact, not only were the electrical and radio communication systems and instruments of the probes not operating to specification, but there were also issues with the launcher interface, and when the photographic system caught fire in a simulation of the Mars flyby sequence it was decided to remove the entire instrument. The optimal launch opportunity had been missed, and the mass that the 8K78 could send to Mars was diminishing with each passing day, but the situation was improved by the removal of the photographic system and was further assisted by the deletion of the spectrometer – possibly after this had failed to detect life on the steppe.[55,56,57]

The first 1M probe arrived on the pad on 8 October, and was launched 2 days later, with the flyby scheduled for 13 May 1961. Unfortunately, the launch vehicle developed a vibration that damaged the electrical components in its attitude control system; the engines stopped after 309 seconds and the probe re-entered over Siberia. In an effort to save even more mass, the entire scientific payload and engine were removed from the second probe, the rationale being that although it could not supply data on Mars, its flight would provide experience of long-range communications and the management of an interplanetary mission. However, when the probe was launched 4 days after its sibling, a liquid oxygen leak froze the kerosene in the pipes and this prevented stage I from firing. If events had gone to plan, this second probe would have made its Mars flyby on 15 May 1961.[58,59] As the launch window was now closing, the third probe was canceled. The telemetry from both launches was intercepted by US electronic monitoring stations in Turkey, thus providing the USA with insight into these missions fully 30 years before the Russians acknowledged their existence. Prior to the collapse of the USSR, a third Mars launch was believed to have been attempted on 24 October, but this was a test of an R-16 intercontinental-range ballistic missile built by Mikhail K. Yangel, a protege of Korolyov, which exploded while it was being serviced, killing engineers, technicians and bureaucrats. A death toll as high as 165 has been reported, including Marshal Mitrofan Nedelin, chief of the Strategic Rocket Forces.[60] Meanwhile, Nikita Khrushchev, the Secretary of the Soviet Communist Party, was in New York to address the United Nations. (It was on this occasion that he famously took off one of his shoes and pounded it on his desk.) According to a sailor who defected from his ship at that time, Khrushchev had with him "spacecraft models" (of the Mars probe, presumably) to display in the event of a successful launch. In retrospect, even if the launch vehicles had successfully placed their probes on course for Mars, the results of the preflight testing suggest that none would have remained operational for the seven months required to reach its objective.[61]

The original intention of the 1V mission was to release a small "cathodic tube-shaped" capsule that would land on the surface of Venus, but this idea was soon discarded in favor of adapting the Martian entry probe for Venus, thus creating the 1VA.[62]* The 1VA was built using a 2.03-meter-long cylinder with a diameter of 105

* Note that although there would appear to be no surviving photographs of the 1M, it presumably closely resembled the 1VA that was developed from it, for which pictures have been published.

cm that was pressurized at 1,200 hPa, which contained all the electronics. Attached to the cylinder was a 2.33-meter-diameter parabolic antenna made of a very fine, almost transparent, copper net with which to receive commands from Earth and transmit data at high speed, plus a pair of solar panels of an irregular shape, each of which was a maximum of 1 meter wide and 1.6 meters tall with a collector area of approximately 1 m^2. A medium-gain antenna was affixed to each solar panel, and an omnidirectional antenna was carried at the tip of a boom. Thermal control was by louvers that exposed a different portion of the main compartment to space according to the internal temperature. The probe had a total mass of 643.5 kg. Its scientific payload included a triaxial magnetometer and two magnetic variometers which, together, were to investigate the interplanetary magnetic field and any field that Venus itself might possess. They were mounted on the omnidirectional antenna boom in order to shield them from interference from the body of the spacecraft. Two ion traps were mounted on the sunward side of the 3-axis-stabilized platform in order to collect data on the solar wind. The spacecraft also had cosmic-ray and micrometeoroid detectors. Pictures show a small parabolic reflecting mirror mounted on one of the solar panels that may have been an infrared radiometer to measure the temperature of Venus.[63] The spacecraft also carried a technological experiment to test the aging of various materials exposed to the deep-space environment.[64] The poor ephemeris was an even greater issue for Venus than for Mars, as the uncertainty in the position of Venus might range up to 100,000 km, which equated to some 15 times the radius of planet! The role of the 1VA was to pass by Venus, but because the nature of the planet's surface (which is totally masked by clouds) was at that time a complete mystery, each probe carried a 70-mm-diameter metal globe of Earth that was designed to float in the event that it should splash down into an ocean. Each globe contained a medallion to commemorate the mission and the country that launched it.

The one-month-long window opened on 15 January 1961, but the first 1VA was not launched until 4 February. Unfortunately the failure of a transformer in stage L prevented power from being supplied to the four ullage rockets that were to settle the propellants in weightlessness for the ignition of the main engine, and it misfired.[65] The probe, which was described simply as a 'Heavy Sputnik' by the Soviets, re-entered the atmosphere 22 days later. A rumor surfaced in the West that the launch had been a failed test of a spaceship carrying a cosmonaut, as the Judica-Cordiglia brothers in Italy insisted that they had recorded a transmission from the doomed person breathing with difficulty.

The second 1VA was launched on 12 February, just before the window closed. After about one hour in low orbit, stage L successfully boosted the probe away from Earth and released it. With this milestone achieved, the Soviets officially named it Venera (Venus), but when later probes in the series were dispatched, this trail-blazer was redesignated Venera 1. With an aphelion of 1.019 AU, a perihelion of 0.718 AU and an inclination to the ecliptic of 0.58 degree, the orbit would yield a 100,000-km flyby of Venus on 19 May. The communication center at Yevpatoria contacted the spacecraft at distances of 30,000 and 170,000 km, and discovered that not only had the thermal control system malfunctioned, but the

A model of Venera 1. This is the Earth-facing side of the spacecraft, but the high-gain antenna is not mounted. (Courtesy of Basil Pivovarov)

attitude control system had also broken down and communications had to be by the omnidirectional antenna. Despite these problems, the spacecraft was able to be contacted again on 17 February, at a distance of 1.7 million km, but it failed to respond on 22 February. The Soviets asked Jodrell Bank to assist. On 4 March this large antenna listened for signals for 3 hours and again for a total of 7 hours over the ensuing days, to no avail. When the Soviets optimistically radioed commands to the probe during the period of its Venus flyby, signals were recorded in Britain but these all proved to be of terrestrial origin. Attempts at re-establishing contact ceased on 10 June.[66,67] Therefore, although Venera was the first spacecraft to make a flyby of Venus, it failed to provide any data on the planet. Its meager results were data on the solar wind inside the Earth's orbit, and limited magnetometer data.[68,69] The post-mortem decided that the malfunctioning thermal regulation system probably caused the Sun sensor to overheat, leading to the loss of attitude control and, as a result, a disruption of both power generation and communications.[70,71,72] A third 1VA was never launched.

Nevertheless, 1961 was a landmark year for the scientific study of Venus. When it was at inferior conjunction on 9 April, American and Soviet teams independently managed to bounce radio waves off its surface and detect an echo. One surprise was that Venus was revealed to have a very slow and retrograde axial rotation. As radar was more precise than optical methods of ranging, such data provided a method of refining the value of the Astronomical Unit and, by reducing the uncertainties in the ephemeris, greatly improved the prospects for future missions. Unfortunately, in the race to be first to announce a result, the team led by Vladimir A. Kotelnikov were misled by "random realizations of noise", and the 149,457,000 km they reported on 12 May was actually some 100,000 km less than the intrinsically more accurate measurement by the Americans. This led to the taunt that perhaps the Soviets had discovered a new planet, as they could not have been pointing their radar at Venus! A more thorough analysis gave the distance as 149,598,000 $\pm$ 3,300 km.[73] It is not known whether the Soviets had hoped to exploit this new information to increase the scientific results of the Venera flyby, but the inaccurate preliminary calculation would have rendered such an effort unproductive.

THE FIRST SUCCESS

A different fate was in store for the first American attempt to dispatch a spacecraft toward another planet. Owing to protracted delays in developing the Centaur stage, whose cryogenic engines challenged NASA's technical capabilities, it began in the spring of 1961 to contemplate canceling the launch of Mariner A to Venus in 1962 in order to devote its efforts to flying the more capable Mariner B at the next window in 1964. Rather than let the 1962 window for Venus pass, and risk the possibility of the Soviets scoring another humiliating 'first', JPL proposed Mariner R as a means of gaining experience of operating in deep space by making a Venus flyby using a cut-down form of its Ranger lunar probe carrying a few kilograms of scientific payload. The preliminary planning for Mariner R would cost several million dollars, but the potential scientific return and prestige seemed to be worth it. Finally on 31 August 1961 NASA canceled Mariner A in favor of Mariner R. In hindsight, the choice appears to have been a considerable risk, since the first probe in the Ranger series had been launched during the same month and failed. (In fact, this effort would deliver no scientific results until Ranger 7 in 1964.) JPL established two engineering groups: one to work out the modifications to enable a Ranger to undertake an interplanetary mission, and the other to improve the Agena stage that would send the spacecraft on its way.[74,75]

The principal structural component of Mariner R was its 102-cm-wide hexagonal base, to which were attached two rectangular solar panels with a maximum output of 222 W. One panel comprised 5,810 solar cells, while the other had 4,900 and, at its tip, a small dacron 'sail' designed to enable the craft to balance out the pressure of the solar radiation. With the panels deployed, the total span was 503 cm. The base contained the attitude control system and the radio communication system. In addition, a 1.22-meter-diameter parabolic antenna was affixed to the underside of the base. Above the base, a pyramidal truss housed the scientific instruments and had an omnidirectional antenna at its apex. The truss took the total height of the craft to 363 cm. Because a high-gain antenna can focus its energy into a narrow beam aimed directly at Earth, Mariner R's communication system would require much less power to transmit data than had been necessary for Pioneer 5 – specifically, 3.5 W against 150 W. A 225-N hydrazine engine was mounted at the center of the base to perform a single course correction. The payload comprised seven scientific instruments, which accounted for some 18.5 kg of the total mass of 203.6 kg.[76,77,78,79] The most important instrument was the 10.79-kg microwave radiometer, which had an oscillating 48.5-cm parabolic antenna to measure the heat radiated at two wavelengths. Since water vapor would absorb energy at only one of these wavelengths, a comparison of their readings would indicate the presence – or indeed the absence – of water in the Venusian (or, as contemporary astronomers preferred to call it, the Cytherean) atmosphere. This instrument was a simplified form of a four-channel device that had been intended for Mariner A.[80,81] A second radiometer was to measure the heat radiated by Venus at infrared wavelengths.[82] Another important instrument was the solar plasma detector. It had been hoped to use an instrument developed by the Massachusetts Institute of Technology for Mariner A, but the principal investigator was visiting China and unavailable when the payload was selected, and it was decided instead to use an instrument built by JPL for Ranger.[83,84] The other instruments were a micrometeoroid detector, a triaxial flux-gate magnetometer to detect magnetic fluctuations, an ionization chamber, and a charged-particle detector. Also, Carl Sagan – more or less informally – suggested that the probe be equipped with a camera to obtain images that could show breaks in the cloud cover or other details that might be visible from nearby that were not discernible from Earth.[85] The primary objective was to measure both the temperature and microwave 'brightness' across as wide a section of the planet's disk as possible, ideally including both the night-side and the day-side. If the 'brightness' was constant from the limb to the center of the disk, then the temperature in excess of 300°C that had been measured by radiometry from Earth was explicable in terms of ionospheric effects, but if it varied across the disk this would indicate that the temperature of the surface was unusually high.[86]

Two identical Mariner R probes were shipped to Cape Canaveral for the 56-day launch window for Venus, lasting from 18 July to 12 September 1962.[87] Mariner 1, was launched on 22 July 1962, but less than 5 minutes into the flight, with only a few seconds remaining before staging of the Agena, the Atlas deviated from its assigned trajectory and threatened shipping routes, and the range safety officer had to issue the self-destruct command. Telemetry from the probe was received for a further minute as it plunged toward the Atlantic. In the ensuing months an investigation determined that the problem was in the Atlas guidance software, and that an error had occurred in the tracking algorithm that was intended to operate on the average velocity. The mathematical symbol to indicate 'average' was a horizontal bar over the quantity; however, in the handwritten equations given to the programmer, the bar had been neglected. On four occasions during the ascent, an antenna on the Atlas that was to receive measured averaged speeds calculated by a ground station had lost 'lock', and the autopilot had invoked its own algorithm, which used the

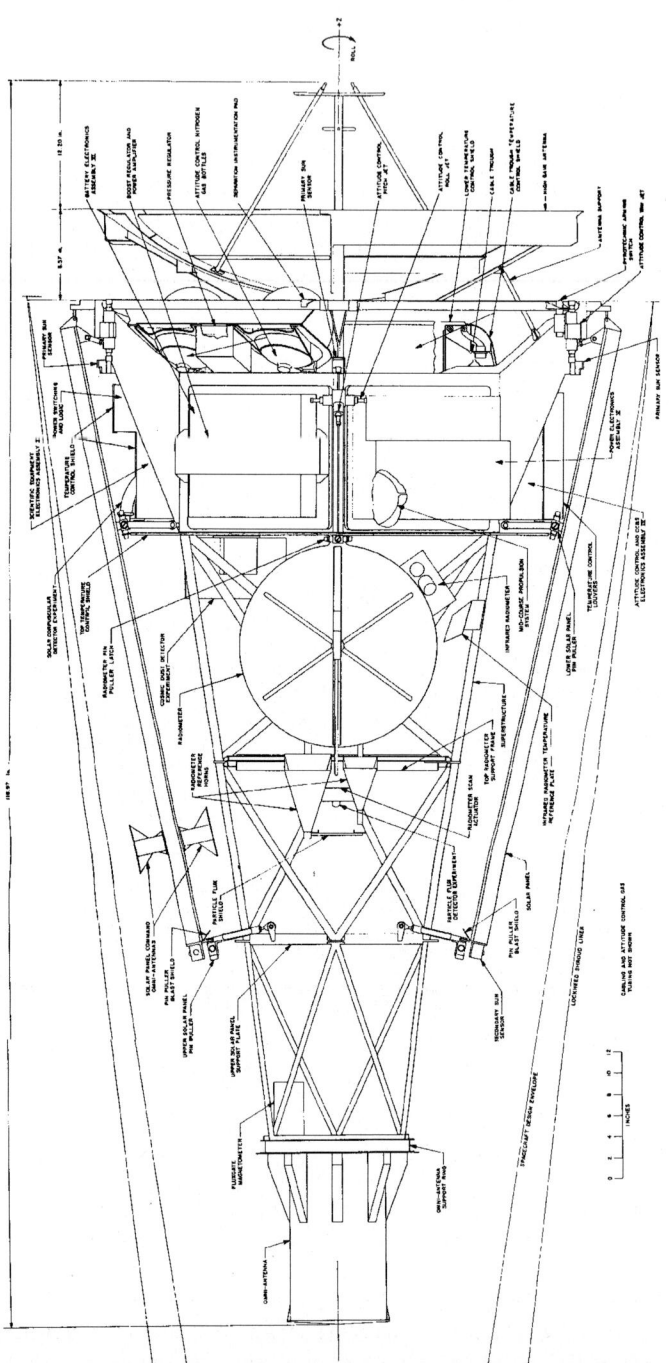

A line drawing of a Mariner R spacecraft in its stowed configuration. Such a craft made the first successful flyby of Venus. (Courtesy of NASA/JPL/Caltech)

A Mariner R spacecraft being prepared for launch. The parabolic antenna is folded beneath the bus, and several scientific instruments can be seen. The radiometers are on the opposite side of the spacecraft. (Courtesy of NASA/JPL/Caltech)

instant velocity from the radar instead of the 'smoothed' velocity. As a result, the program reacted to minor fluctuations in velocity and, in a classic example of a dynamic feedback loop, overcorrected and produced a major departure from the intended trajectory. The faulty program had been present on earlier missions, but this was the first time it had been invoked.[88,89] After almost a month of additional checking and two aborted attempts – one with just seconds remaining – Mariner 2 was able to be launched on 27 August. The flaw in the software had yet to be found, but the flow of tracking data was uninterrupted and the Atlas was not obliged to invoke the software. However, the Atlas lost control of one of its two vernier engines, and after the side-mounted booster system had been jettison the vehicle began to spin around its roll axis – slowly to start with, but then at an accelerating rate until it was rotating once per second, imparting unplanned stress on the structure. After about 70 seconds the roll disappeared, and the vehicle flew on in an almost perfect roll attitude. But at burnout it was pitched up. The Agena recovered from the attitude error and was able to achieve an acceptable parking orbit, and, 26 minutes after lift off, Mariner 2 was injected into a solar orbit ranging between 1.01 and 0.68 AU. Even Radio Moscow issued a brief report of the successful launch.[90,91]

During the first 4 days, the scientific instruments (other than the radiometers) were activated to begin collecting data on the interplanetary environment. The probe was left to roll to optimize the magnetic field data until 3 September, at which time it was instructed to adopt 3-axis stabilization in preparation for a 33.1-m/s course correction the following day to trim the estimated 375,900-km flyby distance to 35,000–40,000 km – some six times the radius of the planet.[92,93] The 109-day cruise was far from nominal: first the spacecraft lost attitude, possibly because of a micrometeoroid impact; then the sensor on the parabolic antenna appeared to lose track of Earth; and after 65 days the power from one of the solar panels ceased. Although the panel was recovered, it failed again 15 days later and could not be reinstated. By now, however, the probe was so close to the Sun that the output from a single solar panel was sufficient to run all the systems and instruments. The problems nevertheless continued: a faulty valve caused an unplanned increase of the pressure in the hydrazine tank; the thermal control system malfunctioned; and several telemetry channels were lost just 5 days before the flyby.[94,95,96] In October the first results were presented. The solar plasma detector had analyzed the solar wind in detail – not only confirming the wind's existence and its solar origin, but also measuring its velocity, which ranged between 400 and 700 km/s for most of the time, but occasionally reached more than 1,250 km/s.[97,98] During the cruise, a detailed analysis of the interplanetary magnetic field was also undertaken, with the probe observing several magnetic storms. Unfortunately, the magnetometer was so heavily influenced by the presence of the probe that it had been necessary to operate it at reduced sensitivity, and its data became increasingly garbled following the interruption of the power supply.[99,100] A temperature problem caused a malfunction of the micrometeoroid sensor as the probe approached Venus, but it detected only two impacts in 1,700 hours in transit. This was a considerably smaller flux than reported by spacecraft in low Earth orbit, indicating that micrometeoroids were as much as 10,000 times more abundant near Earth than in deep space.[101]

The most important instrument carried by Mariner 2 was this microwave radiometer with its very short focal-length antenna. The two horns at the top were pointed into deep space for calibration purposes. (Courtesy of NASA/JPL/Caltech)

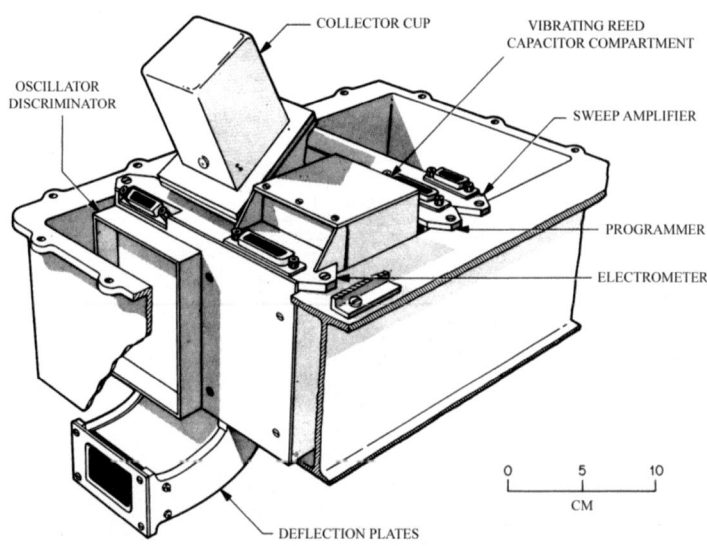

This plasma spectrometer was another important experiment on Mariner 2. Plasma entered it from the aperture at lower left. It provided the first accurate measurement of the solar wind. (Courtesy of NASA/JPL/Caltech)

On the morning of 14 December 1962, the command to configure the instruments for the Venus flyby – in particular, to activate the two radiometers – was sent by the ground because the automatic system had failed. The probe approached the cloud-enshrouded planet from above its equatorial plane, moving from the night-side to the day-side, with the closest point of approach (34,854 km) occurring at 19:59 UTC, and it collected data for a total of 35 minutes passing at a relative speed of 6.7 km/s.[102,103] Despite the uncertainty in the planet's ephemeris, the radiometers were able to make five scans of the night-side (with an average temperature of 217°C), eight across the terminator (322°C), and five on the day-side (238°C). These measurements meant that the 'brightness' across the disk was variable. The inferred temperature of the surface was no less than 425°C, far higher than had been expected for a planet that was often described as the 'twin' of Earth. Some fluctuations of up to 11°C observed by the infrared radiometer at the southern terminator hinted at the presence of either clouds of greater opacity, or high mountains – or even lakes![104,105,106,107] Hypotheses were proposed on the composition of the atmosphere, which seemed to be rich in carbon dioxide and almost devoid of water vapor, with an estimated surface pressure of no less than 20,000 hPa – that is, 20 times that of our own atmosphere at sea level. The magnetometer found that Venus does not have a significant magnetic field, and the data from the ion chamber and plasma sensors supported this conclusion. It was possible that the spacecraft had not actually penetrated the magnetosphere, but if a magnetosphere was present its magnetic field must be so weak as to expose the

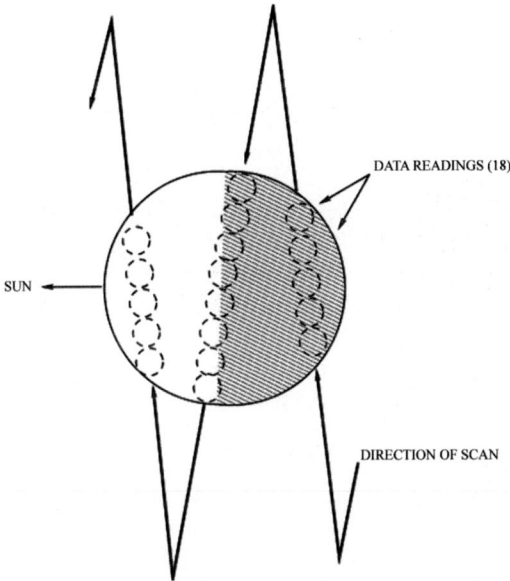

The geometry of the three radiometer scans of the disk of Venus made by Mariner 2, covering the night-side (at right), the terminator and the day-side (at left).

upper atmosphere to the solar wind.[108] One final experiment monitored the Doppler shift on a signal that was sent from Earth and retransmitted by the probe. The degree to which the spacecraft was accelerated during the encounter enabled the mass of the planet to be calculated as 81.485 per cent of that of Earth. In addition, the 22,000 or so positional measurements taken during the entire flight enabled the ephemeris of Venus to be refined, and the closest point of approach was determined with an uncertainty of 15 km – rather than the 1,000 km that applied until several days prior to the event.[109,110,111,112]

Mariner 2 continued to operate for several days beyond perihelion, but telemetry was lost on 30 December because of overheating electronics, and contact was lost on 3 January 1963, at which time it was 86.7 million km from Earth and had been in space for 129 days. Fifty commands were sent without reply on 8 January, and further unsuccessful searches were made on 28 May and 16 August.[113] Despite some gaps in the data – in particular after 17 December – the solar plasma detector had collected data during at least 104 days and produced over 40,000 spectra, all of which showed the presence of the solar wind.[114]

Despite the very meager scientific results (more detailed infrared maps of Venus were obtained at almost the same time using ground-based telescopes) and the long series of technical issues, Mariner 2 demonstrated how to manage and run a simple interplanetary mission. A third probe, carrying an improved microwave radiometer, was considered for 1964, but was deemed superfluous.[115,116,117,118] Mariner R had cost the equivalent of about $100 million at 2000 levels, in addition to $45 million for the development of the abandoned Mariner A.[119,120]

PRODUCT 2MV

As NASA nursed its Mariner 2 toward Venus, the Soviets suffered another string of failures. In the spring of 1961 Korolyov's engineers had set out to design a universal probe named 'Product 2MV' (Mars-Venera). Four different versions were designed, two for Venus and two for Mars. They were to fly in pairs, one for a photographic flyby and the other to deliver a landing module.[121] The common core was based on a 210-cm-long, 110-cm-diameter, 113.3-hPa pressurized cylinder for the standardized flight systems. At one end there was a KDU-414 (Korrektiruyushaya Dvigatelnaya Ustanovka) course-correction engine supplied by Aleksei M. Isayiev. This burned UDMH (Unsymmetrical DiMethyl Hydrazine) and nitric acid, delivered a thrust of 2 kN, and had sufficient fuel for a total of 40 seconds. At the opposite end was either a specialized mission module housing instruments or a spherical landing module. On the sides of the main compartment were two solar panel wings of 2.6 m^2 area, tipped by hemispherical radiators carrying the coolant for thermal control – it had been decided to discard the louver system of the 1M and 1VA probes, since this had not worked as intended. On one side (between the solar panels) there was a 170-cm-diameter parabolic antenna that had a solid inner portion and an 'umbrella-like' periphery, and on the opposite side there was a hemispherical stellar sensor and other sensors to determine attitude. There were also a series of omnidirectional and medium-gain antennas for a communication system operating at four different radio wavelengths. In addition to compressed nitrogen jets, the attitude control system had momentum wheels to maintain the Sun within 10 degrees of the ideal for the solar panels, and to keep Earth in the beam of the high-gain antenna. With its appendages deployed, the 2MV probe was 3.3 meters tall and spanned 4 meters. It was decided that for Venus the landing module would be cooled using ammonia heat exchangers, but for Mars, for which thermal requirements were less stringent, it would be cooled by air circulated by fans.[122,123,124,125]

In 1962, three 2MV probes were launched to Venus in the same window that had been used for Mariner 2. First, on 25 August and 1 September, two landers were launched whose scientific payloads included a suite of atmospheric instruments, a gamma-ray counter to make a preliminary analysis of the composition of the surface and, should there be open water, an ingenious liquid-mercury-switch motion sensor.[126] Unfortunately, on the first of these launches one of the four ullage rockets of stage L failed to fire, and the vehicle started to spin, starving the main engine of propellant and making it cut off some 45 seconds into the planned 240-second burn. In the second launch, a fuel valve failed to open. In both cases the vehicle was stranded in low orbit, and soon burned up. When a flyby probe was launched on 12 September, one of the vernier engines on stage I exploded, but since this occurred 531 seconds into the ascent it managed to continue and achieved parking orbit. However, when stage L ignited to set off for Venus the oxygen pump had a severe problem of cavitation, and the engine shut down after just 0.8 second.[127]

As an interesting aside, in early September 1962 the American press reported a rumor that the launches of a number of Soviet probes had failed. The official

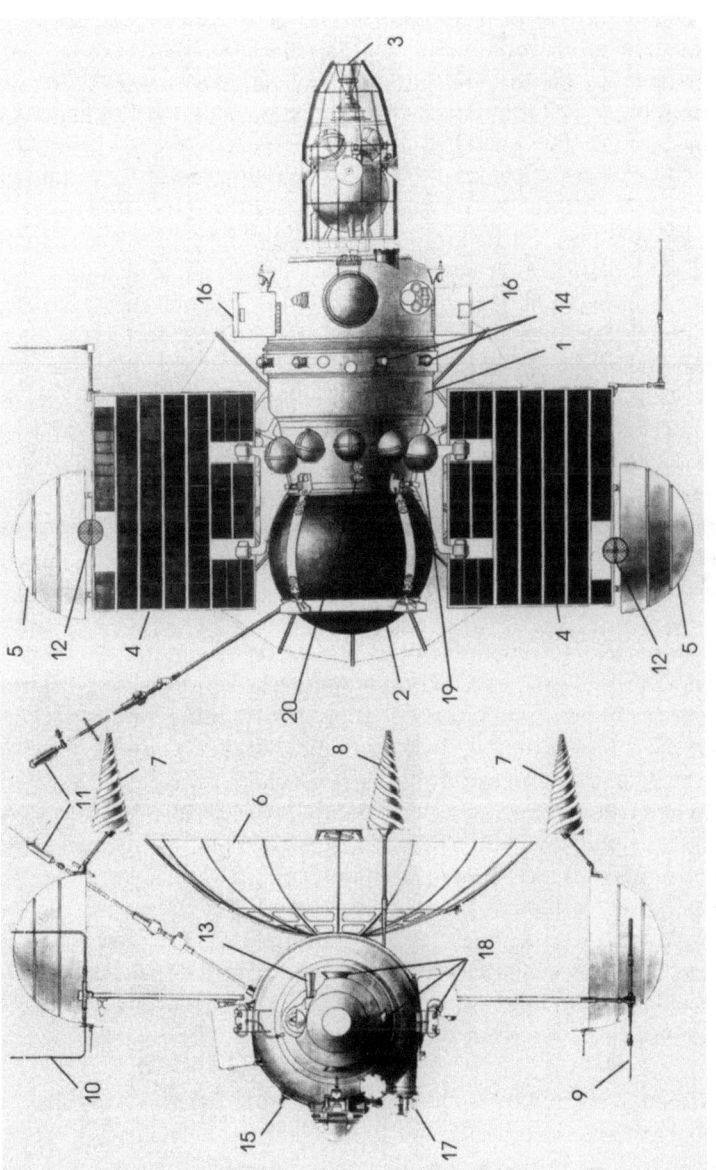

The Soviet 2MV-1 spacecraft equipped with an atmospheric entry probe for Venus: 1, hermetic orbital module; 2, entry capsule; 3, trajectory correction engine; 4, solar panels; 5, radiators of the thermal control system; 6, parabolic high-gain antenna; 7, low-gain antennas; 8, antenna to conduct tests of the entry capsule; 9, metric-range transmitting antenna; 10, metric-range receiving antenna; 11, emergency omnidirectional antenna; 12, antennas; 13, Earth sensor; 14, scientific instruments; 15, solar and stellar orientation sensor; 16, emergency radio system; 17, Sun sensor; 18, attitude control system nozzles; 19, nitrogen tanks for attitude control; 20, Sun orientation sensor. (Courtesy of RSC Energiya)

position was that all Soviet launches were announced, and all were successful. When a Congressman asked NASA's administrator, James E. Webb, to report on the issue, Webb revealed that American intelligence stations had tracked all of the 1962 launch attempts in addition to the Mars and Venus probes of 1960–1961.[128]

The attempts to dispatch missions to Mars in 1962, fared little better. The first, intended for a flyby, lifted off during one of the most tense moments of the Cold War: the Cuban missile crisis. As the 8K78 and its payload stood on the pad, the order came to replace it by an R-7 intercontinental-range ballistic missile carrying a nuclear warhead targeted on the United States.[129] When the Mars flyby probe finally launched on 24 October, it apparently almost caused a major international incident. After a period of coasting in low orbit, stage L fired. Unfortunately, 16 seconds later the kerosene pump lost its lubrication, seized, and caused the stage to explode, which sent a swarm of at least 24 pieces of debris on trajectories that would take them over Alaska, Canada and the United States. When they appeared on early-warning radars, the first impression was that the Soviets must have launched intercontinental-range ballistic missiles. In a short time, however, the computers that processed the tracking were able to report that the objects were in orbit and were not hostile.[130] Another flyby probe on 1 November was faced with difficult circumstances. Five hours earlier, the Soviets had detonated a 300-kiloton atom bomb in space above the weapons-testing site at Semipalatinsk to investigate the effect of the resulting electromagnetic pulse on electrical and radio apparatus, and it caused power shortages at Tyuratam, located several hundred kilometers to the west. Nevertheless, the launch was able to proceed. The 893.5-kg probe was successfully dispatched toward the Red Planet, and announced as Mars 1; because a companion was expected to follow within days, the numeral was made explicit.[131,132] The third probe was launched on 4 November, but severe vibration dislodged a connector in the ignition system of stage L, stranding it in a parking orbit that soon decayed. It had carried a 300-kg spherical Mars lander. Because no first-hand data on the Martian atmosphere was yet available, the descent system again had to be designed using the physical characteristics inferred from telescopic studies. The scientific suite probably included instruments to measure the temperature, density and pressure during the descent, and to determine the chemical composition of the atmosphere. It is also likely that the probe carried instruments to test for life. If it had succeeded in reaching Mars, it would not have fared well, since, as we now know, the atmospheric pressure at the surface is barely one-tenth of that believed at the time, and at best the lander would have returned data for a few seconds before it impacted the planet at a speed of several kilometers per second.[133,134]

The 60-cm-long pressurized compartment of the Mars 1 flyby probe contained an infrared spectrometer to detect organics by way of the 'Sinton bands', an ultraviolet spectrometer to measure ozone in the atmosphere, and a photographic system. The 32-kg camera had lenses of 35- and 750-mm focal length, and its 70-mm film could accommodate 112 frames. Pictures were to be taken using either a 1/1 or 3/1 aspect ratio and be scanned at resolutions of 68 lines (preliminary assessment), 720 lines (medium) or 1,440 lines (high) for transmission to Earth. The film would also record

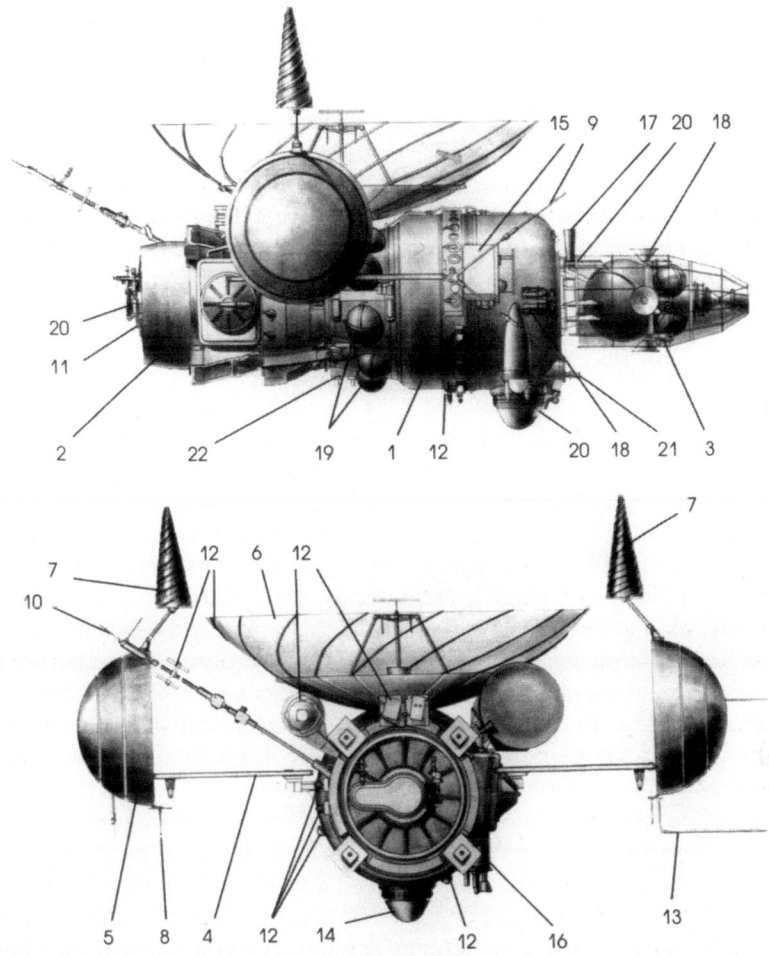

A 2MV-4 flyby spacecraft in the Mars 1 configuration: 1, hermetic orbital module; 2, hermetic imaging module; 3, trajectory correction engine; 4, solar panels; 5, radiators of the thermal control system; 6, parabolic high-gain antenna; 7, omnidirectional antennas; 8, omnidirectional antenna; 9, metric-range transmitting antenna; 10, emergency omnidirectional antenna; 11, camera port and planetary sensor; 12, scientific instruments; 13, metric-range receiving antenna; 14, solar and stellar orientation sensor; 15, emergency radio system; 16, Sun sensor; 17, antenna Earth-orientation sensor; 18, attitude control system nozzles; 19, nitrogen tanks for attitude control; 20, attitude sensor light baffle; 21, rough Sun orientation sensor; 22, Sun sensor. (Courtesy of RSC Energiya)

the output of the spectrometer. The plan was to photograph Mars in a variety of lighting conditions, and cover most of its surface as the planet rotated. It was also hoped to resolve the tiny moons, Phobos and Deimos. Other instruments were mounted on the exterior of the probe, including a boom for a magnetometer, a radio-astronomical package, cosmic-ray detectors, ion and electron detectors, and a micrometeoroid counter.[135,136,137] As soon as Mars 1 had been dispatched, the Crimean Astrophysical Observatory was given sufficient data on its trajectory to calculate how the probe would move across the celestial sphere. Within 24 hours of lift-off, with the probe already half way to the Moon's orbit, the observatory was able to take in excess of 350 photographs of both the probe and the spent stage L as faint rapidly moving points of light.[138] Western speculation about the mission was rampant, suggesting that the probe was to return to the vicinity of Earth after making its flyby (on a trajectory similar to that of Luna 3, the probe which, in 1959, took the first pictures of the far side of the Moon) or that it was to be inserted into Mars orbit.[139]

As Mars 1 departed, it measured the density of charged particles in the vicinity of the Earth. Because it had been launched near the peak of the Taurid meteor stream, it recorded several micrometeoroid hits during its first few days. In its first month it undertook 37 communication sessions with the ground station at Yevpatoria, which uplinked more than 600 commands. As the first probe to travel beyond Earth's orbit of the Sun, it gave unique data on the solar wind, the interplanetary magnetic field, and cosmic rays. In particular, the flux of galactic cosmic rays was more than double that measured in 1959 when the Sun was just past the peak of its 11-year cycle.[140,141] Unfortunately, soon after launch it was realized that one of the tanks of nitrogen for the attitude control system was venting to space, with the implication that sooner or later the probe would lose its ability to control its orientation. On 7 November, the remaining gas was used to spin the probe around an axis perpendicular to the plane of its solar panels at 6 revolutions per hour to keep them facing the Sun in order to generate power. This action obviated a course correction and meant that it would be impracticable to use the camera, because these functions would require the probe to stabilize.[142,143] Starting on 13 December, Yevpatoria held communication sessions at 5-day intervals. In March 1963 Mars 1 surpassed Mariner 2 by transmitting over a distance exceeding 100 million km. When it fell silent on 21 March, it was 106 million km from Earth and 1.24 AU from the Sun; it was also 60 per cent into its flight and had made a total of 61 communication sessions.[144] As the Soviet report stated, "troubles had appeared in the station's orientation system, as a result of which the pointing of the antennas toward the Earth was upset, thereby preventing radio contact during subsequent sessions". Stage L of the launch vehicle had put the probe into a solar orbit ranging between 140 and 250 million km from the Sun. The range of the Mars flyby announced by the Soviets soon after launch was to be between 1,000 and 11,000 km but, since the probe was unable to attempt a course correction, when it flew by the planet on 19 June 1963 it was at a range of 193,000 km.[145] An investigation found the presence of colophony – a residue that can form during the manufacture of the types of valve used in the attitude control system – and this had very likely prevented the valves from sealing properly, enabling gas to leak to space.[146]

In parallel with Korolyov, Vladimir Chelomei's design bureau, whose expertise in the 1950s was the development of cruise missiles, studied the design of a modular spacecraft for flights to the Moon, Mars and Venus. In this grand project, known as Kosmoplan (Spaceplane), a probe was to be propelled by an ion engine mated with a nuclear generator designed to enable it to accelerate slowly out of parking orbit into deep space. After flying past its target, the spacecraft would return to the vicinity of Earth and release a winged craft that would land at an airport. Although the preliminary development was completed in 1962, the project was canceled at the end of 1964 when Khrushchev was deposed. In retrospect, it is evident that the project had been allowed to start only because Chelomei's team included the son of the head of the Communist Party.[147] In fact, the Soviets had even more ambitious plans: a decree signed by the USSR Council of Ministers in June 1960 called for the establishment of automatic spacecraft for the exploration of the outer solar system, in particular the planet Jupiter. The preliminary studies appear to have been carried out, but these aspirations were destined to remain on paper, advancing in status even less than the similar US Navigator program.[148]

THE 'ZOND' PROBES

In 1963, after assessing the performance of the 2MV probes, Korolyov's team made improvements to the solar panels, the hemispherical star sensor, the electronics, the communication system, and the attitude control engines. The first of the new 3MV probes were to be called Zond (Probe) and be launched on interplanetary missions to test the technology prior to dispatching more heavily instrumented probes. The first was launched on 11 November 1963 to test the communication system and provide experience of deep-space navigation. Unfortunately, when stage L fired it was not in the required attitude, and Kosmos 21 (as the probe was named to conceal its actual mission) remained in Earth orbit. Undeterred, the Soviets decided to make use of the launch window for Venus between February and April 1964. They began with an imaging probe that lifted off on an improved version of the 8K78 on 19 February, but it was lost when liquid oxygen leaked from a faulty valve and froze the kerosene pipes. Next was a lander. The launch scheduled for 1 March was delayed to the 27th, but then stage L suffered a power shortage and the probe was abandoned in parking orbit as "the routine artificial satellite Kosmos 27".[149,150,151] When a second lander was successfully dispatched on 2 April it was named Zond 1 and characterized as "an automatic station for further developing a space system for distant interplanetary flights", without mentioning its target – although analysts could readily identify that it was heading for Venus.[152] It was to address several scientific questions. The primary module carried a magnetometer, a micrometeoroid sensor, two types of charged-particle sensors, ion traps, and a Lyman-alpha photon counter for the emission of atomic hydrogen. The lander's payload included atmospheric instruments, a photometer to measure the 'ashen light' of the Venusian night sky (often reported by telescopic observers, but unexplained), and a gamma-ray counter for surface chemistry that was also to be used to detect cosmic rays in interplanetary space. There had been a suggestion for an instrument to detect micro-organisms, but this was not included.[153]

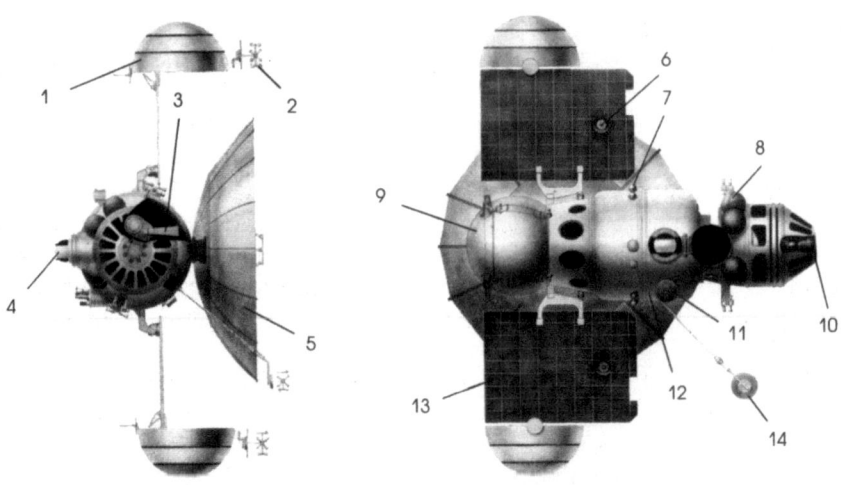

A 3MV-1 spacecraft of the Zond 1 configuration: 1, radiators of the thermal control system; 2, low-gain antenna; 3, Earth sensor; 4, solar and stellar orientation sensors; 5, parabolic high-gain antenna; 6, orientation control sensors; 7, scientific instruments; 8, nitrogen tanks for attitude control; 9, entry module; 10, trajectory correction engine; 11, Sun sensor; 12, orbital module; 13, solar panels; 14, magnetometer boom and entry module test antenna. (Courtesy of RSC Energiya)

Immediately after launch it was discovered that the primary module of the 948-kg probe was slowly depressurizing as a result of the window of the star sensor having been cracked by the rocket vibration. No less than 25 communication sessions were held in the first 2 days, and 60 in the first month. The initial trajectory was evidently very poor, and although two course corrections were made – the first on 3 April at a range of 560,000 km from Earth and the second on 14 May at 13 million km, with the latter providing a change in velocity of 50 m/s – the trajectory would still miss the target by 110,000 km. This discrepancy could have been further refined, but the deteriorating state of the probe's health prompted the cancellation of the course correction that had been scheduled for 30 May.[154] Finally, when the pressure of the compartment reached 6 hPa, the main radio system failed. However, because by this point Zond 1 had been commanded to spin in the manner of Mars 1, it continued to be able to draw solar energy and contact was maintained through the communication system of the lander until 25 May, less than one month from the flyby on 19 July.[155,156]

A commission was established to investigate this latest string of failures. The probes that were under construction were ordered to be subjected to vibration tests and to have their pressure-tight welds examined by X-ray analysis.

Two identical 3MVs were prepared for the launch window for Mars that would open in November 1964. One probe was launched on 30 November, which was later in the window than expected. The 8K78 launch vehicle successfully injected the probe into an orbit that had an aphelion 227 million km from the Sun and a period of 508 days. Western observers had expected the Soviets to use a high-speed trajectory to minimize the transit time in the hope of reaching Mars within the life of the probe, but for the first time they used a transfer orbit that would subtend an angle of more than 180 degrees to encounter its target.[157] This would increase the transit time by one full month, but provided the benefit of a slow approach by decreasing the relative 'asymptotic' speed to the planet to 3.77 km/s (compared to the 3.97 km/s in the case of Mars 1), reaching 5.62 km/s at the point of closest approach. For many years, analysts thought that the objective of the mission must have been to deliver a lander, for which a slower encounter would be a bonus, but following the collapse of the Soviet Union it was revealed that the plan was to take pictures during a 1,500-km flyby.[158,159] In view of this, it is possible that the Soviets wished to test a flight profile intended for landers. The instruments included an infrared spectrometer similar to that of Mars 1, and a new 6.5-kg camera with a lens of 106-mm focal length that was designed to take 40 images on 25.4-mm film to be scanned on board at a maximum resolution of 1,100 lines. As previously, an ultraviolet spectrometer would record its results on the same film.[160] The probe also carried six magnetohydrodynamic engines as an engineering test of a new attitude control system. Built by the design bureau of Aleksandr M. Andrianov, these engines would use plasma obtained from the erosion of solid fluorine that was accelerated to 16 km/s by interaction with a magnetic field and expelled into space to produce a sustainable thrust of a few grams.

When the first communication session on 1 December revealed that a solar panel had not properly deployed – probably because one of the cords designed to pull the panels open after separation from the launcher had broken, raising doubt about whether the probe would survive the trip to Mars – the Soviets announced it as Zond 2 and limited its objectives to "investigations in interplanetary space" to give the impression that it was merely another test. The second solar panel was finally able to be deployed on 15 December, after a series of engine firings, but by then the time to perform the course correction had passed. Three days later, two tests of the plasma engines revealed that although they worked as designed, they were insufficiently powerful to control the attitude of the probe. Meanwhile, other problems developed: for example, the onboard timer failed to activate the thermal control system, and contact became erratic one month into the mission. Jodrell Bank is said to have received signals on three occasions in February 1965 but this may be false.[161,162] Finally on 5 May, a Soviet delegation to a symposium in Chicago on the theme of post-Apollo space exploration, announced: "Transmissions from Zond 2 have stopped. We have been unable to raise it again." The date of the final contact is uncertain, but it may have been 2 May.[163,164] Once again, on 6 August, another inert Soviet probe flew past its target.[165,166,167,168,169]

FAREWELL TO THE 'LITTLE GREEN MEN'

On 8 May 1962 NASA was finally able to attempt the first development flight of the Atlas–Centaur, but it was lost after 63 seconds. In parallel with the investigation into the cause of the explosion, which was attributed to insulation problems on the Atlas, NASA reviewed its schedule for exploring the solar system. It postponed the first Mariner B mission to Mars from 1964 to 1966 (only later to cancel it, because the 1966 window was expected to be used by the first Voyager mission), and postponed the Venus mission to 1965. The agency considered two proposals to exploit the 1964 Mars window. The first, by the Goddard Space Flight Center, was for a spacecraft to deliver a module that would collect data on the physical and chemical parameters of the Martian atmosphere as it descended toward a soft landing. The second proposal from JPL, named Mariner C, called for a pair of identical probes to make a fast flyby to collect basic data and photograph the surface of the planet. Having a mass of only several hundred kilograms, these probes would be able to be launched by the Atlas–Agena. NASA assessed the proposals in terms of the optimal use of the launch window for the development of scientific instruments, and how the mission would contribute to the technologies required by future landings. Another requirement was that the probes use as much existing technology as possible, to minimize risk, development time and cost. The Goddard proposal was appealing, but faced two formidable technical challenges. First, it involved a series of critical events – for example, releasing the capsule, entering the atmosphere at the proper angle, deploying the parachutes and the landing itself – none of which had been demonstrated previously, and raised the issue of reliability. There was also the ethical issue of contaminating Mars with terrestrial biota, which, as NASA put it, would be a "scientific catastrophe", and there was as yet no proven technique for sterilizing a spacecraft. As the Soviets were openly stating their intention to land a probe on Mars as soon as possible, there must have been a temptation simply to 'go for broke', but at the end of 1962 NASA announced that it would mount two JPL flyby missions. While similar to Mariner B in some respects, Mariner C was required to be much lighter.[170,171,172]

To preclude a repetition of the technical difficulties that had beset the Mariner 2 flight, Mariner C was optimized for deep space. The JPL trademark octagonal bus, 138.4 cm wide and 45.7 cm tall, held the sequencer, 10.5-W communication system, batteries, and tanks of propellant. The course-correction engine was similar to that of Mariner R, but with the additional potential of being fired twice. It delivered a thrust of 225 N, and its nozzle, which had four paddles for thrust vectoring, was located at the center of one side panel of the bus. Mariner 2 had had an Earth sensor as part of its attitude determination system, but this had proved unreliable, and was in any case unsuitable for a flight to Mars since Earth would appear close to the Sun in the sky. Mariner C therefore combined a Sun sensor with a star sensor pointed at Canopus (alpha Carinae), selected for its brightness and for being placed almost perpendicular to the plane of the ecliptic. Attitude control was by 12 nitrogen thrusters, and the tank had sufficient gas for a 3.5-year mission. The central structure supported four fold-down 6.5-m^2 solar panels delivering a total of 700 W near Earth. On the tip of each were the thrusters and a 0.16-m^2 'solar sail' which was: (1) to stabilize the probe, since its configuration was unstable for the planned flight attitude; (2) to damp small attitude motions; and (3) to contribute to the attitude control. With its solar panels deployed, the craft spanned 6.79 meters. On top of the octagonal structure was an aluminum honeycomb high-gain antenna weighing less than 1 kg which had an elliptical shape with a 117-cm major axis and a 53-cm minor axis. As a result of the severe mass constraint, this antenna was carried in a fixed position that was optimized for the relative positions of the Sun (for power) and the Earth (for communication) around the time of the flyby. A 2-meter-tall boom with a low-gain antenna stood alongside the elliptical dish. Beneath the octagonal structure was an instrument platform that had limited movement on two axes. On the nominal encounter sequence, a wide-field sensor was to search for Mars, and after the planet had been identified the driving system was to switch to tracking mode, awaiting the planet's entry into the field of a narrow-angle sensor, at which time the imaging sequence would start.[173,174,175,176]

The use of innovative ultra-lightweight bonded structures meant that even though Mariner C was much more capable than Mariner R, it was only slightly heavier. The scientific payload comprised 15.5 kg of the total mass at launch of 261 kg, and was divided in two groups: instruments that did not need to be accurately pointed at the planet were mounted on the body, while those that required accuracy were installed on the movable platform. Eventually, however, the only instrument on the platform was the 5.1-kg camera that was fitted with four alternate red and green filters to maximize the contrast of the dark albedo features against the ochre disk. The camera was coupled to Cassegrain optics with a diameter of 38 mm and a focal length of 305 mm that formed an image on a vidicon screen; each image being converted to an array of 200 x 200 pixels in 64 levels of gray. In order to utilize the camera over wide-ranging levels of illumination as the field of view moved from the limb at local noon to the terminator at local sunset, an automatic gain control was provided to ensure that the image contained at least 15 levels of gray. A tape recorder with a capacity of 22 frames was to store the output of the vidicon, and later replay it for transmission to Earth at the (now amazingly slow, but state of the art in the mid-1960s) rate of 8.33 bps. Each image – which required 24 seconds to be taken and stored, and in excess of 8 hours to send to Earth – was accompanied by engineering data that slowed the effective rate to one frame every 10 hours. Despite this, the pictures would contain only about 0.05 as many pixels as a single Ranger picture of the lunar surface, or about 1 per cent as many elements as an aerial picture of Earth. The images were to be taken in pairs, with a brief time interval in between, and through different filters to enable 2-color composites to be made in cases of overlap. It had been hoped to put another instrument (either an infrared spectrometer or a 3-channel ultraviolet radiometer) on the movable platform in order to measure the composition of the atmosphere, together with profiles to enable the pressure at the surface to be inferred, but although prototypes of these instruments had been built, developmental problems had precluded their being ready in time.[177,178,179] The instruments carried on the spacecraft's body were plasma, radiation, cosmic-ray and micrometeoroid sensors. An ionization chamber and a 3-axis helium magnetometer – to measure the interplanetary magnetic field and determine whether Mars had its own field – were mounted on top of the low-gain antenna boom to protect them from the metallic structure of the probe.

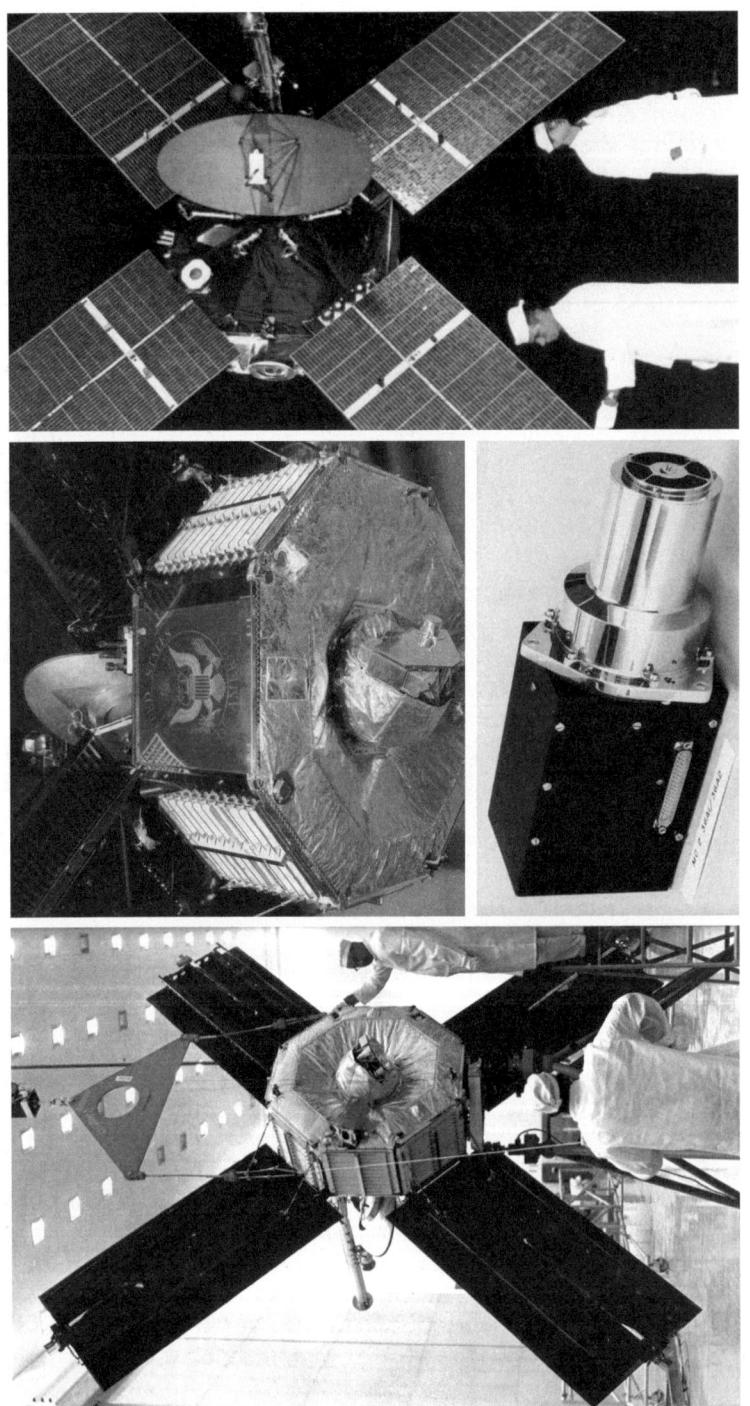

A Mariner C probe is prepared for launch. The 'sails' remain to be installed on the solar panels. Note that some of the thermal insulation blankets are black rather than white (as was used on Mariner 2) because even though black will absorb more solar heat than white, it will do so in a more predictable manner (and, in fact, Mariner 4's temperature varied almost as predicted during its mission). In the center is a view of the rotating platform on the base, and the camera. (Courtesy of NASA/JPL/Caltech)

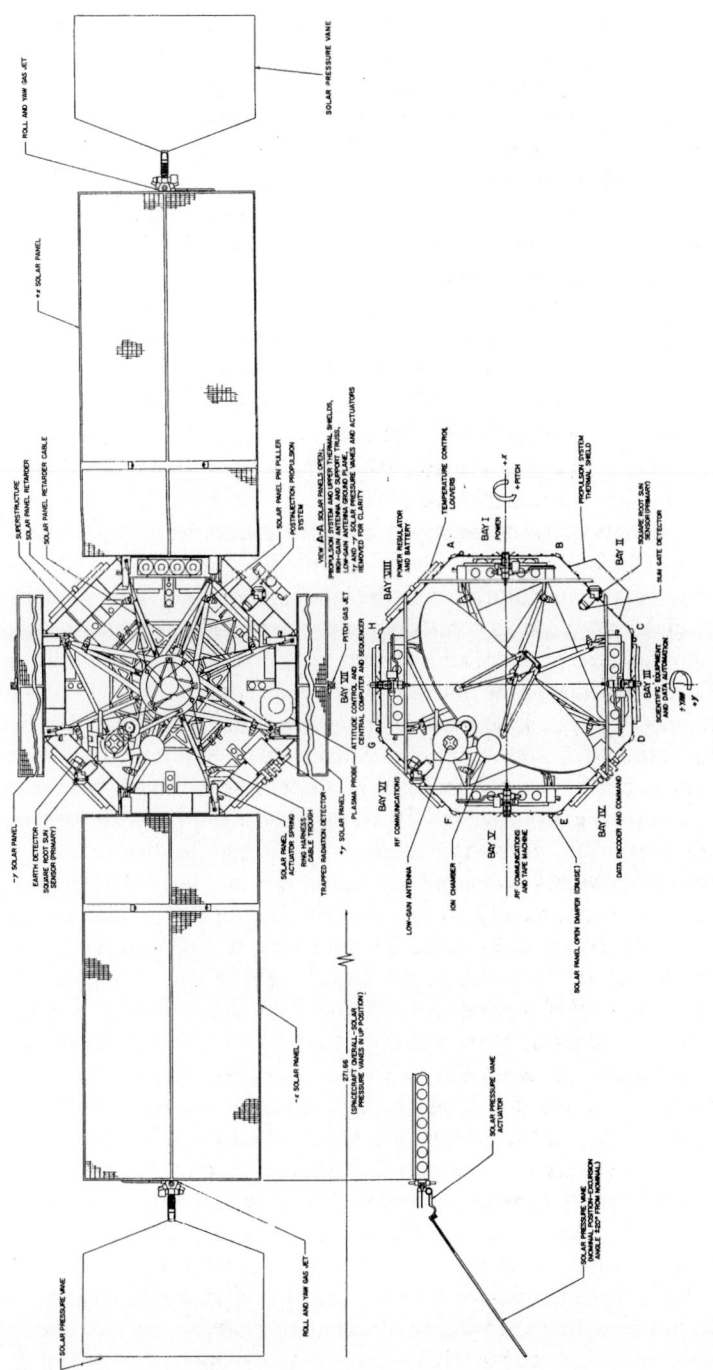

A line drawing of Mariner 4. (Courtesy of NASA/JPL/Caltech)

A unique scientific experiment that did not require an additional instrument was devised by Dan L. Cain of JPL. By planning the flyby to ensure that the spacecraft would be occulted by the planet, the manner in which a simple radio carrier transmitted by the probe was refracted could be used to profile the pressure and temperature of the atmosphere. Engineers were concerned at the loss of communication with the craft for up to an hour during the most critical phase of the mission, but when NASA realized, several months prior to launch, that it would not be possible to include the other platform instruments, this "tricky" experiment was approved as a simple and economical means of obtaining this important atmospheric data.[180,181,182]

With development underway, the Space Science Board of the National Academy of Sciences added its support for the mission, stating that, in view of its biological, physical and geological interest, the exploration of Mars should become the primary goal of NASA's planetary program.[183] The twin Mariner C probes, named Mariner 3 and Mariner 4, were readied for launch in November 1964 – the same window that was used by Zond 2. Mariner 3 lifted off on 5 November and entered parking orbit, and the Agena made the 'escape' burn. However, telemetry from the probe showed that the solar panels had not deployed and its systems were rapidly depleting the battery. It was soon realized that the aerodynamic shroud that had protected the spacecraft during the ascent through the atmosphere was still in place. When a command from Earth failed to belatedly deploy the shroud, engineers suggested that it might be possible to 'blow off' the shroud by momentarily firing the probe's side-facing course-correction engine, but this was not attempted. In fact, the mission was already lost since, owing to the mass of the shroud, the Agena had not been able to achieve the desired transfer orbit for Mars. After 8 hours, the batteries were exhausted and the probe was abandoned in an orbit that ranged between 0.983 and 1.311 AU. An investigation led by JPL took only a few days to identify the root cause of the problem. In order to save mass, NASA had decided to make the shroud of honeycombed fiberglass instead of metal, but this had not been adequately tested. In particular, the pressure difference between the interior of the shroud and the unvented honeycomb cells had caused the inner skin to split and jam the mechanism intended to release the shroud once the launch vehicle was above the atmosphere. Within a fortnight a magnesium shroud that was only slightly heavier than the original was prepared and, 24 hours after passing its tests, was installed on the Atlas–Agena with the second probe.[184]

Mariner 4 left Earth on 28 November, and entered a fast transfer orbit that would pass within 246,378 km of Mars – an orbit designed to minimize the possibility of the unsterilized Agena reaching the planet. The probe performed a course correction on 5 December to pursue a path that would pass the planet at a range of about 9,600 km and produce the desired radio occultation. On this path the probe would neither enter the shadow of the planet (which would deny power at a crucial time) nor block its view of Canopus (required for attitude control), and would ensure that the star sensor could not be distracted by either Phobos or Deimos – in fact, the spacecraft's trajectory was required to approach no closer than 6,000 km to either of the Martian moons.[185]

Preparing and launching the Atlas–Agena with the ill-fated Mariner 3 spacecraft.

Despite the failure of the ion chamber in February 1965 and a component failure degrading the data from the plasma sensor, the probe was able to make observations during its 8-month interplanetary flight. It was discovered that, despite the Sun being near the minimum of its 11-year cycle of sunspot activity, it was far from quiescent, flaring 20 times during the period of observation. However, the attitude control system suffered problems. The Canopus sensor took 2 days to identify its target star, lost it twice in December, probably because it was distracted by sunlight reflected from small fragments of insulation shed by the probe, and ended up viewing gamma Velorum. To preclude similar problems during the flyby, when the probe would require to adopt a specific attitude in order to image Mars, it was decided to open the camera cover in February and rotate the platform into the appropriate orientation.[186] It had initially been hoped to image Syrtis Major, but this was ruled out by the requirement that, at the time of the flyby, Mars had to be above the horizon of the large antenna at Goldstone in California, which was to monitor the occultation. As the historic flyby loomed, astronomers made observations in support of the mission, monitoring meteorological conditions in the regions that were to be imaged. French astronomer Audouin Dollfus, for example, observed 'white clouds' over the Mare Sirenum and Phaetontis regions.[187]

On 14 July a modified encounter sequence was initiated from Earth. The camera and the recorder were powered. Imaging started at 00:18 UTC on 15 July, at a distance of 17,600 km, and continued to 00:43, by which time the range had closed to 12,000 km. The closest point of approach occurred at 01:01, at a range of 9,846 km and a relative speed of 5.12 km/s. An hour and a half later the probe flew behind the trailing limb of Mars, as viewed from Earth, above the region of Electris and Mare Chronium at 55°S and 177°W, where it was late evening, and 54 minutes later crossed the leading limb above Mare Acidalium at 60°N, 34°W, where it was local dawn.[188] The radio-occultation data gave an astonishing result. Since the time of Percival Lowell it had been believed that the Martian atmosphere was primarily nitrogen with a pressure at the surface of 87 hPa, and that the temperature was close to freezing. But Mariner 4 found the atmosphere to be extremely evanescent, with a surface pressure estimated at between 4.5 and 5 hPa at ingress and 8 hPa at egress (the value depending on the composition of the atmosphere, which the probe was not equipped to determine). In 1962 a telescope carried on a balloon to observe from above the bulk of our own atmosphere had obtained similar values for the surface pressure, but this had been presumed to represent only the partial pressure of carbon dioxide, which was thought to be only a minor constituent of the atmosphere, with the remainder being nitrogen, which is difficult to detect. Mariner 4's revelation that this was actually the *total* pressure suggested that the atmosphere was primarily carbon dioxide.[189] (At the 1967 opposition, a telescopic study took spectra which, combined with Mariner 4's occultation data, established an average surface pressure of 5.2 hPa.[190]) At −100°C, the surface was much colder than expected. It was now evident that even if the early Soviet landers or Mariner B had survived the trip, their parachutes would have been ineffective. A lander would require a retrorocket system, and humans would require to wear bulky heated pressure suits. Electron density profiles in the ionosphere were also obtained during the occultation.[191,192]

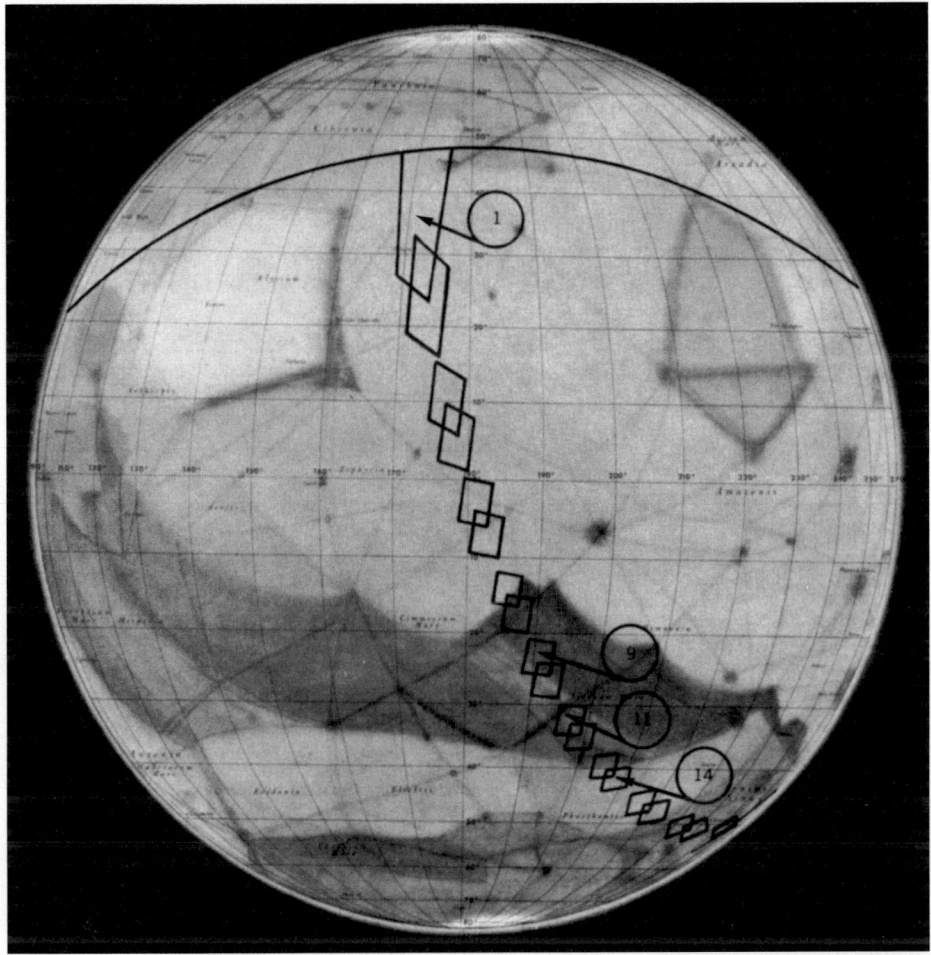

A telescopic map of Mars prepared by R.W. Carder and E.C. Slipher of the Lowell Observatory (comprising a 'canali' network) with the areas imaged during the flyby of Mariner 4 superimposed. The arc marks the limb of the planet from the point of view of the spacecraft. The four numbered frames are shown in the next image.

Doppler tracking from 10 to 20 July enabled the mass of the planet to be determined with an improvement in accuracy of a full order of magnitude.[193] The timing of the occultation provided a measure of the size of the planet. The fact that its radius differed by about 5 km at ingress and egress indicated the presence of major elevation variations, which was surprising as it had been believed that the surface was extremely flat.[194,195] Although the encounter geometry was not well suited to measuring a weak planetary magnetic field, the magnetometer placed an upper limit on the strength of such a field at 0.03 per cent of that of the Earth's field, and the particle sensors detected no trapped radiation.[196] Consequently, the upper atmosphere is exposed to the solar wind.

Some 8.5 hours after emerging from the occultation, Mariner 4 began to relay the pictures from its tape recorder. The long wait for the data was frustrating, and JPL officials endeavored to temper the rising popular expectation of gaining a close-up look at the 'canali'. The press were shown deliberately 'fuzzy' telescopic pictures of the Moon in order to illustrate the best resolution that might be expected. The engineers were concerned, because there were hints in the telemetry that the tape recorder had malfunctioned, and erased some of the pictures. Nevertheless, William Hayward Pickering, director of JPL, cautiously confirmed to reporters that "every explanation that we have attempted to give says we have some pictures".[197] In fact, 21 full images had been taken in addition to 21 lines of another image, together covering almost 1 per cent of the surface. The track began at 37°N, 173°W, ran southeast into the southern hemisphere, across the terminator, and ended in darkness at 50°S, 255°W.

The first image showed a 365-km arc of the limb from a distance of 17,600 km. The presence of a bright halo running parallel to the limb posed a puzzle. It gave the impression of a thick haze, but in such a thin atmosphere that was thought to be unlikely, especially as terrestrial observers had reported no hint of it. Other possibilities were a reflection in the optics or a fault in the imaging system, but ground tests showed that such a flaw would have required the camera to be substantially damaged, which was unlikely. In the third image (once the contrast was corrected to reduce the effects of the Sun being near the zenith) topography features began to be discernible, but only with the seventh image was their nature revealed: craters large and small. The first seven images covered the bright areas of Amazonis and Zephyria. The following six were on the dark Mare Cimmerium and Mare Sirenum. The eleventh image became the most famous of the mission. At 1.4 km per pixel it had the highest resolution of the series, owing to the illumination and viewing geometry. It depicted a crater at 33°S, 197°W, that was 150 km in diameter and flanked by several smaller craters. The International Astronomical Union later named the large crater Mariner, after the probe that discovered it. At the twelfth image, the quality began to decline because the field of view was approaching the evening terminator, and also this was this area in which Dollfus had reported clouds. Some scientists interpreted dark streaks in the fourteenth image as the shadows of clouds, whose altitudes they estimated at 3.5 km. As the field of view approached the terminator beyond the bright Phaetontis region, the gain was to have increased the exposure to compensate for the diminishing light, but this failed, with the result that the seventeenth image was very dark, and by the nineteenth image the camera was viewing the night hemisphere. A total of almost 300 craters were counted, with diameters ranging from 5 km (which was comparable to the camera's limiting resolution) to 150 km.[198,199,200,201]

The presence of large craters begged the question of whether they had ever been seen by terrestrial studies. In fact, at moments of exceptionally clear 'seeing' during the near-perihelic opposition of 1892 William Henry Pickering saw small dark spots where the 'canali' crossed. At about the same time, Emerson E. Barnard reported small dark circular spots on Mars. In 1950 Clyde Tombaugh speculated that Mars had suffered massive impacts, reasoning that the dark spots were indeed

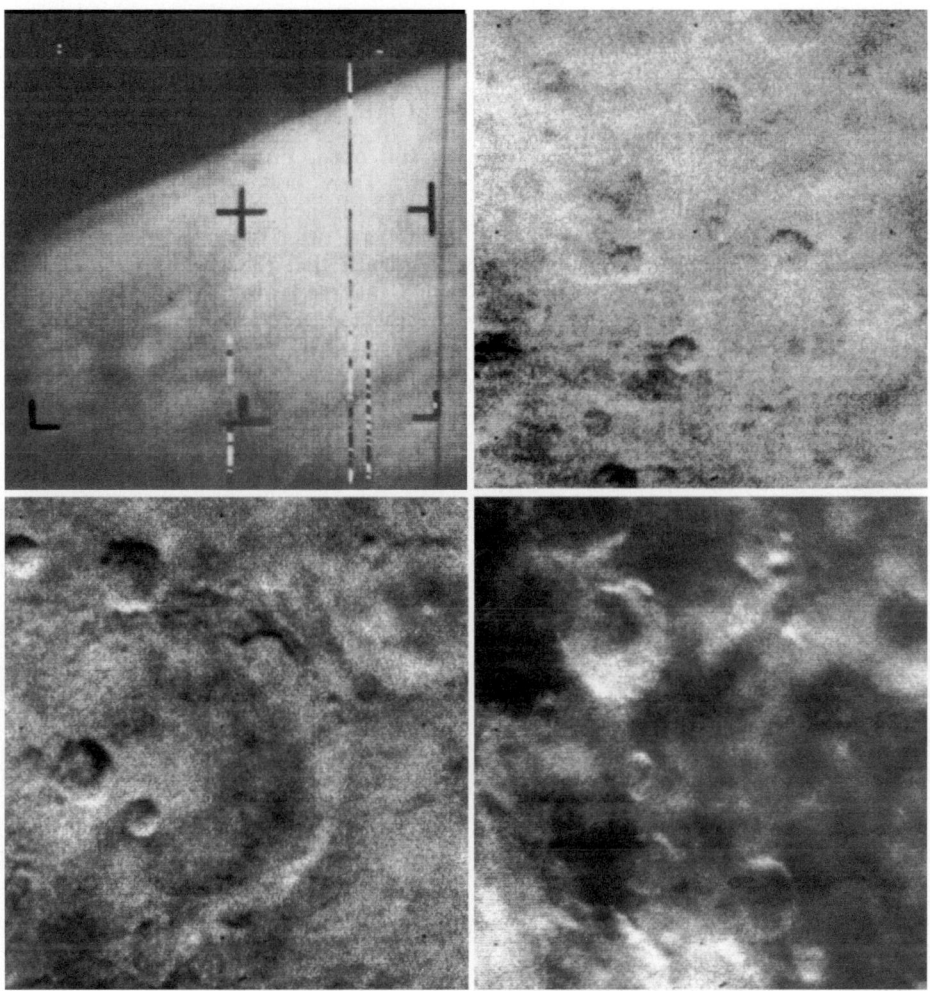

Some Mariner 4 images. Frame 1 (top, left) was taken at 00:18:33.1 GMT on 15 July 1965. It shows the limb of Mars with a hint of haze that future missions would prove to be dust suspended in the atmosphere. In this case, the image is depicted complete with its calibration marks. Frame 9 (top, right) clearly shows the presence of craters. Frame 11 (bottom, left) is the best known and the highest resolution image, showing details as small as 1.4 km. Linear features seen diagonally crossing the lower half of the 150-km-diameter crater were suggested by some as traces of one of the 'canali'. The crater was later named Mariner, in honor of the spacecraft. Some of the craters in frame 14 (bottom, right) show traces of white frost.

impact sites and the 'canali' traced cracks opened by the shock.[202] Ralph Baldwin and E.J. Opik independently offered similar ideas.[203,204,205] The question of whether telescopic observers had seen craters remained open until it was established that not even the Hubble Space Telescope is able to resolve them.[206]

Although Mariner 4's imaging track crossed the lines of some 'canali', they were, as many had maintained, an optical illusion as there was no evidence of structures – or almost no evidence: in analyzing the imagery Eric Burgess drew attention to a pair of parallel linearities that crossed the large crater on the eleventh picture in approximately the position of one of the 'canali', which he interpreted as a down-faulted tract of a 'graben' or 'rift valley'.[207]

In the 26 minutes during which Mariner 4 took pictures and the 54 minutes while it was occulted, the mission overturned centuries of telescopic study of Mars. Not only is the Martian atmosphere more rarefied than expected, but the planet's surface was found to be more like the Moon than like the Earth – a discovery that came as a considerable shock. Physical models based on the surface pressure measured by Mariner 4 indicated that the polar caps were frozen carbon dioxide rather than water ice – a fact that made the prospect of Martian life even less plausible.[208] The reductive model that emerged dominated thinking for the remainder of the decade. It consisted of five main points:

- The geological history of Mars is (in some respects) more akin to that of the Moon than of the Earth.
- The craters are due to impacts, and the fact that the cratering rate is similar to that of the Moon implies that the surface is between 2 and 5 billion years old.
- The state of preservation of the craters indicates that the atmosphere never exercised a significant erosive action, and must have always been very thin.
- With the caveat that only 1 per cent of the surface was inspected, the lack of structures other than craters and the absence of a magnetic field means that Mars has never been subjected to large-scale geological activity.
- Although Mariner 4 could not prove that life has ever existed on Mars, the above points would argue against this possibility.[209]

In the words of science writer Keay Davidson, the flyby that occurred on 14 July 1965 in the US time zones proved to be the Bastille Day of Mars biology![210]

To seek evidence to explain the bright feature seen on the limb in the first picture, and to better calibrate the subsequent images, the camera was restarted on 26 August and 11 pictures of the sky were taken, five of which were later replayed.[211,212] There was no degradation in either the optics or the imaging system, which implied that the effect was a genuine atmospheric phenomenon. Future missions would show it to be due to dust suspended in the atmosphere.

Contact with Mariner 4 was routinely maintained for some months following the flyby; intermittently from March 1966 to early 1967; and continuously again from July 1967 as its orbit took it within 50 million km of Earth. The course-correction engine was fired for a second time as an engineering test, parts of images 16 and 17 were replayed to evaluate whether the tape recorder had degraded, and additional pictures were taken of the sky. In September and December the probe crossed two streams of meteors. In the first case, the meteoroid sensor recorded 17 impacts in less than 15 minutes when Mariner 4 possibly flew within 20 million km of the remains of the nucleus of the long-lost periodic comet D/1895Q1 Swift and through its stream of debris. On the second occasion, a large roll disturbance was detected when the probe was only 10.5 million km from the Leonid meteor stream which, the previous year, had produced the most intense meteor shower on Earth of the entire twentieth century. The probe, however, had run out of attitude control nitrogen on 7 December, and so it is possible that the perturbation was not due to a meteor impact.[213,214] It is worth noting that the small 'sails' mounted at the tips of the solar panels proved to be ineffective: perturbations due to solar radiation pressure were negligible, and the attitude control forces and torques imposed were smaller than the perturbations created by the slow leaking of gas through the nitrogen thrusters. As a result of this finding, Mariner 4 was the last deep-space probe to be equipped with 'sails'. The final contact took place on 31 December 1967. Mariner 4's present orbit is similar to its original one, ranging between 1.107 and 1.561 AU.[215,216] The cost of the Mariner C mission to Mars was the equivalent of $370 million at 2000 levels.[217]

KOROLYOV'S LAST PROBES

Two 3MV imaging spacecraft had been prepared for the 1964 window for Mars, but owing to delays and technical problems only Zond 2 had been launched. The second was launched on 18 July 1965 to test modifications made after the previous failures, and to finish the preliminary reconnaissance of the far side of the Moon that was initiated by Luna 3 in 1959. Although the 959-kg Zond 3 probe was put into a solar orbit that would take it far enough into space to test the endurance of its improved systems, its departure trajectory was designed to take it close by the Moon after some 33 hours, when it snapped 25 images from a distance of 11,570 km. They were stored until the probe was 2 million km distant, and then transmitted to Earth. After a course correction, they were sent again from a distance of 31 million km, and a third time from even further out in long-range tests of the communication system. When contact was lost in March 1966, it was more than 150 million km away. Ironically, this would be the only mission on which a 2MV or 3MV probe performed a meaningful photographic assignment.[218]

Based on this successful flight, four 3MV spacecraft were prepared for the Venus window that opened in November 1965. The mission included some spacecraft that had been built for Mars the previous year and suitably modified. Two were to make flybys, while the others were to penetrate the atmosphere and release landers. The main difference with their Zond predecessors was that the experimental attitude control thrusters had been deleted. The first two launches successfully dispatched Venera 2 (an imaging mission) on 12 November and Venera 3 (a lander) on 16 November. A launch on 23 November encountered problems after 528 seconds when a propellant leak in stage I caused a combustion chamber explosion, and although it was able to reach low orbit it left stage L in an unstable configuration that prevented it from attempting the 'escape' burn. The probe was written off as Kosmos 96; accounts differ as to whether it was an imaging probe or a lander. The final launch was canceled on 26 November after a problem on the pad, and the window closed before the vehicle could be made ready.[219,220] Of the probes in transit, Venera 2 was very similar to Zond 3, but had some new or revised instruments, including an infrared radiometer to measure the temperature of the atmosphere and surface, and a Lyman-alpha photon counter similar to that on Zond 1. It also carried the imaging and communication system that had been 'proven' by Zond 3. Venera 3 had a 337-kg spherical capsule 90 cm in diameter in place of the instrument module, designed to survive a pressure of 5,000 hPa as it performed a parachute descent carrying a scientific payload identical to Zond 1.[221]

When tracking established that Venera 2's departure trajectory would produce a flyby within the required range of 40,000 km above the illuminated hemisphere, and would thus be ideal for imaging, the optional course correction was canceled. If its course was not altered, Venera 3 would pass by the planet at a range of 60,550 km, therefore a course correction of 21.66 m/s was made on 26 December. Some 1,300 distance measures, 5,000 velocity measures, and 7,000 celestial position measures were taken during 26 communication sessions with Venera 2 and 63 with Venera 3, which established that Venera 3 would hit the planet at coordinates computed to an accuracy of 800 km.[222]

At 02:52 UTC of 27 February 1966, Venera 2 flew by Venus at a distance of about 24,000 km. After collecting scientific data and imagery, it was to resume its normal attitude and transmit its data to Yevpatoria, but no such signal was received and attempts to re-establish contact were abandoned on 4 March. In fact, on the last communication session before the probe was to start its flyby sequence, its telemetry showed a sharp rise in the internal temperature of the pressurized compartment. Just to be certain, a command was sent to order the probe to switch on all of its instruments, but this was not acknowledged.[223] It is possible that the probe collected its data but was unable to transmit it to Earth. Meanwhile, after severe thermal problems, Venera 3 had fallen silent on 16 February. At 06:56 UTC on 1 March it penetrated the atmosphere in darkness near the terminator to become the first human artifact to reach the surface of another planet. One source gives the approximate coordinates of its 'landing site' as 0°N, 160°E, but another says it came down in the region between 20°S and 20°N and 60°E and 80°E. Like previous Soviet planetary landers, Venera 3 had been sterilized in order not to contaminate the solar system with terrestrial micro-organisms,[224] and like Venera 1 it carried a globe of the Earth to indicate its point of origin. Veneras 2 and 3 had, like Zond 1, observed the sky in the Lyman-alpha band, showing an increase of emission toward the galactic plane.[225] They also collected data on the interplanetary magnetic field, cosmic rays, charged particles in the solar wind and micrometeoroids. Published particle data for Venera 3 ended on 10 December 1965, while that from Venera 2 ended on 25 January 1966.[226] As regards the thermal stress that apparently disabled both probes, an error may have been committed when applying black-and-white paint to the radiators of the thermal regulation system.[227,228]

Although the Soviets placed more emphasis (and perhaps also greater resources) on the exploration of the solar system than did NASA (17 planetary launches to 1966 against 5) the Soviets had scored only one *total* success – Zond 3, which was an engineering test flight that did not actually encounter a planet. Meanwhile, NASA,

A model of Venera 3 showing (top) its Earth-facing and (bottom) Sun-facing sides. (Courtesy of Dominique Moniez)

despite the constraints imposed by having to use a less powerful launch vehicle, had benefited from the greater reliability of its engineering to make a reconnaissance of both Venus and Mars.[229]

In April 1965, Korolyov, recognizing that his bureau – which was tasked not just with lunar and planetary probes but also manned missions in Earth orbit and the project to beat the Americans to land a man on the Moon – had become seriously overcommitted, handed over responsibility for deep-space exploration to the newly reconstituted Lavochkin design bureau that was under the directorship of Georgi N. Babakin. Because it was meant for Mars, Babakin designated his first probe the 1M. This was a pity, as it was the name of Korolyov's first probe. The 1,000-kg 1M was to be launched in January 1967 by an 8K78 rocket, and release a module to collect in-situ data on the atmosphere as its bus made a flyby. When the probe was redesigned to take account of Mariner 4's results, it was realized that even by using an improved parachute system the entry module would reach the surface within 25 seconds of entering the atmosphere, which was insufficient time to return useful data at the planned data rate of 1 bps. Although such a communication system would be effective for a probe that could survive the descent, it was clear that a lander would need a retrorocket, and this would greatly increase the overall mass. After studying the possibility of a lander based on 3MV technology – and the distraction of another Venus mission using that technology – it was decided in October 1967 to pursue a totally new design.[230,231]

Meanwhile, on 14 January 1966 Korolyov died undergoing what was to have been routine surgery on his colon.

SOLAR PROBES

In 1958 the Ames Research Center started a study of a deep-space probe that would investigate the solar wind unbiased by the presence of Earth, and in November 1962 NASA, having recognized the need to assess the risk that radiation would pose to the Apollo missions that were to venture out to the Moon, approved the project. NASA was also eager for one of its own establishments to acquire deep-space expertise to match that of JPL (which was an independent facility under contract to the agency). The Space Technology Laboratories of Thompson, Ramo and Wooldridge (TRW) received the contract to build five small spacecraft to travel around the Sun on trajectories very similar to that of the Earth. Reincarnating an old name, these solar probes were provisionally designated Pioneer A to Pioneer E. They were small cylinders 89 cm tall and 94 cm wide, whose sides were covered with solar cells. An internal platform held the electronics, scientific instruments and a nitrogen tank, and supported a 152-cm-long tubular high-gain antenna that had an omnidirectional antenna on its end. A trio of 165-cm-long booms projected out from the edge of the platform, one carrying a magnetometer, a second with a nitrogen-gas attitude control thruster, and a third to act as a counterweight. The probe was to spin at a rate of 60 revolutions per minute for stability, with Sun sensors determining the orientation of its axis. The

12.7-kg scientific suite included a 3-axis flux-gate magnetometer (capable of rotating 180 degrees for calibration), two cosmic-ray detectors, and a pair of solar wind plasma sensors (i.e. an electrostatic spherical surface sensor and a Faraday trap sensor). In addition, a simple experiment on the propagation of radio waves between Earth and the spacecraft would provide a measure of the electron density along the line of sight, and the small antenna installed for this experiment would also measure electric fields. Each of these lightweight probes was to be launched by a Thor–Delta in a configuration augmented by three solid-propellant strap-on rockets.[232,233]

Following its launch on 16 December 1965 into an orbit ranging between 147 and 121 million km from the Sun (interior to that of the Earth) the first probe was named Pioneer 6. It was joined on 17 August 1966 by Pioneer 7, which was placed into an orbit ranging between 169 and 150 million km (exterior to that of the Earth). Both of these craft weighed 63 kg. Pioneer 8, slightly heavier at 65.3 kg, was launched on 13 December 1969 into an orbit ranging between 163 and 148 million km. Pioneer 9, a little heavier still at 66.6 kg, was launched on 8 November 1968 into an orbit ranging between 148 and 112 million km. The suite of these latter probes was augmented by a cosmic-dust detector. Pioneer E was destroyed together with its launcher on 27 August 1969. While in parking orbit, the last three missions released small TETR (Test and Training) satellites to test the tracking systems that had been developed for the Apollo missions. Despite their low public profile, the Pioneers were extremely successful, with each operating considerably longer than its design life of six months, and collectively they 'wrote the book' on the nature of the interplanetary medium in the plane of the ecliptic.[234] As a bonus, accurate tracking of the probes provided a means of reliably measuring the Sun–Earth and Earth–Moon mass ratios.[235]

Pioneer 6 monitored continuously until April 1966, then intermittently, and mapped small structures in the solar wind (magnetic field variations, plasma discontinuities, etc.). On 28 January 1967 the triangulation of simultaneous observations from Earth and Pioneers 6 and 7 enabled the probable location of the site of flare activity on the Sun to be determined. Pioneer 6 became the first probe ever to be tracked while passing behind the Sun, with the occultation occurring between 21 and 24 November 1968. When solar interference reduced the signal-to-noise ratio below that at which useful data could be obtained, the manner in which the probe's carrier wave was affected provided a method of studying the solar corona.[236,237,238,239,240] In April 1969 it detected solar disturbances that went on to cause the shut down of all radio communications in the Arctic. In 1971 Pioneers 6 and 8 were in antipodal relation to one another, and Pioneer 7 was contacted in September 1972, when in conjunction with the Earth at a distance of 312 million km.[241] In 1973 Pioneer 8 collected data along a line of sight near comet Kohoutek's position, and Pioneer 6 did likewise near the position of the comet's ion tail some 100 million km from its nucleus.[242] In 1977 Pioneer 7, and later Pioneer 8, passed through the Earth's magnetotail (the part of the magnetosphere that has been 'blown' downstream by the solar wind) and were able to sample it out to a distance of 19 million km. In October 1982 Pioneers 8 and 9 passed within 2.4 million km of

A Pioneer solar wind spacecraft as it would appear in orbit.

each other (the previous closest encounter between such craft had been 13 million km) and the event was used to recalibrate Pioneer 8's plasma instrument, which had been damaged shortly after launch. When Pioneer 7's orbit took it within 12.1 million km (0.08 AU) of Halley's comet on 20 March 1986, this marked the closest that a US spacecraft came to this most famous interloper to the inner solar system. It measured how the flux of cometary material disturbed the solar wind.[243,244] The most amazing fact about these four spacecraft is that they continued to work for

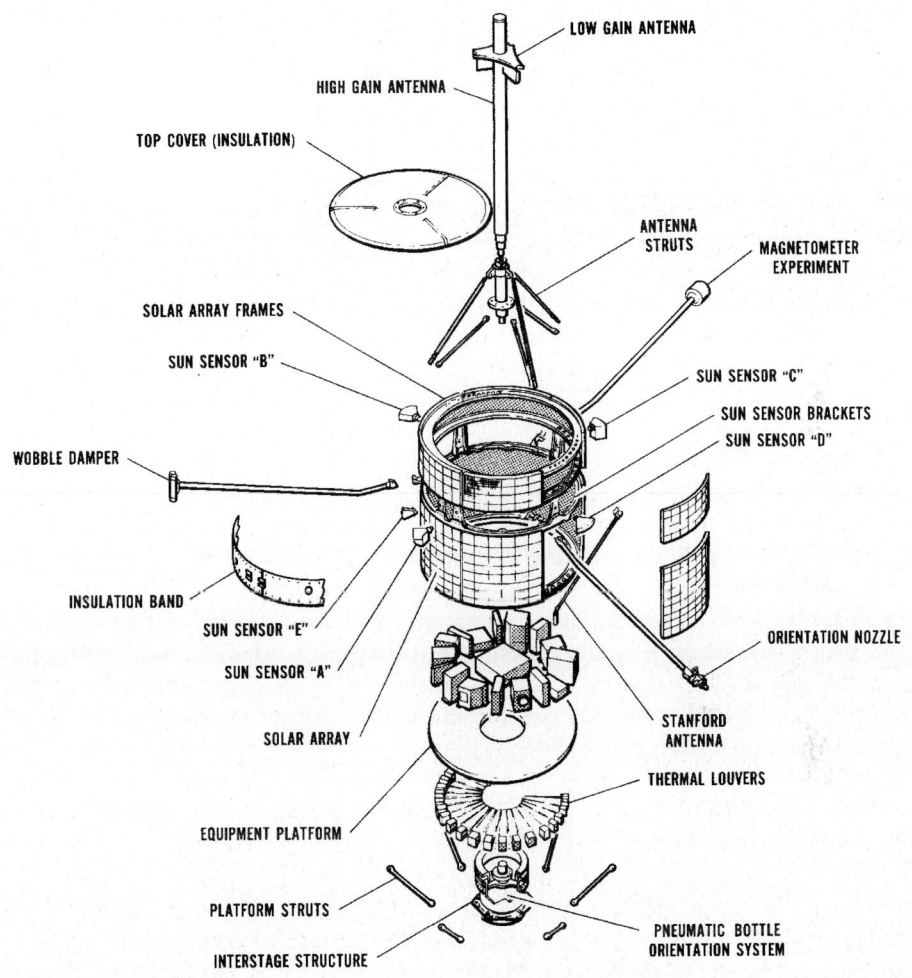

LOW GAIN ANTENNA

HIGH GAIN ANTENNA

TOP COVER (INSULATION)

ANTENNA STRUTS

MAGNETOMETER EXPERIMENT

SOLAR ARRAY FRAMES

SUN SENSOR "B"

SUN SENSOR "C"

SUN SENSOR BRACKETS

SUN SENSOR "D"

WOBBLE DAMPER

INSULATION BAND

SUN SENSOR "E"

SUN SENSOR "A"

ORIENTATION NOZZLE

SOLAR ARRAY

STANFORD ANTENNA

THERMAL LOUVERS

EQUIPMENT PLATFORM

PLATFORM STRUTS

INTERSTAGE STRUCTURE

PNEUMATIC BOTTLE ORIENTATION SYSTEM

An 'exploded' view of a Pioneer solar wind spacecraft.

decades. Ailing, Pioneer 9 was last heard from on 3 March 1987. Pioneer 7 was still returning data from its cosmic-ray detector and plasma analyzer in February 1991. (In effect, it was providing information on how materials and sensors degraded over time in the space environment.) Pioneer 7 was last contacted on March 1995, and Pioneer 8 was last heard from on 22 August 1996. In 1996 Pioneer 6's primary radio system failed, and in deciding that it was too expensive to provide further support, NASA declared the spacecraft's mission complete on 31 March 1997.[245] Nevertheless, some of its instruments were reactivated several months later in a training session for Ames' Lunar Prospector probe. The most recent contact was on 8 December 2000, in order to mark the 35th anniversary of its launch. Pioneer

6's longevity probably owed much to the fact that it did not have an onboard computer.[246,247,248]

In the mid-1960s, NASA worked with the Massachusetts Institute of Technology on a project named Sunblazer. The idea was to design a very small probe (the initial target mass being 4.5 kg) able to be launched by an inexpensive solid-fueled rocket into an orbit to enable it to reach perihelion at 0.53 AU at the same time each year – this being timed to occur at superior conjunction. While close to the Sun in the sky, the probe's 500-W peak-power dual-frequency radio transmitter would broadcast to enable receiving stations to measure the rotation of the linearly polarized radio waves to study the solar magnetic field, determine the electron density of the solar corona, and employ a coded signal to accurately measure the 'time of flight' of radio waves along that line of sight. The preliminary design envisaged the probe as a cylindrical capsule that would use a 4-bladed solar sail to keep it facing toward the Sun, and have a solar panel rated for 30 W at perihelion and 12 W at aphelion. It would not require an active thermal control system, even at perihelion, and would have no appendages apart from a pair of whip antennas. A series of probes with conjunctions occurring at different times of the year would facilitate ongoing coordinated solar studies.[249] The addition of other instruments soon increased the mass of the probe to 18 kg. The first Sunblazer was scheduled for launch by an unprecedented five-stage form of the all-solid-propellant Scout C rocket from the infrequently used station on Wallops Island on the Atlantic coast of Virginia in 1968, but was postponed first to 1972, then to 1973, and NASA's Office of Science and Applications canceled the project prior to the first launch.[250,251,252,253]

TOGETHER TO VENUS

After assessing the scope to further develop the technology of Korolyov's 3MV, in March 1966 Babakin's team set to work on a new design for a spacecraft capable of performing a variety of missions to Mars and Venus. They would have the luxury of being able to use the recently introduced 8K82 'Proton' launcher, which Vladimir Chelomei had developed for a manned circumlunar mission but could propel 4 or 5 tonnes into deep space (the actual mass depending on the launch window). This was more than enough for the flyby, orbiting, atmospheric sampling and landing probes. Babakin's plan was to begin in 1969 with a pair of Mars orbiters that would release atmospheric capsules. But in May the government ordered that spacecraft should be sent to Venus in 1967, the window for which was only 13 months away, and so the studies of the new probe were put on hold while the team devoted itself to adapting the 3MV for the task. A major aim of this redesign was to improve the reliability of the systems (in fact, all subsequent failures would be due to the launch vehicle rather than the spacecraft), and in the process the mass was increased by almost 200 kg. In an effort to concentrate on beating NASA to the Venusian surface, the requirement for an imaging variant was deleted. Babakin named the new Venus probe the 1V, once again reusing one of Korolyov's designations.

A mockup of the NASA–MIT Sunblazer small solar probe. The four sail blades were to stabilize the craft with its solar panels facing the Sun. (Alternative configurations included a ring-shaped solar sail.)

The standard second-generation Venera bus made by the Lavochkin design bureau. The principal external difference from Korolyov's design was the absence of the hemispherical radiators for thermal control. (Courtesy of Patrick Roger-Ravily)

Wholly new Earth, Sun and star (Canopus) sensors were developed and, against the advice of Korolyov's team, the thermal control system was revised to utilize gas instead of liquid, thereby eliminating the hemispherical radiators. Instead, a radiator was installed in behind the (enlarged) high-gain antenna. The radiator was tested in a thermal-vacuum chamber at the Lavochkin factory, the first such apparatus in the USSR. The atmospheric capsule was redesigned to withstand a deceleration of 350 g and a pressure of 10,000 hPa (the surface pressure on Venus was thought to be about 7,200 hPa). A vibration-damping system and a radar altimeter were also added. The landing module was initially tested in a press, and then in April 1967 with the first centrifuge in the world capable of testing spacecraft massing up to 500 kg at accelerations of up to 450 g. The tests showed that electrical components and cable brackets had a tendency to crack when stressed, so, with less than two months remaining until the launch window opened, all such items had to be replaced and some structures reinforced. To give the project the best chance, many engineers ignored the May Day holiday! After unprecedented rigorous testing, two identical spacecraft were sent for launch and a third was used to simulate the flight on the ground.[254,255,256]

The main spacecraft of the new 1,106-kg probe housed a triaxial magnetometer, four cosmic-ray detectors of three different types, four ion traps, and a spectrometer to detect both atomic hydrogen in the Lyman-alpha region and oxygen. Mariner 2 had not found a planetary magnetic field during its 35,000-km flyby, but the Soviet

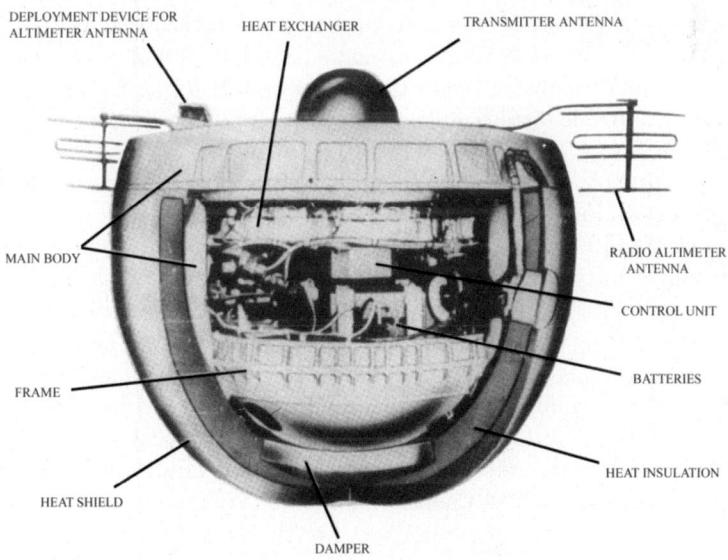

A cutaway showing the Venera 4 capsule with its radar altimeter antennas deployed. (Adapted from Vinogradov, A. P., Surkov, Yu. A., Marov, M. Ya., "Investigation of the Venus Atmosphere by Venera 4, Venera 5 and Venera 6 Probes", paper presented at the XXI International Astronautical Congress, Constance, 1970)

probe's primary module would report data until its destruction on diving into the atmosphere and therefore was likely to detect any weak field there might be.[257] The atmospheric capsule had a diameter of 1 meter and a mass of 383 kg. It included an external heat shield and the probe itself, which was connected to the shield by a layer of insulating material to protect it from shock and vibration. The probe was made 'bottom-heavy' for 'weathercock' stabilization. It had two compartments: one for the parachutes, the communication antenna, and the two radio-altimeter antennas;

The most important instrument on the Venera 4 capsule analyzed the composition of the planet's atmosphere. (Reprinted from Vinogradov, A. P., Surkov, Yu. A., Marov, M. Ya., "Investigation of the Venus Atmosphere by Venera 4, Venera 5 and Venera 6 Probes", paper presented at the XXI International Astronautical Congress, Constance, 1970)

and the other for the scientific instruments and their support systems (i.e. batteries, radio, accelerometers, etc.). In the event of its splashing down into water or a liquid of similar density, the hermetic capsule was designed to float, and a second communication antenna was to be released by the dissolving of a 'sugar lock'. The mechanical telemetry system of the atmospheric capsule had a data rate of 1 bps.[258] A 2.2-m^2 drogue parachute was to be opened when the external pressure was 500 to 600 hPa, and be followed by the 55-m^2 main parachute; both canopies being made of a fabric capable of withstanding a temperature of 450°C. Six scientific instruments were mounted inside the capsule. The most important was a 1-kg atmosphere analyzer to measure the partial pressures of nitrogen, carbon dioxide, oxygen and water vapor. In fact, terrestrial spectroscopy had shown the presence of vapor above the enshrouding clouds, but the amount of vapor in the lower atmosphere was not known and, despite the microwave radiometry that implied that the surface of the planet was extremely hot, some scientists still thought it was possible that most of the water was located at or near the surface – as on Earth. The instrument contained ampoules divided by a septum. Atmospheric samples were to be introduced to both sides, and then the gas of interest chemically removed from one side so that its partial pressure could be inferred from the differential across the septum. A similar instrument would detect the same gases in case they were present only in trace quantities. However, the suite was not capable of detecting neon, which was believed to be the third most significant constituent. It was also equipped with a barometer to measure pressures in the range 130 to 40,000 hPa, two platinum wire thermometers, a densitometer and the radio altimeter. The altimeter was to measure the difference in frequency between the transmitted and surface-reflected signals. By the nature of its design, however, there would be a 30-km ambiguity in its readings – that is, at a true altitude of 40 km the output from the altimeter could be interpreted variously as 10 km, 40 km, 70 km, etc.[259,260,261,262,263]

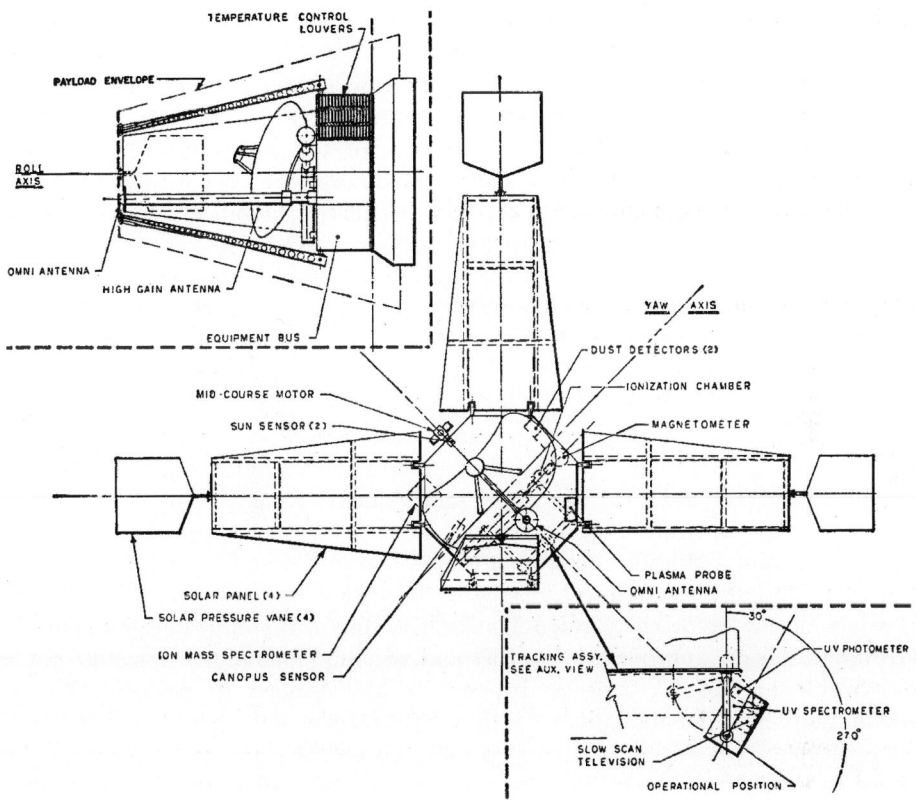

A proposal to adapt the Mariner 4 backup spacecraft to conduct a cometary flyby. (Courtesy of Space Systems/Loral)

On the American side, JPL requested to fly the spare Mariner C probe that it had built to the specification of Mariners 3 and 4. One proposal, made in collaboration with the Philco laboratories, was to launch it on an Atlas–Agena on a trajectory to make a flyby of the periodic comet Pons–Winnecke in 1969 in order to study the nucleus using an improved vidicon camera and other appropriate instruments. However, the ephemeris of the comet was poorly defined, and the fact that it was intrinsically faint would make it difficult to optically track its position from Earth, which meant that the spacecraft would have to make observations. To arrange a 5,000-km flyby, the probe would have to make at least two course corrections: one soon after launch to correct the 'escape' maneuver, and the final one, based on an ephemeris developed using its own observations, no more than a month before the encounter. As an alternative, the joint study proposed a mission in 1967 to the somewhat brighter comet Tempel 2.[264] A contemporary proposal by the Space Technology Laboratories of TRW envisaged sending a 200-kg spacecraft based on the solar Pioneer probes to comet Encke. Its instrument suite would include a camera with a resolution of about 150 meters, and as the probe neared its objective it would slow its spin to a rate conducive to imaging.[265] Although NASA decided not to attempt a cometary flyby at this time owing to the many 'unknowns', those in favor of such a mission continued to conduct studies.

In December 1965 NASA finally authorized JPL to send the third Mariner C to Venus. It was to make a fly by at one-tenth the range achieved by Mariner 2, and carry a larger scientific payload. It was also extensively modified. The solar panels were reduced in area from 6.54 m^2 to 4 m^2 because Venus is closer than Mars to the Sun, and owing to the geometry of the trajectory they were installed with the cells facing the opposite direction to the high-gain antenna. The 'solar sails' (which had proved unnecessary) were replaced by the antennas of a double-frequency radio-occultation experiment. In addition, a new data-acquisition system was built, and an actuator was installed to move the high-gain antenna from its flyby position to the optimal post-encounter position for data relay. Also, because the probe would be exposed to a greater degree of solar heat, an aluminized teflon thermal shield was designed to unfurl under the bus as the solar panels were deployed.[266]

The primary scientific objective of the mission was to measure the pressure at the surface, and a study to determine the best way of doing so recommended the use of an atmospheric capsule to provide pressure and temperature profiles, possibly all the way to the surface. However, owing to the short development time (18 months from approval to launch) and the limited budget, it would not be practicable to make a capsule. Instead, it was decided to perform a radio-occultation experiment similar to that of Mariner 4 – but this time more sophisticated.[267] As the probe crossed the limb of the planet, it was not only to transmit a radio carrier to Goldstone for later analysis to determine how the signal was refracted in passing through Venus's atmosphere, but by using its double-frequency radio system it would simultaneously receive a signal from an antenna operated by Stanford University, which it would analyze and store on tape for subsequent replay to Earth. This method was expected to yield a measurement of the density profile of electrons in the high atmosphere and various other data, including the percentage of water vapor in the lower atmosphere. There would also be a plasma experiment (very similar to that of Mariner 4) to study the solar wind, an ultraviolet photometer to detect atomic hydrogen and oxygen in the upper atmosphere, a thin-window Geiger–Mueller counter to test for the presence of Van Allen radiation belts, and a magnetometer. Since the magnetometer readings provided by Mariner 2 had been corrupted by interference from the metallic mass of the spacecraft, an effort was made on this occasion to minimize that source of error. Finally, the mission was to yield celestial mechanics data to improve knowledge of both the mass of Venus and the value of the Astronomical Unit.[268,269,270,271,272] The decision not to have a camera was controversial. Even by using a powerful telescope Venus is bland to the naked eye, but as early as 1927 a study using film sensitive to ultraviolet had hinted at atmospheric structure. An analysis of ultraviolet observations published in France in 1961 gave a rotational period of about 4 days in a retrograde manner.[273,274] Only later did radar studies confirm the direction of rotation, and reveal the amazing fact that the spin is so slow

An early NASA Marshall Space Flight Center proposal for a Venus entry probe that would be delivered by a flyby bus.

that a 'day' on Venus is longer than the planet's 'year' (i.e. the time that it takes to complete an orbit of the Sun). The upper atmosphere, in contrast, travels around the planet much more rapidly. Close-up pictures through ultraviolet filters would have been very welcome, but the data-acquisition system could not have dealt with the combined output of a camera and the radio-occultation experiment, and the latter took priority. Since there would be no imaging system, it was possible to delete the scan platform and reduce the capacity of the tape recorder by 80 per cent. Mariner 5 massed 244.8 kg (one-fourth of Babakin's modified 3MV, and substantially less than the entry capsule), and with its solar panels deployed it spanned 5.56 meters. Adapting an off-the-shelf craft reduced the cost of the mission to such an extent that it was no more than $165 million at the 2000 level.

Two modified 3MV spacecraft were to be launched for the 'V-67' mission, as it was referred to. In order to ensure that Venus would be above Yevpatoria's horizon when a probe made its atmospheric entry, the launches had to occur within narrow time intervals. The first lifted off on 12 June and, after a brief time in parking orbit, stage L placed Venera 4 on a trajectory that would take it to within 60,000 km of its target. Two days later, an Atlas–Agena D put Mariner 5 on a trajectory that would take it to within 75,000 km of Venus. For the first time, an

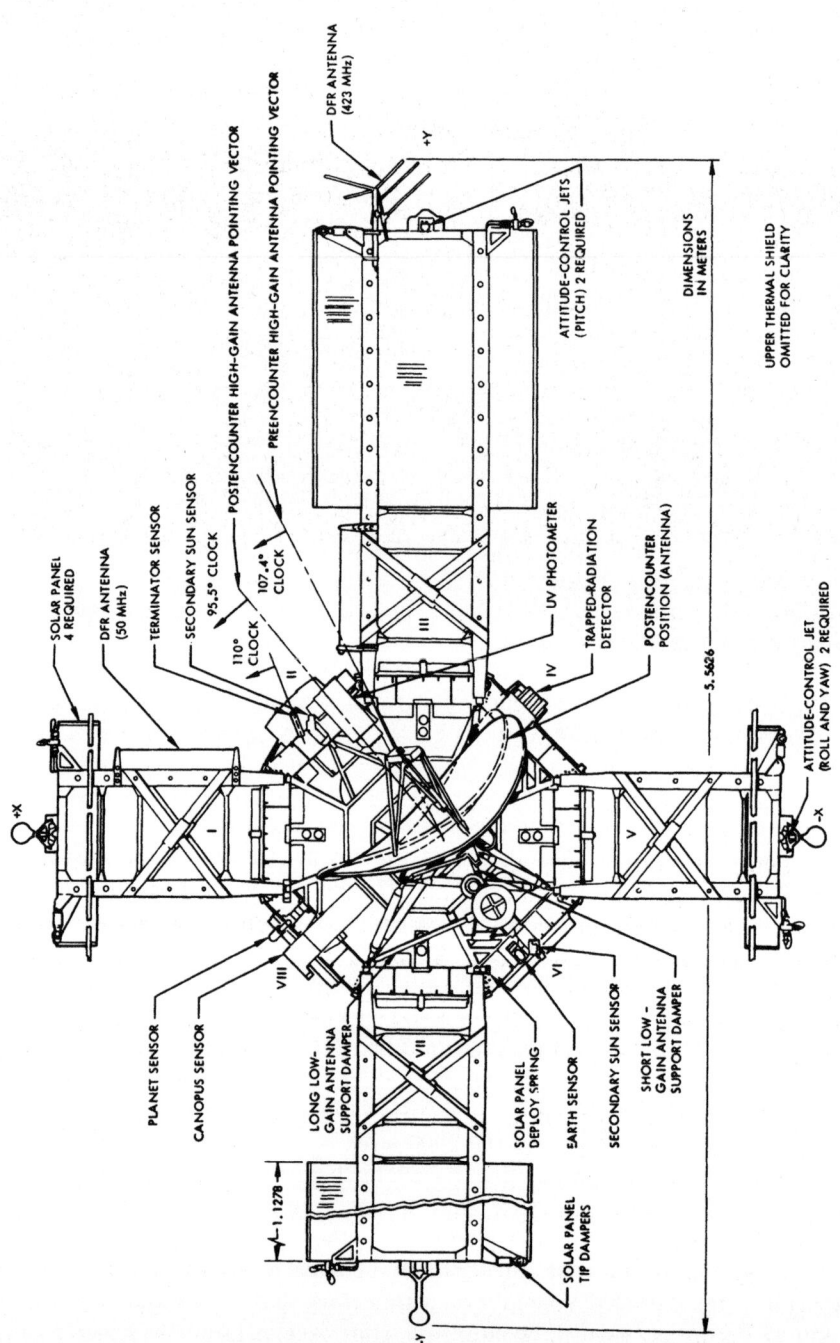

A line drawing of Mariner 5.

Although Mariner 5 used the same hexagonal bus and propulsion system as Mariner 4, its solar panels were smaller and its base had a sunshade owing to the fact that it was to venture toward rather than away from the Sun.

international flotilla was in transit for a single planetary target. When the second Soviet probe was launched on 17 June, a fault in the fuel pump of stage L prevented this engine from firing, and the stranded probe (named Kosmos 167) re-entered the atmosphere 8 days later.[275]

Venera 4 made its course-correction maneuver on 29 July, by which time it was 12 million km distant from the Earth, and this targeted it precisely to the night hemisphere, 1,500 km from the terminator. During the flight, a decrease of cosmic rays relative to Venera 2 was noted, and was probably due to the approaching solar maximum. Approaching its target, Venera 4 gave the Soviets their first remote-sensing data for another planet. This set the upper limit of the planetary magnetic field at 0.03 per cent of Earth's. On closing to within 14,000 km of the planet, the magnetometer and ion traps detected the 'bow shock' in the interplanetary magnetic field caused by the presence of Venus. A tenuous envelope made up almost entirely of atomic hydrogen was noted by the spectrometer. Its density increased 10-fold between 20,000 km and 6,000 km, and 100-fold at 1,000 km. This envelope hinted that the planet may once have had a large amount of water, that this had been dissociated by solar radiation, and that the hydrogen was leaking to space. However, the instrument did not detect any corresponding emission from atomic oxygen.[276,277,278,279]

At 04:34 UTC on 18 October 1967 the four metallic straps that had restrained the atmospheric capsule were released and it accompanied its bus into the atmosphere at an

The preparation and launch of Mariner 5 on 14 June 1967.

entry speed of 10.7 km/s. While the bus was disintegrating, the capsule's heat shield reached a temperature of 11,000°C as it endured a deceleration of 350 g. After it had slowed to 300 m/s, the cover of the parachute compartment was jettisoned, the chute was opened and, at 04:39, as the capsule descended at 10 m/s, its instruments started to collect data. After 94 minutes, by which time the descent had slowed to 3 m/s, the transmission ceased – not because the battery had reached its predicted lifetime of 100 minutes but because the hull had been crushed by the pressure. The wreckage should have fallen at 19°N, 38°E in an area that is now known as Eistla Regio. In excess of 70 pressure measurements and 50 temperature measurements had been reported; the initial pressure was 500 hPa and the value increased to 18,000, during which time the ambient temperature rose from 25°C to 270°C. All remaining doubts about the high temperatures reported by Mariner 2 were silenced. The only item that could be determined was the altitude at which the parachute had opened, because the depth of atmospheric penetration and the vertical profiles could only be calculated from knowledge of the capsule's aerodynamic properties. The most interesting result was the first in-situ analysis of the atmosphere of another planet. The capsule had collected two gaseous samples: the first corresponding to an external pressure of 700 hPa, and the second to a pressure of 2,000 hPa. Some 90 per cent of the gas was carbon dioxide; nitrogen was less than 7 per cent, and oxygen was 0.7 per cent. Overall, the results provided an intriguing insight. On Earth water is efficient at removing carbon dioxide from the atmosphere and 'fixing' it in carbonate rock. In fact, the amount of carbon dioxide in the atmosphere of Venus is comparable to that in terrestrial rocks. If the carbon dioxide could be removed from Venus's atmosphere, it would leave an atmosphere made primarily of nitrogen that was comparable to the atmosphere of Earth. Water vapor is indeed present in Venus's atmosphere, but at less than 1 per cent. Remarkably, this result could neither prove nor disprove that the visible clouds are made of water droplets or icy crystals. The initial Soviet reconstruction of the probe's descent said that it started to collect data at an altitude of 26 km, and that the loss of signal occurred when it reached the surface and turned over, in doing so masking the antenna.[280,281,282,283,284,285,286,287,288]

Mariner 5 had performed a course correction on 19 June to reduce the range of its flyby to within 4,100 km, and also to place it onto a trajectory that would create the all-important occultation. During its cruise, it had returned data on the interplanetary medium, and when its observations of shock fronts in the solar wind were combined with observations by Explorer 34 (a satellite in a highly eccentric Earth orbit) it was possible to infer a mean shock speed of 940 km/s.[289] On the morning of 19 October, the day after Venera 4 arrived, Mariner 5 began to collect data as it approached the dark hemisphere of the planet from the north of its orbit. At 17:34 the carrier signal disappeared. Although in geometric terms the spacecraft had passed beyond the limb 7 minutes earlier, the dense atmosphere had been refracting the signal. The probe reached its closest point of approach to the planet 14 seconds later. After an almost diametrical occultation, the signal reappeared over the day-side at 17:55. As a result of the flyby, the craft's trajectory was deflected by more than 100 degrees, in toward the Sun. As a result of this gravitational 'slingshot', its orbit ranged between 0.579 AU and 0.735 AU, and no longer intersected Earth's orbit.

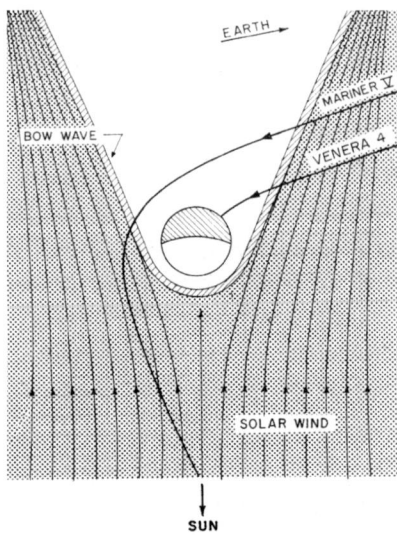

Venera 4 crossed the bow shock in the solar wind caused by Venus once just before it entered the atmosphere on the night-side. Mariner 5 had its trajectory deflected, in the process crossing the bow shock twice. The radio occultation occurred as it passed behind the dark limb as seen from Earth. (Reprinted from Jastrow, R., "The Planet Venus: Information Received from Mariner V and Venera 4 is Compared", Science, 160, 1968, 1403–1410)

The Doppler measurements, collected over a 10-day period centered on the flyby, enabled the ephemeris to be refined. Furthermore, the mass of Venus was able to be measured as 81.5003 per cent of Earth, and its shape (in particular, its oblateness) was inferred.[290,291] Meanwhile, the instruments had collected a variety of interesting data. The magnetometer put an upper limit for the magnetic field at 0.2 per cent of the terrestrial field, and it found the bow shock in the solar wind at a distance of 50,000 km. The radiation sensors did not detect any trace of Van Allen belts – the charged-particle flux densities being much lower than those around Earth. The ultraviolet photometer did, however, confirm the presence of atomic hydrogen, and the absence of oxygen in the upper atmosphere. As the probe receded from the planet 20 minutes after closest approach, it crossed the bow shock wave for the second time. The most interesting data, of course, was from the occultation experiment. The first indication of signal attenuation was detected at an altitude of 77 km over the night-side, where the pressure was estimated at 1 hPa. When calibrated with the compositional data of Venera 4, the pressure and temperature profiles implied a temperature at the surface of about 530°C and a pressure of 75,000 to 100,000 hPa! The fact that the profiles on the day-side were very similar to those on the night-side indicated that the rapidly rotating atmosphere was essentially uniform – it did not have time to cool off while crossing the night-side. An ionosphere was detected extending to an altitude of about 3,000 km over the night-side and to 500 km over the day-side. Although the lowest frequency of the double-frequency occultation experiment was severely affected by ionospheric phenomena, the other observations gave similar results – even revealing a transition zone in the 45–50 km altitude range, very probably caused by warm and cold air layers. The ionosphere also explained how Venus interacts with the solar wind. Earth has a substantial magnetosphere that interacts with the solar wind by way of a 'detached' shock wave, where the solar wind piles up against the sunward side of the magnetosphere. Having no global magnetic field, and hence no magnetosphere, the Moon simply absorbs plasma into its surface while the magnetic field passes through it. Although Venus has no magnetic field, the dense day-side ionosphere creates the bow shock that causes the solar wind to deflect around the planet. The enormously different characteristics of the atmospheres of Venus and Mars are strikingly evident from the fact that, during the Mariner 5 occultation, the frequency of the carrier signal was shifted more than 3,000 times as much as for Mariner 4.[292,293,294,295,296]

The results of the two missions were clearly contradictory: the Soviet probe had indicated a surface pressure of 18,000 hPa, while the American probe had measured a value at least 5 times as great. Accepting the Mariner 5 figure, the Soviets briefly considered that Venera 4 might have landed on an extremely tall mountain, and then conceded that it must have started to return data at an altitude of about 55 km (i.e. 26 km as initially announced, plus some 30 km of altimeter ambiguity) and fallen silent at an altitude in the range 23–27 km owing to the increasing pressure.[297] Combining the results from the missions suggested a radius for the planet that was some 20 km greater than that measured using terrestrial radars.[298]

Mariner 5 had continuously monitored solar X-rays after the encounter.[299] In November 1967, it was commanded to make roll maneuvers to enable the ultraviolet photometer to scan the sky in search of ultraviolet and Lyman-alpha emissions. It detected stars that radiate intensely in this part of the spectrum, the galactic plane, and the Large Magellanic Cloud.[300] On 21 November communications were switched to the omnidirectional antenna, and on 4 December the spacecraft was placed into hibernation. On 26 April 1968 an unsuccessful attempt was made to re-establish contact. A carrier signal was received on 14 October, but attempts to receive telemetry or to have commands acknowledged were fruitless, and the mission was officially terminated on 5 November 1968.[301]

The 1967 Venus missions provided the occasion for an unusual and long overdue 'space first': in March 1968 American and Soviet scientists gathered at the Kitt Peak National Observatory for the Second Arizona Conference on Planetary Atmospheres to compare their scientific results. The consensus that was reached at the conference was that *neither* probe had been successful in detecting the surface of the planet, nor in measuring its radius. When they published their results, each team made great use of its rival's results.[302,303]

A VOYAGER WITHOUT SAILS

JPL intended the Mariner spacecraft to be just its first step in the exploration of the solar system, making a reconnaissance for the advanced Voyager probes that would

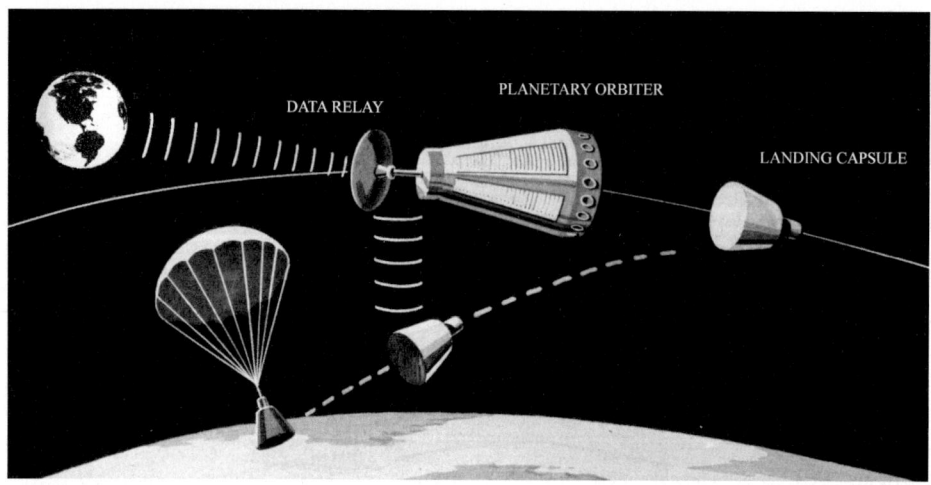

DATA RELAY

PLANETARY ORBITER

LANDING CAPSULE

A drawing of the Voyager Mars–Venus orbiter and lander as envisaged by NASA in the early 1960s.

land on Venus and Mars. The initial plan was to dispatch the first of these spacecraft with an incredible scientific payload of 230 kg as early as 1966. Whereas JPL had previously built its spacecraft in-house, NASA instructed it to switch to an industrial supplier. In April 1963 the Missile and Space Vehicle Division of General Electric, and AVCO, a designer of re-entry vehicles for the warheads of intercontinental-range ballistic missiles, were awarded study contracts. They considered two options for Voyager: one for a spacecraft of between 2,700 and 3,175 kg that would require the Saturn IB (a launch vehicle being developed for the Apollo program), and a lighter 1,800-kg probe in case only the smaller Titan III that was being designed by the US Air Force was available. The studies had similarities and differences. Both proposals were for a solar-powered orbiter that would serve as a relay for its lander, and have a scientific payload of 98 kg in the General Electric proposal and 61 kg in the AVCO proposal. General Electric planned two cone-shaped entry and landing capsules, each of which would have a 70-kg scientific payload that would include instruments to address the issue of whether there was life on Mars. AVCO proposed a single lander with a 91-kg payload. In both cases, the landers would be powered by RTGs (Radioisotope Thermal Generators) that would ensure an operational life on the surface of Mars of at least 180 days. Both companies also reviewed what was known of Mars on the basis of its telescopic appearance, and cited potential landing sites. Both studies listed the dark Syrtis Major, where astronomers had often observed a 'wave of darkening' that, before the shocking results from Mariner 4, was believed to be due to the seasonal growth of vegetation. In contrast to the detail of the proposals for Mars, the exploration of Venus was vague.[304]

In addition, the Philco–Ford Corporation independently offered NASA a 173-cm-diameter spherical lander weighing 544 kg which, once on the surface of Mars, would deploy four spring-loaded 'petals' to right itself. (The Soviets would later use this scheme on their landers.) It would have a camera, instruments to study the atmosphere and surface, and an 'Automated Biological Laboratory'.[305] In fact, NASA had made an early start in funding the development of instruments to test for life on Mars. Built by Hazelton Laboratories, Gulliver was named after Lemuel Gulliver of Jonathan Swift's 1726 tale *Gulliver's Travels*, who discovered unusual forms of life in distant lands. It was to deploy two tethered projectiles coated with silicone grease to collect samples of dust that would be introduced to a culture medium tagged by a radioisotope. If this prompted metabolic processes similar to those of terrestrial life, then the sample would release isotope-enriched carbon dioxide, whose beta particles could be detected by using a Geiger–Mueller counter. The Wolf Trap was named after its principal proponent, Wolf V. Vishniac of the University of Rochester, New York. Like the other instruments, it was not to detect life directly, but was to seek evidence of biological processes. In this case, a sample of dust would be put into a culture fluid in order to study its turbidity and acidity (i.e. pH), both of which could be influenced by biological activity. Multivator and Minivator were variants of a concept under study at Stanford University and at JPL in which Martian dust and aerosols would be analyzed for evidence of activity by enzymes. Unlike Multivator, Minivator could collect samples of dust using a motor blower. In addition to these instruments, the 'laboratory' would probably have included such instruments as a gas chromatograph.[306,307,308]

At the same time, General Electric was working on a study for a concurrent Mars mission for NASA's Office for Space Science. This was called Beagle, after the ship on which Charles Darwin made his scientific investigation into the origin of species. It called for four probes, two for the 1969 launch window, with the others following in 1971. As each spacecraft would have weighed at least 25 tonnes,

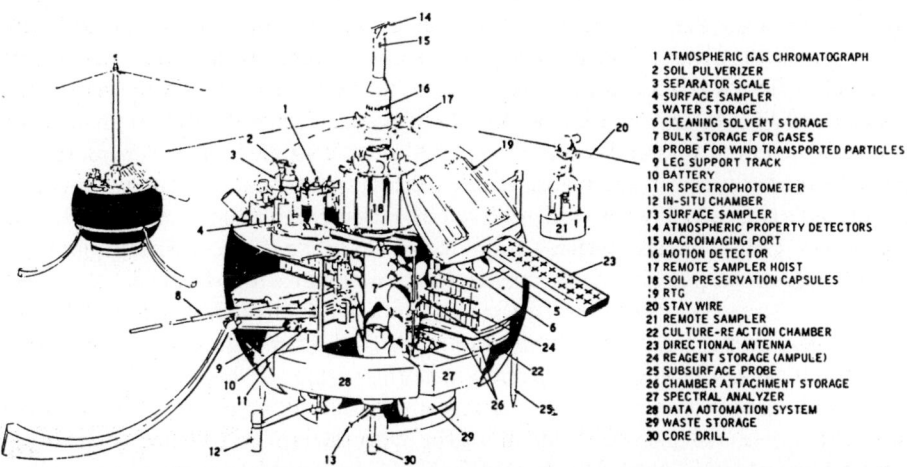

1 ATMOSPHERIC GAS CHROMATOGRAPH
2 SOIL PULVERIZER
3 SEPARATOR SCALE
4 SURFACE SAMPLER
5 WATER STORAGE
6 CLEANING SOLVENT STORAGE
7 BULK STORAGE FOR GASES
8 PROBE FOR WIND TRANSPORTED PARTICLES
9 LEG SUPPORT TRACK
10 BATTERY
11 IR SPECTROPHOTOMETER
12 IN-SITU CHAMBER
13 SURFACE SAMPLER
14 ATMOSPHERIC PROPERTY DETECTORS
15 MACROIMAGING PORT
16 MOTION DETECTOR
17 REMOTE SAMPLER HOIST
18 SOIL PRESERVATION CAPSULES
19 RTG
20 STAY WIRE
21 REMOTE SAMPLER
22 CULTURE-REACTION CHAMBER
23 DIRECTIONAL ANTENNA
24 REAGENT STORAGE (AMPULE)
25 SUBSURFACE PROBE
26 CHAMBER ATTACHMENT STORAGE
27 SPECTRAL ANALYZER
28 DATA AOTOMATION SYSTEM
29 WASTE STORAGE
30 CORE DRILL

A cutaway of the Philco–Ford Voyager lander. (Courtesy of Space Systems/Loral)

they would have required the powerful Saturn V booster. Designed to send an Apollo mission to the Moon, this launcher was capable of dispatching over 35,000 kg to Mars. Each probe would comprise an orbiter and a pair of 9,500-kg landers, with the landers carried on opposite sides of the rectangular orbiter. In addition to a 6-meter-wide heat shield, each lander would have four braking petals with a maximum diameter of 10 meters. A lander would carry 2,250 kg of instruments and an RTG power supply. In addition to relaying the data from its landers, the orbiter was to observe the planet using its own instruments. However, in view of the very high cost of this project (estimated at $1 billion in 1964 terms, and equivalent to around $5.5 billion in 2000) it is not surprising that Beagle did not advance beyond a paper study.[309]

Voyager suffered two setbacks within 12 months. In late 1964 NASA further postponed the first Mars mission from 1969 to 1971. Then in July 1965 Mariner 4 revealed the planet's atmosphere to be more tenuous than even the most pessimistic estimate. In view of the weight that the lander would undoubtedly gain when it was redesigned, it was announced in October 1965 that the Saturn V would be used, and that the spacecraft would be sent in pairs. Ironically, since flying missions in parallel would create management problems, this decision resulted in yet another postponement to 1973. As a result of these changes, each orbiter/lander would have a launch mass of about 12 tonnes. The orbiter would include a central cylindrical bus housing the electronics, systems and tankage for 6,500 kg of propellant. On top of the bus would be a circular solar panel array with a total area of 22.5 m^2, to which would be attached the high-gain antenna and a scan platform for 180 kg of scientific instruments. The main engine (based on the Apollo Lunar Module descent engine) would project from the center of the solar panels. The 2.5-tonne entry capsule and lander would be carried on the far end. The lander would weigh about 350 kg 'dry', would have its own landing engine and three legs, and would be equipped with: an imaging system that could operate during the final phase of the descent and after the landing; a high-gain antenna for direct communication with Earth; a robotic arm to collect samples of soil, move rocks and excavate trenches; and about 135 kg of scientific instruments, some of which would test for the presence of life. In this configuration, the RTG had been deleted, and the lander would be powered by a chemical battery. Although a single Saturn V would launch two identical spacecraft, they would adjust their trajectories in order to reach Mars about 10 days apart. They would enter orbits ranging in altitude between 1,100 and 22,500 km. Following its release, the lander would make the deorbit burn and enter the atmosphere at a speed of approximately 4 km/s. Once atmospheric braking had slowed it to a speed of between 140 and 335 m/s, which would occur at an altitude of about 6 km, the lander would shed its 6-meter-diameter heat shield and deploy parachutes. When it was a few meters above the surface, it would use its engine to come to a virtual stop, then fall the remainder of the way. Since the spacecraft would require more propellant to enter orbit with the lander attached than without it, there was an alternative plan in which the lander would be released just prior to the insertion maneuver, and the mass thereby saved would enable the payloads of both the orbiter and the lander to be increased. More Voyagers would follow in

The definitive Voyager Mars orbiter and lander. They were to be launched in pairs on a single Saturn V. (Courtesy of NASA/JPL/Caltech)

1975 with RTG-powered landers that would have a surface lifetime of two years, and others in 1977 and 1979, possibly carrying a rover.[310,311] Because a landing posed such a challenge, 13 tests of the proposed landing system were made in the context of the Planetary Entry Parachute Program (PEPP). After a series of tests using sounding rockets, heat shield models were released by balloons at an altitude of 50 km and accelerated to supersonic speeds by clusters of rockets to return to Earth by various combinations of parachutes and other kinds of decelerators.[312]

However, the revisions greatly increased the estimated cost, which would exceed $1 billion for a single two-probe mission. In late 1965 it was decided that JPL would work on the lander, with the assistance of the Langley and Ames Research Centers. On 27 January 1967 responsibility for the project was formally transferred from JPL to the Marshall Space Flight Center, which would be able to use its familiarity with the Saturn V to coordinate the integration of the spacecraft. However, this move was rendered obsolete because on that same day three astronauts died while checking out the first Apollo capsule, and this led to a major reorganization of the space program. In June, with the war in Vietnam costing in excess of $2 billion per month, Congress cut the agency's budget for fiscal year 1968 (which would start in October 1967) by slowing the pace of work on a nuclear rocket and some science programs, including Voyager. Even the Apollo funding was reduced. Although the Senate then deleted Voyager from the budget, a joint

The performance of the Saturn V as a launcher for planetary missions

Target	Orbit Inclination (degrees)	Time of Flight (days)	Saturn V Performance (kg)	Saturn V + Centaur Performance (kg)
Mars	2	150	35,400	39,000
Asteroid Ceres	10	200	3,160	10,000
Asteroid Ceres	23	80	None	2,700
Jupiter	0	750	10,900	16,300
Comet Encke	12	100	10,800	15,400
Comet Schwassmann–Wachmann 3	9.5	500	None	7,500
Solar Probe – 0.2 AU	0	80	None	6,650
Solar probe – 0.12 AU	0	76	None	2,500
Out of the Ecliptic at 1 AU	25	–	None	5,800
Out of the Ecliptic at 1 AU	35	–	None	580
Solar System Escape	0	–	None	5,900
Solar System Escape	10	–	None	3,850
Earth–Moon Lagrangian Point	0	4.2	34,500	38,500

Adapted from: Bromberg, J.L., Gordon, T.J.: "Extensions of Saturn", paper presented at the XVII International Astronautical Congress, Madrid, 1966

House–Senate Committee restored $42 million to the agency's 1968 budget to keep the program alive. However, it was canceled on 22 August, immediately after the Manned Spacecraft Center invited proposals for a Mars Sample Return probe to be deployed by piloted spacecraft that would either fly past the planet or enter into orbit around it.[313]

A REPEAT MISSION

Encouraged by Venera 4's success, the Soviets prepared two more spacecraft for the January 1969 window for Venus, as the V-69 mission. The short development time meant that the atmospheric capsule could not be redesigned to survive the predicted surface pressure. However, the mass saved by deleting water-buoyancy enabled its hull to be strengthened to withstand a pressure of 25,000 hPa and a deceleration of 450 g, and by replacing the parachute with one that had an area of just 18 m^2 it was possible to increase the rate of descent in order to extend the temperature, pressure and density profiles more deeply into the atmosphere.[314,315] The bus of the new 2V probe was almost identical to that of the 1V, with a launch mass of 1,130 kg, of which 405 kg was the capsule. The instruments in the capsule were revised: it had a completely new densitometer; a more accurate radio altimeter would report when it reached altitudes of 35, 25 and 15 km, although still with an ambiguity of 30 km; the atmospheric analyzer was modified to accurately measure the partial pressure of

A test related to the Martian landers was the SPED (Supersonic Planetary Entry Decelerator) in which a metallic parachute was tested on a sounding rocket.

gaseous oxygen; it had six barometers and three thermometers; and a photometer to measure the ambient light in the Venusian clouds.[316] At the time, the Soviets were in exploratory talks with French scientists to supply instruments for Soviet probes in the 1970s and 1980s, and they took this opportunity to ask the French to provide the thermometers and barometers for the V-69 probes in the hope that such instruments would operate better than those on Venera 4, whose measurements were ambiguous and difficult to interpret. The French agreed, but as a result of the national strikes of May 1968 they were unable to complete their instruments until September, by which time it was too late to integrate them into the probes.[317]

Venera 5 was successfully launched on 5 January 1969. Although Tyuratam was hit by a blizzard on 10 January, the launch of the second probe could not be delayed because the pad was required for Soyuz 4, which was to lift off 4 days later for a joint mission with Soyuz 5, and the reliable Semyorka proved itself by riding out the storm and dispatching Venera 6.[318] The two spacecraft flew transfer orbits designed to reach Venus in four months. In the case of Venera 5, the initial flyby distance was 25,000 km, and for Venera 6 it was no less than 150,000 km. Course corrections in mid-March changed the velocities by 9.2 m/s and 37.4 m/s respectively, successfully targeting each probe to the planet's night-side. Venera 5 was to enter the atmosphere some 2,700 km from the terminator at an incidence angle of 62 degrees. Venera 6 was to do so the next day, at a point 300 km away. The Soviets recognized that it would have been very useful to have both probes in the atmosphere at the same time, as data could be obtained on wind speed and direction, but the lack of a second deep-space tracking antenna had made this experiment impossible. However, to all intents and purposes the observations by the onboard instruments could be deemed "almost simultaneous".[319] In transit, the instruments on the spacecraft found that the intensity of galactic cosmic rays was lower than that measured by Venera 4 two years earlier – this correlating with the fact that the Sun was nearing the maximum of its 11-year cycle of activity. As they closed on Venus, both spacecraft collected data on the planet's envelope of atomic hydrogen.

At 04:08 on 16 May Venera 5 started its automatic entry sequence at a distance of 50,000 km from Venus, released its capsule at 06:01 at a distance of 37,000 km, and was destroyed when it penetrated the atmosphere at a speed of 11.18 km/s. Once the capsule's speed had been reduced to 210 m/s, which occurred at an altitude of 60 km, it deployed its parachute and began to collect data. Some 53 minutes later, 36 km deeper and descending at 5 to 6 m/s, the capsule imploded when the ambient temperature was 320°C and the pressure was 27,000 hPa. The wreckage would have crashed to the surface at 3°S, 18°E. Venera 6 entered the atmosphere at 06:03 the next day and collected data for 51 minutes until it reached a measured height of between 10 and 12 km, its wreckage falling at 5°S, 23°E.

The probes were programmed to make atmospheric composition measurements at different altitudes to maximize the overall yield. Venera 5 measured the composition of the atmosphere at external pressures of 600 and 5,000 hPa, while Venera 6 did so at 2,000 and 10,000 hPa. Carbon dioxide was confirmed to be the main constituent, with a concentration of up to 97 per cent. The remainder (at least above an altitude of 32 km) was mostly nitrogen, with a trace of oxygen. The concentration of water

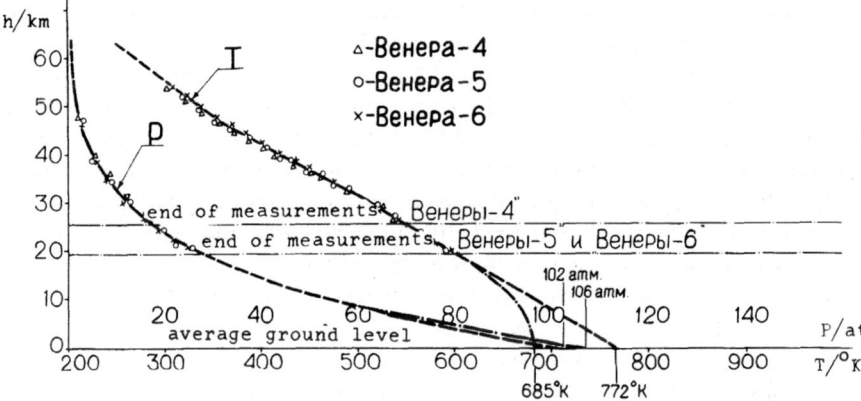

An extrapolation of the temperatures and pressures measured by Veneras 4, 5 and 6 in order to predict conditions at the surface of Venus. (Reprinted from Vinogradov, A. P., Surkov, Yu. A., Marov, M. Ya., "Investigation of the Venus Atmosphere by Venera 4, Venera 5 and Venera 6 Probes", paper presented at the XXI International Astronautical Congress, Constance, 1970)

vapor was found to be five times greater in the upper atmosphere than at lower altitudes. The temperature and pressure profiles were similar to those of Venera 4, but the density profile showed marked differences, probably as a result of using a different instrument. An early extrapolation of Venera 5's measurements gave a surface pressure of 140,000 hPa and a temperature of 530°C, and for Venera 6 the figures were 60,000 hPa and 400°C. In an effort to explain this discrepancy, Soviet scientists suggested that the second capsule had descended over a mountain range that had corrupted the altimeter data, but later missions showed that the landing site on Guinevere Planitia not only lacks mountains but lies substantially below the mean elevation of Venus. The next explanation was a malfunction in one or other of the altimeters (the one on Venera 5 in particular had persistently given altitudes 10 km higher than actuality) but when the atmospheric models were recalibrated the two datasets were found to be consistent with a surface pressure of about 100,000 hPa.

The most sensational discovery made by the V-69 probes was made by Venera 5. Four minutes before being crushed, the photometer recorded a high light level for a few seconds, suggesting an outbreak of lightning. In this case too, however, the most likely explanation was a sensor malfunction. The corresponding sensor on Venera 6 never returned any data.[320,321,322,323,324,325]

MARS AGAIN

Several weeks after the launch of Venera 4, Babakin's team turned its attention back to Mars. The first plan for the 2M spacecraft was to adapt the E-8 heavy lunar probe that was to be launched in several variants starting in 1969. The M-69 mission called for the spacecraft to release an atmospheric probe with instruments to collect basic data. Once it had decelerated to Mach 3.5 in the local conditions, it would deploy its parachute and begin to transmit. The uncertainty in the ephemeris then available to the Soviets meant that the angle at which the probe would hit the atmosphere might be anything between 10 and 20 degrees, which in turn meant that the descent time would range between 230 and 900 seconds. The main spacecraft was to enter orbit around the planet, thereby achieving another 'first' at least 2 years before NASA's contemporary schedule envisaged it. Seven months into the study of a configuration based on the E-8, it was realized that it imposed several shortcomings, including a large displacement of the center of mass as propellant was consumed. Even although there was little more than one year to the launch window, it was decided to abandon the design and start a completely new one. It was built around a large spherical tank that incorporated an internal bladder membrane to enable it to contain both the fuel and the oxidizer for the engine that would make the orbit-insertion maneuver. Above was a cylindrical equipment compartment and a 2.8-meter-diameter high-gain antenna. The tank also supported the radiators for the thermal control system, and two wings of solar panels providing a total collector area of 7 m². A pair of external pressurized cylindrical compartments contained the scientific payload. The conical atmospheric capsule was carried on top. The orbiter was to have 13 instruments weighing a total of 99.5 kg: a magnetometer; a micrometeoroid detector; a charged-particle detector; a cosmic-ray and radiation belt detector; a spectrometer for low-energy ions; a low-frequency radiation detector; a gamma-ray spectrometer; a mass spectrometer for hydrogen and helium; an X-ray spectrometer; an ultraviolet spectrometer; an X-ray photometer; a radiometer; and an infrared photometer. In addition, it had three cameras with 35, 50 and 250-mm focal lengths and red, green and blue filters. As usual for a Soviet spacecraft, the cameras used a photographic film that could be developed on board, scanned and transmitted to Earth. The imaging system had a capacity of 160 frames in a 1,024 × 1,024-pixel format. A wide-angle image taken near the low point of the orbit was expected to cover an area 1,500 km on a side, and a telephoto shot would cover an area 100 km on a side. The best resolution was expected to be about 200 meters per pixel. The atmospheric capsule had an instrument suite similar to that of the Venus probes: pressure, density and temperature sensors, and an atmospheric composition analyzer. As the capsule was not expected to survive the landing it had no instruments for surface science.

According to an assessment by Soviet engineers, this was an excellent example of how *not* to design a spacecraft: not only were the electronics obsolete even by Soviet standards, but the internal arrangement of the pressurized compartment did not allow for easy access to the apparatus if it had to be replaced. Worse, when it was realized that the spacecraft was too heavy for the Proton launcher, it was decided to delete the atmospheric capsule and attempt what was expected to be a "simple" mission involving three months of orbital investigations.[326,327]

Even before it launched Mariner 4, NASA had begun to consider further missions to Mars using this same technology. One option was to deliver a small atmospheric

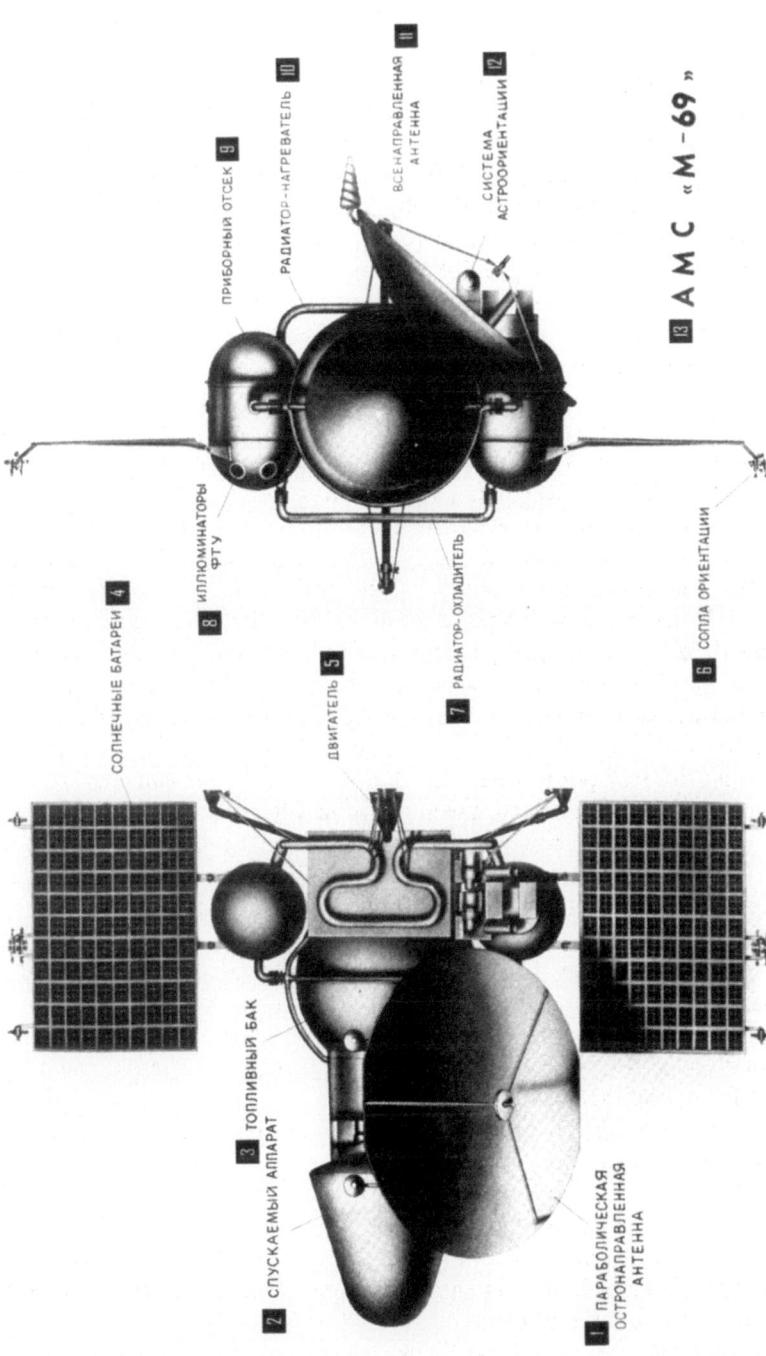

The M-69 Mars spacecraft: 1, parabolic high-gain antenna; 2, entry capsule; 3, fuel tank; 4, solar panels; 5, orbit insertion engine; 6, nozzles of the attitude control system; 7, cooling radiator; 8, camera ports; 9, instrument module; 10, thermal radiator; 11, omnidirectional antenna; 12, stellar orientation system. In fact, the missions were flown without the atmospheric entry capsule.

capsule. The Raytheon Company studied a simple 60-cm-diameter spherical capsule that would be carried inside a cylindrical sterilization shroud and have a collapsible boom in order to establish aerodynamic stability as it entered the atmosphere. Based on pre-Mariner 4 knowledge, it was expected to take about 40 seconds to reach the surface (where it would not survive the impact). During this time, it would report basic data with its minimal suite of instruments: a 3-axis accelerometer, a barometer, a thermocouple, and an infrared radiometer with a field of view 'forward' through a sapphire window in the thick beryllium heat shield to measure the constituents of the atmosphere.[328] General Electric's Missile and Space Division investigated a capsule that had four 'lunes' of metal foil attached to the inside of the heat shield to serve as an omnidirectional antenna to send this data to the flyby bus.[329] Although an atmospheric probe would be much simpler than a lander such as was proposed by the Goddard Space Flight Center prior to NASA's decision to focus its efforts on Mars, and would be a logical follow-up to Mariner 4, JPL was not enthusiastic, and in any case funding constraints meant that it would not be possible to take advantage of the 1966 window for Mars. When the October 1965 decision to use the Saturn V for Voyager led to the postponement of the first such mission to 1973 at the earliest, NASA opted to send a pair of improved Mariners in 1969 to fly past the planet at a distance of just 3,200 km.[330,331,332]

The hexagonal frame of the 1969 Mariner spacecraft was almost unchanged, but every system was modified: the course-correction engine was upgraded, the attitude control system was improved, the solar panels were enlarged to 7.75 m^2 (increasing their overall span to 5.79 m) to provide 449 W in the vicinity of Mars, the high-gain antenna was increased to a diameter of 102 cm; and the low-gain antenna was mounted on a 223-cm-tall boom that took the overall height to 3.35 meters. The new scan platform carried a battery of instruments that increased its overall weight to 57.6 kg. The principal instrument was a pair of vidicon cameras. The 3.5-kg wide-field camera had a focal length of 52 mm for a field of view of 11 × 14 degrees, and would have a surface resolution of 3 km if the flyby occurred at the planned altitude. The 14-kg narrow-angle camera had a focal length of 508 mm for a field of view of 1.1 × 1.4 degrees, and would have a maximum resolution of 300 meters. An image spanned 704 × 945 pixels. Because the optics were boresighted, the differing focal lengths meant that each narrow-angle frame provided a closer look at an area within a wide-angle frame. The wide-angle camera was fitted with blue, green and red filters, but the narrow-angle had a single filter (referred to as 'minus-blue'), chosen to penetrate haze. Instead of predicting the lighting conditions, it was hoped that by reading the brightness of some of the pixels of the most recent wide-angle image the exposure would automatically be corrected. The maker, however, was three months late in delivering the camera package to JPL, and when the software was found not to work there was insufficient time to overcome the fault, and the function had to be disabled.[333] Also on the platform were a spectrometer to analyze two ultraviolet bands to determine the atmospheric composition; an infrared radiometer with two sensors to measure the temperature of a narrow track across the same field of view as the cameras; and a 2-channel infrared spectrometer to measure the concentrations of hydrogen, nitrogen, carbon and sulfur oxides in the atmosphere and also study the

An artist's impression of a spacecraft based on Mariner 4 releasing a spherical entry probe at Mars. Nearby is the jettisoned hemispherical cover of the sterile container. (Reprinted from "Astronautics & Aeronautics", a publication of the American Institute of Aeronautics and Astronautics, December 1964)

composition and temperature of the surface. The infrared spectrometer was devised by George Pimentel, the scientist who, in 1965, had recognized that the 'Sinton bands' in the spectrum of Mars, which had been thought to prove the presence of vegetation, were actually due to deuterated water vapor in the Earth's atmosphere.[334] There were two separate data-recording systems: one an analog tape recorder for the imagery, and the other a digital recorder to store the data from the other instruments. The upgraded communication system increased the data rate to 16,200 bps. As each spacecraft passed behind Mars as viewed from Earth, the Martian atmosphere would be studied by the radio-occultation technique. Although the modifications increased the launch mass to 413 kg, this did not pose a problem because the Agena stage used for Mariner 4 had been superseded by the more powerful Centaur.[335,336,337,338]

On 14 February 1969, soon after Mariner 6 was installed on its launch vehicle, an unusual mishap occurred. The Atlas was designed to maintain its rigidity by internal pressure, and when the pressurization valves were accidentally opened this caused it to start to deflate! Fortunately two technicians managed to close the valves before any damage was done. The Centaur stage and the spacecraft were removed from the Atlas and installed on the vehicle designated for the sister probe, and on 24 February Mariner 6 was successfully launched and inserted directly into a solar orbit with a perihelion of 0.99 AU and an aphelion of 1.588 AU, heading for Mars. The mishandled Atlas was replaced by one assigned to the fifth Application Technology Satellite (ATS-E), and on 27 March Mariner 7 was successfully dispatched in a solar orbit ranging between 0.971 and 1.568 AU. For the first time, NASA had two spacecraft on course to another planet.[339,340,341]

The Soviets transported the twin 4,850-kg spacecraft for the M-69 mission out to Tyuratam in January 1969. It was a particularly cold winter, and to make matters worse the explosion of the first N-1 lunar rocket on 20 January smashed the windows of buildings within a radius of many kilometers. Preparations were hindered by the fact that M-69 was given a lower priority than the robotic and human programs related to the Moon. In fact, one of the Lavochkin lunar probes was being prepared for launch in the same building. In view of the rushed development, the engineers held out little prospect of the probes successfully completing their assigned tasks.

After stage D of the Proton had been inserted into parking orbit, this stage was to fire its engine to achieve a very high apogee, at which time the spacecraft would fire its own engine for the 'escape' burn to head for Mars. A course correction 40 days after launch would trim the flyby range to within 10,000 km, and another correction 10 to 15 days from Mars would reduce this to about 1,000 km. For course corrections the probe would be 3-axis stabilized, but otherwise on the six-month interplanetary cruise it would be spun around an axis that would maintain its solar panels facing the Sun. On reaching their destination, both probes were to enter similar orbits, inclined at 40 degrees to the Martian equator, with altitudes ranging between 1,700 and 34,000 km and a period of about 24 hours. After a preliminary phase, the lowest point would be reduced to 500 km in order to undertake imaging. In the absence of the atmospheric capsules, the objective was to investigate the

Preparing Mariner 6. The scan platform on its base was more sophisticated than that of Mariner 4. (Courtesy of NASA/JPL/Caltech)

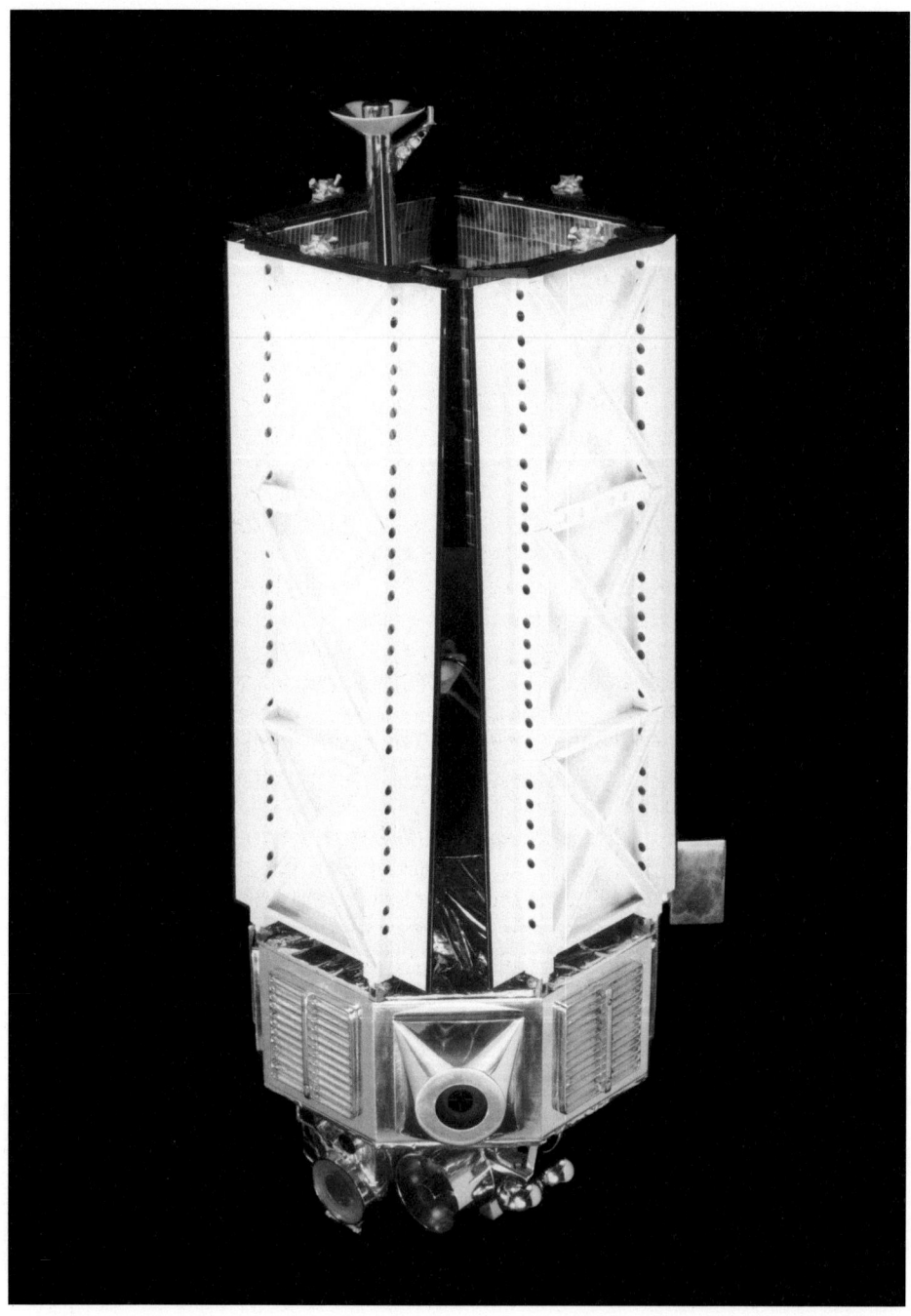

One of the Mariner 69 spacecraft folded for launch. As in the case of Mariner 4, the course correction engine was on the side of the bus. (Courtesy of NASA/JPL/Caltech)

atmosphere by remote sensing, both by using the onboard instruments and by radio occultations.

The launch on 27 March (the same day as Mariner 7 was dispatched) started well, but the fuel pump in the third stage seized and caught fire, and 436 seconds into the flight the engine shut down. The debris of the probe fell on the Altai mountains. The launch of the second probe on 2 April was an even worse disaster: one of the six engines of the first stage exploded 0.02 second after lift-off, and 41 seconds later the underpowered vehicle fell to Earth about 3 km from the pad. In the opinion of Soviet engineers, destiny and the unreliable Proton launch vehicle had saved the probes from inevitable failure![342,343]

Mariners 6 and 7 made course corrections on 1 March and 8 April, respectively. Other than some celestial mechanics and engineering tasks, they were not to conduct scientific observations on the interplanetary cruise. The only anomaly while in transit was that the star sensor on Mariner 6 lost its lock on Canopus as the scan platform was unlocked. As a result, this task for Mariner 7 was modified. It was also decided that during the flyby the probes should use gyroscopes instead of the star sensor for attitude determination. As Earth celebrated the triumph of the Apollo 11 mission, on 28 July Mariner 6 began its 'far encounter' sequence, taking its first image from a range of 1,255,000 km. Although the cameras alternated every 42 seconds, the tape recorder stored only one narrow-angle view every 37 minutes, the intention being to gain a sense of the planet's rotation, with a resolution better than that of any terrestrial telescope. Some 20 hours later, with the range now reduced to 725,000 km, the 33 images on tape were transmitted to Goldstone at the rate of one frame every 5 minutes. As the range reduced from 561,000 to 175,000 km, 16 more images and part of another were returned the next day. But just as Mariner 6 was about to start its 'close encounter' sequence, within a 1-minute interval the signal from Mariner 7 first faded and then was lost. While most of the engineers monitored the progress of the first probe, others set out to attempt to regain contact with the second, which was running several days behind its mate. After 7 hours a command was sent to switch to the low-gain antenna. With communications restored, it was found that 15 telemetry channels had been lost, including the one to show the position of the scan platform. From the fact that the spacecraft was slowly spinning and, as tracking revealed, its closest point of approach to Mars had been displaced by 130 km, the first theory was that the craft must have been hit by a meteoroid. However, it was later realized that the silver–zinc battery had exploded; in venting its electrolytes it had acted like a thruster and not only caused the probe to drift in attitude (thereby swinging the high-gain antenna away from Earth) but also slightly deflected the trajectory.[344,345] In the meantime, Mariner 6 had successfully taken 24 close images of the Martian surface: 12 wide-angle and 12 narrow-angle, over an interval of 17 minutes. Its closest point of approach was at 05:19 UTC of 31 July, at an altitude of 3,429 km above a point just south of the equator between Sinus Meridiani and Sinus Sabaeus. The picture track had been chosen to cover a broad longitude range near the equator in order to inspect several transitions between light and dark areas, two dark 'oases' (Juventae Fons and Oxia Palus) and the variable Deucalionis Regio. The

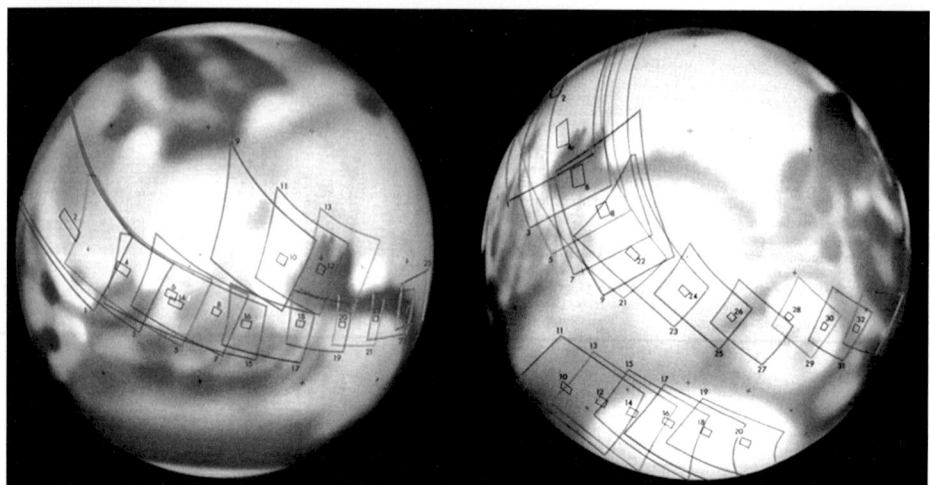

A map of Mars showing the imaging tracks of Mariner 6 (left) and Mariner 7 (right).

spacecraft passed over the limb above Sinus Meridiani at 4°N, 4°W, and emerged over the night-side above Boreosyrtis at 79°N, 276°W.[346] Once all of the Mariner 6 data had been downlinked, procedures were put in place to enable Mariner 7 to undertake its flyby. The first task was to regain the scan platform. Exploiting the high data-rate, this was swung in small increments until Mars entered the field of view. Once the planet was centered, the system was able to be recalibrated. On 2 August Mariner 7 returned 33 full-disk images, and a further 58 the next day.[347,348,349]

In both cases the resolution of the initial far-encounter images was no better than through the best terrestrial telescope. Owing to the axial tilt of the planet, the view was primarily of the southern hemisphere, where the polar cap extended as far north as 60°S; the north polar region was not visible. As the probes closed in, it became possible to verify the reality of some features glimpsed from Earth, and to banish the last lingering hope that there might be 'canali'. These images provided the first link between the albedo features observable from Earth and actual surface structures. The region of Edom was found to be a large crater (later named in honor of Schiaparelli) and the bright feature known as Nix Olympica (Snows of Olympus), which had been believed to be a high plateau, seemed to be a 500-km-diameter crater that possessed an unusually bright rim. Differences in the images from the probes showed evidence of clouds forming at high northern latitudes, and there were bright clouds over Tharsis, Candor, Tractus Albus and Nix Olympica itself. Several months after the encounter, a detailed analysis of the images revealed that three Mariner 7 pictures included the larger of the two moons, Phobos. The best such image was taken on 5 August from a distance of 138,000 km. Although Phobos spanned just 40 pixels, it was possible to infer that it had an elliptical shape with a major axis of 22.5 km and a minor axis of 17.5 km, from which it was possible to determine that it had a low albedo indicating that its surface was extremely dark – in fact, the darkest object then known in the solar system. Because

Phobos appeared on several images, it was possible to determine its orbital parameters quite accurately. It proved impossible to locate the outer, even smaller moon, Deimos.[350]

Immediately before the flyby, Mariner 7 used the filters on its wide-angle camera to take several images from a range of 44,000 km. These were transmitted directly to Earth without being stored on tape, and then combined to produce a 'color' image that clearly showed a light blue haze over the south pole, confirming a phenomenon noted by terrestrial observers.[351,352,353]

As the Mariner 6 far-encounter images had shown interesting detail on the fringe of the south polar cap, the scientists requested that Mariner 7 be told to take 8 more images than the planned 25, and to draw the track as far south as possible. It became the first probe to be reprogrammed in flight. To gain storage for the extra images, it was necessary to transmit data from other instruments 'live'. After taking 9 images of the equatorial region in the vicinity of Sinus Meridiani, Mariner 7 was to take 11 images of the southern polar region, then finish with the vast bright circular Hellas. A total of 33 close-up images were taken: 17 wide-angle and 16 narrow-angle. The closest point of approach to Mars occurred at 05:01 UTC on 5 August, at a range of 3,430 km. As viewed from Earth the spacecraft crossed the limb 19 minutes later, at 58°S, 330°W, over the Hellespontus, made an almost diametrical occultation lasting 30 minutes, and reappeared at 38°N, 148°W, over Amazonis. On the assumption that the atmosphere was entirely carbon dioxide (a fair first approximation), the four radio occultation measurements gave a pressure at the surface ranging from 4.2 hPa over Hellespontus to 7.3 hPa over Amazonis.[354,355]

The first Mariner 6 images covered the Sinus Aurorae and Pyrrhae regions, and showed new and unexpected features. On the limb, there was a thin haze similar to that seen by Mariner 4. A major discovery was a new 'chaotic terrain' of irregularly shaped and chaotically jumbled ridges and depressions, typically some 2 to 10 km in length and 1 to 3 km in width. The absence of recognizable large craters implied that it was relatively young, although there was no way of knowing its absolute age. On the far-encounter images it had appeared as dark spots and lines. This was the first case of a type of terrain on another planet for which there was no analog on either the Earth or the Moon. When the cameras turned to Sinus Meridiani and imaged some of the albedo transitions, it was realized that there was no morphological difference between areas that appear bright telescopically and those that appear dark. On Sinus Meridiani there were several craters of two types: large ones with flat floors that were probably ancient and had been modified by tectonic processes and aeolian erosion, and small 'bowl-shaped' ones whose 'fresh' appearance implied a much younger age. The focus of attention switched to Pyrrhae, then to Sinus Sabaeus, and finally to the Deucalionis region, all of which were heavily cratered, after which the field of view crossed the terminator into darkness. The first images of Mariner 7 were also in the vicinity of Sinus Meridiani. These first pictures covered the very limb of the planet, and revealed horizontal layers of haze 10 km thick at various altitudes in the range 15 to 25 km. Images of the limb taken later showed haze at an altitude of 40 km. With this evidence, it was decided that the bright glow on the limb of the first image taken by Mariner 4 was haze, not the result of a

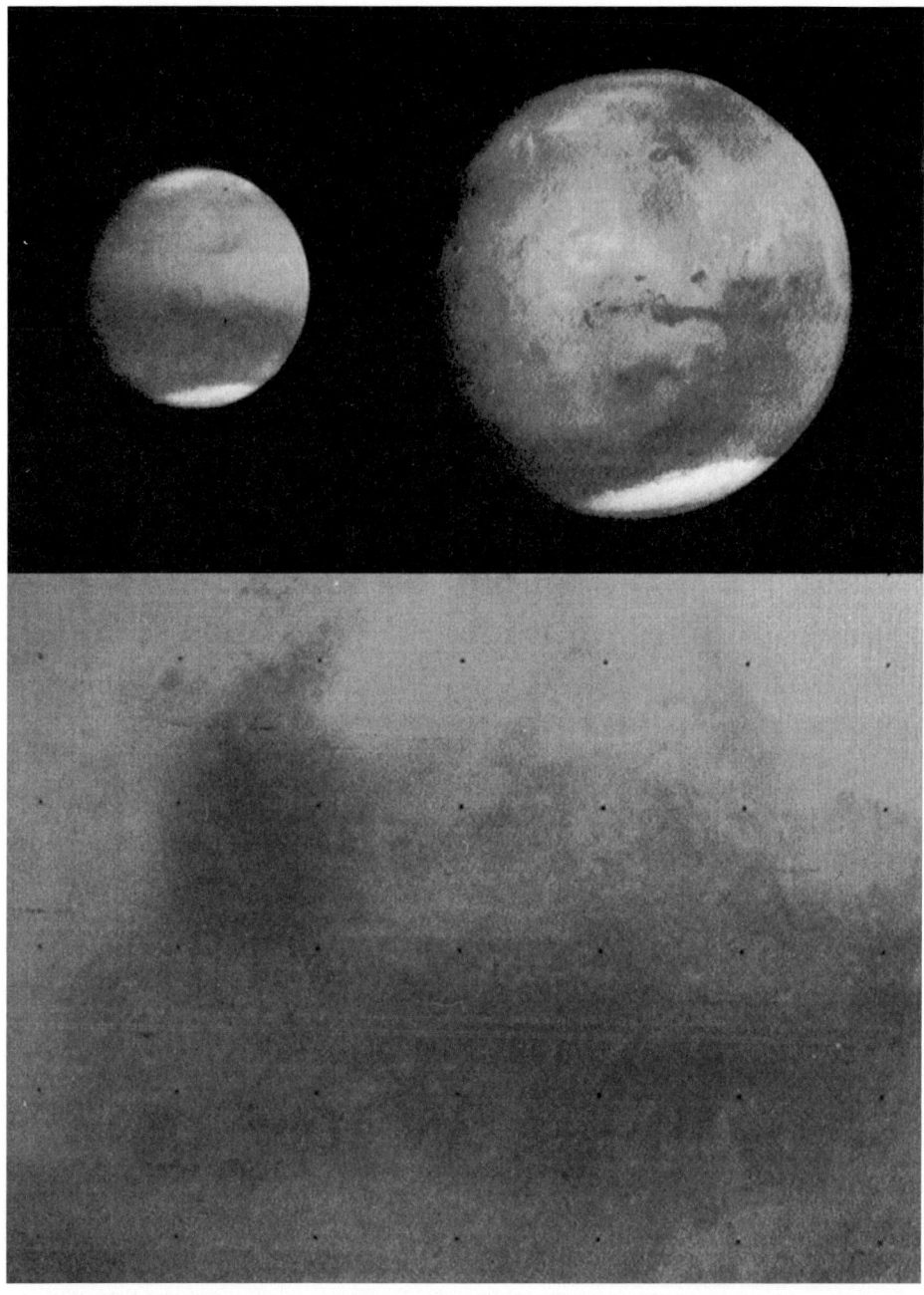

Some Mariner 6 and 7 far-encounter images: Mariner 6 F15 (top left), Mariner 7 F69 (top right); and Mariner 7 F86 (bottom) showing the dark region of Syrtis Major.

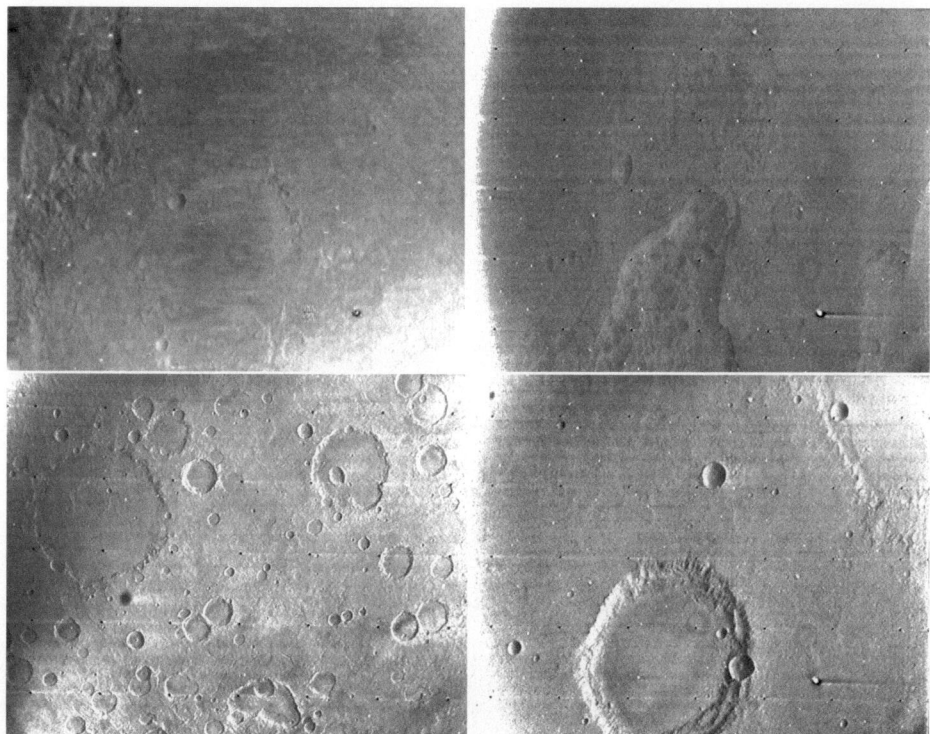

Some near-encounter images by Mariner 6: (clockwise from top left) frames 6N08 and 6N14 showing 'chaotic terrain', and frames 6N18 and 6N21 showing craters in Sinus Sabaeus.

malfunctioning or damaged camera. The next 11 images, which showed the fringe of the south polar cap, were by far the most interesting. The first hint of the cap was noted at latitude 59°S, and by 62°S the coverage was uniform. There was no sign of a thin dark line around the cap such as had been frequently reported by telescopic observers. Many craters had frost coating their inner walls. One of the best-known images from the mission showed two craters in the southern area – one 80 km in diameter and the other 50 km – juxtaposed to form a structure nicknamed 'the giant's footprint'. The final sequence was of the heavily cratered Hellespontus, sloping gently down to Hellas, which appeared featureless to the limiting resolution of 300 meters.[356,357,358] The only hint of clouds was in several near-encounter images showing the pole.

Although one of the channels of the infrared spectrometer on Mariner 6 failed, all the other instruments functioned properly. By measuring the absorption of carbon dioxide, and therefore the height of the 'gas column' above the surface, the infrared spectrometers provided data on the topography of the imaged areas. The results were in agreement with the surface pressures inferred from the radio occultations. It was evident that there was no consistent difference in elevation between the areas that

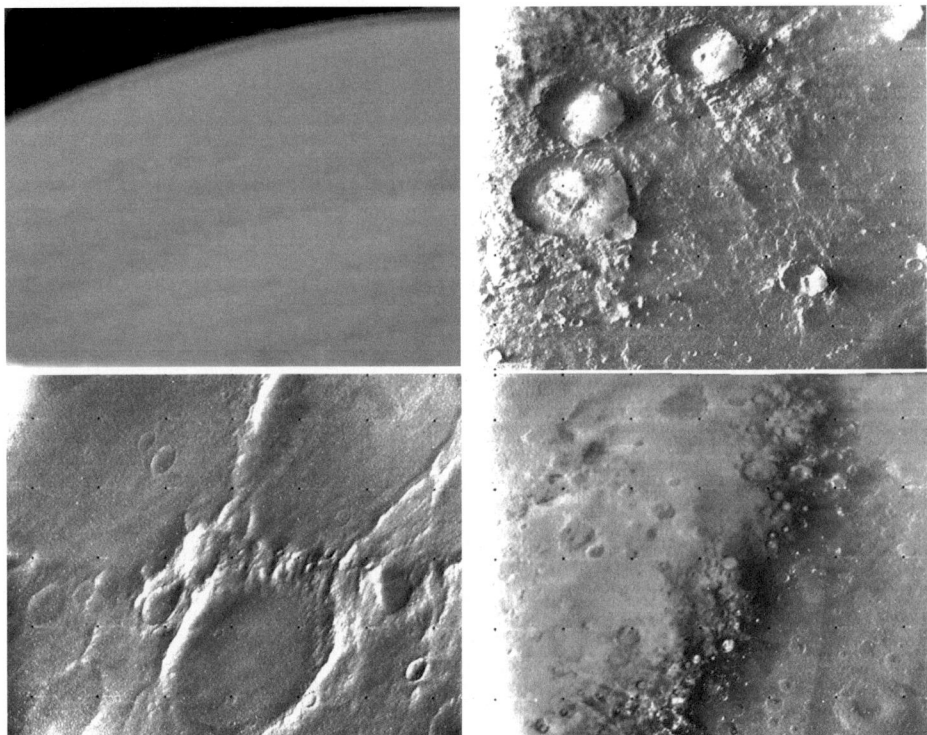

Some near-encounter images by Mariner 7: (clockwise from top left) 7N01 showing the limb with a haze similar to that detected by Mariner 4; narrow-angle frame 7N12 showing frost covered craters at the edge of the southern polar cap; wide-angle frame 7N13 showing part of the polar cap; and narrow-angle frame 7N20 showing a pair of large juxtaposed craters nicknamed 'the giant's footprint' – the larger of the two was dedicated to the late astrobiologist Wolf V. Vishniac.

appear light or dark telescopically, and Hellas was revealed to be a vast basin whose floor lies 5.5 km below its rim. At first, it was reasoned that any structures in Hellas must long-since have been buried by dust blown into the basin, and hence it must be as featureless as it had appeared, but the results of later missions would reveal that at the time Mariner 7 flew by the floor was obscured by cloud.[359] A very preliminary analysis of Mariner 7's infrared spectra revealed the presence of nothing less than ammonia and methane, both of which had great implications for the presence of life, but a deeper analysis determined that the data had been misinterpreted.[360] The instruments detected only three atmospheric constituents: carbon dioxide, carbon monoxide and water vapor. There was no trace of ammonia, nitrogen oxides or hydrocarbons down to the instrument sensitivity of better than 1 part per million.[361] Even today, the data from the infrared spectrometers on the 1969 Mariner missions remains unique for its wavelength.[362,363] The infrared radiometers measured the temperature, which ranged from –27°C at local noon on the equator down to –73°C at Syrtis Major in darkness. The most interesting measurements were taken over the south polar cap, where the temperature was $-125°C$. The combination of low pressure and temperature meant that the cap was primarily frozen carbon dioxide, rather than the water ice that had been expected.[364] The infrared spectral data misinterpreted as ammonia and methane was then recognized to have been due to sunlight reflecting off carbon dioxide ice. The ultraviolet spectrometers found emission features of atomic oxygen, and also of ionized forms of carbon dioxide and carbon monoxide in the upper atmosphere. One of the main objectives of the experiment was to search for traces of nitrogen (which, being unreactive, is difficult to detect at visual wavelengths), but no emissions were found to show its presence. Although Mariner 7 detected some ozone over the south pole, the results confirmed the atmosphere to be virtually pure carbon dioxide.[365,366]

The main result of the mission was fittingly described in this way: "Before the space era, Mars was thought to be like the Earth; after Mariner 4, Mars seemed to be like the Moon. Mariners 6 and 7 have shown Mars to have its own distinctive features, unknown elsewhere within the solar system."

One important discovery that was not recognized at the time was that the craters that could just barely be discerned in the far-encounter images were primarily in the southern hemisphere – this being the first hint that the northern hemisphere was very different. The near-encounter images were also used to map the distribution of crater sizes, and the results gave a clear indication that Mars and the Moon are not as similar as had been presumed after Mariner 4 found the planet to be cratered.[367]

Apart from the damage caused to Mariner 7 by the explosion of its battery, both spacecraft survived their encounters with the Red Planet and were deflected by its gravity into slightly different orbits, ranging from 1.14 to 1.75 AU for Mariner 6 and from 1.11 to 1.70 AU for Mariner 7. In August Mariner 6 collected 10.5 hours of ultraviolet spectra of the Milky Way.[368] Between the end of April and mid-May 1970, both spacecraft were on the far side of the Sun as seen from Earth, and it was possible to measure the deflection of their radio carriers as a result of passing close to the Sun (in effect, the radio beam was bent, with the 200-microsecond increase in the time of flight of the signal being consistent with General Relativity) and the way in which the signal was altered yielded insight into the inner solar corona.[369] Useful data was also provided by a cavity radiometer – an engineering instrument carried to measure the spacecraft's thermal budget. At this time the Sun was near the peak of its 11-year cycle of activity, and the data from these instruments served to indicate that the total luminosity of the Sun varies very little over a timescale of several weeks.[370] Contact with both probes was maintained for almost 2 years after launch, but the dates of their last communications do not appear to have been published. Overall, NASA's missions to Mars in 1969 had cost the equivalent of $570 million at 2000 levels.

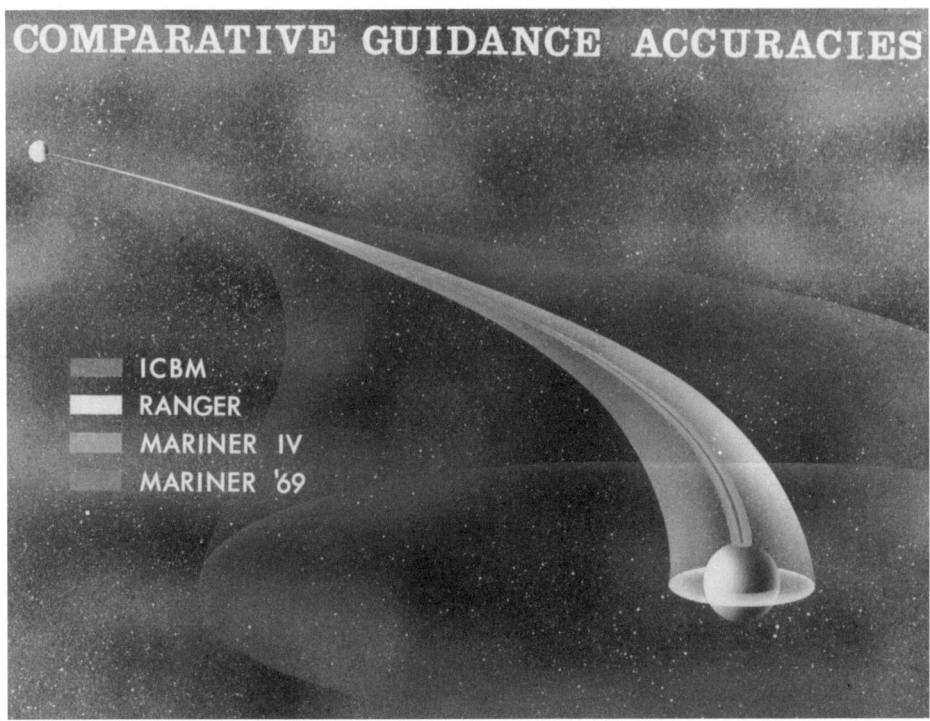

The 1960s saw a great increase in the accuracy of astronautics guidance. This figure compares the relative accuracies of an ICBM, Ranger lunar impactor, Mariner 4 and the 1969 Mariners at the distance of Mars. (Courtesy of NASA/JPL/Caltech)

OTHER PLAYERS

In the 1960s and early 1970s, as the Americans and Soviets were starting to explore the solar system, other nations were launching their first artificial satellites and thus becoming 'space powers' – most notably France, Japan, China and Great Britain. In addition to cooperating in space research with NASA, and to a lesser degree with the Soviets, in 1962 several European nations formed ESRO (European Space Research Organization). One of the early objectives was to send a probe to the Moon, but the establishment of a deep-space program was also suggested. In 1963 plans for a lunar mission were abandoned, in part because of the lack of a suitable European launcher but also because it would be expensive and would most likely simply duplicate what the Americans and Soviets were well on the way to undertaking. ESRO therefore focused on solar system exploration, with the intention of identifying areas of study and objectives that were *not* being pursued by the others. It studied three possible missions. One was a Jupiter flyby probe. Another was the polarimetric investigation of Venus and Mars. Since astronomical polarimetry was being pursued by America only on

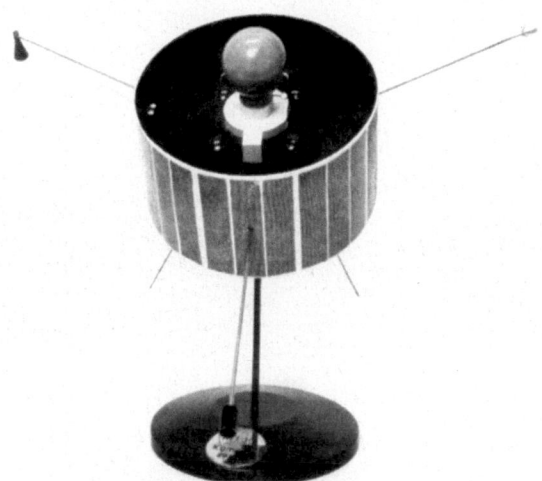

A model of the cometary probe that was under study at ESRO in the mid-1960s. The baseline plan was to intercept periodic comet 41P/Tuttle–Giacobini–Kresak in 1973. Had the probe encountered the comet, it could have studied the mysterious outbursts that made it 10,000 times brighter than expected at perihelion. (Courtesy of Henning W. Scheel)

an occasional basis, this offered the prospect of Europe making a unique contribution. Moreover, the Meudon Observatory in Paris, having initiated physical observations of the planets as early as 1941, was the foremost center for polarimetric studies of the solar system. Unfortunately, the European space scientists were mostly interested in cosmic rays and high-energy astronomy, and dismissive of missions to Mars and Venus. Another European specialism was cometary research, and so studies were made of a mission to fly past a comet. The initial work involved a long-period comet, but such a flight posed too many 'unknowns', and the effort was soon switched to a periodic comet. The basic scientific payload and objectives of the probe were defined, and a preliminary study was carried out in Germany, but the mission was deferred.

In early 1968 several new ideas were put forward as candidates for ESRO's next big scientific project, one of which was for a mission to Mercury. This probe was to take the first close-up images of the planet's surface, and determine whether it had a magnetic field and a thin atmosphere. It would also provide an opportunity to study the solar wind within 0.38 AU of the Sun, and when the probe was on the opposite side of the Sun from Earth its radio carrier could be used to investigate the nature of the solar corona. The project was named MESO (MErcury SOnde). The spacecraft would have a launch mass of 450 kg, of which 69.4 kg would comprise instruments. It would be launched by an Atlas–Centaur supplied by NASA, and have 10 scientific instruments which, in addition to those to study particles and fields, would include a variety of cameras, an integrated photometer and polarimeter, an infrared radiometer and a microwave radiometer. However, when put forward for approval in May 1969 it was deemed both too expensive and beyond ESRO's technical capabilities.[371]

A drawing of the European MESO Mercury flyby spacecraft that was proposed and refused in 1969. (Courtesy of ESA)

While solar system exploration has always been – and still is – a government-led activity, during the 1960s corporations in the United States were proposing planetary missions of their own, often in competition with NASA. In 1961 General Dynamics studied a Solar Monitor probe. The large probe would be launched by a Saturn I into an elliptical solar orbit that had its perihelion at 0.16 AU, its aphelion between 0.43 and 0.93 AU, and a period as short as 60 days. Two possible architectures were proposed: the first resembling a spinning top, with eight small solar panels and a parabolic antenna of wire-mesh; and the second, dubbed the 'Wright Flyer II' owing to its resemblance to a pioneering aircraft, with two large solar panels. Power would be produced both by solar cells and by a 6-meter-diameter mirror that would use solar heat focused on liquid metal to generate power.[372] Another interesting case was Lockheed's 1967 proposal for Starlite, to be launched using a purpose-built rocket named Starlet. Many aspects of this project were highly innovative or revolutionary: (1) the Starlet would use highly corrosive fluorine and liquid hydrogen as propellants, the benefit being sheer power, and be launched from a mobile, military-like transporter; (2) Starlite would use an inflatable aluminized reflector to generate power, to communicate with Earth, and to provide radiation-pressure stabilization; and (3) carrying just over 11 kg of scientific payload, Starlite was to be able to undertake a wide variety of missions of durations up to 10 years, including very close solar flybys, Jupiter flybys, and multiple missions to Jupiter, Saturn, Neptune and Pluto.[373] As yet however, no such 'private venture' has been mounted.

The 'spinning top' configuration of the 1961 Solar Monitor that was to be launched by a Saturn I rocket. (Courtesy of General Dynamics Corporation)

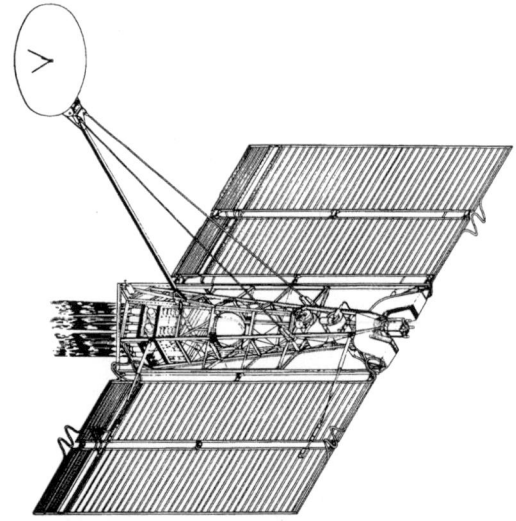

In the 1960s NASA studied alternative propulsion systems for its deep-space probes. One proposal for missions beyond the orbit of Mars was electrical propulsion using a pair of 35 kW SNAP-8 nuclear power units; the two 'wings' are not solar panels but radiators. (Original image copyright of General Electric)

SOURCES

Citations refer to the chapter references section at the end of the book.

1 Bille-2004	20 Ness-1970	39 Koppes-1982b
2 Marcus-2006	21 Smith-1960	40 Ezell-1984b
3 Ulivi-2004a	22 Lewis-1960	41 AWST-1961
4 Hufbauer-1991a	23 Fan-1960	42 Ezell-1984c
5 Goddard-1920	24 Greenstadt-1963	43 Varfolomeyev-1998a
6 Portree-2001	25 Greenstadt-1966	44 Harvey-2006b
7 Turner-2004	26 Cahill-1963	45 Perminov-1999a
8 Burgess-1953	27 AWST-1960a	46 Semenov-1996a
9 Kronk-1984a	28 AWST-1960b	47 Harvey-2006a
10 Kronk-1984b	29 AWST-1960c	48 Vladimirov-1999
11 Kronk-1999	30 Powell-2005b	49 Chertok-2007
12 Crocco-1956	31 Melin-1960	50 Perminov-1999a
13 DeVorkin-1992	32 Melin-1960	51 Mitchell-2004a
14 AWST-1957	33 Butrica-1996a	52 Lardier-1992a
15 Wilson-1987a	34 Melbourne-1976	53 Semenov-1996a
16 Powell-1984	35 Clark-1960b	54 Varfolomeyev-1993
17 Powell-2005a	36 Ezell-1984a	55 Semenov-1996a
18 Corliss-1965a	37 Koppes-1982a	56 Lardier-1992a
19 Clark-1960a	38 Stone-1961	57 Chertok-2007

58 Siddiqi-2002a
59 Varfolomeyev-1998b
60 Siddiqi-2000a
61 Oberg-1981
62 Semenov-1996b
63 Mitchell-2004b
64 Gatland-1964
65 Varfolomeyev-1998b
66 Gatland-1964
67 Grahn-2000
68 Axford-1968
69 Ness-1970
70 Lardier-1992b
71 Siddiqi-2002b
72 Chertok-2007
73 Butrica-1996b
74 Koppes-1982c
75 NASA-1965a
76 Wilson-1987b
77 Schneiderman-1963
78 NASA-1965b
79 Forney-1963
80 Barath-1964
81 Corliss-1965b
82 Corliss-1965c
83 Neugebauer-1997
84 Hufbauer-1991b
85 Davidson-1999a
86 Sonett-1963
87 NASA-1965c
88 Nicks-1985a
89 van der Linden-1994
90 Nicks-1985b
91 Hill-1962
92 NASA-1965c
93 Coleman-1962
94 Scheinemann-1963
95 Forney-1963
96 Nicks-1985c
97 Neugebauer-1962
98 Hufbauer-1991c
99 Coleman-1962
100 Ness-1970
101 Sonett-1963
102 Nicks-1985d
103 NASA-1965c
104 Chase-1963
105 Barath-1963
106 Barath-1964

107 Pollack-1967
108 Frank-1963
109 Sonett-1963
110 NASA-1965d
111 NASA-1963
112 Kolcum-1963
113 NASA-1965e
114 Nicks-1985e
115 Wilson-1987b
116 Gatland-1972a
117 Wilson-1982a
118 Westphal-1965
119 ESRO-1966
120 ESRO-1969
121 Semenov-1996c
122 Flight-1963
123 Perminov-1999b
124 Mitchell-2004c
125 Siddiqi-2002c
126 Mitchell-2004c
127 Siddiqi-2002c
128 Knap-1977
129 Harford-1997
130 Gatland-1972b
131 Zheleznyakov-2001
132 Lissov-2004
133 Siddiqi-2002d
134 Harvey-2006a
135 Mitchell-2004b
136 Mitchell-2004d
137 Murray-1966
138 S&T-1963
139 Murray-1966
140 Flight-1963
141 Gatland-1972c
142 Siddiqi-2002d
143 Perminov-1999b
144 Gatland-1972c
145 Clark-1986
146 Perminov-1999b
147 Siddiqi-2000b
148 Siddiqi-2000c
149 Siddiqi-2002e
150 Flight-1964a
151 Varfolomeyev-1998c
152 Flight-1964b
153 Mitchell-2004c
154 Knap-1977
155 Siddiqi-2002f

156 Lardier-1992c
157 A&À1964
158 LePage-1993
159 Murray-1966
160 Mitchell-2004d
161 Flight-1965a
162 Grahn-2000
163 Flight-1965b
164 Harvey-2006a
165 Lardier-1992c
166 Siddiqi-2002g
167 Semenov-1996e
168 Huntress-2002
169 Clark-1986
170 Koppes-1982d
171 Ezell-1984d
172 Murray-1966
173 Wilson-1966
174 NASA-1964
175 Wilson-1987c
176 Stone-1963
177 Corliss-1965d
178 Leighton-1965
179 Murray-1966
180 NASA-1964
181 Koppes-1982e
182 A&A 1964
183 Murray-1966
184 Koppes-1982f
185 Null-1967
186 NASA-1967a
187 Chapman-1969
188 Anderson-1965
189 Danielson-1964
190 Barth-1974
191 Kliore-1965
192 Kliore-1973
193 Null-1967
194 Kliore-1973
195 Dollfus-1998a
196 Ness-1979a
197 Sullivan-1965
198 Chapman-1969
199 Leighton-1965
200 NASA-1967a
201 Herriman-1966
202 Tombaugh-1950
203 Baldwin-1949
204 Opik-1950

205 Opik-1951
206 Sheehan-1996a
207 Burgess-1966
208 Murray-1989a
209 Ezell-1984e
210 Davidson-1999b
211 AWST-1965a
212 AWST-1965b
213 Beech-1999
214 Wiegert-2007
215 NASA-1967a
216 Wilson-1987c
217 ESRO-1969
218 Ulivi-2004b
219 Varfolomeyev-1998a
220 Varfolomeyev-1998b
221 Mitchell-2004b
222 Gatland-1972d
223 Harvey-2007a
224 Siddiqi-2002h
225 Kurt-1971
226 NSSDC-2004a
227 Perminov-2002
228 Lardier-1992d
229 Murray-1966
230 Perminov-1999c
231 Lantranov-1999
232 Corliss-1965e
233 Stone-1964
234 Turnill-1984a
235 NASA-1971a
236 Burlaga-1968
237 Bukata-1969
238 Levy-1969
239 Goldstein-1969
240 Merat-1974
241 Flight-1972
242 Brandt-1974
243 Borrowman-1983
244 Mihalov-1987
245 Flight-1997
246 Pioneer-2004
247 Kelly Beatty-2001
248 Wolverton-2004a
249 Harrington-1965
250 The Tech-670221
251 Hollweg-2004
252 Hollweg-1968
253 Wilson-1979

254 Wotzlaw-1998
255 Perminov-2002
256 Mitchell-2004e
257 Ness-1970
258 Kerzhanovich-2003
259 Vinogradov-1970
260 Jastrow-1968
261 Wilson-1987d
262 Lardier-1992e
263 Gatland-1972e
264 Wetmore-1965
265 Park-1964
266 Wilson-1987e
267 Koppes-1982g
268 Snyder-1967
269 Fjeldbo-1971
270 Lazarus-1970
271 Van Allen-1968
272 Ness-1970
273 Boyer-1961
274 Sheehan-1999
275 Varfolomeyev-1998b
276 Gringauz-1976
277 Wilson-1987e
278 Gatland-1972e
279 Ingersoll-1971
280 Vinogradov-1970
281 Jastrow-1968
282 Wilson-1987e
283 Gatland-1972e
284 Lardier-1992e
285 Marov-1978
286 AWST-1967a
287 AWST-1967b
288 AWST-1967c
289 Lazarus-1970
290 Anderson-1968a
291 Anderson-1968b
292 Fjeldbo-1971
293 Kliore-1968
294 Wilson-1987d
295 Hunter-1967
296 Snyder-1967
297 Jastrow-1968
298 Anderson-1968c
299 Van Allen-1968
300 Barth-1970a
301 Wilson-1987d
302 Jastrow-1968

303 Mitchell-2004e
304 Ezell-1984f
305 Neumann-1966
306 Ezell-1984g
307 Corliss-1965f
308 Quimby-1964
309 AWST-1964
310 Ezell-1984h
311 NASA-1967b
312 Murrow-1968
313 Ezell-1984i
314 Perminov-2002
315 Lardier-1992f
316 Vinogradov-1970
317 Blamont-1987a
318 Reeves-2003
319 Vinogradov-1970
320 Marov-1978
321 Wilson-1987f
322 Gatland-1972f
323 Siddiqi-2002i
324 AWST-1969a
325 AWST-1969b
326 Lantranov-1999
327 Perminov-1999d
328 Giragosian-1966
329 Beuf-1964
330 Koppes-1982h
331 Ezell-1984j
332 Murray-1966
333 Leighton-1971
334 Murray-1989b
335 Wilson-1987g
336 Koppes-1982i
337 Siddiqi-2002j
338 Ezell-1984k
339 Ezell-1984l
340 Wilson-1987g
341 Powell-2004
342 Lantranov-1999
343 Perminov-1999d
344 Koppes-1982j
345 Wilson-1987g
346 Kliore-1973
347 NASA-1969a
348 NASA-1969b
349 NASA-1969c
350 AWST-1970
351 Collins-1971

352 Leighton-1971
353 Schurmeier-1970
354 Kliore-1973
355 Wilson-1987g
356 Leighton-1971
357 Collins-1971
358 Schurmeier-1970
359 Herr-1970

360 Murray-1989c
361 Barth-1974
362 Forney-1997
363 Kirkland-1998
364 Neugabauer-1971
365 Barth-1971
366 Barth-1974
367 Leighton-1971

368 Barth-1970b
369 Anderson-1975
370 Foukal-1977
371 Ulivi-2006
372 Mari-1962
373 LMSC-A847990

2

Of landers and orbiters

A NEW DECADE

In July–August 1970 the US National Academy of Sciences' Woods Hole Summer Study Conference held a symposium at which space scientists put forward ideas for planetary exploration in the next 10 to 15 years that would, in part, directly shape NASA's program and, indirectly, the Soviet one. It was suggested that orbiting and landing probes should explore Mars and Venus in detail, orbiters should study Jupiter and Mercury, flyby probes should reconnoiter the outer planets, the Sun should be approached to at least 0.3 AU, and preferably 0.05 AU, and should also be observed from high latitudes. Finally, there should be preliminary investigations of comets, asteroids and other minor bodies of the solar system.[1]

TO THE SURFACE!

The first deep-space probe of the 1970s was dispatched by the Soviet Union, in the launch window for Venus during the summer of 1970. Since the new type of heavy spacecraft was still in development, Babakin prepared two 3V carrying modified landing capsules for the V-70 mission. In order to improve the capsule's stability on reaching the ground its diameter was reduced, and to increase the structural integrity of the pressure hull the number of apertures was reduced. Its shell was built of two forged titanium halves which were machined to a high precision and covered with shock-absorbing material. This way, the capsule was expected to survive for up to 90 minutes in temperatures and pressures as high as 540°C and 180,000 hPa, and to reach the surface in working order. The capsule was tested in the only hyperbaric chamber in the world able to simulate such conditions. To ensure a reasonably short descent time and a survivable impact, a 'variable geometry' parachute was designed. It was to be kept partially closed by nylon straps that would melt at a temperature of 200°C in the lower Venusian atmosphere, whereupon the area of the canopy would increase from 1.8 to 2.4 m^2. Even with a minimal payload of a barometer and the platinum-wire thermometer, the structural modifications had increased the mass of the capsule by about 100 kg, making the overall spacecraft too heavy for the 8K78 Molniya launch vehicle. To compensate, the main spacecraft was lightened and its instrument suite reduced, and some of the telemetric systems were deleted from the launcher – an option that would complicate a postmortem if the launch vehicle were to fail.[2,3] The first probe lifted off on 17 August, and less than a hour later was in a solar orbit ranging between 0.69 and 1.01 AU, heading for Venus. The total mass of the spacecraft, including the 495-kg atmospheric capsule, was 1,180 kg. The second probe was less fortunate. When it was launched on 22 August an electrical problem caused stage L to shut down 25 seconds into the 244-second 'escape' maneuver. The stranded probe was named Kosmos 359 and left to burn up on 6 November.[4]

After two course corrections, Venera 7 approached Venus in December. On 10 December its instruments recorded an intense solar flare that was noted 2 days later by Lunokhod 1, a Soviet lunar rover. On 12 December the atmospheric capsule was powered up, and a system chilled its interior to –8°C to further improve the length of time it would be able to operate in the hot Venusian atmosphere. On 15 December, as the spacecraft penetrated the atmosphere at a speed of 11.5 km/s and an angle of 60–70 degrees below the local horizon, it released the capsule. The entry site was on the night-side, some 2,000 km from the terminator. The capsule was decelerated at 350 g and endured a heat-shield temperature of 11,000°C. By the time it was at a height of 60 km, its speed had slowed to 200 m/s. The parachute now unfurled, and the instruments started to collect data. The external pressure at that altitude was 700 hPa. Later in the descent, the mechanical commutator of the telemetric system suffered a major failure and fixated on providing temperature data. The descent was nominal for the first 6 minutes, then, at an altitude of 20 km, the parachute canopy began to melt and rip, and the rate of descent increased. At 5:34:10 UTC, 35 minutes after entry instead of the expected 60 minutes, the probe fell silent. After stating that the signal had ceased when the probe impacted the surface, the Soviets reported that almost 2 months of analysis of the telemetry had established that the probe had sent data for an additional 22 minutes 58 seconds at a signal strength barely 1 per cent of its earlier level. The only data that was directly reported was the temperature, which showed a constant 475°C. The Soviets concluded that the main parachute tether had been so weakened by the intense heat that it had snapped, whereupon the probe had fallen through the final 3 km of the dense atmosphere at a terminal velocity of 16.5 m/s, then rolled over on impact with its antenna facing away from Earth. The impact was at 5°S, 9°E on Tinatin Planitia. Using mathematical models of the atmosphere, the ground pressure was determined to be 90,000 ± 15,000 hPa.[5,6,7,8,9] Venera 7 was the first probe from Earth to survive a landing on the surface of another planet. Moreover, this provided the first indication of the mechanical properties of the surface. The fact that the vertical speed had diminished to zero in under 0.2 second suggested a solid dust-free surface with a bearing strength estimated to be between lava and wet clay – admittedly very broad limits, but the first direct measurement nevertheless.[10]

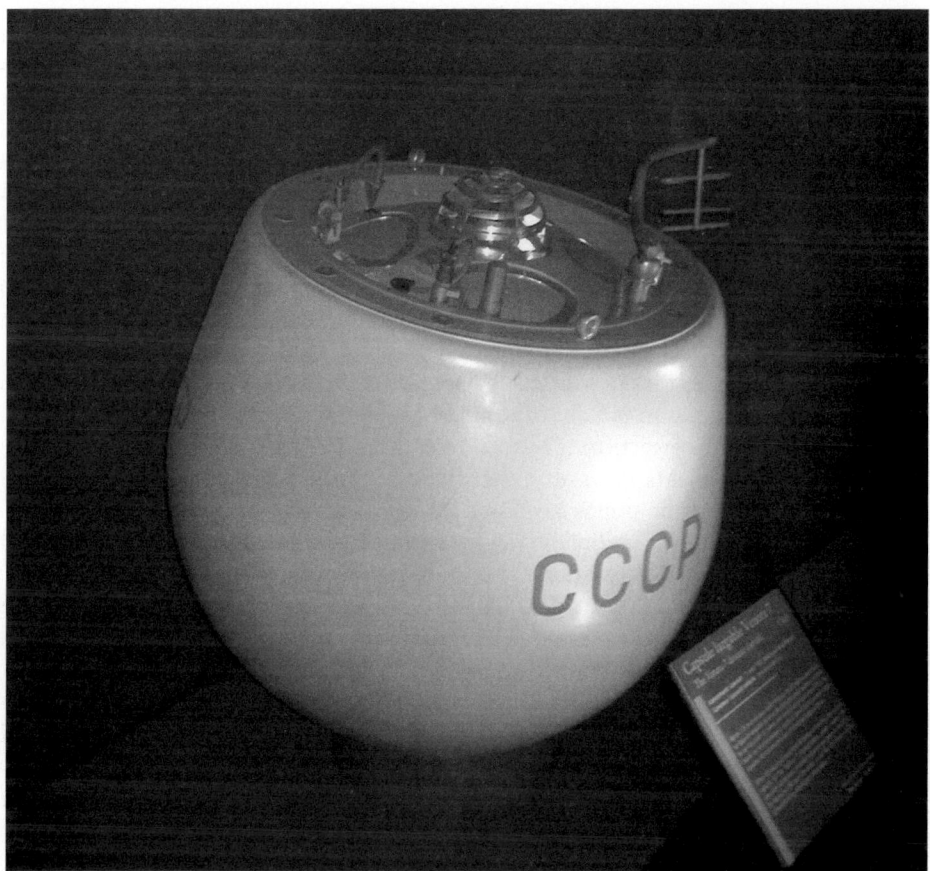

The Venera 7 capsule was the first probe to survive a landing on another planet.

INTO THE STORM

Immediately after the M-69 debacle, the Soviets began to prepare for the next launch window to Mars. The initial idea was to dispatch orbiters to improve the ephemeris and capsules to characterize its atmosphere, and then at the next opportunity to send more orbiters with soft-landing probes. However, it was then noted that the window in the late spring of 1971 opened within a few months of a 'great opposition' – an event that occurs only once in every 15 to 17 years, when Mars reaches perihelion while at opposition relative to Earth, and the separation is at the minimum possible distance. At such a time, the energy requirements of an interplanetary transfer are minimized, and the payload for a given launch vehicle is maximized. It was therefore decided to advance the soft landers from 1973 to 1971. The engineers initially studied probes based on the M-69 design, but when they realized that this had no 'growth potential' the idea was abandoned. The M-69 is therefore the only Soviet deep-space probe never to make even a partially successful mission. In fact, until the truth was revealed by Russian sources, western analysts had presumed that the 1969 attempts were early flights of the 1971 bus.

The 3M bus designed by Babakin's engineers at the Lavochkin design bureau was destined to become the standard for deep-space missions until the mid-1980s. A central module included a pressurized 180-cm-diameter cylinder, within which was a propellant tank. The base of the module enlarged into a 'mushroom-shaped' structure that was 2.35 meters in diameter and held a toroidal tank built around the gimbaled KTDU-425A (Korrektiruyushaya Tormoznaya Dvigatelnaya Ustanovka) course-correction and braking engine. To the sides of the cylinder were two rectangular solar panels, each 2.3 meters tall and 1.4 meters wide, with a total span of 5.9 meters. On one side between the solar panels was a 250-cm-diameter parabolic high-gain antenna, and on the opposite side was a large radiator for the cooling fluid of the thermal control system. The Sun and star sensors, the navigation, attitude control and communication systems, and some scientific instruments, were mounted beside the engine. On the opposite end of the bus could be carried either a scientific payload for the orbital study of the objective planet, or a landing capsule protected by a conical heat shield that was 320 cm in diameter and had a vertex angle of 120 degrees. The total height of the vehicle was 4.1 meters. The spacecraft had three stabilization regimes: (1) the axis perpendicular to the plane of the solar panels could be pointed to the Sun, and the orientation left to drift around this axis; 2) this axis could be pointed to the Sun, and the spacecraft set spinning every 11.4 minutes; or (3) the spacecraft could be stabilized in all three axes.[11] With a mass in excess of 4,500 kg, the new spacecraft required the 4-stage version of the 8K82 Proton launch vehicle which, since its introduction with the ill-fated M-69s had successfully dispatched several lunar missions.[12,13,14]

Two versions of this spacecraft were prepared for the 1971 launch window. The M-71S (Sputnik, orbiter) had a launch mass of 4,549 kg, including 2,385 kg of fuel. Its payload was: a magnetometer; three photometers (one visible, one infrared and one ultraviolet); two radiometers (one infrared); a spectrometer; a charged-particle spectrometer; a cosmic-ray sensor; and an FTU (FotoTelevisionnoye Ustroistvo) photo-television system incorporating two cameras. One camera, named Vega, used optics having a focal length of 52 mm, and the second, Zufar, had a focal length of 350 mm for a field of view of 4 degrees. They were each equipped with three color filters, and used 25.4-mm-wide photographic film. Imaging was to be undertaken primarily from the low point in the spacecraft's elliptical orbit of Mars, with pictures being taken in cycles of 12 at intervals of 35 or 140 seconds, and from the planned altitude the best resolution of the surface was expected to be in the range 10 to 100 meters.[15] The M-71P (Pasadka, lander) had a launch mass of 4,650 kg, including the 1,210-kg lander. The orbiter was equipped with its own 2-camera imaging system; an infrared radiometer to scan the temperature of the Martian soil; two ultraviolet photometers to study the atmosphere; an infrared photometer; an instrument to detect hydrogen in the upper atmosphere; a 6-channel visible-light photometer; a combined radiometer and radar to study the characteristics of the atmosphere and the surface to a maximum depth of 50 meters; an infrared spectrometer to determine

A mockup of the Soviet Mars-71P spacecraft. The conical heat shield of the lander is at the top. The black and white striped object is the lander's thermal radiator for the interplanetary cruise. (Courtesy Patrick Roger-Ravily)

topographic relief; cosmic-ray and plasma sensors; and a 3-axis magnetometer mounted on a boom positioned below the solar panels.

The M-71 spacecraft were also to carry the French STEREO-1 solar radio-astronomy experiment – the first non-US, non-Soviet experiment to be flown on an interplanetary probe. The opportunity arose from the 1966 Soviet–French agreement on space science, under which French scientists had flown a laser retroreflector to the Moon on Lunokhod 1 in November 1970. Over a period of at least 6 months, this new instrument was to measure the directivity of solar radio bursts by carrying out simultaneous solar observations from Earth and space with an increasing angular separation due to the different orbital motions of Earth and the spacecraft. In particular, it would study the type I (brief, circularly polarized and of a limited bandwidth) and type III (longer, unpolarized and of a larger bandwidth) solar radio bursts. The experiment consisted of two almost identical receivers, one based at the Nançay Radio Astronomy Station in France, and the other a low-noise receiver and a 3-element Yagi antenna on a 1.2-meter-long boom mounted on one of the solar panels of the Mars-bound probe. Although neither drawings nor written descriptions of the spacecraft interfaces were permitted to be issued to the French scientists, the hardware was successfully built. Nor were the French informed of the spacecraft that

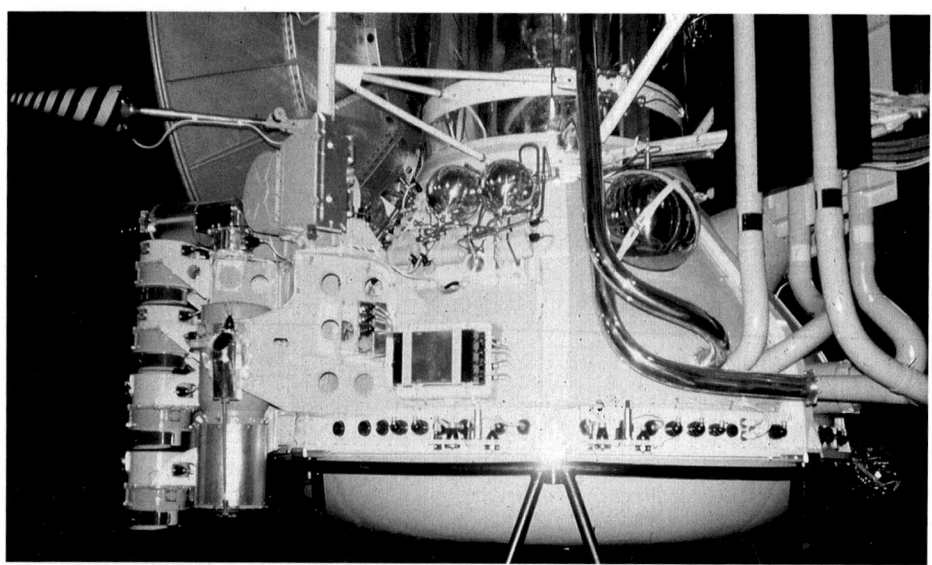

Detail of the Mars-71P mockup. The base of the bus housed a toroidal fuel tank and the orbit insertion engine. Instruments and sensors such as those seen at left were mounted on its periphery. Note the pipes of the thermal control system. (Courtesy Patrick Roger-Ravily)

would carry their experiment: one went on the single M-71S and the other on one of the M-71P.[16,17,18,19]

The landing module was to be released when the spacecraft made its approach to Mars. It had a solid-rocket motor with which to put itself into the entry corridor, a stabilization system for attitude control in free flight, a heat shield, a toroidal compartment with a parachute, and the spherical lander, which had a diameter of 1.2 meters and a mass of 358 kg. The landing system had a 13-m^2 drogue parachute and a 140-m^2 main parachute. The lander was based on the design of the E-6 'egg' probe that had made the first successful lunar soft-landing in 1966, but in this case was enclosed by a 20-cm-thick plastic shell to cushion the impact. Four spring-loaded petals were to open to orient it in an upright position on the surface. It carried a mass spectrometer to analyze the composition of the atmosphere during the descent; a meteorological package comprising a barometer, a thermometer and an anemometer; two cameras to take 500 × 6,000 pixel panoramas of the site; and a gamma-ray spectrometer and an X-ray spectrometer at the tips of different petals in order to place these sensors as near as possible to the ground and enable the composition of the material on the surface to be analyzed. As the capsule was powered by a battery, it had an operational life of at best a few days. Four short whip aerials were to send data to directional antennas on the mothership, which would record it for later relay to Earth. Every component of the lander had been sterilized using germicidal lamps to preclude terrestrial bacteria contaminating the Martian environment.[20,21,22]

Detail of the Mars-71P mockup. Another view of the bus, showing the cylindrical portion of the spacecraft that housed propellant, and the white toroidal parachute container of the lander. (Courtesy Patrick Roger-Ravily)

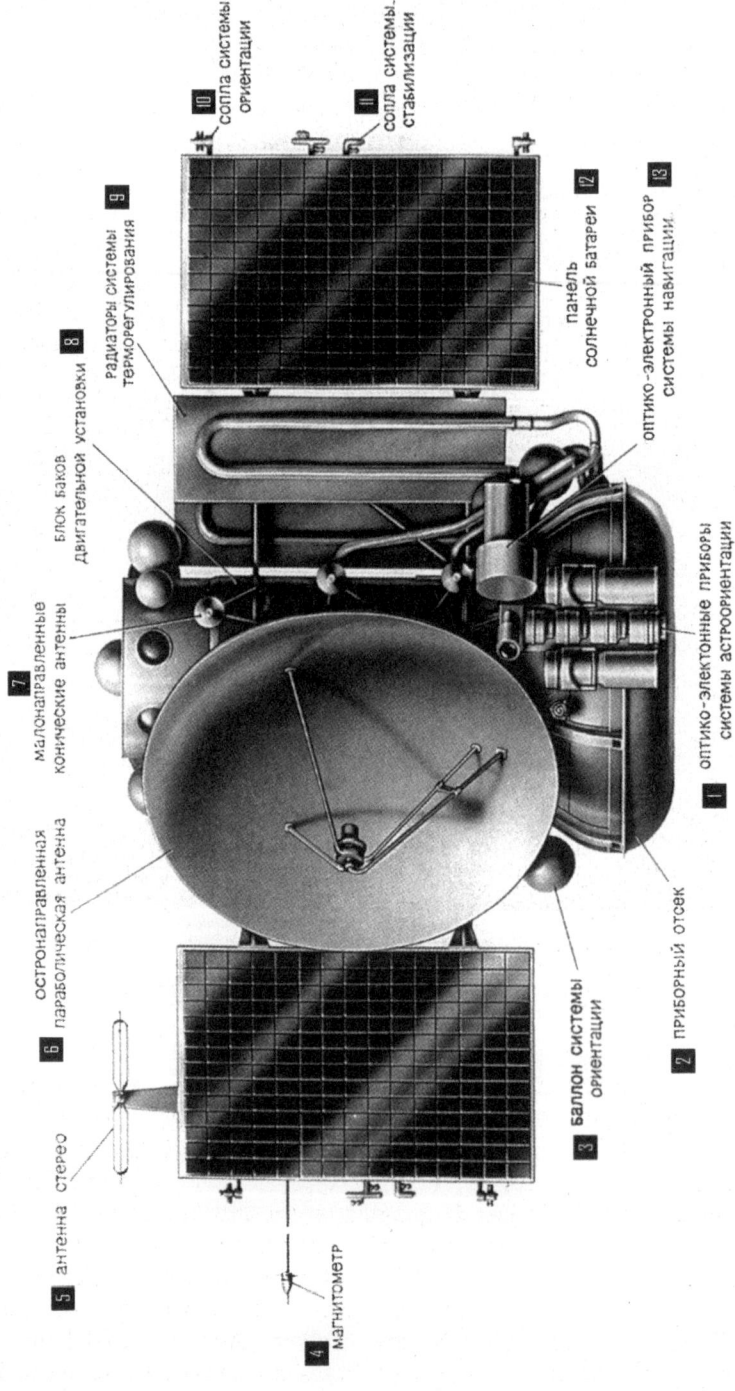

The M-71S configuration: 1, attitude determination sensors; 2, instrument module; 3, attitude control system tank; 4, magnetometer; 5, antenna of the 'Stereo' experiment; 6, high-gain antenna; 7, low-gain antenna; 8, propulsion system tank; 9, thermal control system radiator; 10 and 11, attitude control system nozzles; 12, solar panel; 13, navigation system sensor.

A mockup of the M-71 lander that was the first probe to land on Mars. In the event of coming to rest on its side, it had the capability to right itself. The black object on top is the PrOP-M 'walker'. (Courtesy of Olivier Sanguy/Espace Magazine)

It was also announced that data from the landers was to assist in the development of the future 'Planetokhods', or planetary rovers.[23] This statement assumed a new meaning when the Russians revealed in the early 1990s that each lander carried an amazing experiment – a 4-kg mini-rover equipped to measure soil properties. This had been designed and built in just 18 months by the institute of mobile vehicle engineering VNII Transmash of Leningrad, which had already designed Lunokhods for use on the Moon. It was named PrOP-M (Pribori Otchenki Prokhodimosti-Mars, instrument for cross-country characteristics evaluation on Mars), 'Marso-khodik' or 'Micromarsokhod' (small Martian walker). After a 6-fold boom had placed it on the ground, the 21.5 × 16 × 6-cm box-shaped robot was to use skids in

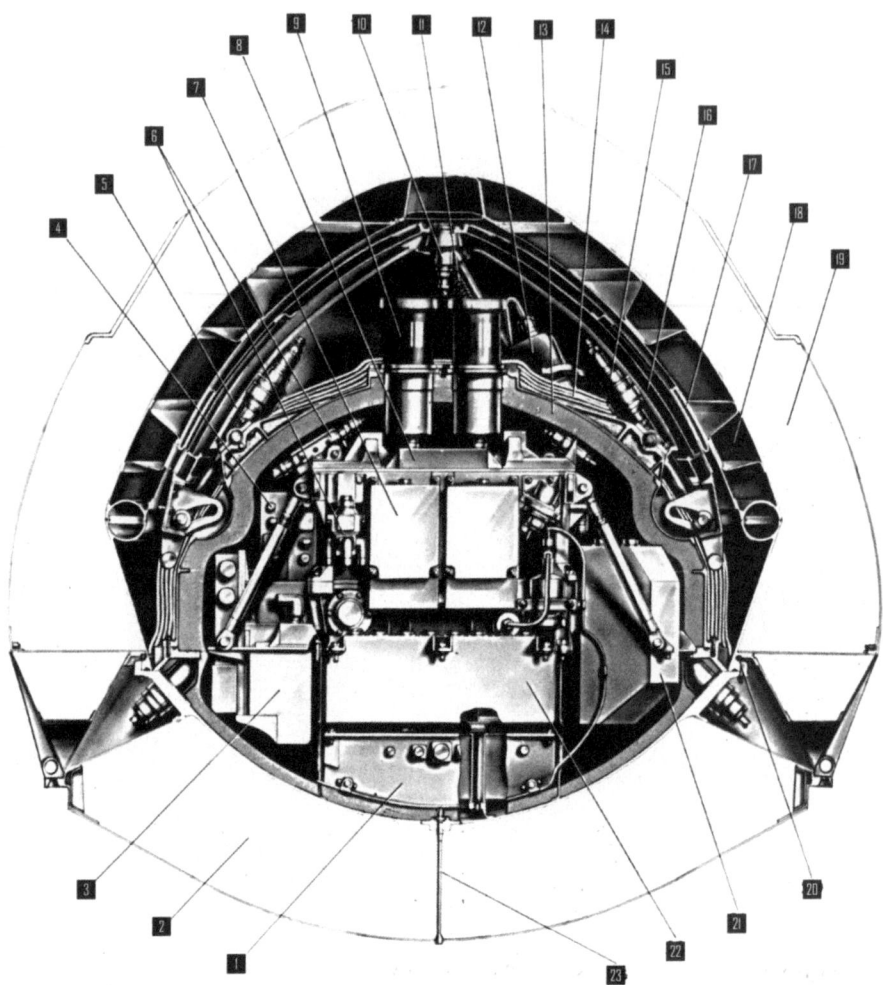

The M-71 lander in its stowed configuration: 1, radar altimeter electronics; 2, lower shock absorber; 3, telemetric units; 4, radio unit; 5, antennas; 6, radio system; 7, radio system units; 8, scientific instruments; 9, stereo cameras; 10, lander raising petal lock; 11, scientific instrument deploying device; 12, scientific instruments; 13, thermal insulation system; 14, thermal insulation blanket; 15, pyroactuators to raise the probe after landing; 16, petals; 17, shock absorber opening bag; 18, aeroshell cover; 19, upper shock absorber; 20, cartridge for release of the shock absorber; 21, control electronics; 22, batteries; 23, descent atmospheric pressure sensor.

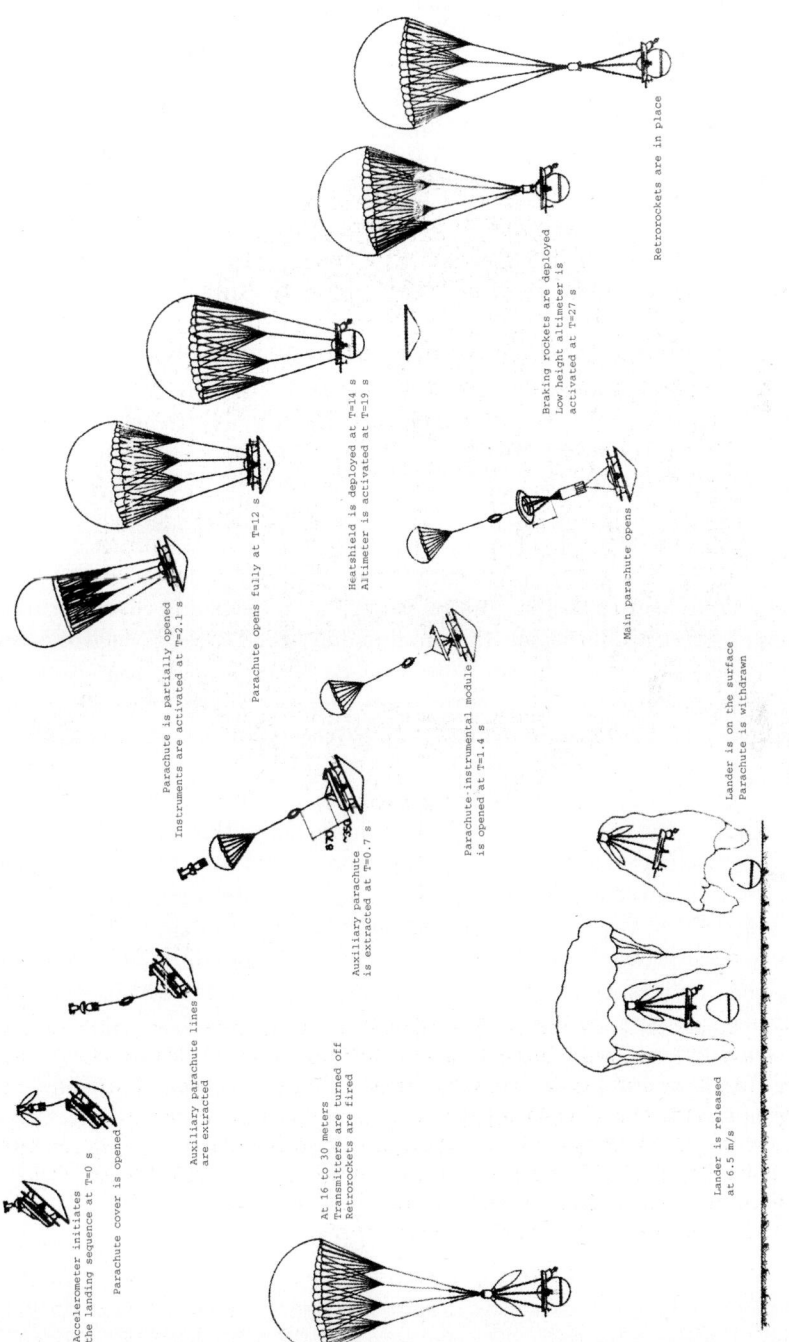

The M-71 landing sequence.

a parallelogram configuration to walk at a speed of 1 meter per minute out to a maximum of 15 meters from the lander, its range being limited by the umbilical that would supply power (it used just 1 W), commands and telemetry. On the front was a bumper with two levers that worked as an obstacle sensor. The rudimentary artificial intelligence of the robot was to enable it to identify the position of the obstacle, and then take several steps backward and attempt to walk around it, steering by moving the skids in opposite directions. It was to halt every 1.5 meters to undertake soil-mechanics measurements using its two instruments: one a penetrometer supplied by Transmash, and the other a densitometer provided by the Institute of Geochemistry of the Soviet Academy of Sciences. More data was to be gleaned from analysis of the pictures of the traces left in the soil by excrescences of various sizes imprinted on the underside of the vehicle. The Soviets planned to use the data obtained to design the locomotion system for more sophisticated Mars rovers. This approach was the same as that used for lunar missions, in which data collected by the penetrometer and densitometer of the Luna 13 lander in 1966 was exploited in the design of the Lunokhod rovers of the 1970s.[24,25,26,27,28,29]

The plan was to launch the M-71S first, to enter orbit around Mars and refine the ephemeris. When the two M-71P made their approach shortly thereafter, they were to release their landers 46,000 km from the planet, and 900 seconds later the landers were to fire their own rocket motors to adopt the atmospheric entry corridor for their assigned targets. During the 4 hours it would take to reach Mars, the landers would be spin stabilized. Some 100 seconds after it detected atmospheric deceleration, and still traveling at more than Mach 3.5 (measured in terms of the local environment), a lander would deploy its drogue parachute. The main parachute would be opened by a timer that would also jettison the heat shield and power on the radar altimeter. As the speed was reduced to 65 m/s, the instruments were to start to take pressure data through a capillary hole at the bottom of the protective shell. At a height somewhere between 16 and 30 meters, the altimeter was to command a cluster of braking solid-rocket motors to fire and to jettison the parachute, leaving the probe to impact at a speed of about 12 m/s. After compressed air had jettisoned the two halves of the protective shell, the four stabilization petals would open to right the lander. Its first task would be to take a panorama of the site, which would be sent to the orbiter at a data rate of 72,000 bps. As landing sites, the Soviets chose two areas in the southern hemisphere that were thought to be lowlands and would provide a deeper column of air to increase the efficiency of the parachutes.[30] The entire descent sequence was tested by 15 flights of M-100B sounding rockets which released scale models of the lander and landing system from heights of 130 km. Ground tests demonstrated the ability of the probe to survive impacts at up to 28.5 m/s, imparting decelerations of up to 180 g.[31]

On 14 November 1968 NASA had announced that in 1971 it would send probes to Mars on a mission that would form an intermediate step between the Mariners 6 and 7 flybys and the landings which, despite the demise of the Voyager program, it still hoped to make in the mid-1970s. Once again, JPL was given the mission. Two identical probes were to enter orbit around Mars and study its surface for at least 90 days, and possibly half a local year. The first probe, initially designated Mariner H

Two views of the tiny PrOP-M walking robot. It is a pity that this never showed its worth! The bottom view shows the underside of the 'Marsokhodik', showing the conical penetrometer. (Courtesy of VNII-TransMash)

but intended to become Mariner 8, was to enter an orbit with a periapsis of 1,250 km and an apoapsis of 17,300 km inclined at 80 degrees to the Martian equator with a period of 12 hours. It was to map 70 per cent of the surface at 1,000-meter resolution and 5 per cent of it at 100-meter resolution. This orbit would allow the mapping of a strip between 60°S and 40°N in its first 90 days, with opportunities to re-examine interesting sites at 20-day intervals. Mariner I was to enter an orbit with a periapsis of 850 km and an apoapsis of 28,600 km inclined at 50 degrees with a period of 20.5 hours that would cause it to repeat its ground track every 5 days. It was primarily to study the atmosphere, and transient phenomena in the atmosphere or on the surface. One objective was to improve the coordinate system for the Martian globe, which at that time incorporated inaccuracies as large as 50 km, to enable the locations of features to be determined to within about 1.5 km.

The Mariner orbiter used the same basic architecture as had proved itself on JPL's earlier missions to Mars. At the center was a 1.39-meter-wide, 0.46-meter-tall magnesium octagonal bus that housed all the electronics and flight systems. This supported four 2.14-meter-long rectangular solar panels that gave the spacecraft a total span of 6.89 meters. Each panel had an area of 1.92 m^2, and together they could supply up to 800 W in the vicinity of Earth and between 450 and 500 W at Mars. In fact, the only major difference to the flyby probes was the orbit-insertion system, which was placed above the bus structure. It included the single 1,340-N engine that was capable of being fired as many as six times and the 0.75-meter-diameter propellant tanks, one for monomethyl hydrazine and the other for nitrogen tetroxide. The tanks had a total capacity of 463 kg, and were protected from micrometeoroids by a fiberglass cover. Sun and Canopus sensors provided attitude determination, and the attitude was controlled by 12 nitrogen jets positioned at the tips of the solar panels. It had a tubular 10.2-cm-diameter, 1.45-meter-long low-gain antenna and a 1.02-meter-diameter high-gain dish antenna mounted on a 2-position joint. When using the 64-meter-diameter antenna at Goldstone, the peak data rate would be 16,200 bps, but only 2,025 bps when using the 26-meter antennas of the Deep Space Network at Canberra in Australia, Johannesburg in South Africa and Madrid in Spain. Because the high-gain antenna would not be able to point at Earth in the attitude required for the Mars orbit-insertion maneuver, a medium-gain horn antenna was fitted beneath one of the solar panels for this purpose. The scan platform was installed on the base (opposite the engine), could be oriented in azimuth (215 degrees total range) and elevation (69 degrees), and carried all the scientific instruments. One camera had 12.5-mm-diameter optics with a focal length of 50 mm for a wide-angle view, and the other had a 251-mm-diameter, 508-mm-focal-length Cassegrain system for a narrow-angle view. Both had an 8-position wheel for color and polarizing filters. At the intended periapsis, the wide-angle camera was to provide a surface resolution of 1,000 meters per pixel, and the narrow-angle camera a resolution of 100 meters. The scan platform also carried a 2-channel infrared radiometer that was almost identical to those of the 1969 flyby probes, to measure the surface temperature with a field of view half that of the narrow-angle camera; an ultraviolet spectrometer to collect data on the pressure, temperature and density of the atmosphere and to

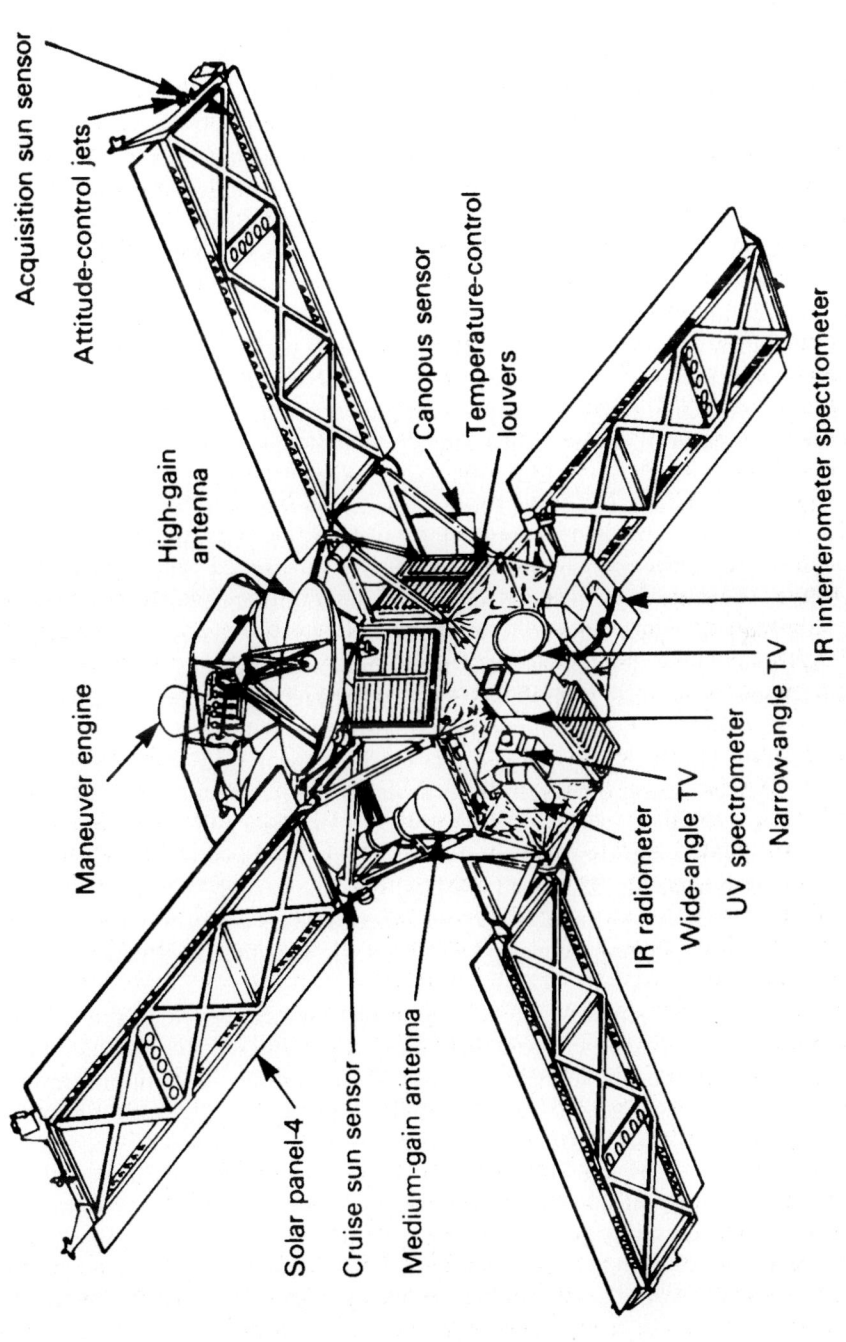

Acquisition sun sensor

Attitude-control jets

Canopus sensor

Temperature-control louvers

High-gain antenna

Maneuver engine

IR interferometer spectrometer

IR radiometer

Wide-angle TV

UV spectrometer

Narrow-angle TV

Solar panel-4

Cruise sun sensor

Medium-gain antenna

A line drawing of NASA's 1971 Mariner spacecraft.

investigate whether there was ozone; and an interferometric infrared spectrometer to identify the secondary chemical species in the atmosphere – there was a particular interest in testing for water vapor, because its presence would require a re-evaluation of the prospects for Martian life. (Despite some problems with the development of the hardware for this instrument, it was ready in time for tests.) And, of course, once the probes were in orbit, tracking and the radio-occultation events as they repeatedly disappeared and reappeared around the planet's limb would yield a wealth of data for mapping not only the atmosphere but also the gravitational field – in particular to evaluate the shape ('geoid') of the planet. When opportunities presented themselves, the small moons, Phobos and Deimos, were also to be studied. Of the orbiter's 998-kg launch mass, the scan platform and its suite of instruments accounted for 82.2 kg.[32,33,34,35,36,37,38]

The first launch was Mariner 8, on 9 May. The Atlas performed flawlessly, but an unstabilized pitch oscillation on the Centaur stage starved the engines of propellants and it shut down prematurely, 365 seconds into the mission, dumping the spacecraft into the Atlantic some 1,500 km down range.

The first Soviet launch on 10 May was also a write off. The M-71S was inserted into parking orbit, but stage D of the Proton failed to perform the 'escape' burn, and the probe (named Kosmos 429) re-entered 2 days later. A programming error had prevented ignition of the engine. The Soviets had thus lost not only the chance to become the first to place an artificial satellite around another planet, but also the opportunity to update their ephemeris for the lander-carrying spacecraft that were to follow. All would now depend on how their automated systems coped with the situation. When an M-71P was 50,000 km from Mars, it was to determine the location of the planet optically, then adopt a specific orientation. An hour later it was to release the lander, which would orient itself for the maneuver that was to put it on course to enter the atmosphere. When the main spacecraft was 20,000 km out, it was to make a second measurement of the planet's position and calculate the best time to execute the orbit-insertion maneuver, which would hopefully result in a closest point of approach of 2,350 ($\pm$1,000) km.[39,40] The two M-71P spacecraft were dispatched faultlessly on 19 and 28 May and named Mars 2 and Mars 3; the first entered a solar orbit ranging between 1.01 and 1.47 AU, and the second an orbit ranging between 1.01 and 1.50 AU.

Meanwhile in America an investigation had rapidly traced the loss of the Centaur carrying Mariner 8 to a failed integrated circuit in the pitch guidance module, which was readily corrected. To enable a single orbiter to make observations appropriate to both of the planned missions, it was decided that Mariner 9 should enter an orbit similar to that designed for Mariner 8 but at the lower inclination of 65 degrees.[41,42] On 30 May (12 days later than initially planned) Mariner 9 was boosted on a direct-ascent trajectory into a solar orbit ranging between 0.99 and 1.57 AU. Although it set off later than its Soviet rivals, the fact that it was on a faster trajectory meant that it would arrive first.

The ill-fated Mariner 8 spacecraft is prepared for launch. The high-gain antenna and the small horn antenna for communications during Mars orbit insertion are visible.

The STEREO-1 apparatus was activated 3 days after launch, and almost every day thereafter. The Soviets exploited the fact that the French had not been informed of the probes that carried their apparatus to cover the loss of the M-71S, and said that a malfunction of the Soviet electronics of Mars 2 had prevented the instrument being powered on, and that, as a result, they would receive data only from Mars 3. On 1 September 1971 most of the French ground equipment for this experiment was destroyed by fire – possibly due to sabotage. Nevertheless, some instruments were able to be made operational within days, and the radio array was finally reconnected 5 weeks later.[43]

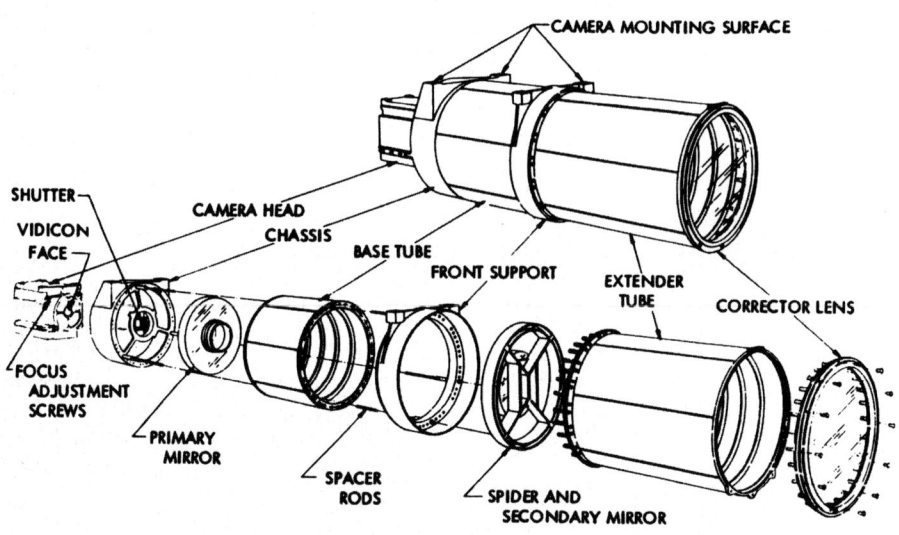

The narrow-angle camera of the 1971 Mariner spacecraft.

In the days after launch, all three probes performed a course correction: Mariner 9 on 4 June, Mars 2 on 5 June (or according to another source, 17 June) and Mars 3 on 8 June. The primary and secondary high-gain transmitters failed on both Soviet spacecraft on 25 June, but after several days of communication using their low-gain antennas the high-gain systems came back on line. The cause of the malfunction was never determined, but it is possible that solar heat was concentrated by the parabolic dishes and damaged some components of the antenna feeds. (As a precautionary measure it was decided to install thermal protection blankets on the antennas of future deep-space probes.) On 3 August Georgi N. Babakin, who had led the Soviet programs of automatic lunar and planetary exploration since taking over the Lavochkin bureau in 1965, died of a heart attack. He was succeeded by Sergei S. Kryukov, who, although he had only recently joined Lavochkin, had earlier been one of Sergei Korolyov's deputies.[44] The event that would determine the fates of these missions occurred at the end of September. On 22 September a South African astronomical observatory that was monitoring the state of the Martian atmosphere in advance of the arrival of the flotilla of probes noted the appearance of a 'bright spot' in the Noachis region of the southern hemisphere. This soon developed to form a 2,400-km-long streak, then masked Hellas and went on to rapidly spread around two-thirds of the circumference of the planet. A second dust cloud appeared 6 days after the first, this time in Eos. At the same time, observations through blue filters indicated that the entire atmosphere had increased in opacity. By mid-October it was apparent that the dust storm was becoming global in extent. Dust storms are not uncommon, particularly when Mars is at perihelion, but this 'storm of the century' masked the entire planet – an event last seen at the great perihelic opposition of 1909.[45]

In October and early November, while in interplanetary space, Mariner 9 used its

An Atlas–Centaur lifts off with Mariner 9.

ultraviolet spectrometer to observe the Lyman-alpha emission of our galaxy, and to make photometric measurements of several classes of variable star. After the camera had been calibrated by taking long-range pictures of Saturn, it began to image Mars on 8 November, but the plan to monitor the slow retreat of the south polar cap during the local summer was frustrated by the fact that no detail could be discerned owing to the dust storm.[46] Images taken over 11–13 November as the spacecraft closed in were featureless apart from a hint of the south polar cap and the presence of four mysterious dark spots, one of which matched the location of Nix Olympica, a region that terrestrial radar had shown to be elevated. It was inferred, therefore, that the spots were the summits of mountains. When further study revealed that all of the dark spots had a large crater, it became apparent that the mountains were volcanoes. On 14 November Mariner 9 fired its engine for 915.6 seconds, consuming 431 kg of propellant as it entered an orbit ranging between 1,398 and 17,916 km. The engine was fired again on the fourth revolution to refine the orbit to 1,394 × 17,144 km at an inclination of 64.34 degrees and a period of 11 hours 57 minutes, which would revisit any given longitude every 35 revolutions (17 local days). Having arrived, Mariner 9 now had to wait for the dust to settle before it could start its main imaging mission. On the other hand, radio-occultation data provided invaluable temperature profiles during the storm, as well as welcome insight into the complex dynamics of Mars's atmosphere. Because dust at altitudes of up to 30 km was absorbing solar radiation and heating the upper atmosphere, while blocking the Sun and cooling the usually warmer lower atmosphere, the atmosphere was revealed to be essentially of uniform temperature – i.e. it was 'isothermal'.[47] The infrared spectrometer measured a reduction of the temperature difference between the day-side and the night-side from 95°C to 40°C. This instrument also showed that the mountain summits that projected above the storm were indeed at very high altitudes. Three days after orbit insertion, imagery showed a hint of craters, and a bright linear feature crossing the Coprates region where radar had indicated there to be low-lying terrain – subsequent Mariner 9 observations would show this to be a canyon system some 500 km wide and 4,000 km long; much larger than any terrestrial counterpart. Within just two weeks, Mariner 9 had transformed our view of Mars. Between them, the flyby probes had imaged just 10 per cent of the planet's surface and, it was now realized, had given the misleading impression that Mars was merely a land of impact craters; but the volcanoes and canyons that were now revealed were evidence that it had, in fact, a rich geological history.[48,49,50,51] The canyon system was later named Valles Marineris (Valleys of Mariner).

In late November, the first detailed images of the two Martian moons were taken. Distant images of Phobos had first been taken by Mariner 7, and then by Mariner 9 prior to orbit insertion. On 26 November the outer satellite, Deimos, was imaged, and on 29 November Phobos was imaged from a range of 5,270 km. Both proved to be irregular bodies, approximately described by triaxial ellipsoids of 28 × 23 × 20 km (Phobos) and 16 × 12 × 10 km (Deimos), and both had impact craters. The largest crater on Phobos was about 10 km in diameter, and was later named in honor of Angeline Stickney, the wife of Asaph Hall who discovered the satellites in 1877. It was also discovered that both satellites always show the same face to Mars, their rotations having become tidally locked.[52]

Mars 2 corrected its trajectory for the second time on 20 November, and after a final refinement on 27 November it released its lander. Approximately 4.5 hours later it entered a 1,380 × 25,000 km orbit inclined at 48.9 degrees with a period of 18 hours (the plan had called for a higher apoapsis to provide a period of 25 hours). Meanwhile, the lander had independently fired its own engine to head for the planet, but unfortunately the entry angle was too steep, with the result that it entered the atmosphere at 5.8 km/s and hit the surface before the parachute had time to deploy. It crashed near 44°S, 47°E (or according to an alternative source, 45°S, 58°E) in Noachis, the first artifact to reach the planet's surface. The fault was attributed to the optical navigation system, which had overcompensated trajectory errors during the final targeting maneuver.[53] Despite the existence of a 'hot line' between Moscow and JPL for timely exchanges of news related to the Martian missions, the Soviets postponed the announcement of the crash for more than 48 hours. When the Soviet press finally released the news, the scientists at JPL were left to ponder why the 'hot line' had been installed.[54]

Mars 3 managed to enter orbit on 2 December, but for some unknown reason the insertion burn was cut short, producing an extremely eccentric orbit ranging between 1,530 and 214,500 km inclined at 60 degrees with a period of 307 hours (just a few hours short of a fortnight).[55] Meanwhile, 4 hours 35 minutes after its release, the lander entered the atmosphere at almost precisely the requisite angle of incidence and deployed its parachute. When the radio altimeter indicated that the lander was 25 meters above the surface, the parachute was jettisoned and the retrorocket was fired to soften the impact. Nevertheless, when the 450-kg probe made contact at 13:50:35 UTC it did so at the relatively high speed of 20.7 m/s. The landing site was at 45°S, 158°W, between the regions of Electris and Phaetontis. Thirty minutes after achieving orbit, Mars 3 was to adopt a 3-axis-stabilized attitude appropriate to relay the early data from its lander to Earth. The first attempt at stabilization failed because the optics of the primary Sun sensor had become coated by the efflux of the main engine, but control was regained using a backup sensor in time for the lander's transmission. The lander's first task was to provide a panorama of the site, then a meteorological report. The camera was similar to that of the lunar surface probes, in that it used a photometric sensor and a mirror that nodded up and down and rotated horizontally to produce scan lines. After 14.5 seconds, however, the signal was lost, and never recovered. The data recorded by the orbiter was relayed to Earth several times between 2 December and 5 December, but revealed nothing extra. Even when 'processed' on Earth, the 79 image columns of the panorama remained mostly 'noise', and the most that could be inferred was a very low contrast view of a section of the horizon. While the cause of the problem is not known, the suggestions include that the capsule was destroyed by high winds; that the capsule sank into very fine dust; and that the lander functioned but there was a fault in the relay system on the orbiter. The most likely explanation, however, is that the lander's transmitter was damaged by electrostatic discharges on the whip antennas induced by the dust in the rarefied atmosphere. Western observers suspected that the story of a section of panorama being received was an invention to represent what had been a total failure as a partial success, and to substantiate the Soviet claim to be the first to soft land a probe on Mars.[56,57,58]

By 28 November Mariner 9 had completed its preliminary mapping of the craters on the summits of Nix Olympica and the other volcanoes. Images of the limb taken in early December clearly showed layers of dust at altitudes of 30 and 50 km. While waiting for the atmosphere to clear, the spacecraft continued to provide regular radio occultations and the celestial mechanics experiment collected data about the planet's gravitational field. On 7 December the probe suffered the only major anomaly of its mission, when the temperature of one component of the radio system unexpectedly increased and the output power began to fall slowly. The spacecraft was commanded to switch over to the backup, and although tests showed that the temperature of the primary system would eventually stabilize, it was decided not to attempt to switch it on again. The infrared radiometer took measurements of Phobos for 52 minutes while the moon passed through the planet's shadow on 12 December, and for a time following its egress. Unfortunately, while the moon was in darkness it was not possible to take images to confirm that the radiometer was pointing at it. These observations (plus 12 more made later in the mission) provided a direct measurement of the thermal inertia of the moon's surface, showing there to be a layer at least 1 mm in thickness of fine material similar to the lunar regolith, which was consistent with polarimetric observations by terrestrial telescopes.[59,60] On 30 December, with the dust storm gradually abating, the spacecraft made a trim maneuver to adopt the optimal mapping orbit: the period was increased by a little more than 1 minute, and the altitude at periapsis was made 1,654 km to ensure that adjacent images would overlap. Finally, on 2 January 1972 the mapping mission started. The latitudes between 65°S and 20°S were covered on the first 20 days, and latitudes between 25°S and 30°N during the next 20 days. The coverage of the second band was particularly welcome because this was where sites would have to be selected for the Viking landers that were due to arrive in 1976. The primary mission was concluded on 11 February, at which time, although the storm had subsided, bands of haze were still evident in pictures of the planet's limb.

The scientific output of the Mariner 9 mission was truly excellent.[61,62,63,64] First, it revealed a dichotomy between the intensely cratered and evidently ancient southern hemisphere and the much less cratered and probably younger northern hemisphere. Due to the geometry and timing of their encounters, the coverage of the flyby probes had been essentially confined to the cratered terrain. In the northern hemisphere, the most interesting features were the volcanoes. With a base diameter of some 600 km, Olympus Mons (as Nix Olympica was renamed) was the largest, and apparently also the most recently active. According to the spectrometers, which enabled altitudes to be inferred from atmospheric pressure, the summit stood about 15 to 30 km tall and its flank was marked by a 2-km-high scarp. The other three dark spots that had stood out from the dust storm proved to be large volcanoes in the Tharsis area, some way off to the southeast of Olympus. A volcanic edifice with a 1,600-km base was found in nearby Solis Lacus (Lake of the Sun) and named Alba Patera, but because it was extremely shallow it was not discovered until the dust storm had cleared. A group of smaller volcanoes was found on the other side of the planet in the Elysium region. All these structures were shield volcanoes formed by low viscosity lava. It was as if they stood over 'hot spots', where columns of magma rising from the

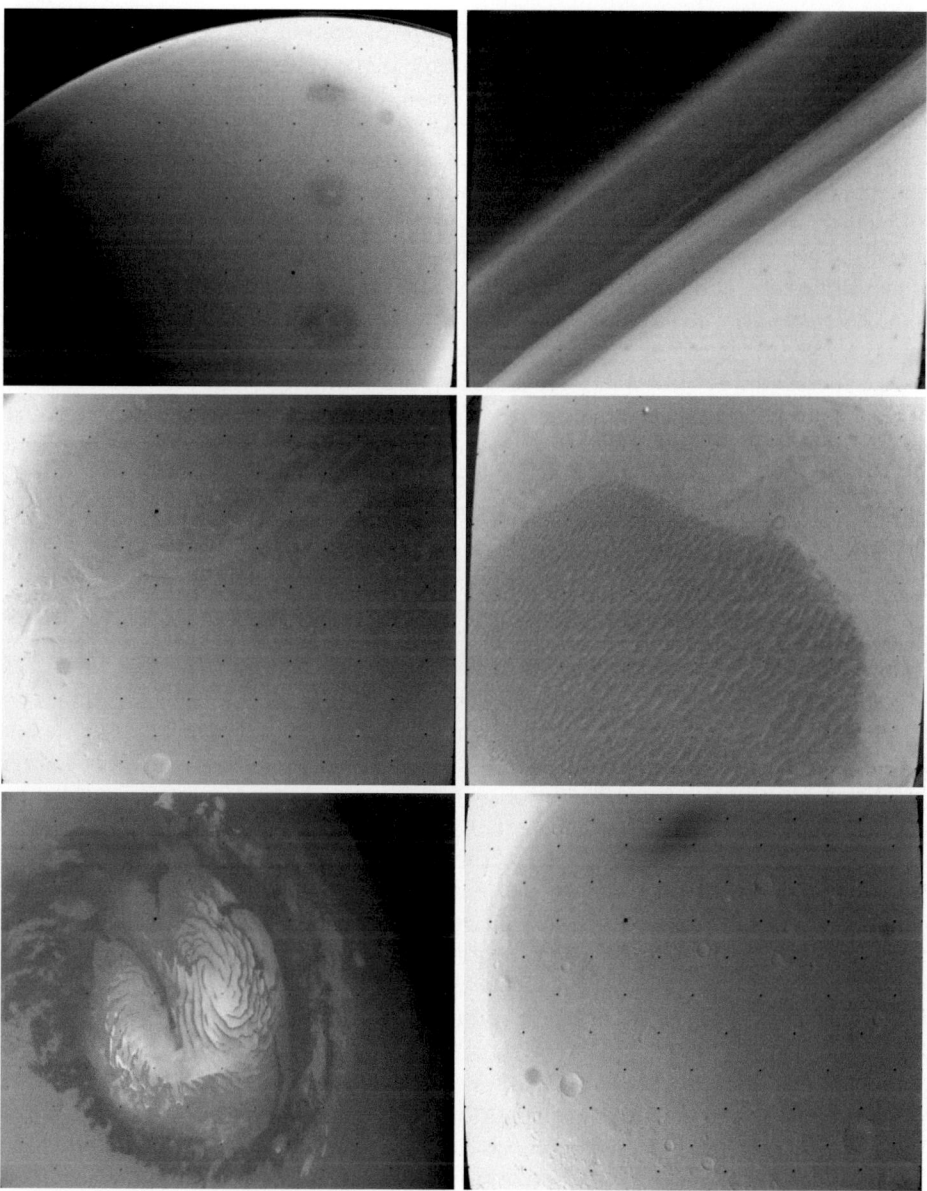

Some Mariner 9 images. Top: the volcanoes (north to south: Ascraeus, Pavonis and Arsia Mons) on Tharsis emerge from the dust storm on 15 December 1971, and a limb view on 1 March 1972 showing atmospheric layering. Center: a portion of the Valles Marineris canyon complex, and a 70-km-wide field of dunes inside a 150-km-diameter crater in the Hellespontus region. Bottom: the northern polar cap seen on 12 October 1972, and the elongated shadow of Phobos crossing a field of small craters.

interior had welled up through the crust. This phenomenon also occurs on Earth, but because our lithosphere is divided into drifting 'plates', the magma from a 'hot spot' produces a chain of volcanoes, the best example being the Hawaiian islands. On Mars, however, where 'plate tectonics' does not appear to have been active, the magma from each 'hot spot' made a single massive edifice. Interestingly, the total mass of the Hawaiian chain is similar to that of Olympus Mons. Orbital tracking and gravimetry confirmed that both Tharsis and Elysium coincide with gravitational anomalies created by the plumes of material. In fact, the three volcanoes in the Tharsis region reside on a 'bulge' some 5,000 km across that rises to an altitude of 7 km. Interestingly, the altitudes of the summits of the edifices on this bulge are comparable to that of the summit of Olympus Mons, suggesting that the internal pressure was simply unable to force any more magma up through their central vents. The ages of the volcanoes is a matter of debate, but a study of cratering on their flanks suggested that the Tharsis volcanoes were active 200 million years ago, and some of the flows on Olympus Mons appear to be considerably younger, raising the interesting possibility that volcanic activity might still be ongoing. The southern hemisphere was mapped at a time when dust was still abundant in the atmosphere, which reduced the resolution of the images, but very subdued volcanic edifices that were difficult to discriminate from impact craters were identified, and named Tyrrhenum, Hadriaca and Amphitrites Paterae. These features indicated that volcanism played a part in the planet's earliest history.

The second most interesting feature revealed by Mariner 9 was Valles Marineris. Although one of the 'canali' had been named Coprates, even the best telescopes had given no hint of the canyon system. The suspicion that this structure must have been made by the erosive action of running water resurrected the idea that a considerable amount of liquid water must have existed on the surface at some time in the past; but for liquid water to have been stable on the surface, at that time Mars must have had a much denser atmosphere, since its current surface pressure is below the triple point of water – meaning that if liquid water were to be placed on the surface of Mars, it would not remain in that phase, it would promptly either freeze, or evaporate, or do both simultaneously. However, it was possible that the water that cut this system of canyons was subsurface ice that was melted by geothermal heat associated with the Tharsis volcanoes – the canyon system does, after all, run down the eastern flank of the bulge. Interestingly, cratering studies suggested that the canyon system might be relatively young. There were other 'outflow channels' to the north that appeared to originate in large patches of the chaotic terrain first glimpsed in 1969, and ran north across the line of dichotomy to drain onto the low-lying Chryse Planitia, where there were examples of elevated features that had been etched into 'tear drop' shapes by vigorously flowing water. A study of the distribution of the craters and their sizes, indicated that the surface of Mars is less heavily cratered than our own Moon, which in turn indicated that either erosive phenomena had been at work during at least part of the planet's history, or that the impact flux was less than that in the case of the Moon, which is unlikely.[65] Indeed, with Mars being closer to the asteroid belt, the impact flux could be expected to be rather greater. Over its operating life, and using both its wide-angle and narrow-angle cameras, Mariner 9 took a total of 80 images of Phobos and Deimos before and after orbit insertion. When processed, this data enabled their orbital parameters to be accurately determined; the results confirming that tidal effects are causing Phobos to slowly spiral inward, and that one day it will be broken up and the debris will fall and make more craters.[66,67]

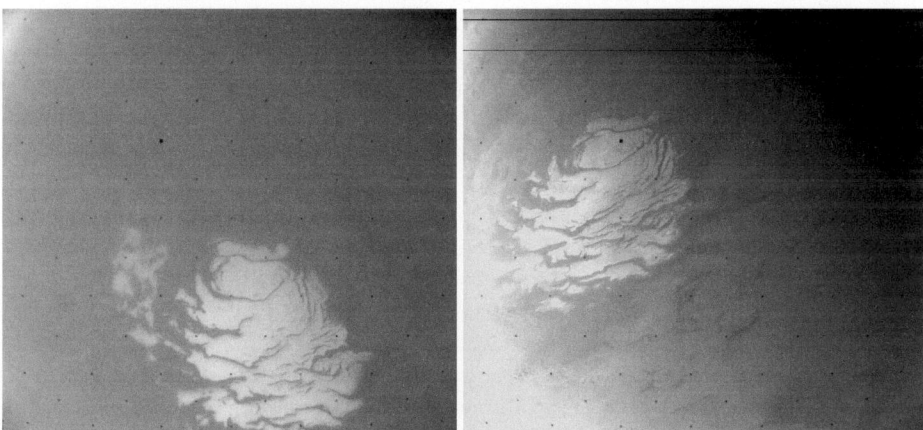

Two images of Mars taken by Mariner 9 showing seasonal changes in the south polar cap.

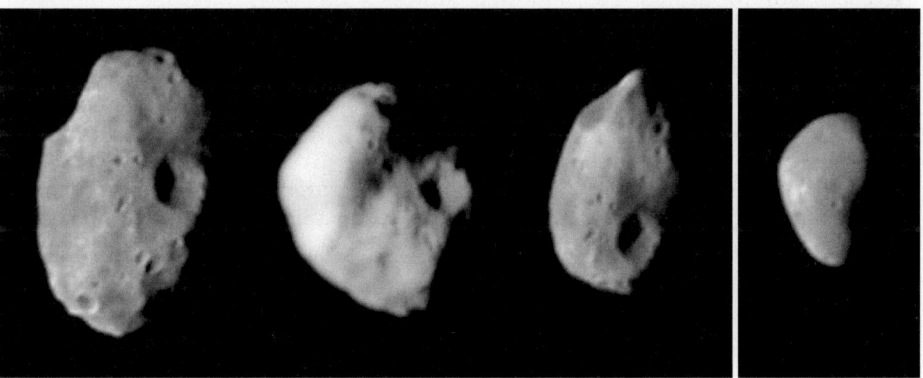

Images by Mariner 9 of the two small moons of Mars: Phobos (left) and Deimos (right).

Mariner 9 also observed both polar caps, witnessing the recession of the southern cap with the progress of the southern summer. The fact that the southern cap held its basic shape as it shrank – even when the evaporation of carbon dioxide should have been at its maximum – proved that it was a combination of 'dry ice' and water ice. The presence of water vapor was confirmed by the infrared spectrometer, which observed absorption features at characteristic wavelengths in the polar clouds. The terrain surrounding the caps was classified as 'laminated' because it comprised many layers (each several meters thick) that were offset from each other in a pattern that suggested that they probably formed as layers of dust intermixed with carbon dioxide ice. The shape of the caps and surrounding terrain showed that the direction of the planet's rotational axis had varied over time. Initial calculations revealed that the tilt of the axis relative to the perpendicular to its orbit can vary from a minimum of 15 degrees to a maximum of 35 degrees with a 160,000-year period.[68] However, more recent studies have suggested that the axial variation might be chaotic. Each 12-hour orbit produced up to two occultation pairs per day, and a total of 260 radio-occultation events were monitored. Entry and exit latitudes ranged over almost all of the planet, ranging from as high as 86°N to as low as 80°S (i.e. essentially over the poles). The temperature profiles in the day-time were found to be quite sensitive to the presence of dust in the atmosphere. In fact, it looked as if the profile obtained by Mariner 4 in 1965 indicated the presence of substantial dust in the atmosphere at the time – which would also explain the low contrast in that probe's pictures. Over the northern polar cap, temperatures dipped as low as –95°C while surface temperatures measured by the infrared radiometer were –123°C, indicating the existence of strong thermal gradients in the lowest several kilometers that were unable to be resolved by the occultation experiment. The interferometric infrared spectrometer measured not only the temperatures over the polar regions, but monitored the evolution of the temperature profiles with the changing seasons. In early summer the temperature profile was almost isothermal, but later in the season a temperature inversion developed near the 10-km level. In all instances, the temperature of the atmosphere exceeded that of the surface. At high latitudes the early morning temperature reached the condensation temperature of carbon dioxide at an altitude of 10 km, indicating that the morning clouds were made of crystals of carbon dioxide ice.[69] The infrared spectrometer took in excess of 20,000 spectra of Mars. During the dust storm, it took measurements of the composition and size of the airborne particulates, as well as the air temperatures and temperature gradients. After the storm, it measured the ground pressures over a wide range of surface features,[70] and this data provided a surface elevation from which it was possible to produce a preliminary topographic map that showed that the lowest lying terrain was in Hellas – where a pressure of 8.9 hPa was measured – and that the most elevated areas were in the Claritas–Tharsis region. By coincidence, one of the measurement points was very close to the summit of Pavonis Mons, where the pressure of 1 hPa indicated an elevation of 12.5 km. The entire southern hemisphere (including the polar cap) was revealed to be several kilometers higher than the low-lying northern plains, thereby posing one of the major mysteries about the planet's history.[71]

Overall, Mariner 9's pictures proved that the dark and bright albedo features that so fascinated telescopic observers are only rarely correlated with actual topography. One of the few correlations is the circular bright region in the southern hemisphere, Hellas, which Mariner 7 showed to match the location of a large impact basin and (as noted) Mariner 9 found to be the lowest point on the planet. Most of the features on telescopic maps were simply impossible to recognize in close imagery, while for others the interpretation was incorrect. For example, the characteristic dark 'tongue' of Syrtis Major, long thought to be a dried lake bed and hence low lying, is in fact an elevated plateau.[72] In 1973, the International Astronomical Union published a map that focused on the newly revealed structures rather than the variation in albedo, and introduced a new nomenclature.[73]

Meanwhile, an attitude control failure had prompted the Mars 2 orbiter to forgo 3-axis stabilization. The spacecraft was unable to take images while spinning for stability, but was able to take other data.[74] Although the dust storm initially masked the surface, the fact that Mars 3 was in an unplanned and extremely eccentric orbit meant that it reached periapsis only once every two weeks, and therefore was able to wait for the atmosphere to clear. Imaging sessions were held on at least four dates: 12 and 14 December 1971, 28 February 1972, and 12 March 1972. The full-disk photographs obtained on 28 February from 18,000 km were of particular interest because the phase angle was one at which Mars had not been previously observed. A haze layer was detected on the limb, as well as clouds consisting of large particles at altitudes of up to 30-40 km.[75] The STEREO-1 apparatus on Mars 3 was turned off on 25 February after a total of 185 hours of operation over 208 days, during which it had collected a total of about 1 megabyte of raw data – which may now sound ludicrously small but was at that time significant for an experiment of its kind. After the spacecraft had entered orbit around Mars, the time available to the instrument had been drastically reduced as planetary observations took priority. Despite the data being marred by errors that required considerable effort to correct using the computing facilities available at that time, STEREO-1 gave the first evidence that the radiation of a type I solar burst is highly directional in space, with the correlation between Earth- and space-detected events declining at Earth–Sun–spacecraft angles near 10 degrees, and vanishing above 20 degrees. In contrast, the time profile of type III events was the same at angles up to 80 degrees.[76,77,78,79]

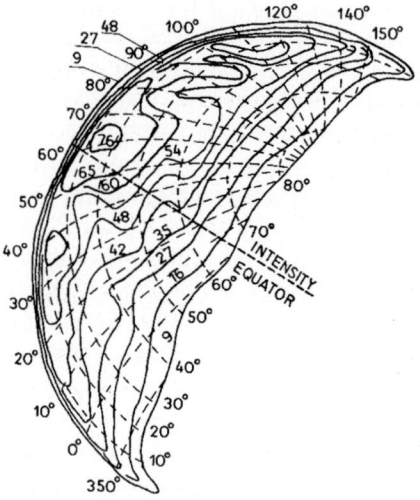

A photometric map of Mars obtained from images taken by Mars 3 on 28 February 1972, shortly before the northern spring equinox. The area covered includes Mare Acidalium. (Courtesy of the International Astronomical Union)

Although Mars 3 returned few images, the Soviet probes provided interesting data on the planet and its environment. In particular: they measured the density of electrons in deep space, noting a marked decrease shortly before reaching Mars; they detected atomic hydrogen, atomic oxygen and carbon monoxide in the atmosphere above an altitude of 100 km; and they measured the temperature on the day-side at the subsolar point as 13°C, and near the antipodal point as –110°C. Near the north pole it was –45°C (probably measured at night, before sunrise). The darker albedo areas of the surface were warmer than the brighter areas and had a greater thermal inertia, indicating that such surfaces were rather less dusty. Furthermore, the inferred temperatures to a depth of about 35 cm proved to be relatively constant not only over the diurnal cycle, but also across a wide range of latitudes, indicating that the near-surface material is a poor heat conductor. The radio occultation measurements by Mars 2 using a double-frequency system were in good agreement with the Mariner 9 results, with atmospheric pressures at the surface ranging between 5.5 and 6 hPa.[80,81,82,83] The infrared photometer on Mars 3 observed a minimal amount of precipitable water vapor in the atmosphere, and even less over the polar region. This contrasted with Mariner 9, which showed 50 times as much precipitable water over the north polar cap in spring, as well as at lower latitudes.[84] But the most intriguing data returned by the Soviet probes concerned the planetary magnetic field. After Mariner 4 reported that Mars had no magnetic field, NASA did not pursue further investigations. The Soviet probes measured the radiation levels at various distances from the planet, and on several orbits a bow shock, a magnetopause and a magnetic field were clearly detected over the sunlit side at strengths 10 times the interplanetary field – the maximum magnetic field strength being approximately 0.015 per cent that of Earth. But the scant data was insufficient to determine other characteristics of this tenuous field, such as its tilt relative to the planet's rotational axis. Unfortunately, the orbits of the probes were less than ideal for such an investigation, and indeed on some orbits no magnetosphere was identified. Furthermore, the Soviet scientists disagreed over the interpretation – some saying it was a planetary magnetic field, and others arguing that the field was of solar origin and that what had been observed was simply an ionosphere on the sunlit side, similar to that found at Venus.[85,86] The dates of the last contacts with the two probes have not been officially reported. One source says that this occurred in March 1972; another that they fell silent almost simultaneously in July.[87,88] Both dates are probably incorrect, as Mars 2 is reported to have collected data relating to the propagation of radio waves through the solar corona from April until August as Mars approached conjunction.[89] The end of the mission was announced on 24 August 1972, after Mars 2 had made 362 orbits and Mars 3 had completed only 20 of its much more eccentric orbits.[90]

After concluding its primary mission, Mariner 9 was still healthy and continued to map at a reduced pace, and by the time Mars entered solar conjunction most of the surface had been documented at a resolution of 1–2 km, and 2 per cent had received scrutiny at a resolution of 300 meters per pixel. The only anomaly during this time was a minor issue with the computer memory. It was powered down from 2 April until early June, because (1) the relative geometry of Mars, the Sun and the probe's orbit would produce a long season of eclipses, and (2) the Deep Space Network was involved in tracking a more bandwidth-hungry mission: the flight of Apollo 16 to the Moon. Mariner 9 was powered up again on 4 June for an 'extended mission', and by early August it had mapped the areas around the north pole on which the Sun had finally risen. Commanding of the spacecraft had to be discontinued from August until mid-October, while the planet was too close to the Sun as seen from Earth for reliable communication. After conjunction, it resumed its mapping of the northern regions, although, owing to the distance from Earth, a maximum of only 12 pictures per week could be returned. On 27 October, during its 698th orbit, it exhausted its supply of gaseous nitrogen for attitude control, and perturbations caused it to enter a slow spin. By the time of the next contact (later that day) it was spinning at a rate of one revolution every 51 minutes. While this might sound slow, it reduced the illumination of the solar panels sufficiently to prevent the batteries from recharging – so much so, in fact, that Mariner 9 was never heard from again. Frustratingly, there were still 15 images stored on its tape recorder. Taking into consideration planetary perturbations and the slow drag of Mars's atmosphere, in about 2020 the spacecraft will enter the atmosphere and its wreckage will crash to the surface. It is believed that 50 years of exposure to the space radiation environment should have completed the thorough sterilization process carried out prior to launch, ensuring that it will pose no threat to the environment. Mariner 9 was an outstanding success. In addition to 167 days of cruise in transit to Mars, it functioned in orbit of Mars for 349 days, as against the planned 90 days, and executed a total of 45,960 commands. It returned 7,329 images and other scientific data for a total of 54 gigabytes – fully 27 times as much as all of the previous missions combined.

Although the Soviets had managed to beat NASA to the Martian surface with the disappointing Mars 2 and Mars 3 landers, the American program had scored two important 'firsts' with the Mariner 9 mission: the first Mars orbiter and the first full mapping of another planet. Moreover, the scientific results of Mariner 9 far surpassed both qualitatively and quantitatively those from the Soviet orbiters. For NASA, it was now time to prepare for the follow-on Viking orbiter/lander missions, whose candidate landing sites were chosen on the basis of Mariner 9's data and publicly announced on 6 November 1972, less than two weeks after that spacecraft had fallen silent.[91]

A FIRST LOOK BEYOND THE ASTEROIDS

After the success of the Pioneer solar probes, scientists and engineers at NASA's Ames Research Center started work on two follow-on probes that would monitor the solar wind and study interplanetary and cosmic-ray phenomena as far as 4 AU from the Sun; they were given the preliminary designations of Pioneer F and Pioneer G. At the same time the Goddard Space Flight Center was studying a 'Galactic Jupiter Probe' that, in addition to performing a similar mission, would explore the asteroid belt and the environment around Jupiter. The spacecraft envisaged by Goddard was to be a simple spinning probe that was powered by a Radioisotope Thermal Generator (RTG) in which the heat from the radioactive decay of small amounts of plutonium was transformed into electric power by using thermocouples. This would mark their first use on a deep-space probe. The use of solar panels would not be feasible because as the energy in insolation decreases with the square of the distance from the Sun, the area of solar panel required to generate a given power increases such that a spacecraft operating at 2 AU from the Sun needs a panel that is four times as large as if it were at Earth's distance from the Sun. Although the Galactic Jupiter Probe was ambitious, it was simpler than Navigator, which had been proposed by JPL, and also Navigator's successor, the Grand Tour in which (as will be described in the next chapter) a spacecraft was to pass in turn by each of the giant planets of the outer solar system. Goddard proposed building four 300-kg probes (plus a spare) and dispatching them in pairs over two consecutive launch windows in a program that would cost an estimated $100 million.[92] In 1967 NASA transferred the Pioneer F and Pioneer G missions from its solar science to its planetary exploration division, which had just had its main project (the Mars-landing Voyager) canceled. In addition, since the National Academy of Sciences deemed investigating Jupiter to be a very high priority objective for US planetary exploration, it was decided to add a Jupiter encounter to the Pioneer mission and, in turn, to cancel the Galactic Jupiter Probe – the rationale for this choice being that the Pioneer mission would be less costly, together with a recognition that Goddard was already busy supporting the Apollo missions. In view of the fact that its spacecraft would venture much further from the Sun than originally envisaged (Jupiter is at 5.2 AU), Ames deleted the solar panels in favor of RTGs. The Pioneer Jupiter project was officially approved by NASA on 8 February 1969, and TRW, which had built the Pioneer solar probes, was awarded the contract to make the new spacecraft.[93,94]

The 258-kg Pioneer resembled the Mariners and Rangers designed by JPL, in that it was built around a hexagonal bus, 35.5 cm deep and 142 cm wide, that contained the electronics and equipment. At the center of the hexagon was the 42-cm-diameter spherical tank for 27 kg of hydrazine fuel, sufficient for a total change in velocity of 187.5 m/s. On top of the bus was the 2.74-meter-diameter high-gain antenna made of an aluminum honeycomb sandwich. On top of the three-legged feed for the high-gain antenna was a medium-gain antenna that took the total height of the probe to 2.9 meters. An omnidirectional antenna was mounted beneath the central bus for use in the early post-launch phase of the mission. There were two 8-W transmitters, only one of which would be used at any given time. Depending on the distance to Earth, data rates up to 2,048 bps could be selected. The bus also supported a pair of triple-segment telescoping booms, each of which had mounted at its far end a stack of two SNAP-19 (System for Nuclear Auxiliary Power) RTGs capable of converting 5 to 6 per cent of the heat generated by the decay of plutonium in order to provide 155 W at launch, and 140 W by the time of the Jupiter encounter. With the booms in their extended position, the RTGs were situated 3 meters from the center of the bus. Another boom, 5.20 meters in length, held the magnetometer away from the spacecraft's metallic mass. Unlike the 3-axis-stabilized JPL vehicles, the Jupiter-bound Pioneers were to be spin stabilized. Attitude determination was by Sun and Canopus sensors, and attitude control would be by three pairs of 1.8–6.2-N

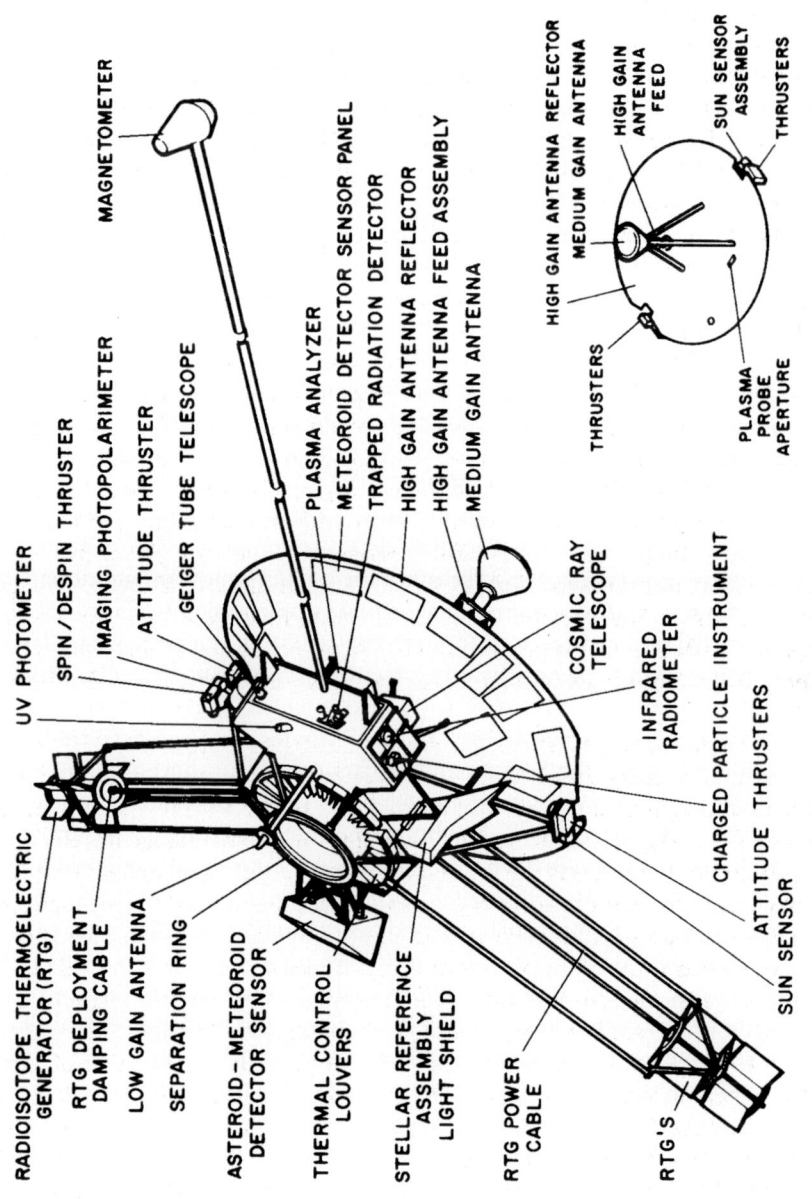

RADIOISOTOPE THERMOELECTRIC GENERATOR (RTG)

UV PHOTOMETER

SPIN / DESPIN THRUSTER

RTG DEPLOYMENT DAMPING CABLE

IMAGING PHOTOPOLARIMETER

MAGNETOMETER

LOW GAIN ANTENNA

ATTITUDE THRUSTER

SEPARATION RING

GEIGER TUBE TELESCOPE

ASTEROID - METEOROID DETECTOR SENSOR

PLASMA ANALYZER

METEOROID DETECTOR SENSOR PANEL

THERMAL CONTROL LOUVERS

TRAPPED RADIATION DETECTOR

HIGH GAIN ANTENNA REFLECTOR

STELLAR REFERENCE ASSEMBLY LIGHT SHIELD

HIGH GAIN ANTENNA FEED ASSEMBLY

MEDIUM GAIN ANTENNA

RTG POWER CABLE

COSMIC RAY TELESCOPE

INFRARED RADIOMETER

CHARGED PARTICLE INSTRUMENT

RTG'S

ATTITUDE THRUSTERS

SUN SENSOR

HIGH GAIN ANTENNA REFLECTOR

MEDIUM GAIN ANTENNA

HIGH GAIN ANTENNA FEED

SUN SENSOR ASSEMBLY

THRUSTERS

THRUSTERS

PLASMA PROBE APERTURE

A line drawing of the Pioneer Jupiter spacecraft.

monopropellant hydrazine thrusters installed on the rim of the parabolic antenna. The thrusters could be pulsed or fired continuously to control the spin rate (in order to maintain the ideal 4.8 rpm), to control motions about the yaw and pitch axes, or to make trajectory corrections. For continuous communication, the spin axis was to be maintained pointing toward Earth. In order to ensure this orientation, the axis of the feed of the high-gain antenna was slightly offset from the spin axis so that if the spin axis drifted the signal received by the probe would oscillate in intensity as it rotated, and an automatic system called CONSCAN (Conical Scan) would adjust the axis to minimize this modulation, thus ensuring that the antenna remained pointing within 0.3 degree of Earth. However, to save propellant, attitude corrections were always to be commanded by Earth. Like the solar Pioneers, the Jupiter-bound probes did not include a computer, relying instead for their operation on commands sent by Earth. There was an onboard cache capable of storing five commands for more complex or time-critical sequences. The trade-off of this commanding strategy was the need for extensive planning, and, in fact, the planning of the Jupiter encounter began 2 years prior to the actual event.[95,96,97,98] A flat hexagonal compartment carrying most of the scientific payload was positioned on the opposite side of the bus from the RTGs.

The suite included no fewer than four instruments to study charged particles and high-energy radiation.[99] The charged-particle detector comprised two systems, one to operate in interplanetary space and the other for use in the Jovian radiation belts. A cosmic-ray telescope was to monitor galactic and solar cosmic rays in transit, and high-energy particles in the vicinity of Jupiter. The radiation trapped in the planet's magnetosphere was to be analyzed by an instrument sensitive to particles across a wider range of energies. A University of Iowa instrument to study charged particles in the Jovian system utilized seven Geiger–Mueller counters that had been calibrated prior to launch to allow for the background of gamma rays produced by the RTGs and Radioisotope Thermal Heaters – the latter being pellets of radioactive material employed to warm the electronics.[100] The boom-mounted helium gas magnetometer operated in eight different ranges to enable it to measure both the interplanetary field and the predicted much stronger Jovian field. A plasma analyzer based on the instrument carried by the Pioneer solar probes was to analyze the solar wind beyond the orbit of Mars for the first time. An ultraviolet photometer was to report on neutral hydrogen and helium in interplanetary space, and also seek evidence of the boundary of the heliosphere (the volume of space immersed in the supersonic solar wind). It would also measure the amount of hydrogen and helium in Jupiter's upper atmosphere, and investigate the possibility of auroras at Jupiter's poles. It was mounted in a fixed orientation that would ensure two scans of Jupiter, the first at 50 Rj (Jupiter radii) and the second at 10 Rj, plus observations of the Galilean satellites in the 5 days leading to the encounter.[101] To investigate why the planet emits more heat than it receives from the Sun, a 2-channel infrared radiometer viewing through a 7.6-cm-diameter Cassegrain telescope was to measure the temperature field across the planetary disk at wavelengths unobservable from Earth.

A vidicon camera of the type developed by JPL cannot be employed on a spinning spacecraft, but an alternative system was devised to secure the first close-up view of

Pioneer 10 about to be enclosed by the aerodynamic shroud of its launch vehicle. The trapezoidal object protruding from the spacecraft to the right is the pod containing the 'Sisyphus' meteoroid–asteroid detector.

Jupiter. The University of Arizona managed the development of the imaging photo-polarimeter, which was to serve three roles: (1) At high sensitivity, it would be used to analyze the zodiacal dust in the inner solar system, and the integrated starlight of the galactic plane, marked by the Milky Way. (2) The polarimetric function would be used during the Jovian encounter to gain data on particles in the clouds of the atmosphere under various phase angles. In particular, 'rainbows' would be sought, because these are characteristic of spherical cloud particles. (3) Photometry using red and blue filters and a mechanism to slowly draw the scanner across the planet's disk would enable Ames to assemble images of the cloud tops. The resolution would be inferior to what a vidicon could have achieved, but better than that attainable by a terrestrial telescope. The instrument had a 2.5-cm-diameter Maksutov optical system whose 'look angle' could be slewed by up to 151 degrees. For imaging, the spacecraft's spin would enable the telescope to collect photometric data across a swath 14 degrees in length, after which the telescope would be slewed by a small increment to enable it to scan an adjacent swath on the next pass. As an alternative, the telescope could be held fixed when the spacecraft was very close to Jupiter and the high relative speed would cause the planet to drift across the field of view. The starting point for an imaging swath could be commanded from Earth in terms of the spin position angle, or be triggered automatically when a light sensor detected the presence of the planet in the field of view. Although there were filters for only two colors (red and blue), a third color (green) could be artificially added later to provide the 'aesthetically pleasing' images that the human eye expected. Significant processing of the raw photometry would be required to produce images, not only when the scanning technique introduced severe distortion along the limb, but also when the relative speed was sufficiently significant to cause large gaps between the individual scan lines. Photopolarimeter prototypes were tested using high-altitude balloons.[102,103,104]

Because these spacecraft would be the first to venture into the asteroid belt, they each had a 'Sisyphus' detector mounted on an external pod. This utilized four 20-cm telescopes, each of which had an 8-degree-wide field of view (with a slight overlap) and fed a photomultiplier. The non-imaging system was designed to detect sunlight

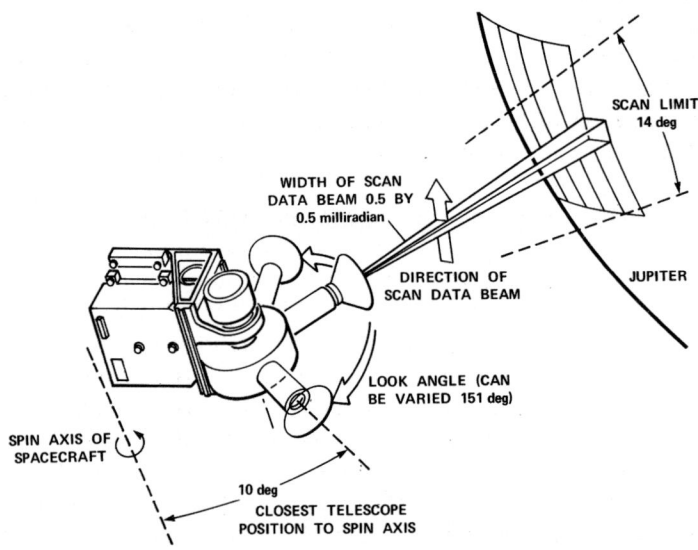

As the Pioneer spacecraft rotated, the imaging photopolarimeter 'nodded' up and down to build up an image of an object by the 'spin–scan' technique.

reflected by particles in the asteroid belt between Mars and Jupiter. An event would be recorded whenever a source was simultaneously detected by any three telescopes. The instrument was sufficiently sensitive to detect kilometer-sized asteroids at large distances, or small particles floating nearby. A related instrument was the meteoroid detector mounted on the outside of the parabolic antenna. It comprised 13 panels with a total area of 0.605 m^2, each having 18 argon and nitrogen-filled cells which would leak when punctured. The detectors on Pioneer F were sensitive enough to record the slow outgassing resulting from a breach caused by a micrometeoroid with a mass of 10^{-9} grams. The cells on the experiment on Pioneer G had thicker walls to investigate larger particles.

And, of course, the spacecraft's radio transmitter would allow celestial mechanics and occultation experiments to be undertaken. In addition to having both probes pass behind Jupiter to investigate the planet's atmosphere, the plan was to arrange for one of them to be occulted by Io to find out whether this satellite (which was evidently able to modulate the emission of radio waves from the planet) possessed an atmosphere.

But the most widely known items on the twin Pioneer probes were gold-anodized aluminum plaques, 15 × 23 cm, showing information on the probe, its mother planet, and the species that designed it. These were included in case of recovery by an alien civilization in the far future, because during their encounters with Jupiter the probes would be accelerated to speeds sufficient to escape the Sun's gravitation and in doing so would enter the interstellar realm and pursue independent orbits around the center of the galaxy. The inclusion of such a plaque was proposed by the science writer Eric Burgess, and the design was provided by Carl Sagan and Frank Drake in December 1971. To provide a scale, it had a schematic neutral hydrogen molecule, whose hyperfine transition gives an accurate measurement of frequency (and thus of time) and length; respectively 1,420 MHz and 21 cm. Immediately below, a diagram with lines showed the locations of 14 pulsars relative to the Sun, with their pulsation frequencies expressed in binary notation with respect to the characteristic frequency of hydrogen, and another line pointed at the galactic center. If correctly interpreted, this diagram could provide the alien civilization with information concerning the position of the Sun at the time of launch, plus the time of launch. A third diagram gave the origin of the spacecraft as the third planet from the Sun. Lastly, there was a sketch drawn by Linda Salzman, Sagan's wife, of the Pioneer probe, a naked man with a raised hand as if in greeting, and a naked woman. Two reference marks on the rightmost side showed the height of the human figures to be approximately 8 times the 21-cm unit. This drawing drew considerable criticism because the human figures were naked, because (despite an attempt to make them 'pan-racial') they appeared to be white and of Caucasian descent, and because the woman's pose implied that she was passive. But these were purely human issues, unlikely to be of significance to an alien. The 'message' was designed to be easily interpreted, but when it was shown to a number of leading scientists some were mystified! The plaque on each spacecraft was attached to the struts supporting the high-gain antenna dish, with its engraving facing inward to protect it from micrometeoroids and other sources of erosion.[105,106]

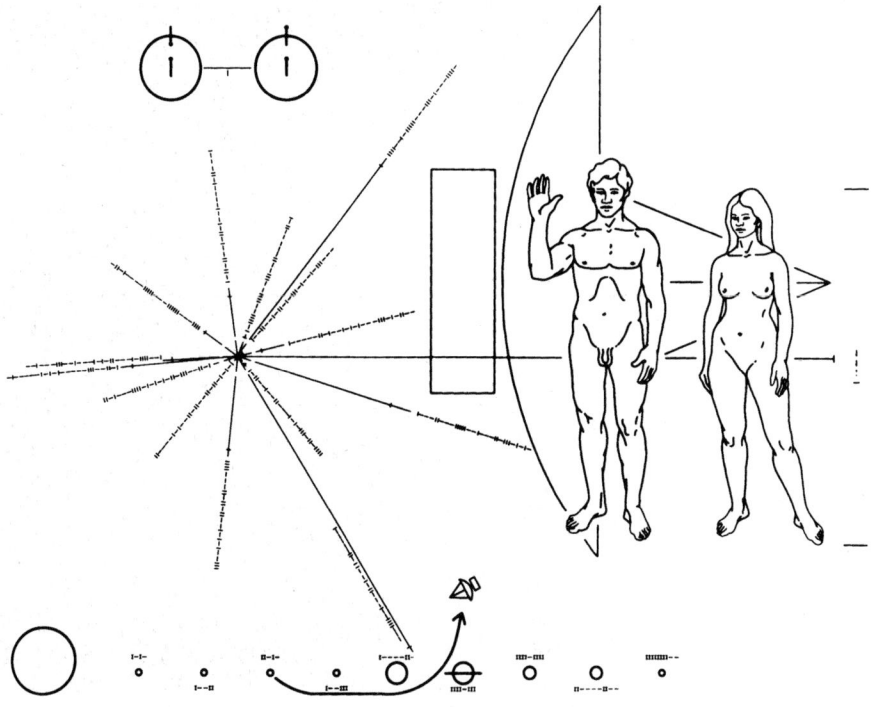

The Pioneer plaque was designed as a message to any extraterrestrials that might recover the spacecraft in the distant future. In addition to a man and a woman and the outline of the vehicle, the plaque defines the position of the solar system relative to a number of radio pulsars, the position of Earth in the solar system, and the trajectory. The hydrogen molecule (at top) is used as a standard measure of distance and time.

In December 1971 an Atlas–Centaur launcher was erected at Cape Canaveral, and in January 1972 Pioneer 10 was airlifted from California, integrated with its RTGs and fueled. In order to attain the speed required to reach Jupiter, it was to receive a final boost by a TE-M-364-4 solid-fuel 'kick stage' giving a thrust of 66.7 kN. This was derived from the engine that was used to brake the Surveyor craft that landed on the Moon in 1966–1968.[107] The launch window ran from 25 February to 20 March. The first attempts on 27 and 28 February and 1 March were scrubbed owing to high winds in the upper atmosphere. Then at 01:49 UTC on 2 March the first spacecraft designed to venture beyond the orbit of Mars was successfully put on a direct-ascent trajectory. Prior to firing the third stage, the vehicle was spun up to 21 rpm to even out any irregularity in the solid rocket burn. The speed of 14.356 km/s made Pioneer 10 the fastest craft yet to leave Earth. Once it had been released, it deployed the booms with the RTGs and the magnetometer, which, owing to the conservation of angular momentum, had the effect of slowing its spin to 5 rpm. Its heliocentric orbit ranged between 0.99 and 5.97 AU, and was inclined 1.92 degrees to the plane of the ecliptic. In the following 10 days it powered on and tested all its instruments, and

In order to protect the plaque from micrometeoroids, it was mounted facing inward on the struts that supported the high-gain antenna.

began to report data on the environment. As had often happened in the past, the Canopus sensor took a while to find its target star. A course correction of 14 m/s on 7 March refined the trajectory for Jupiter, and a second correction of 3.3 m/s on 26 March arranged the occultation by Io.[108] During the first 20 months the primary task was to be an analysis of interplanetary space between 1 and 5 AU, including the 'terra incognita' beyond the orbit of Mars. Galactic cosmic-ray fluxes at increasing distance from the Sun were compared to those measured in the vicinity of Earth by Explorer 35 in lunar orbit and the IMP (Interplanetary Monitoring Platform) 5 and 6 probes.[109] Four powerful solar storms occurred between 2 and 7 August 1972, and Pioneer 10 collected data in cooperation with is predecessors orbiting near 1 AU. At 2.2 AU, which was Pioneer 10's position at that time, the speed of the solar wind was half of that recorded close to Earth by Pioneer 9.[110,111]

One concern for this pioneering mission was whether the spacecraft would be able to pass safely through the asteroid belt. In mid-July 1972 it reached the inner edge of the belt, and, to everyone's relief, in February 1973 emerged unscathed. During the crossing, it approached within 9 million km of two known asteroids: on 2 August it passed a 1-km-body discovered during the course of the Palomar–Leiden survey and as yet unnamed, and on 2 December it flew by asteroid (307) Nike, which was about 58 km in size. Neither asteroid was within the detection range of the probe's optical sensors.[112,113,114] The asteroidal experiments gave mixed results. The optical sensors reported a total of 283 events, but a thorough analysis failed to produce any meaningful orbit solutions. The principal investigator for Sisyphus proposed that his instrument had discovered a hitherto unsuspected population of dark interstellar dust grains which he called 'cosmoids', and that these not only outnumber dust native to the solar system but might explain the mystery of the so-called 'missing mass' of the universe.[115] But most scientists maintained that the instrument had malfunctioned, and most of the events it noted were 'noise'. Whereas the brightness of the zodiacal light implied by optical transits of particles was much greater than that indicated by the imaging photopolarimeter, observations of celestial objects such as Jupiter and the star alpha Centauri confirmed that Sisyphus was working properly. Another result of the experiment was that an almost constant event ratio was recorded from 1 to 3.5 AU from the Sun, a slight increase near 2.6 AU, and an abrupt cut-off beyond 3.5 AU, with the particles detected being mostly in the 0.1 to 1 mm size range, but a few in the 10 to 20 cm size range. The photopolarimeter proved the asteroidal origin of the zodiacal light, because its brightness decreased sharply between 2.8 and 3.3 AU, becoming negligible relative to the background of galactic starlight beyond that distance. The spacecraft also established beyond doubt that the Gegenschein, or counterglow, is a phenomenon originating in the asteroid belt and not, as some had argued, dust residing at the Lagrangian point of the Sun–Earth system lying beyond Earth.[116,117] Free of interference from the zodiacal light, the photopolarimeter was able to obtain maps of the background starlight. Structures of the galactic disk such as the nearest spiral arms were readily recognizable.[118,119] An electronic fault in the data acquisition system of the meteoroid detectors meant that the data from 126 of a total 234 cells was lost. The remaining cells recorded only 55 hits between launch and the initial approach to Jupiter, but the density of particles in the asteroid belt proved that the dust in this region of space does not pose a physical threat, which was good news for the more ambitious missions that JPL was planning to the outer planets. Ten cells were punctured over an interval of 61 hours during the Jovian encounter, but the reason for this was not immediately evident.[120]

As Pioneer 10 receded from the Sun, the magnetic field strength, the solar wind density and the particle fluxes were seen to decrease as expected, but the predicted increase of lower energy galactic cosmic rays did not occur, which indicated that the radius of the heliosphere was much larger than believed. Some theories had posited that the boundary with interstellar space might be somewhere between the orbits of the outer planets, but this new data suggested that it could possibly be as far away as 100 AU.[121,122]

The first hint of Jupiter's presence was noted about 360 Rj from the planet, when radiation sensors detected an increase in particle fluxes. In fact, Jupiter proved to be the source of bursts of high-energy electrons that were detectable as far 'inward' as the orbit of Mercury. The encounter activities officially began on 6 November 1973, with a photopolarimeter imaging session from a range of 25 million km. Two days later, Pioneer 10 crossed the orbit of Sinope, which at that time was the outermost of Jupiter's known satellites, although the moon was nowhere near that point. At 20:30 UTC on 26 November, when the spacecraft was at 109 Rj (equivalent to 6.4 million km), it reported a decrease in the speed of the solar wind from 420 to 250 km/s and a

A Pioneer 10 'spin–scan' image of Jupiter, with the Great Red Spot at the terminator.
The satellite Io is at upper right and is casting its shadow on the planet's cloud top.

100-fold increase in its temperature in passing through the bow shock front where
the solar wind piled up against the sunward side of the planet's magnetosphere.
One day later, after traversing the magnetosheath region beyond the bow shock, it
found itself in the Jovian magnetosphere. The existence of this region of space, and
the fact that it contained a radiation belt, had been known since the 1950s when
Jupiter was realized to be a source of radio noise and bursts. In fact, the Jovian
radiation belts were discovered before the Earth's, which were not identified until
the United States put up its first satellite in 1958. If the Jovian magnetosphere was
in proportion to Earth's, then it would have been expected to extend to 53 Rj at its
subsolar point, but Pioneer 10 established it to be almost twice as large. It was
nevertheless dynamic, expanding and contracting in response to gusts in the solar
wind. Indeed, when the spacecraft was at 55 Rj it found itself back in the
magnetosheath region, and it had closed to 48 Rj before the magnetosphere
inflated out beyond it again.[123,124,125]

On 2 December, 2 days prior to the flyby, the photopolarimeter images began to
yield a better resolution than terrestrial telescopes. As the spacecraft penetrated
deeper into the Jovian magnetosphere, the energetic charged particles saturated the
detectors of some of the instruments and induced erratic behavior in the electronics
of the photopolarimeter, disrupting its imaging, but it was reinstated in time to take
images of the planet's terminator. Although at some 200,000 km the point of closest
approach had been expected to be safe, conditions in the magnetosphere prompted
concern that Pioneer 10 might not survive the intense radiation belt. Extrapolations
showed that if the charged-particle fluxes continued to increase at the observed rate,

then 2 hours before periapsis its electronics would be totally disabled. Sixteen hours before Jupiter, it passed within 1.42 million km of the outer Galilean moon Callisto, followed by Ganymede, the largest moon, 4 hours later at a range of 440,000 km. It passed Europa at a range of 330,000 km 6.4 hours before the Jovian encounter. The photopolarimeter was able to image them all. The best satellite image taken by this spacecraft was of Ganymede from 750,000 km at a resolution of 400 km, but apart from showing the existence of dark terrain in the south polar and equatorial regions, and a bright north polar region, it did not reveal anything about the largest moon of the solar system. Europa was too distant to reveal interesting markings, but overall it resembled Ganymede.[126,127,128,129] Ultraviolet photometry indicated the presence of a cloud of neutral hydrogen and oxygen molecules in space at Europa's distance from Jupiter. As Europa was known to be covered by water ice, it seemed likely that the cloud consisted of molecules that had been ejected ('sputtered') from the surface by micrometeoroid impacts.[130] The spacecraft flew by Io at a range of 340,000 km 3.3 hours prior to periapsis, but the photopolarimeter succumbed to the radiation and was unable to image it at decent resolution.

A little more than 2 hours from periapsis, the disk of Jupiter entered the field of view of the infrared radiometer – the only instrument to have been dormant during the interplanetary cruise – and during the 82 minutes that the planet took to cross the field of view, the instrument collected data to make a temperature map at each of its two wavelengths.[131]

The small red spot on Jupiter imaged by Pioneer 10.

The magnetic field strength had been almost constant during the approach, but at 3.5 Rj it began to rise dramatically, and at periapsis it attained a value comparable to that on the surface of Earth. To a first approximation it was a dipolar field inclined at an angle of just less than 15 degrees to the planet's axis of rotation.[132,133] Several instruments had been monitoring energetic particles and radiation belts. At distances beyond 20 Rj, the particles seemed to be confined to a thin disk coincident with the plane of the magnetic equator, with this 'magnetodisk' appearing to wobble (from the spacecraft's perspective) as the planet followed its 10-hour axial rotation. Within 20 Rj, the particle fluxes increased rapidly, as would be expected for such particles trapped inside a dipolar field. Local decreases in the flux of particles corresponding to crossing the orbits of the two inner Galilean satellites were also noted.[134,135,136,137] Although the radiation peak approached the design maximum for the spacecraft, most instruments survived essentially unscathed. Two cosmic-ray detectors were briefly saturated, and the photopolarimeter endured glitches, but the only instrument to be written off was the non-imaging Sisyphus detector, whose optics were totally fogged by the charged particles.

At 02:26 UTC on 4 December 1973, Pioneer 10 passed 132,000 km over Jupiter's cloud tops at a planetocentric distance of 203,240 km, or 2.85 Rj, while traveling at a relative speed of about 35 km/s. It was 1 minute ahead of the predicted schedule. An analysis based on tracking from mid-October to the end of December established that the estimated mass of the planet based on how it perturbed the orbits of several asteroids was slightly less than the true figure.[138] Some 10 minutes after the point of closest approach, the spacecraft passed within 30,000 km of the small inner satellite Amalthea, although without crossing inside its orbit. Six minutes later, when some 550,000 km 'beyond' Io as viewed from Earth, the spacecraft was occulted for about 1 minute. The radio occultation revealed the moon to possess both an atmosphere

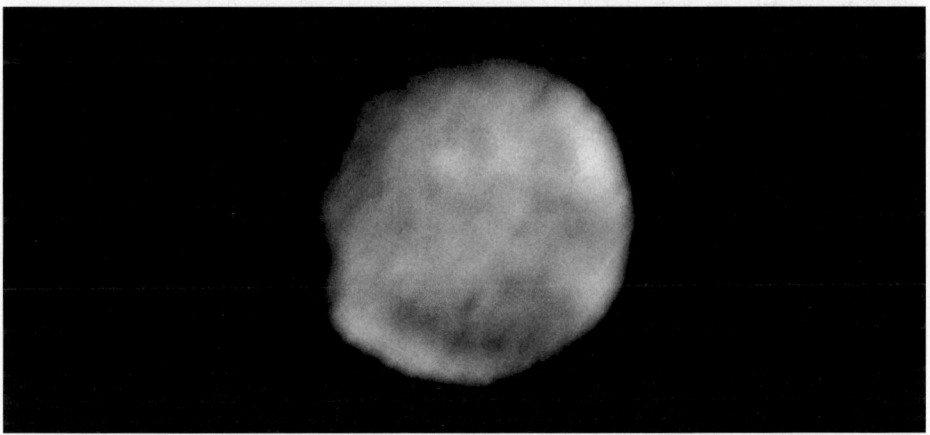

A 'spin–scan' image taken by Pioneer 10 of Jupiter's satellite Europa. Just six years later, far better Voyager images would reveal this to be one of the most fascinating moons in the solar system.

and an ionosphere that reached to an altitude of 700 km, and were probably made of gaseous and atomic sodium, the spectral signatures of which had been discovered by terrestrial telescopes in 1972. The evanescent atmosphere was at most 1/20,000th as dense as Earth's, with a surface pressure in the range 0.1 to 0.01 Pa.[139] Furthermore, the ultraviolet spectrometer identified a cloud of hydrogen atoms drawn out around Io's orbit for about 120 degrees, like the tail of a comet, and this overlapped with a cloud of atomic sodium that had recently been discovered from Earth, and which, while much less dense, traced out the full orbit. The discovery of an ionosphere around Io offered an explanation for why the Jovian radio emissions appeared to be modulated by the period of this moon's orbital motion, because a conducting ionosphere might be connected by electrical currents to the planet's polar regions. However, the source of the tenuous atmosphere and the torus of sodium remained a mystery. One theory resulting from the flyby suggested that because Io orbits within the most intense part of the planet's radiation belt, the impact of the electrons and protons might sputter sodium atoms from the surface of the moon with sufficient energy to escape its weak gravity.[140,141,142]

Some 78 minutes after the moment of closest approach, Pioneer 10 passed behind Jupiter's limb, and this radio occultation was used to investigate the ionosphere and atmosphere of the planet. It was occulted near the evening terminator at a latitude of 27.7°N, where the Sun was 9 degrees above the horizon, and reappeared almost an hour later near the morning terminator at 58°N, where the Sun was 4 degrees below the horizon. Unfortunately, Jupiter rotates so rapidly that its polar diameter is 0.94 of that across its equator, and small errors in its assumed geometrical shape could result in significant errors in the temperature and pressure profiles. Moreover, owing to the distance of the planet from Earth, small errors in the measured position of the probe could also introduce large additional uncertainties. There was a concern, in fact, that the occultation technique that had proved so successful in sounding Mars and Venus would not be useful for investigating the atmospheres of the giant outer planets. But the experiment was able to investigate the dynamics of the atmosphere, revealing the existence of a jetstream at the 150-hPa level, in addition to vertical shear winds and convective regions. The infrared radiometer took better data, giving a very detailed atmospheric temperature profile that conflicted with the one inferred from terrestrial measurements and from how Jupiter occulted stars. The data also confirmed that the planet radiated more heat than it received from the Sun. The fact that the same amount of heat was radiated in the polar regions as at the equator meant that the temperatures were the same for a given pressure level; in fact, at any given level the temperature differences did not exceed 3°C. The temperature varied with altitude, however. The atmosphere is divided into a series of latitudinal 'belts' and 'zones'. The darker belts were found to be warmer and hence at lower altitudes, while the brighter zones were cooler and higher. The albedo difference was because the ammonia clouds that formed in the ascending flows were white and their absence in the descending flows provided a deeper view. Furthermore, the day and night hemispheres exhibited the same temperatures. Finally, the radio occultation also established that the ionosphere of the planet extended to an altitude of 3,000 km.[143,144,145,146,147,148,149] Helium could not be detected by conventional spectroscopic techniques from Earth, but its presence in the atmosphere was confirmed by the spacecraft's ultraviolet spectrometer, thereby proving that the planet has a bulk composition similar to that of the Sun, as Rupert Wildt had argued four decades previously.

Between 6 November and 31 December, Pioneer 10's polarimeter took in excess of 500 images of Jupiter's atmosphere, the best of which had a resolution of 320 km per pixel. These revealed the finer structure of the belts and zones, and of the Great Red Spot, and also facilitated viewing the disk of Jupiter at perspectives that are impossible from Earth. The anticlockwise motion of the Great Red Spot was well observed, detailing its 6.5-day rotation. Some 18 months earlier, terrestrial observers had noted the appearance of a smaller red spot, and this was present in several of the probe's images. Bright nuclei, plumes and festoons at the boundaries of the belts and zones were observed in much greater detail than was possible from Earth. Despite suffering radiation-induced problems, the instrument was able to operate in its polarimetric mode and take some data on the shape, size and refraction index of the cloud particles. By tracking no fewer than 87 distinct features in the atmosphere, it was possible to create a complete velocity profile for features in the latitude range 50°N to 55°S. In addition to documenting the relative velocities of the jetstreams, this established that there was no meridional component, meaning that features that developed at a given latitude remained at that latitude.[150,151,152,153,154,155] The imaging and radiometric data suggested that the bright zones and the Great Red Spot were convective regions that were ascending in the atmosphere, and the dark belts marked descending flows. One early idea was that the Great Red Spot was a manifestation of a 'Taylor column' (i.e. a cylinder of stagnant air stationed over a surface feature), although it was evident that in this case the surface must be far beneath the clouds. The new data showed the Great Red Spot to be a high-pressure vortex with a central

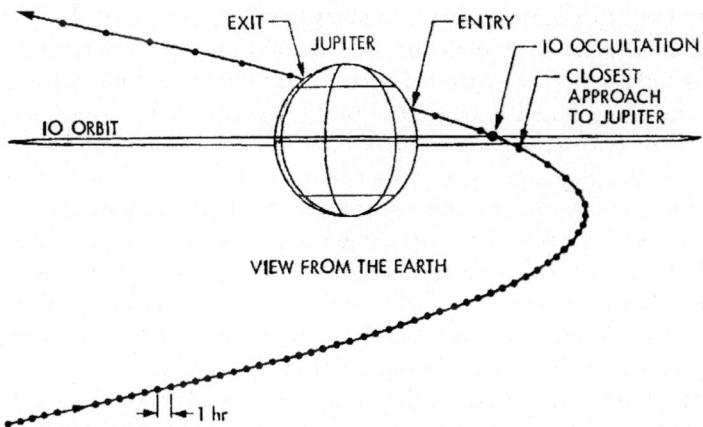

The geometry of Pioneer 10's flyby as viewed from Earth. Minutes after making its closest approach to Jupiter, the trajectory provided radio occultations by Io and the giant planet.

rising column that tends to subside around its periphery.[156,157] In some respects, it is similar to a terrestrial hurricane, although the center of a hurricane is a low-pressure region. That is, a hurricane forms over a tropical ocean when solar energy causes evaporation, and as the heated air rises it creates a low-pressure region into which the surrounding air is drawn at sea level. When a storm makes landfall and loses its source of warm moisture, it weakens. The longevity of the Great Red Spot would seem to derive from the fact that on Jupiter there is no 'land' to dissipate its energy.

As Pioneer 10 receded it continued to take imagery and other measurements, but the encounter was declared to be complete on 2 January 1974 when the spacecraft exited the magnetosphere by crossing the shock wave at 189 Rj.[158] The flyby had drastically modified the probe's solar orbit, and its new eccentricity of 1.737 was sufficient for it to escape the gravitational influence of the Sun. On its new trajectory, it would enter a galactic orbit that was more circular than that of the Sun.[159] The solid-fuel rocket stage that had augmented the Atlas–Centaur had followed its payload to Jupiter, and was similarly ejected from the solar system. At the time of launch, it was estimated that the Deep Space Network would be able to track Pioneer 10 until mid-1980, at which time it would still be within the orbit of Neptune, but the ongoing program of upgrades to that facility increased the spacecraft's 'trackability', first to its maximum projected lifetime of 10 years (when it would be more than 30 AU from the Sun) and then to its actual life, which would depend on how long it could continue to operate on the diminishing power delivered by its RTGs.[160]

While Pioneer 10 was flying through the asteroid belt, its twin was being prepared for launch. Pioneer 11 was identical to its predecessor with the exceptions that its Geiger–Mueller counter was greatly improved and it carried an additional flux-gate magnetometer for the measurement of strong magnetic fields. Its launch on 9 March

As Pioneer 10 looked back after the encounter it provided a view of Jupiter from a perspective never before seen.

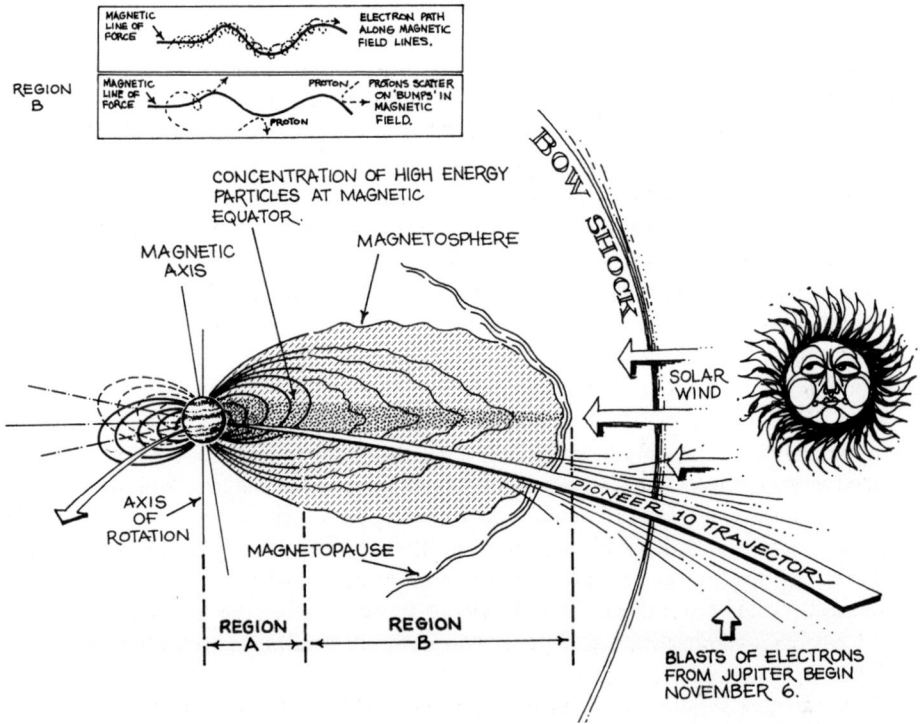

The model of the Jovian magnetosphere inferred from the Pioneer 10 data.

1973 marked the introduction of avionics to enable the Centaur to process wind-speed data in real time so as to improve its trajectory.[161] The spacecraft was injected into a heliocentric orbit ranging between 1.0 and 6.12 AU, inclined at 2.97 degrees to the ecliptic. If something were to go amiss with Pioneer 10 during its Jovian flyby, Pioneer 11 would back it up; but if the first encounter was successful then Pioneer 11 was to venture even closer to the planet on a polar flyby that would deflect it above the ecliptic and back across the solar system to encounter Saturn in 1979. In response to criticism that such an adventure would jeopardize funding for the more advanced Mariner Jupiter–Saturn (later to be named Voyager) mission that JPL was planning, Ames pointed out that its spacecraft would serve as a pathfinder to that mission – just as Pioneer 10 had in the asteroid belt and at Jupiter – and in particular would determine the degree to which Saturn's ring system posed a risk to a spacecraft. As JPL was hoping to use Saturn as a stepping stone to the planets beyond, the prospect of 'proving' the necessary route through the rings ensured its support. But Pioneer 11 first had to reach Jupiter safely. No sooner was it on its way than it faced a problem when one of the two RTG booms failed to extend, which, by conservation of angular momentum, made the spacecraft rotate at a faster rate than desired. The boom was partially extended by pulsing the thrusters, and the probe was then reoriented to expose more of the boom to the Sun, whereupon its extension was abruptly completed.

Next, the malfunction of a spin-control thruster prompted concern that if this were to be used it would not shut off. It was therefore decided to allow Pioneer 11 to spin at a higher rate than planned. And then a critical component in the primary transmitter failed, and its backup had to be activated. In addition, shortly after launch one of the photomultiplier tubes of the asteroidal dust detector cracked and failed, with the implication that the instrument would now have a much higher rejection ratio for 'sightings'. On 11 April 1973 Pioneer 11 executed its first course correction to refine its aim for Jupiter, but the nature of the flyby remained to be decided. The spacecraft emerged unscathed from the asteroid belt in mid-March 1974. The thicker-wall gas cell instrument had recorded 20 punctures since launch, seven of which occurred in the belt. This number of impacts was about half that recorded by the first probe, implying that an almost equal number of small and large particles were present. Apart from spurious radiation-induced events, the instrument went on to record only two hits near Jupiter.[162,163] On 19 April 1974, following Pioneer 10's success at Jupiter, Pioneer 11 consumed 7.7 kg of propellant in making a 42-minute 36-second course correction for a velocity change of 63.7 m/s to adopt a trajectory that would take it three times closer to the planet than its predecessor. Although this would require Pioneer 11 to penetrate more deeply into the radiation belts, the polar flyby meant that, by rapidly crossing the equator from south to north, it would spend the minimum of time in the most intense radiation. In fact, notwithstanding the closer flyby, it was expected that Pioneer 11 would receive a smaller radiation dose than its predecessor. As an additional bonus, the polar flyby would provide an opportunity to investigate the planet's polar regions, where the meteorology was expected to differ from that near the equator.[164,165]

Inbound to Jupiter, Pioneer 11 took this image of the planet and the Great Red Spot.

On 7 November, as Pioneer 11 approached the Jovian system, it crossed the orbit of Sinope, and on 18 November it began to take photopolarimetric data and images of the planet. At 3:39 UTC on 25 November it crossed the bow shock at 109.7 Rj, entering the magnetosphere at 97 Rj. But the magnetosphere proved to be even more dynamic than the previous year, and the bow shock washed over the spacecraft for several hours at 92 Rj, and then again at 77.5 Rj on 28 November. The data taken by Pioneer 11 at high Jovian latitudes, together with that from its predecessor near the equatorial plane, provided a 3-dimensional map of the magnetosphere, showing it to have an almost hemispherical shape on the sunward side, in contrast to the flattened shape that had been inferred from the Pioneer 10 data.[166,167] The imaging sequences had been designed to document the Great Red Spot when it was near the center of the visible hemisphere, where it would be evenly illuminated and not distorted by perspective. A computer simulation to recreate the timing of the spot's transits was prepared to assist in planning. Owing to this improvement, Pioneer 11's pictures of the Great Red Spot had four times the resolution of the best by Pioneer 10 – which, in any case, were degraded by radiation. The blue filter showed the vortex structure, while the red filter showed the finer details. Pioneer 11 was also able to photograph one of the 'white ovals' in the southern hemisphere, and confirm the disappearance of the small red spot imaged by Pioneer 10 in the northern hemisphere.[168,169] There were glitches as the stepping motor of the imaging telescope occasionally jammed and produced a series of identical scans, but owing to the different trajectory and instructions sent from Earth to reject them, very few radiation-induced spurious commands affected the instruments. On 2 December, the day before the flyby, the spacecraft began to cross the orbits of the Galilean moons, although of course it was far south of the plane in which they travel: the closest approach to Callisto being at a range of 772,000 km, 21 hours prior to its closest approach to the planet; Ganymede (674,000 km, 7.5 hours); Io (327,000 km, 1.7 hours); and Europa (590,000 km, 1.1 hours).[170] It took images and polarimetric data of all except Europa, which was in the planet's shadow when the opportunity arose. As a radiation glitch had prevented Pioneer 10 from imaging Io, Pioneer 11's observations were very welcome. In total, over 200 images of the Galilean moons were taken by the two probes, many of them showing the moons alongside the planet, a detailed analysis of which enabled their ephemerides to be improved to an accuracy of several hundred kilometers to help in planning the Mariner Jupiter–Saturn (Voyager) missions that were intended to observe them in close-up.[171,172] Pioneer 11 penetrated within the orbit of Amalthea, but although it flew within 120,000 km of this satellite 36 minutes after the Jovian flyby, it did not study it. The encounter provided a new perspective on the planet's radiation belts, and the discovery that the counts of particles at high latitudes were comparable to those at lower latitudes challenged the 'thin magnetodisk' model.[173] Shortly before the point of closest approach to the planet, it flew within 6,000 km of the torus of sodium that marks Io's orbit, and recorded the highest electron flux in a particular range of energies encountered by either Pioneer spacecraft.[174] At 1.7 Rj an anomalous depletion of energetic particles was detected that could not be linked to any known satellite. A suggestion was put forward by researchers that Jupiter may possess some kind of 'dark' – and hence difficult to observe from Earth – ring that absorbed such particles.[175]

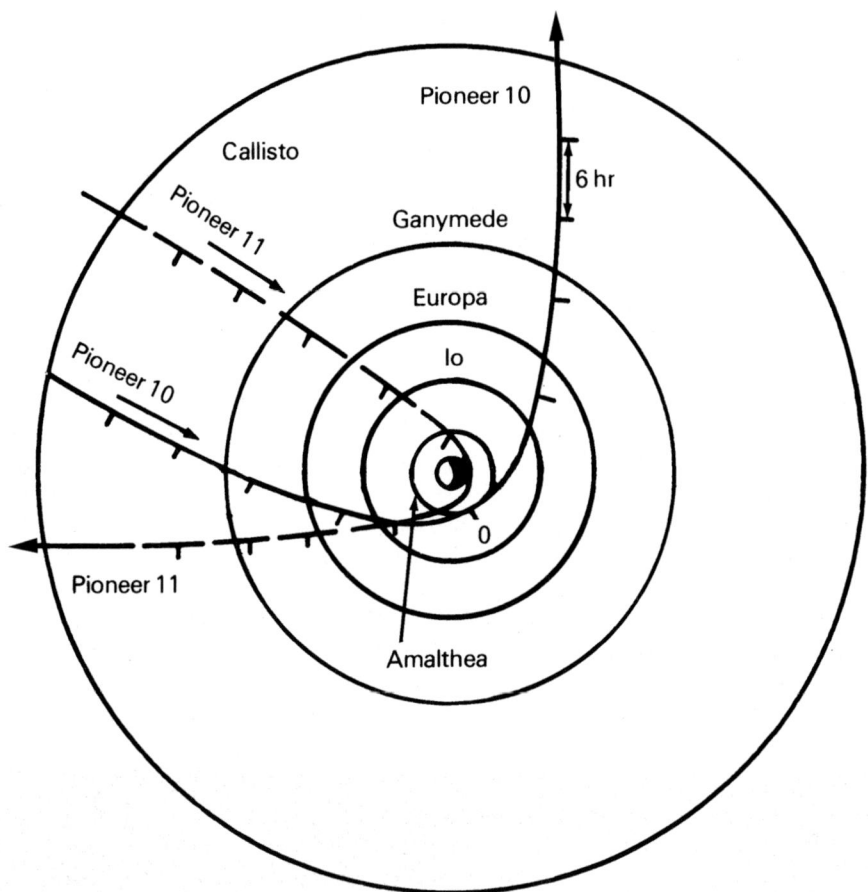

The trajectories of Pioneers 10 and 11 in the Jovian system as viewed from 'above' the planet's orbit. In the case of Pioneer 11, the trajectory was designed to deflect it on to Saturn, which at that time was on the opposite side of the solar system.

At 05:02 UTC on 3 December Pioneer 11 passed behind Jupiter's disk at 79.5°S. It was traveling so fast that the signal gave just 2 minutes of data on the ionosphere and upper atmosphere, but this was sufficient to verify the temperature profile by its predecessor. At 05:22 UTC the vehicle reached its closest point of approach, passing 42,500 km above the planet's cloud tops at a planetocentric distance of 113,850 km (1.59 Rj).[176] After 43 minutes in occultation it reappeared at 20°N, but the radiation had affected a critical component of the radio system and the egress data had to be ruled unusable.[177]

Images taken shortly after Pioneer 11 had begun to withdraw documented the north polar region at a best resolution of 270 km/pixel. Working toward the pole from the North Temperate Belt, the regular belt and zone structure yielded to a mottled aspect of an irregular collection of vortices, bright ellipses, rings and spots that were more or less aligned along the parallels. At the pole the cloud tops were much lower, and were covered by a thicker atmosphere that was transparent in red light and opaque in blue light.[178] The infrared radiometer had obtained an almost complete map of the planet at each of its two wavelengths, one inbound and the other while withdrawing – although part of this was lost to spurious commands.[179] Measurements obtained by Pioneer 11's two magnetometers during the closest part of the flyby recorded many fine details of the structure of the magnetic field, and revealed that the field is offset from the center of mass.[180] Departing on a trajectory steeply inclined to the ecliptic, the spacecraft reported a miscellany of bow shocks before finally emerging into the realm of the solar wind on 9 December at a distance of 95 Rj. Its orbit now ranged between 3.72 AU (north of the ecliptic) and 30.14 AU (south of the ecliptic), with an ecliptic crossing in 1979 that would enable it to intercept Saturn. With the exception of the Sisyphus detector (which had begun to generate spurious data and was turned off) most of the instruments were fully functional, and the engineers had every confidence that their spacecraft would reach its next objective.

Meanwhile, Pioneer 10 was crossing the outer regions of the solar system. In mid-1975 the variable Cepheid stars gamma Pegasi and delta Ceti happened to be in the field of view of the ultraviolet photometer, which took a 'light curve' for both.[181] In March 1976 the probe encountered the 'tail' of the Jovian magnetosphere, revealing this to extend at least as far out as Saturn's orbit.

After a small burn on 2 December 1975 to calibrate the performance of its engine, Pioneer 11 made a 30 m/s maneuver on 18 December 1975 to correct its trajectory

This Pioneer 11 view of Jupiter ranges from the equator almost to the north pole (the pole itself is in darkness) and shows a wealth of atmospheric structure. This is one of the most scientifically important images taken by the Pioneers of this planet.

for any errors amplified by the Jovian flyby. It reached perihelion on 3 February 1976. A preliminary Saturn-targeting maneuver of 16.6 m/s was performed on 26 May 1976. In August 1976 the probe reached its maximum elevation of 1.1 AU above the ecliptic (equivalent to an ecliptic latitude of about 15 degrees) but the opportunity to investigate the solar wind in this region of space was lost because the plasma analyzer had malfunctioned following the Jupiter encounter. However, good data was returned by the magnetometers, which revealed the interplanetary magnetic field out-of-ecliptic to be nearly always pointing away from the Sun (although it would reverse its polarity with each 11-year cycle). The field near the ecliptic can point either toward or away from the Sun at any time during the cycle, depending on the position of the 'current sheath' that separates the regions of opposite polarity.[182] After several attempts, the plasma analyzer was successfully revived on 3 December 1977.[183]

Before Pioneer 11 could be precisely targeted for the encounter, the scientists and engineers had to decide where they wished to pass through the plane of Saturn's ring system. There were two possibilities. One involved an inbound crossing between the planet and the C ring. This would enable the instruments to collect important data on the magnetic field close to the planet and on possible interactions between charged particles and the rings, but in 1969 French astronomers had found a very faint ring (the D ring) extending from the C ring almost to the atmosphere, and terrestrial radar had recently shown that the rings were not made solely of smoke-like micron-sized particles, as had been thought, but included centimeter-to-meter-sized icy bodies that would pose a major risk to a spacecraft.[184] Nevertheless, some scientists argued for such a close crossing as it would yield much better data of the gravitational field and provide insight into Saturn's oblateness, and (in combination with future Voyager data) might even allow the mass of the rings to be estimated. Finally, a pass near to the planet could be used to set up a close flyby of Saturn's mysterious largest moon, Titan; this would not only provide imaging and radiometry but would also enable its mass to be estimated.[185] The second possibility was to pass through the ring plane near a planetocentric distance of 2.87 Saturn radii (Rs), just beyond the A ring. While this would require traversing the faint outer ring (the E ring) that American astronomers had found in 1966, the risks would be less, but so too, unfortunately, would be the likelihood of scientific discoveries. However, the case for such a ring-plane crossing was that this was the distance at which one of the Voyagers would require to cross the ring plane if it were also to visit Uranus. If Pioneer 11 survived, then the door for the 'Grand Tour' of the outer solar system would be wide open. Hence, the choice was between an immediately rewarding inner crossing, and an outer crossing that promised scientific rewards from later missions. The scientists with instruments on Pioneer 11 held a meeting at Ames on 1 November 1977, and all but one of the 13 members voted for the inner crossing. NASA headquarters, however, had reserved the right to make the final decision, and decided that Pioneer 11 was a pathfinder for Voyager.[186] The final trim maneuver on 13 July 1978 that set up the outer ring-plane crossing also advanced the Saturn flyby by 2 days, to 1 September 1979, to reduce solar interference from the conjunction of Saturn that would begin in mid-September.[187]

The 2-month Saturn encounter sequence began on 2 August 1979, with the Deep Space Network devoting more and more tracking time to the mission. On 15 August the spin axis of the probe was slightly adjusted to allow the ultraviolet photometer to view the planet in order to search for hydrogen and helium in its atmosphere and observe the rings and satellites. On 20 August the axis was adjusted again to enable the photopolarimeter to initiate routine imaging of the target, which at that time was 10 million km away. Because the axis of the Saturnian system is inclined to the orbit of the planet, the Sun sometimes illuminates the north face of the ring system, and at other times it illuminates the south face. Pioneer 11 was approaching from north of the ecliptic and the south face of the ring system was in sunlight, which meant that it viewed the rings in silhouette. When the first long-range observations were made on 20 November 1975, test photometric scans had reassured the scientists that useful observations of the shadowed face of the ring system would be feasible.[188,189] On 24 August the first of three polarimetry scans of the bright region of Iapetus suggested a highly reflective particulate such as ice or snow. Unfortunately, the enigmatic dark side could not be targeted.[190] On 27 August the spacecraft had its closest approach to Phoebe at a range of 9.2 million km, and the next day it flew by Iapetus at a range of just over 1 million km. Meanwhile, ultraviolet scans were made of Hyperion, Tethys and Dione, and a polarimetry scan of Rhea. The bow shock was first detected on 31 August at a planetocentric distance of 24.1 Rs, and then twice again later in the day, indicating that the magnetosphere was reacting to gusts in the solar wind in a similar way to the much larger Jovian magnetosphere.[191,192] Photometric and ultraviolet scans were taken of all of the large moons except Mimas, the innermost. It was also discovered that Dione, Tethys and Enceladus caused a 50-fold decrease in proton fluxes inward of 7.5 Rs, the most intense radiation being out near the orbit of Rhea. Overall, however, the radiation belts were much weaker than those of Jupiter.

At 14:29 UTC on 1 September, 2 hours prior to the moment of closest approach, Pioneer 11 survived the south-bound ring-plane crossing at a planetocentric distance of 2.82 Rs. The infrared radiometer now had an opportunity to scan the southern part of Saturn's disk and the illuminated face of the rings. The atmosphere was colder than Jupiter's, with some small latitudinal variations. There was a 15°C difference in temperature between the illuminated and shadowed faces of the ring system, and a lesser difference between the illuminated face and the section on which the planet's shadow fell. These observations enabled the size and thermal characteristics of the bodies that constitute the rings to be inferred, indicating that (given assumptions) many had to be at least 10 cm across.[193] At 2.53 Rs, 23 minutes after the ring-plane crossing, the rates of the magnetospheric particle instruments abruptly fell by 99 per cent for 9 seconds, then recovered. The most plausible explanation was that the spacecraft had crossed the orbit of an unknown object that had to be at least 150 km in diameter, and was probably larger, immediately after this had 'swept' the particles from its orbit. The miss-distance was estimated at just 2,500 km. Later studies could not decide whether 'Pioneer rock', as it was dubbed, was Janus or Epimetheus, two small co-orbiting moons discovered telescopically in 1966 when the ring system was viewed from an edge-on perspective. Unfortunately, the object was too small for its mass to be determined by the manner in which it

A series of pictures taken by Pioneer 11 during its Saturn encounter. Top left: the dark side of the rings and the line of their shadow on the planet. Some structure in the rings can be recognized. The vertical lines at the bottom of the picture are due to the manner in which the photopolarimeter operated. Top right: a wider view of the rings in silhouette. The blob at left is the moon Tethys. Bottom left: the shadow of the rings cast across the terminator, viewed as Pioneer 11 departed from the planet. Bottom right: Titan, the largest of Saturn's satellites. Distortions in the pictures, in particular the first pair, are again due to the manner in which the photopolarimeter operated.

perturbed the spacecraft's trajectory.[194,195,196] The signatures of several additional objects were found in radiation data, most of which still remain unconfirmed.[197,198] All the radiation detectors reported that the charged-particle flux was zero inward of the outer edge of the A ring, implying that the particles had been absorbed by the ring material.[199,200,201,202]

At 16:31 UTC, Pioneer 11 flew 21,000 km over the cloud tops at a planetocentric distance of 80,982 km, or 1.35 Rs. At that time it was 9,000 km south of the C ring and traveling at 31.7 km/s relative to the planet.[203] Some 90 seconds later the radio carrier was occulted at 11.5°S, reappearing after 1 hour 19 minutes over the morning terminator at 9.6°S. The signal was weak owing to the fact that the Earth was on the opposite side of the Sun, and further degraded by solar interference, making the data difficult to analyze. Nevertheless, the electron density in the planet's ionosphere was obtained and the atmospheric temperature profiles were in good agreement with the results from the infrared radiometer.[204] The data from the magnetometers indicated the presence of a magnetic field with a strength at the 'surface' comparable to that at Earth's surface. Intriguingly, the axis of the magnetic field appeared to be almost precisely aligned with the spin axis, which posed a challenge to the standard theory

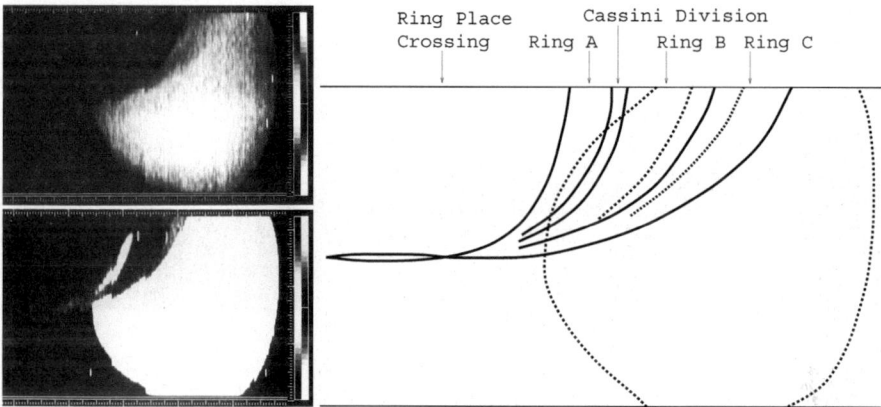

Two maps of Saturn's infrared radiation taken by Pioneer 11's spectrometer: the top image shows the temperature variations across the planet disk, with the rings system underexposed, and the bottom image shows the planet overexposed to reveal details in the colder rings. The motion of the spacecraft in the 2.5 hours it took to collect the data introduced distortions that made the planet's disk egg-shaped. As the diagram explains, in the leftmost portion of the image the dark face of the rings was in view; as the ring-plane crossing approached the rings became progressively narrower, and afterwards the illuminated face became visible and the perspective gradually opened up, giving the impression of the rings 'opening' toward the right of the image.

of a dipole field. There was no clear periodicity related to the rotation of the planet, undoubtedly owing to the co-alignment of the magnetic and spin axes. There was no trace of a Jovian-style magnetodisk.[205,206] Imaging was impractical at the time of closest approach because the spacecraft was traveling at too great a relative speed for the spin–scan system to operate effectively. However, low-resolution polarimetric data was used to assemble a single crude image of the illuminated face of the ring system. At 18:24 UTC the spacecraft made its north-bound crossing of the ring plane at 2.78 Rs, some 4,000 km beyond the outer edge of the A ring, again surviving unscathed. During the 4.5-hour interval centered on the time of closest approach to the planet, which included both ring-plane crossings, the meteoroid experiment had at least four punctures, including two particles from the E ring that impacted when Pioneer 11 was 900 km above the ring plane. Although this instrument had serious operational limitations, having been designed for the asteroid belt rather than Saturn's rings, it was significant that these were the only impacts to be detected within 25 million km of the planet.[207] While speeding away, images were taken of Saturn at a one-third-illuminated phase – a perspective that is not possible from Earth.[208]

A total of 440 images had been taken by the photopolarimeter, several of which, with a resolution of 90 km, were better than those achieved by terrestrial telescopes. Five images with a resolution of about 180 km were taken of Titan on 2 September from a range of 356,000 km, but they showed only a featureless orange fuzzy disk. Polarization measurements provided an estimate of the size of the particulates in the moon's atmosphere, while the infrared radiometer measured its heat balance and cast doubt on whether a greenhouse effect was at work. The ultraviolet instrument discovered a cloud of hydrogen accompanying the satellite along its orbit, undoubtedly produced by dissociation of methane in the atmosphere by solar radiation.

When the unilluminated face of the ring system was viewed, the denser rings were dark because their material blocked sunlight, but the gaps were brilliant as a result of microscopic particles forward-scattering sunlight. The measurements indicated that although the ring particles were typically centimeters to meters in size, there were also many that were much smaller, probably eroded from larger bodies by collisions. No trace of the tenuous inner D ring was found. A new thin ring, named the F ring, was spotted in the pictures just outside the known ring system at 2.33 Rs. It was only about 800 km wide, with a hint of local concentrations and clumps. While Pioneer 11 is usually credited with its discovery, this ring had very likely been observed several times from Earth, the first such report being by the Swiss astronomer Emile Schaer in October 1908.[209] The 3,400-km gap between the A ring and the F ring was named the Pioneer division by the project scientists, but this was not ratified by the International Astronomical Union. A new 200-km-sized satellite was spotted slightly beyond the F ring, but only as a tiny speck of light spanning a mere three pixels in two images, and was assigned the provisional designation of 1979S1. There was insufficient coverage to compute a reliable orbit, but it is now believed to have been Epimetheus, the moonlet that is co-orbital with Janus. Although the imager saw very few details on the disk of Saturn, it was possible to discern weak activity near the equator suggestive of jetstreams, and also at high northern latitudes. Meanwhile, the polarimetry facilitated mathematical models of the composition and structure of the atmosphere that showed there to be an ammonia haze at high altitudes.[210] During the encounter, the ultraviolet instrument swept across the rings and the planet, and observed a cloud of atomic hydrogen associated with the rings, thereby proving that the ring particles are mostly made of water ice. Emissions from auroras on Saturn were also detected around the poles.[211]

Owing to problems with the Deep Space Network antenna at Goldstone in California and the weather at Canberra in Australia there were periods during the encounter when the data rate had to be halved from a nominal 1,024 to 512 bps, which meant that only about 20 pictures at high resolution were able to be received.[212] Moreover, while the US had sought exclusive use of the frequency band used by Pioneer 11 for 1–2 September, it had not requested 3 September, and a conflict arose with Kosmos 1124, a Soviet military OKO early warning satellite built by the same design bureau (i.e. Lavochkin) that specialized in planetary probes. Although noisy and difficult to interpret, 25 minutes of Titan radiometric data were able to be received, from which an average global temperature of −193°C was calculated for this strange enshrouded satellite.[213,214,215] Continuous Deep Space Network tracking between 17 August and 4 September provided insight into the gravitational fields of Saturn and its satellites, from which it was inferred that the planet has an 'inner core' rich in rock and metal and an 'outer core' of ammonia, methane and water, surrounded in turn by a layer of high-pressure metallic hydrogen, a layer of liquid hydrogen and an atmosphere rich in hydrogen and helium. While estimates of the masses of the largest moons were obtained, the mass of the ring system proved impossible to determine. The computed densities of Iapetus, Rhea and Titan implied that all three were primarily ices, with little or no rock or iron.[216] On the outbound leg, the bow shock was encountered on no fewer than nine occasions, the final one on 8 September at a distance of 102 Rs. However, owing to interference from the onset of solar conjunction this distance was only an approximation.[217]

The Saturn flyby had put Pioneer 11 into a hyperbolic orbit with an eccentricity of 2.15 that was inclined at 16.6 degrees to the ecliptic, giving the spacecraft sufficient heliocentric velocity to escape the Sun's gravity. In the 1970s consideration had been briefly given to arranging the post-Saturn trajectory to intercept Uranus in December 1985 (just a month before JPL intended one of its Voyagers to arrive) but there were doubts that the spacecraft and its instruments would remain functional for 12 years, and that the scientific output from the extended mission would justify the additional cost. Other theoretically alternative extended missions (ignoring the spacecraft's state of health) included sending it back to Jupiter, to Neptune, or to one of a selection of 33 periodic comets.[218] In the event, it was decided to leave Pioneer 11 to depart from the solar system. The next major events came when Pioneer 10 crossed the orbit of Neptune on 13 June 1983; and Pioneer 11, which was traveling more slowly, did likewise on 23 February 1990.[219,220] During the remainder of their missions, the two spacecraft undertook scientific observations of phenomena in the outer solar system, with the prospect of finding the boundary of the heliosphere where the solar wind would slow down, become subsonic and cause the particles to form a 'termination shock'. It was expected that beyond this boundary there would be a wall of plasma and magnetic fields (the heliopause) and, beyond that, a bow shock in the interstellar medium produced by the motion of the solar system in the galaxy. Pioneer 10's ultraviolet photometer could observe the glow of sunlight backscattered by hydrogen and helium out near the heliopause, and its cosmic-ray telescopes were able to monitor how the flux of galactic cosmic rays was modulated with the solar activity cycle. Such measurements indicated that the heliopause was at least 50 AU, and possibly 100 AU, from the Sun.[221,222,223] Having 'doubled back' across the solar system after Jupiter to reach Saturn, Pioneer 11 was traveling in the approximate direction of the solar apex (i.e. the direction of the Sun's motion) and might live long enough to reach the heliopause, but as Pioneer 10 was heading in the opposite direction it was unlikely to reach this region.[224,225] Reporting over two full solar cycles, the Pioneers found that at heliocentric distances of up to 20 AU the average speed of the solar wind was almost constant, although there was still some structure.[226] By the solar minimum of 1986, Pioneer 11 was again at a high ecliptic latitude, and crossed a region of low-speed, low-temperature and high-density solar wind.[227] It passed within 6.5 AU of Uranus in January 1986, and its data on the state of interplanetary space was used in support of a flyby of that planet by one of the Voyager spacecraft.[228]

Like their predecessors in the inner solar system, the Pioneers exploring the outer solar system lived long lives, although some of their instruments suffered at

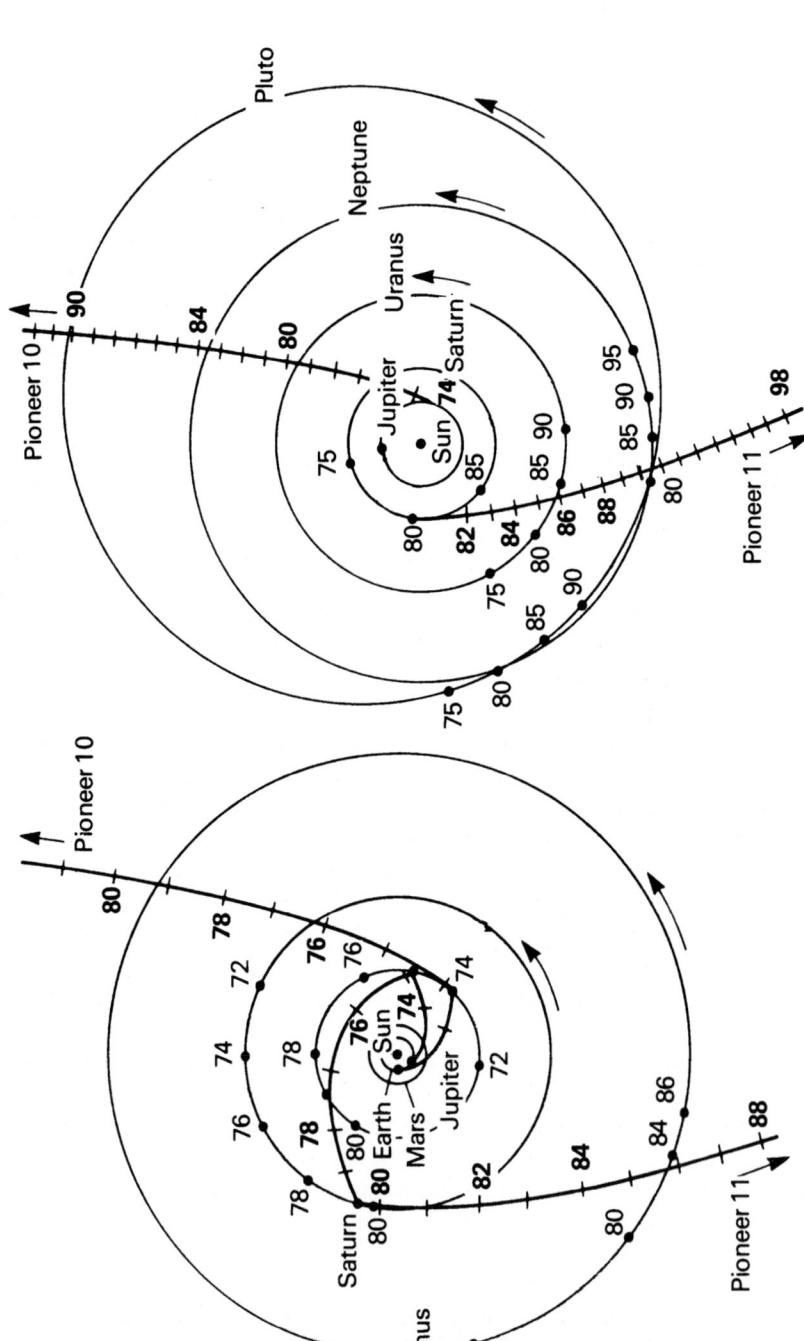

The trajectories of Pioneers 10 and 11 in the solar system, considered from a vantage point above the ecliptic plane.

an early stage. Among the first to fail were the meteoroid dust detectors. The one on Pioneer 10 lost one of its two channels soon after launch, and succumbed to the low temperature in May 1980. By the time Pioneer 11's instrument was switched off in September 1983, one channel had recorded many more hits than the other. Between them, the two dust detectors had recorded a total of 225 penetrations. There was an almost constant flux of dust beyond Jupiter that was thought likely to have derived from comets in highly eccentric orbits that spend most of their time in the outer reaches of the solar system (e.g. Halley's comet), comets in low-eccentricity orbits beyond Jupiter (e.g. Schwassmann–Wachmann 1), and icy bodies in a disk in the plane of the ecliptic somewhere beyond Neptune.[229] In addition to studying the heliopause, one of the major objectives as the probes departed the solar system was to perform accurate celestial mechanics measurements to seek perturbations from possible unknown bodies in the outer solar system, and search for evidence of low-frequency gravitational waves (as predicted by General Relativity). The Pioneers were particularly well suited for such research because they were spin-stabilized. In contrast, a 3-axis-stabilized spacecraft (such as the Voyagers that were to join the Pioneers in leaving the solar system) made many more maneuvers that complicated such fine measurements. Pioneers 10 and 11 were tracked during several oppositions. A search for very-low-frequency waves (down to 0.0001 Hz) that would produce small periodic variations of the distance between the Earth and the spacecraft found only noise.[230,231,232] However, astonishing results were obtained from Doppler tracking – although the system of Pioneer 11 had failed in October 1990 at a heliocentric distance of about 30 AU. Even after data processing had removed 'noise' arising from planetary perturbations, radiation pressure, the interplanetary medium, General Relativity and a miscellany of other influences, both spacecraft were found to have been subjected to a constant sunward acceleration of about 0.00000008 cm/s^2. No trace of this acceleration has been reliably found in the tracking data of other spacecraft, or in a high-precision ephemeris of Mars derived from long-term tracking of the craft on its surface. Putative causes of the anomalous acceleration include gravity from Kuiper belt objects, gravity from the galaxy, errors in the planetary ephemerides and errors in determining the orientation, precession or nutation of the Earth, but no single mechanism is satisfactory. More mundane suggestions include slow gas leaks and the radiation of heat by the RTGs, although there is no reason why either should manifest itself as an acceleration in the direction of the Sun.[233,234,235,236] Other explanations have ranged from minor effects predicted by General Relativity to completely new theories of gravitation.[237] By any measure, the 'Pioneer Anomaly' has proved to be a long-standing mystery.

In early September 1995, with its RTG almost exhausted, Pioneer 11 was told to cease routine scientific observations. On 30 September the Deep Space Network lost the communication and could not regain it. It seems likely that the spacecraft was no longer able to maintain an attitude to hold Earth in its high-gain antenna beam.[238] By that time, the spacecraft was at a heliocentric distance of 44.1 AU. Pioneer 10 was also having its share of problems. In particular, it was experiencing difficulties in measuring its attitude. After 1997 there was insufficient power to operate both the transmitter and the thrusters, and tracking had to be interrupted for periodic maintenance of the attitude. Even with most of the instruments switched off, the power was so limited that it was not possible to run the remaining instruments constantly, and eventually only one of the charged-particle detectors and the Geiger–Mueller counter were on, together returning a handful of data at the slow (but power efficient) speed of 8 bps. As with the Pioneers in the inner solar system, Pioneer 10's mission was officially terminated at 19:35 UTC on 31 March 1997, shortly after the 25th anniversary of its launch, by which time it was at a heliocentric distance of 67 AU. In fact, contact was intermittently maintained thereafter, and Pioneer 10 was used to train controllers for the Lunar Prospector low-cost lunar orbiter that Ames built. Later, contact was again re-established when Pioneer 10 was made the subject of a NASA study on receiving weak signals from distant spacecraft, and as a calibration target for a SETI (Search for Extraterrestrial Intelligence) study by the Arecibo radio telescope. A factor that worked against further such use of Pioneer 10 was its obsolescence, as its commands were still being generated by one of the last surviving 1970s vintage punched-card computers. A two-way lock was not achieved during the periodic antenna-pointing maneuver in the summer of 2000, but Pioneer 10 was again contacted in April 2001, when it became evident that a component of its transmission system had failed, and the spacecraft had activated the back-up.[239] The final telemetry from Pioneer 10 was received on 2 March 2002 (the 30th anniversary of its launch) after 8 months of silence, by which time it was at a heliocentric distance of 79.7 AU. Despite its age it was relatively healthy, and the Geiger–Mueller counter was still working. The probe was contacted again on 22 January 2003 at 82.1 AU, but only the carrier signal was received. An attempt at communication on 7 February was unsuccessful, probably because the RTGs were then delivering no more than a trickle of power, and NASA announced that it would make no further attempts. Nevertheless, the agency approved a request sponsored by a private advocacy group, the Planetary Society, to try to gain a few more data points on the Pioneer Anomaly between 3 and 4 March 2006, but there was no carrier signal. Pioneer 10 is leaving the solar system at 2.38 AU/year in a direction that is several degrees from the Hyades star cluster in Taurus and is almost opposite to the motion of the Sun around the center of the galaxy. In about 33,000 years it will pass within 3.3 light-years of the red dwarf Ross 248. It is worth mentioning, however, that this star is currently located in the constellation of Andromeda at a distance of about 10 light-years, and the encounter will occur not so much because the spacecraft transits the intervening space but because the motion of Ross 248 through space will cause it to travel across the sky into Taurus and approach within about 3 light-years of the Sun. Pioneer 11 is the slowest of the spacecraft currently departing the solar system, traveling at just 2.21 AU/year in the direction of the constellation Aquila, and even although it is moving in the opposite direction from its partner, it too will pass within 2.7 light-years of Ross 248 in about 35,000 years.[240,241]

Star encounters by the Pioneer spacecraft

Spacecraft	Date (Year)	Star Name	Distance (light-years)
Pioneer 10	32608	Ross 248	3.27
Pioneer 11	34511	Ross 248	2.67
Pioneer 11	42405	AC + 79 3888	1.65
Pioneer 11	157464	DM + 25 3719	2.46

A third spacecraft, known as Pioneer H, was assembled, and in 1971 Ames made a feasibility study on the use of it for an Out-of-the-Ecliptic mission. In the baseline scenario it would be launched by a powerful Titan IIIE–Centaur rocket and use the same TE-M-364-4 'kick stage' as its predecessors. After a little over 1 year it would reach Jupiter, and undoubtedly continue the previous studies. A polar flyby at 3 Rj would maintain the perihelion near 1 AU, but deflect it into an orbit inclined at 92.5 degrees to the ecliptic to provide opportunities to pass directly over the north pole of the Sun, and then the south pole, at a range of about 2 AU in order to investigate not only the solar wind, magnetic fields and solar and galactic cosmic rays at high solar latitudes, but also the processes that cause the expansion of the solar corona. All the particles and fields instruments used by Pioneer 10 and 11 were suitable for such a mission. There was also the prospect of adding instruments to measure cosmic dust, electric fields and radio noise at high solar latitudes. A pole-to-pole radio occultation of the Sun would provide unique data. As the availability of the Titan IIIE–Centaur rocket was questionable, options were investigated for employing the somewhat less powerful Titan IIIC. The worst

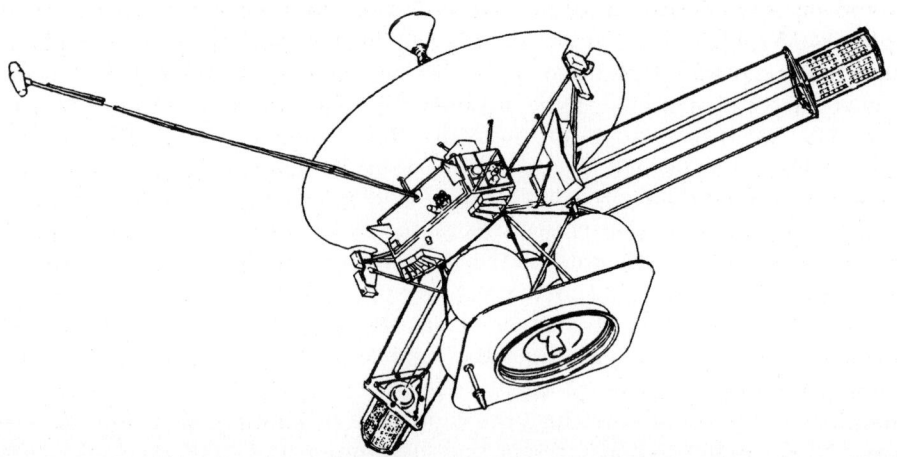

A line drawing of the proposed Pioneer Outer Planets Orbiter.

option was the previously used Atlas–Centaur, as the angle of inclination would not have been able to exceed 42 degrees. As a point in favor of this Out-of-the-Ecliptic mission, the launch windows were not particularly strict: the only real constraint was to reach Jupiter, for which a window opens at 13-month intervals.[242] Pioneer H was proposed to NASA in 1973 with a view to launch in July–August 1976, but was rejected.[243] In its place, Out-of-the-Ecliptic missions were invited in the context of international cooperation with Europe. A variety of other missions for the Pioneer–Jupiter bus were also studied, including the addition of a propulsion module, more powerful RTGs and a better communications system to produce a Pioneer Outer Planets Orbiter that would be put in orbit around either Jupiter or Saturn in the 1980s to continue the work of its predecessors.[244] As an alternative, the basic flyby bus could be modified to deliver an atmospheric probe to Jupiter, Saturn or even Uranus.[245] Although none of these proposals was pursued, they were all instrumental in defining Jupiter-Orbiter-with-Probe, which became the Galileo spacecraft that was sent in the 1990s to conduct a sustained exploration of the Jovian system.

THE TASTE OF VENUS

While a new 'heavy' Venus lander was being developed, the Soviets decided to use the 1972 window to follow up the Venera 7 results using a revised 3V bus. To assist in the design of the cameras for the next generation of landers, one of the objectives was to measure how the illumination varied during the descent and on the ground. This required the probe to enter the atmosphere over the day-side of the planet, which posed two different problems. The relative positions of Earth and Venus in their solar orbits meant that the planet was in a crescent phase when a probe arrived. Earlier probes had penetrated the atmosphere on the night-side for the simple reason that this readily provided a line of sight to Earth to transmit their data. An attempt to land on the crescent would require a very accurate trajectory, and from a site on the surface in this area Earth would be at an elevation of, at most, 30 degrees above the horizon, rather than near the zenith as previously. To overcome this, the antenna on the lander was redesigned to provide a 'funnel-shaped' rather than a 'pear-shaped' directionality profile. Having knowledge of the surface pressure, the engineers were able to build the lander for 105,000 hPa (about 15 per cent greater than the pressure measured by Venera 7), and use the mass that was saved in the construction of the pressure hull to accommodate a stronger parachute, an active thermal control system with a compressor and heat exchanger, a second communication antenna, and three new instruments. The second antenna would be carried in the parachute bay, and as soon as the capsule landed the parachute would be jettisoned and the new antenna, connected by a cable, would be ejected onto the surface. The disk-shaped unit had a spiral aerial on each side, and a gravity switch to select the one that faced upwards. Hence, even if the lander were to roll over and point its main antenna away from Earth, the deployed omnidirectional antenna would ensure that a signal was able to be received. In addition to cadmium-sulfide photometers to measure the illumination at altitudes below 50 km, the lander's payload included an instrument to detect ammonia in the atmosphere by measuring the color variation of a chemical reactant, and a gamma-ray spectrometer to measure the composition of the surface. The latter was housed in the lower half of the hull in order to be as close as possible to the ground (unless the capsule were to roll over, of course) and had been calibrated first on Earth by placing it in a tunnel of rock that was known for its low content of radioactive elements, and then, during the parachute descent, by three measurements of the composition of the heat shield to provide a point of reference.[246,247,248] Since this capsule would be the first to enter the atmosphere well away from the center of the planet's disk as viewed from Earth, measurements of the Doppler shift on its transmission during the parachute descent would provide wind speed profiles. Such an experiment had been intended for Venera 7, whose entry point was 10 degrees from the 'sub-Earth point', but the erratic behavior of the spacecraft meant no data could be extracted from the telemetry.[249] The modified parachute system was tested in a wind tunnel filled with carbon dioxide at 500°C. The first of the two identical spacecraft was to deliver its 495-kg lander over the day-side, and the second to the night-side.

On being successfully dispatched by a Molniya launch vehicle on 27 March 1972, the first probe was named Venera 8. The second achieved parking orbit 4 days later, but when stage L attempted the 'escape' maneuver its engine shut down 125 seconds into the 243-second burn and the spacecraft, announced as Kosmos 482, was left in an orbit with an apogee of 9,805 km. This is not only the last planetary probe to be launched by a Molniya, but also the last failed attempt by the Soviet Union to dispatch a planetary probe. At the end of June the spacecraft detached a fragment, possibly the capsule. The main craft re-entered on 5 May 1981, but the fragment remains in space.

A course correction on 6 April targeted Venera 8 for the illuminated crescent so accurately that a second correction was deemed unnecessary. During its 117 days of flight, the bus performed faultlessly – making 86 communication sessions, collecting information on the deep-space environment and, in particular, detecting a powerful solar flare. Approaching Venus, measurements were taken of various radiations and of the density of hydrogen and deuterium in the upper atmosphere. Several days before the encounter, the lander's internal temperature was reduced to −15°C and on 22 July its restraining straps were released. A hour later it entered the atmosphere at an angle of 77 degrees below local horizontal at a speed of 11.6 km/s. After 1 minute, by which time the rate of descent had slowed to 250 m/s, the parachute opened and telemetric and scientific data began to stream back. The ammonia concentration was sampled twice on the 53-minute parachute descent. It was 0.1 per cent at an altitude of 46 km and an order of magnitude less at 33 km (or, more specifically, at the altitudes where the pressures were 2,000 and 8,000 hPa respectively). The 56 minutes of data from the photometer indicated that the light flux decreased 3-fold between 50 and 32 km, at which altitude there was a change of gradient (possibly where the cloud layer was thickest), and decreased another 4-fold by the surface.[250] According to atmospheric models based on

Parachute

External Antenna
before release

Antenna

Light Level
Sensors

Pressure and
Temperature
Senors

External Antenna
after release

CCCP

A drawing of Venera 8 on the surface of Venus. (Courtesy of NPO Lavochkin)

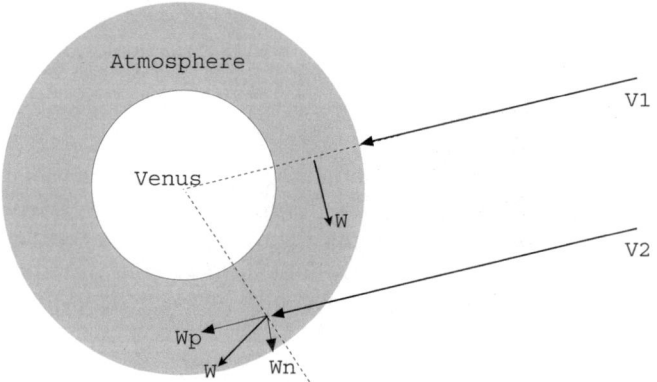

Unlike its predecessors, Venera 8 was to enter the atmosphere of Venus away from the sub-Earth point. Whereas previously the wind speed W was perpendicular to the velocity of the probe V1 and therefore did not introduce a Doppler shift, in this case the wind would introduce a Doppler shift proportional to its velocity component Wp parallel to the velocity of the probe V2. (Not to scale)

previous missions, the thickest cloud layer had been expected to be present at an altitude of about 48 km. At the landing site it was just after dawn, with the Sun having risen just 5 degrees above the horizon, and barely 1.5 per cent of the incident sunlight was reaching the ground. During the descent the winds carried the capsule some 60 km horizontally. The anemometric Doppler experiment measured a wind speed of 100 m/s at high altitude, decreasing to some 50 m/s at an altitude of 40 km, and to just 1 m/s below 10 km. The winds followed the rotation of the planet – because its axis is almost inverted, the direction of rotation is east to west – and exceeded the angular rate of the surface, thereby giving the first in-situ confirmation of the 4-day 'super rotation' of the atmosphere. During the descent, a radar antenna was aimed over the side of the capsule. In addition to monitoring the decreasing altitude, the antenna was able to measure the dielectric characteristics of the surface, yielding a value much lower than that measured by terrestrial radars – possibly indicating that the area over which Venera 8 was descending had a porous surface. As the capsule drifted, the radar noted its passage over hills of up to 1,200 meters in elevation. The second stage of the parachute's opening sequence occurred at an altitude of 30 km. At 09:29 UTC the capsule touched down at a speed of 8 m/s. It landed at 10.7°S, 335.25°E, on Navka Planitia. During its first 13.3 minutes on the surface, the probe sent pressure, temperature and illumination data using its main antenna: the pressure was about 90,000 hPa and the temperature was 470°C. The fact that this was almost the same as reported by Venera 7 on the night-side suggested that conditions were essentially independent of the diurnal cycle.[251,252] The backup antenna then sent a 20-minute period of gamma-ray data, and the main antenna resumed transmitting until the capsule eventually succumbed after some 50 minutes on the surface. The gamma-ray spectrometer took 42 minutes to finish its analysis. The percentages of potassium, uranium and thorium were in amounts similar to some terrestrial alkaline basalts. Future orbiters equipped with imaging radars would show sites with structures like flat volcanic shields, and it is likely that the material analyzed by Venera 8 was such a structure.[253,254,255,256,257,258,259]

The year 1972 was important for studies of Venus for another reason, because two astronomers in the US independently found that the polarization of the upper level clouds indicated the presence of a sulfuric acid aerosol, and thereby solved the long-standing mystery of the substance that forms the droplets in the visible clouds. Furthermore, it was later realized that this highly corrosive acid would have reacted with the ammonia-tracer on Venera 8 and falsified the results from that experiment, because the ammonia levels are actually 1,000 times lower than those measured by Venera 8.[260]

THE CURSE OF THE TRANSISTOR

After the spectacular success of Mariner 9, NASA set about preparing its ambitious Viking missions, in which orbiters would dispatch landers, but budgetary constraints obliged it to skip the 1973 window and aim to launch in 1975. The Soviets decided to attempt to use 1973 to recover from the partial or total failures that had befallen their 1971 flotilla, and so steal a march on Viking. The Lavochkin engineers faced two major issues. First, because the window would open just 1 year after the M-71 missions were scheduled to finish, it would be difficult (if not impossible) to apply the lessons learned to make substantial modifications to the design. Moreover, since the opposition of Mars in 1973 would be less favorable, a spacecraft would have to be accelerated to a speed 300 m/s faster than before in order to reach Mars, which meant that not even the Proton launch vehicle would be able to dispatch a combined orbiter and lander. It was therefore decided to split the mission. A pair of spacecraft were to arrive first and enter orbit around Mars. Trailing a month behind would be two others that would release their landers and then fly past the planet; their reduced propellant loads would enable them to be launched by the Proton. The designations were retained: the vehicles intended to enter orbit were designated M-73S, and those ferrying landers were M-73P.

The M-73S spacecraft had a launch mass of 3,440 kg, and was equipped with no fewer than 15 instruments. The most important instruments were an imaging system, an infrared radiometer whose large antenna occupied the top of the spacecraft where a lander would be carried, and a 256-channel gamma-ray spectrometer mounted on a boom. On this mission the 52-mm-focal-length Vega camera had four filters and the 350-mm-focal-length Zufar had only an orange filter. There was sufficient 25.4-mm film for 440 images that could be scanned at 10 possible resolutions, including 235 × 220 pixels for preview, 940 × 880 pixels for normal resolution, and 1880 × 1760 pixels for high resolution. Furthermore, a digital 'push-broom' linear camera was to build up panoramic images of indefinite length that viewed 15 degrees to each side of the ground track; it had been tested in lunar orbit prior to being assigned to Mars. Also carried were an ultraviolet photometer, a charged-particle detector, a magnetometer, and five photometers for a variety of wavelengths, including those

A drawing of the M-73S Mars configuration: 1, high-gain antenna; 2, infrared radiometer; 3, fuel tank; 4, thermal control system radiator; 5, solar panel; 6, instrument module; 7, magnetometer.

characteristic of carbon dioxide and water. One of the spacecraft carried several photopolarimeters, together designated VPM-73 (Visual Polarimeter-Mars). This had been designed by the French expert on planetary polarimetry, Audouin Dollfus of the Meudon Observatory in Paris, in cooperation with the Georgian astronomer Leonid Ksanfomaliti who, after working at the Abastumani Observatory, joined the Soviet Institute for Cosmic Research, IKI (Institut Kosmicheskikh Isledovanii); and built jointly by IKI and the National Space Studies Center in France, CNES (Centre National d'Etudes Spatiales). The prototype was assembled in the Soviet Union, calibrated in France, and two units were made and mounted on a single spacecraft in order to view the surface at two different angles and in concert with the camera and the infrared radiometer. It was to collect polarimetric data at phase angles that were inaccessible to terrestrial telescopes. At periapsis it would have a 20-km footprint and would be used to investigate both the atmosphere and the texture of the planet's surface.[261,262,263,264,265,266,267]

The M-73P spacecraft had a launch mass of just 3,260 kg. Because each bus was to perform a flyby it carried a minimal suite of instruments, which were designed to study the deep-space environment: a magnetometer, a plasma sensor, a cosmic-ray sensor, a micrometeoroid detector, and an instrument to study charged particles in the solar wind. They also carried a new form of the French STEREO experiment. STEREO-2 to -4 were development units, and the STEREO-5 version that was flown had a 10-meter-long dipole antenna and a 2.5-meter-long monopole antenna. A final French experiment, also developed jointly with IKI, was a Lyman-alpha photometer based on an instrument of the D2A-Tournesol (Sunflower) scientific satellite. Its function was to collect data on the emission of hydrogen atoms in the solar system and to characterize the local interstellar medium.[268,269,270] The M-73 lander was almost identical to the previous design, but had in addition a second transmission system to return in real time the data taken during the atmospheric descent – the temperature and pressure, the light levels measured by a photometer with colored filters, and the current drawn by the pump that supplied atmospheric gas to the mass spectrometer. The data reported during the descent would ensure that at least something was received in the event of the lander not surviving the touchdown.[271,272,273,274] If the landing was successful, a more detailed report would be made. As the Soviets had asked NASA to supply Mariner 9 data for 43°S, 42°W and 24°S, 25°W, these areas were widely expected to be the sites of new landing attempts.[275]

A serious problem emerged less than 4 months before the window opened, when all the spacecraft systematically suffered malfunctions in their power systems during ground tests. The culprit was easily identified as a simple transistor, named 2T-312, built in Voronezh. In order to save the national gold reserves, someone had ordered that the gold-plated contacts of the transistor be substituted by aluminum, even though this would reduce their lifetime to approximately 2 years. The obvious solution was to replace the 2T-312 transistors, but this would probably take so long that the 1973 window would be missed. Instead, a reliability analysis, very likely one of the first such exercises carried out in the Soviet Union, indicated that if the missions were to proceed using the substandard transistors there was a 50 per cent

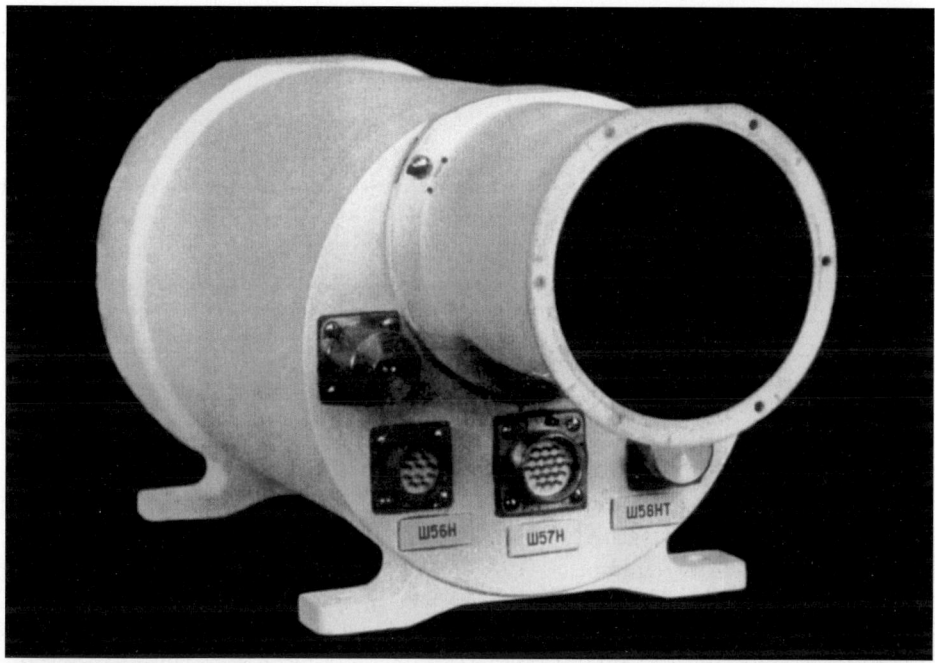

The French–Soviet VPM-73 photopolarimeter carried by Mars 5. (Courtesy of Audouin Dollfus)

probability of success. While this was distinctly low by aerospace standards, it was decided to proceed. First to set off were the two orbiters, Mars 4 and Mars 5, which were launched on 21 and 25 July into transfer orbits with periods of 556 and 560 days and aphelia at 1.63 and 1.64 AU respectively. The landers, Mars 6 and Mars 7, followed on 5 and 9 August and flew slower transfers with periods of 567 and 574 days and aphelia at 1.67 and 1.69 AU. Several days after launch, each spacecraft successfully made its first course correction. However, the lifetime of the transistors started at the time of their manufacture, and at the end of September Mars 6 suddenly lost all the telemetric channels that reported the status of its systems, leaving only the channels for command transmissions and positional measurements. In a remarkable (for that time) display of openness, Roald Sagdeev, director of IKI, announced that one of the probes had suffered a telemetry failure, although this went largely unnoticed in the West.[276] To the delight of the engineers, Mars 6 autonomously determined its position with respect to its objective and made the second course correction in February. Next to suffer was Mars 4, whose main computer lost some of its capabilities, including the ability to command the engine, with the implication that it would be unable to enter orbit around the planet. Finally, Mars 7 lost some of its transmission systems. Only Mars 5 was apparently working as designed, and executed calibrations of its instruments at increasing distances from Earth.[277,278,279,280] As Mars 4 flew by Mars at a distance of 1,844 km on 10 February, the engineers managed to power up the imaging system a bare 2 minutes prior to the moment of closest approach, and over the next 6 minutes the probe took 12 images and two panoramas of a swath of the southern hemisphere, including the site chosen for the Mars 6 landing. Then dual-frequency radio observations were made as the spacecraft was occulted. In addition to yielding data on pressures and temperatures, Mars 4 measured the electron density over the day- and night-sides of the planet, providing the first indication of an ionosphere over the night-side.[281] The spacecraft continued to report data on the interplanetary environment for some time thereafter, but in essence its mission was over.

Mars 5, the only fully functional orbiter, performed its second course-correction maneuver on 2 February, and 10 days later successfully entered an orbit around the planet that ranged between 1,760 and 32,586 km and was inclined to the equator at 35.33 degrees with a period of 24 hours 53 minutes – several minutes longer than a local day. Given the loss of Mars 4, Mars 5 would have to provide a relay for both of the landers. However, shortly after orbit insertion it was discovered that the main compartment of the spacecraft was slowly losing pressure, which in turn meant that within 3 weeks its apparatus would become unusable. An accelerated scientific program was therefore initiated in order to collect as much scientific data as possible within this short period.[282,283,284] Five imaging sessions yielded five panoramas and a total of 108 images, of which 43 were deemed usable. It also confirmed the indication from the Mars 4 radio-occultation of a night-side ionosphere, and identified a second possible electron density maximum. The photopolarimeters of the VPM experiment recorded the degree of polarization along ground tracks extending over 100 degrees in longitude in the Thaumasia and Argyre regions. All of the Martian areas that were analyzed had finely grained soils ranging from lunar-like dust (in Mare Erythraeum) to larger grains similar to those present in terrestrial sand dunes (in Claritas Fossae, Thaumasia Fossae and Ogygis Rupes). The VPM also performed small-scale texture analyses, and proved able to detect details such as rugged terrain at the rim of the large craters Lampland and Bond, where there were probably large and dust-free boulders. These results were consistent with thermal inertia data from the infrared radiometer, which also showed a maximum temperature at the surface of –1°C in the afternoon, decreasing to –43°C along the line of the terminator, and –73°C over the night-side.[285,286] There had been major dust storms in June and October 1973, but by this point the atmosphere was generally transparent and the VPM data provided little insight. However, white clouds and yellow clouds were both observed, often in the same area and sometimes persisting for several days. The white clouds were cirrus-like, and made of ice crystals. No large water droplets were detected that could have formed rainbows. The yellow clouds were a very fine dust. The instrument collected both limb and terminator data, and detected layers of very small particles at altitudes from 40 to 60 km.[287,288] Some of the most interesting observations were made by the gamma-ray spectrometer during six periapsis passes at altitudes below 2,000 km. These were the first measurements of the composition of the Martian soil, and covered the regions of Thaumasia, Argyre, Coprates, Lacus Phoenicis and Sinus Sheba. Radioactives such as thorium, uranium and potassium were detected, and the amounts of oxygen, silicon, aluminum and iron in the soil were measured.

One of the low passes was over Tharsis, and gamma-ray spectra were gained of Arsia Mons, the southernmost of the volcanoes, showing its composition to resemble terrestrial alkaline basalts.[289] Magnetic and plasma data were obtained during nine communication periods. Once again, Soviet scientists claimed that they had found "necessary and sufficient arguments for the existence of the Martian intrinsic magnetic field". By re-evaluating the data from Mars 2 and Mars 3 it was calculated that the north magnetic pole was in the southern hemisphere, and that the dipole-axis was within 15 degrees of the axis of the planet's rotation.[290]

The last contact with Mars 5 occurred on 28 February, at which point, lacking an orbiting relay station, the Soviets lost the possibility of a sustained landing mission. Mars 7, having pursued a faster trajectory than Mars 6, arrived on 9 March. It had suffered such a series of malfunctions that only one of its radio systems was active. The first command to separate the lander was not acknowledged by the spacecraft but it was issued again, accepted and executed. The lander had several failed 2T-312 transistors. These prevented the engine from firing to adopt a trajectory to make a landing at 50°S, 28°W, and the helpless probe accompanied its carrier on a flyby at a range of 1,300 km. The main spacecraft went on to have the longest lifetime of the group, and provided data during a solar occultation from July to September 1974. Of the two STEREO-5 experiments, the one on Mars 6 failed to return any data, while that on Mars 7 provided data right through to May 1974. Moreover, the Lyman-alpha photometer on Mars 7 provided measurements of the temperature of interstellar gas in the immediate vicinity of the Sun.[291,292,293,294] The date of its last transmission is not known.[295]

Despite having been the first spacecraft to suffer from transistor failures, Mars 6 had struggled on and dramatically demonstrated the reliability and ruggedness of its advanced control system working without input from Earth. Not only did it make its second course correction, it also computed the correct attitude and timing to release its lander, and as it flew by the planet at a range of 16,000 km it acted as a relay station as its probe made its atmospheric descent, then provided a double-frequency radio-occultation that indicated a surface pressure of between 4 and 10 hPa.[296] The lander was released when 48,000 km from the planet on 12 March, and it entered the atmosphere 3 hours later (at 09:06) at a speed of 5.6 km/s. It was fortunate that a second transmission system had been added to provide data in real time during the descent. After the parachute deployed, the telemetry immediately indicated that the probe was oscillating much more than expected, thereby rendering the early data unintelligible. Unfortunately, after 148 seconds the probe fell silent – immediately after the braking rocket was fired to soften the touchdown at 09:11. The cause of the loss of signal was never determined. It is possible that the communications system malfunctioned, or the probe landed in a rough area. A more mundane explanation is that the mothership passed below the local horizon and broke the relay link; if that is so, then the timing was terrible! Although the site is accepted to have been on Mare Erythraeum, the coordinates are disputed, with one source giving 23.9°S, 19.4°W, and another 24°S, 25°W. The lander provided interesting atmospheric data for the final 8 km of its descent. The extrapolated surface temperature was about –43°C, the pressure was 6 hPa, and the wind was blowing at between 8 and 12 m/s. The light

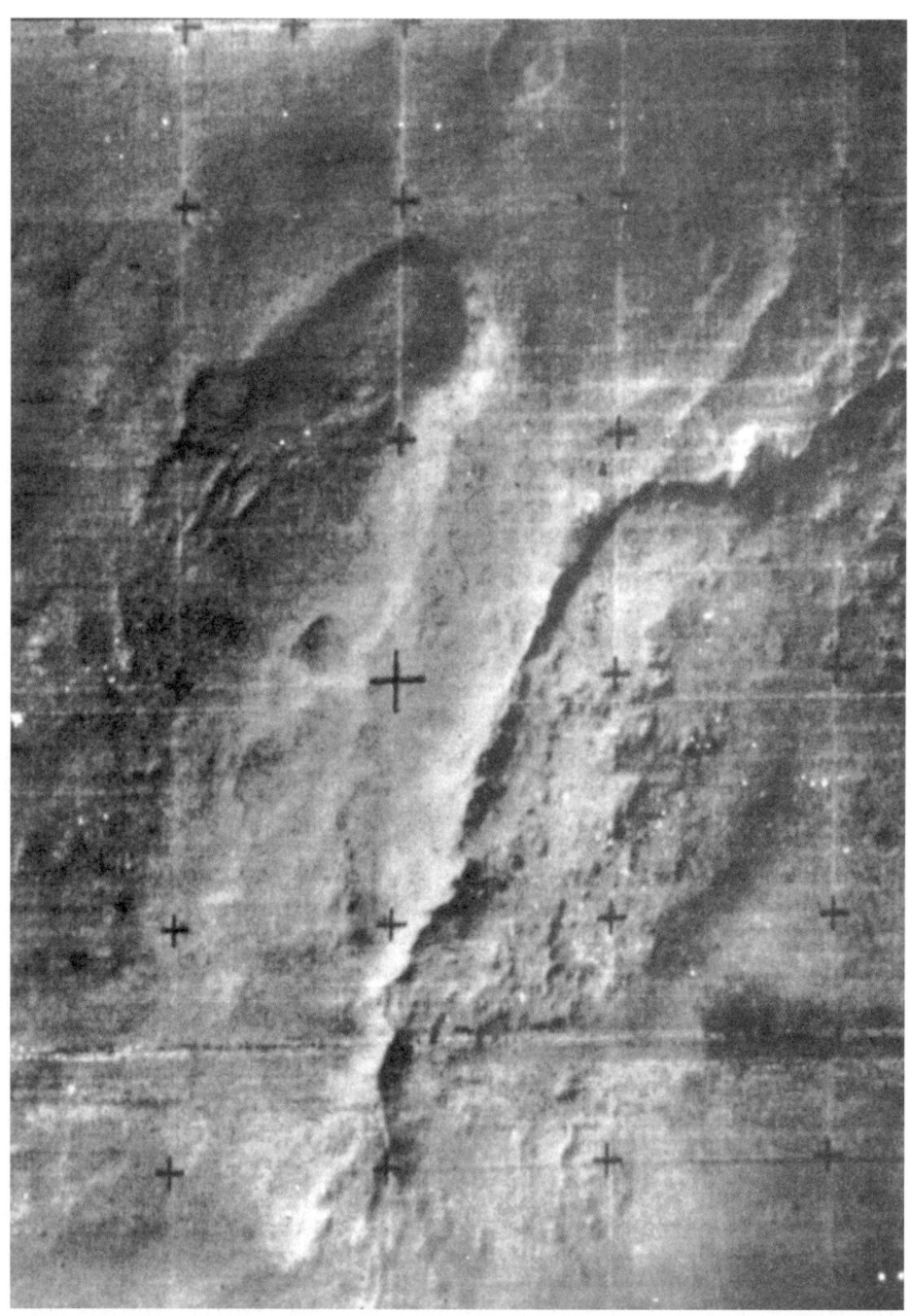

An image of the Martian surface near 35°S, 38°E that was taken by Mars 5.

levels measured by the photometer indicated that visibility was good. Contrary to exaggerated press reports, it seems unlikely that the instrument was able to return color images![297] The telemetry from the current drawn by the pump of the mass spectrometer showed that this was pumping more air than expected, which the Soviet scientists interpreted as evidence that the atmospheric composition was one-third inert gas, probably argon – a finding which, if true, would have implications for future missions, because the presence of so much argon would greatly increase convective heat transfer at hypersonic speeds and would necessitate a revision of the heat shields and entry profiles. However, it would later be proved that this finding was incorrect.[298,299,300,301]

SOVIET SOIL FROM THE RED PLANET

With the total or partial failure of the four M-73 spacecraft, the Soviet aspiration to achieve a Martian spectacular in advance of NASA's Vikings had been dashed. However, the Lavochkin design bureau was working on a bold concept that, if developed, would eclipse a successful Viking mission. In early 1970 Georgi Babakin had ordered his designers to exploit their experience of the E-8 lunar sample-return probes to design a spacecraft to land on Mars, collect a sample and return it to Earth for chemical, geological and biological analyses. The first preliminary project, named 5NM, was completed during the summer of 1970. The 20-tonne spacecraft was to be launched by the N-1 rocket that Sergei Korolyov's bureau was developing for a manned lunar landing mission. The schedule called for a launch in September 1975, and a transfer orbit that would take 1 year to reach Mars. The 3,600-kg bus would incorporate both the spherical propellant tank developed for the M-69 design and the toroidal tank of the M-71 design. On approaching Mars, the lander would be released and the bus would make a flyby, in the process relaying to Earth the data from the lander. The 16-tonne lander would have a 6.5-meter-diameter heat shield, on the periphery of which were 30 petals to be deployed on entering the atmosphere to increase its diameter to 11 meters and increase its aerodynamic drag. Rather than use a parachute system, it would be equipped with a block of four tanks and four throttleable landing engines. Its surface mission would last no more than 3 days, during which it would take a sample identified in a panoramic image of the landing site. The sampling system is not known, and was probably never decided, but could have included a robotic arm, perhaps equipped with a drill. Once a 200-gram sample had been stored in the return capsule, a two-stage rocket would lift off and enter orbit, where it would wait for the July 1977 Earth window to open. It would then release a 750-kg Venera-based spacecraft that would boost out of orbit for a return flight lasting 291 days, concluding with the release of a 15-kg spherical capsule that would land on Soviet territory. In order to test elements of the 5NM mission, a 4NM pathfinder was planned that would deliver a large rover instead of the ascent module to the surface of Mars. However, the project soon stalled, partly owing to the low reliability of Soviet technology, which did not instill confidence that a spacecraft would continue to operate for a 3-

year period, and partly because there was no way to ensure that the sample would not become contaminated by, or contaminate, Earth if the capsule were to be damaged on landing.

Following the dismal M-73 missions, this ambitious plan was resuscitated in 1974 as 5M. By this time of course, the N-1 launch vehicle was no longer available – its development had been canceled after four successive failed launches between 1969 and 1972, and after the plan to send cosmonauts to the Moon had been shelved. The 5M mission was therefore to use the much less powerful Proton. As the spacecraft would have a mass of at least 8,500 kg, it would be launched in two parts, with the fourth stages of the launch vehicles rendezvousing in parking orbit and docking prior to setting off to Mars. The bus would include a simple toroidal propellant tank and be fitted with two solar panels. On approaching Mars the bus would set up the desired entry trajectory, release the lander, and execute a deflection maneuver so as to perform a flyby. The large conical entry module would not use a parachute, but would instead rely on its aerodynamical design to generate some lift in the rarefied atmosphere and follow a complex trajectory designed to progressively slow it sufficiently for the two-stage retrorocket to settle it on the surface at a vertical speed not exceeding 3 m/s. After a sample had been collected, the 2,000-kg ascent stage would lift off and enter Mars orbit, then rendezvous and dock with another vehicle (launched by a third Proton) that would depart with the sample. To overcome the contamination issue, the return spacecraft was to enter Earth orbit, where a Soyuz spacecraft would rendezvous with it and collect the sample. As with the 5NM plan, the 5M was to be preceded by a 4M mission in which the bus and lander would deliver a rover to the surface of Mars. To this end, in 1974 VNII Transmash started to work on a large rover, and over the next 4 years investigated a number of different locomotive solutions. The first prototype (named 4GM) used four tracks on independent suspensions; the second (KhM) had a pair of large 'walking skids' like those of the PrOP-M, but now for a rover that weighed 240 kg; the third (KhM-SB) resembled the Lunokhods; and the fourth (EOSASh-1) used a locomotive system that combined the advantages of six wheels with a 'walking' articulated bogie.

After another major redesign, the Mars sample-return project was approved in January 1976. The use of aerodynamic lift during entry was discarded in favor of the original idea of a variable-geometry heat shield, now having an 'all out' diameter of 11.35 meters, in addition to the use of parachutes. The 9,135 kg probe now included a 1,680-kg orbiter and a 7,455-kg lander comprising the ascent-to-orbit stage and the return capsule. The mission was designated M-79 because launch was scheduled for 1979, with Mars arrival in June 1980; and to shorten the 3-year flight, a Venus flyby was included in the plan. The lander was to be released after the spacecraft had entered orbit around the planet, with the insertion burn being made by one of the two stage D modules. At the suggestion of the planetary geologist Alexandr Vinogradov, it was decided to maintain the Martian sample at a constant high temperature during the return flight in order to sterilize it, thus eliminating the threat of any biological contamination of the Earth's biosphere in the event of the capsule being damaged on landing. The office of I.V. Barmin

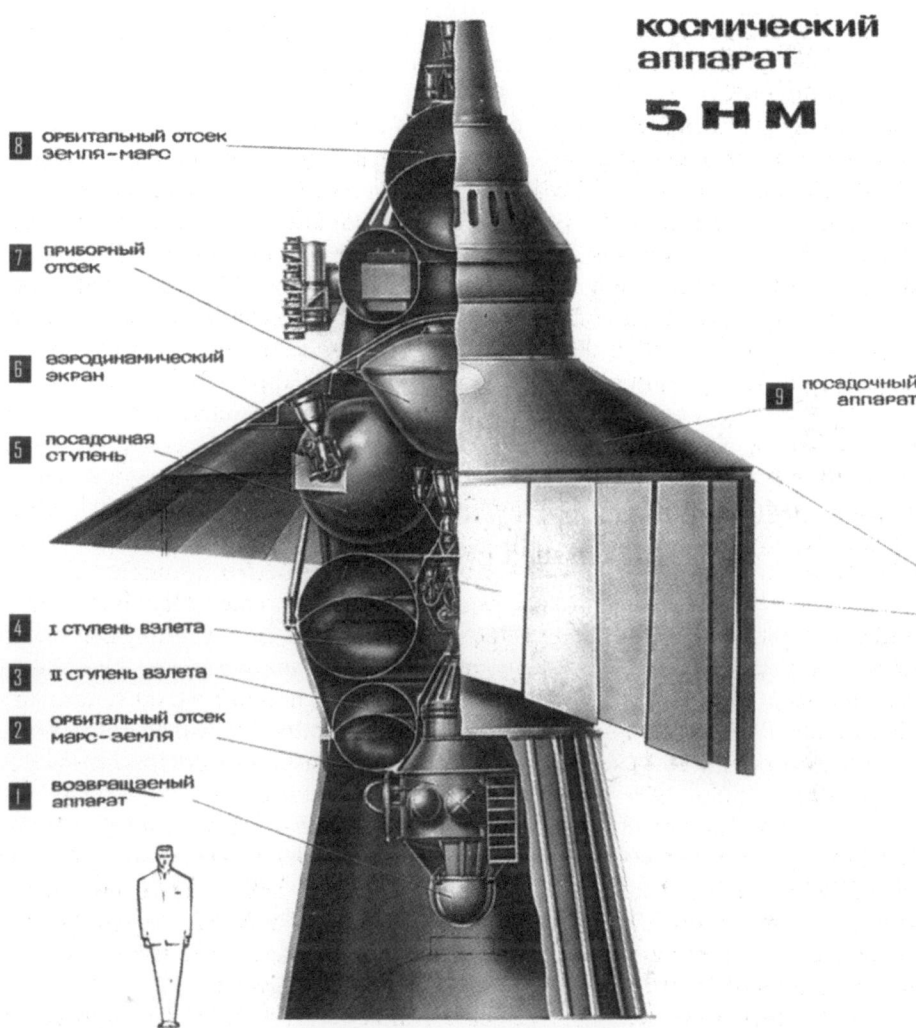

The 1970 version of the 5NM Mars sample-return probe: 1, sample-return capsule; 2, Mars-to-Earth module; 3, second stage of the Mars ascent rocket; 4, first stage of the Mars ascent rocket; 5, landing stage; 6, heat shield; 7, instrument module; 8, Earth-to-Mars module; 9, Mars landing module.

started work on a drill able to reach a depth of 3 meters on Mars. In 1977, as the first components for the spacecraft were being fabricated, the government ordered a reliability analysis of many space projects following a series of technical failures of the Igla (Needle) docking system of the Soyuz manned spacecraft – this system was to be used in the joining of two Proton stages in Earth orbit at the start of the M-79 mission, and later of the sample-return capsule and the orbiter in Mars orbit.

The 4GM prototype for a tracked rover that was developed by the Soviets in 1974 in connection with the 4M project. (Courtesy of VNII-TransMash)

The analysis found that the M-79 project had such a low probability of success that it was cancelled, and Sergei S. Kryukov was dismissed as director of the Lavochkin design bureau.[302,303,304,305] In fact, at this time the Soviet program for the exploration of the solar system was in the midst of a reorganization. To the scientists and engineers involved in the program, the largely secret debate was known as the 'war of the worlds'. The main outcome was the realization that Soviet deep-space exploration should not mindlessly address the same agenda as the US program, but should focus on scientific targets of its own. In particular, as the spate of studies had shown, and in view of the complete success of the Viking missions, the Soviets had neither the technology nor the resources to make a significant contribution to the exploration of Mars. It was therefore decided to concentrate Soviet efforts on Venus which, although it had a harsher environment, was nearer, would be less demanding on the limited reliability of Soviet technology and, significantly, was evidently being ignored by NASA.[306] When their technology had improved, there would be an opportunity to resume exploring Mars and, indeed, possibly also other destinations. In 1979, dismissing past failures, Viatcheslav M. Kovtunenko, the newly appointed director at Lavochkin, ordered the development of a new bus: the UMVL (Universalnyi Mars, Venera, Luna; Universal for Mars, Venus and the Moon).

A depiction of a 3M bus equipped with solar panels and carrying two heavy Mars landers mounted in tandem. This could represent an unmanned precursor to the proposed *Aelita* human expedition. (Courtesy of the TsNIIMash museum)

THE PLANET OF CONTRADICTIONS

Even after the invention of the telescope, the innermost planet of the solar system, Mercury, remained one of the most mysterious. As it never strays far from the Sun in the sky it was difficult to observe. Although the size of its disk indicated that it was fairly small, little else was known for certain – not even basic data such as the length of its day or the composition of its atmosphere, if any. In the early 1960s, advances in radar technology provided a new tool for studying the planets, and in 1962 the Soviets became the first to bounce radar off Mercury. When JPL did this in 1963, the results seemed to confirm what telescopic observers had inferred: the axial rotation of the planet was synchronized with its orbital period, in which case one hemisphere would be permanently illuminated and the other would be in permanent darkness. However, new radar observations in 1965 revealed that the situation was not so simple, and that the planet turned on its axis once every 58.6 days. Since this period is exactly two-thirds of the 88-day orbital period, the planet makes precisely three rotations for every two circuits of the Sun. As the intervals at which the planet was well placed for visual viewing were so restricted, it had appeared to present the same hemisphere at a given illumination, and this had misled astronomers into concluding that its rotation was tidally locked. Taken together with its marked eccentricity, the spin–orbit resonance will present an observer standing on the surface with a number of amazing phenomena, such as the halting and reversing of the passage of the Sun across the sky.[307]

Although Venus was an early target for interplanetary probes, it was apparent that Mercury would pose a much more difficult challenge. While it is true that Mercury's orbit occasionally brings it as close to Earth as Mars can approach, a spacecraft on a mission to Mercury requires a very energetic transfer orbit to 'drop' its perihelion so close to the Sun, and this, in turn, means that on reaching the planet its relative velocity will be large, which will either limit the opportunity to make observations during a flyby or make an orbit-insertion maneuver very expensive in terms of fuel. In fact, a probe leaving Earth's vicinity for Mercury must be braked almost as much as a probe bound for Jupiter must be accelerated. However, in the case of Mercury there is a shortcut. In the early 1960s Michael Minovitch, then a student at JPL, discovered that if a craft heading for Venus on a relatively low-energy transfer were to make a close flyby, the perturbation of its trajectory by the planet's gravity could deflect it on to Mercury – for free! Hence, whereas a direct launch to Mercury would require a large rocket such as the Titan III, a mission to Mercury via Venus would be feasible using a smaller (and cheaper) Atlas–Centaur. In effect, the 'gravity assist' at Venus would act as a 'third stage'. Calculations showed that launch windows would be available for such a mission in 1970 and again in 1973, and then a wait until the 1980s.

In fact, three Mercury missions were under study in the late 1960s. In the Soviet Union the principal research and development institution of the space and rocketry industry, TsNIIMash, proposed adapting the technology of the Venus–Mars craft for a dual mission to Venus and Mercury. According to the 'National Intelligence Estimates' drawn up by the CIA, such a mission could have been launched as early as 1970; however, it did not proceed beyond the concept phase. Furthermore (as related earlier), in 1969 the European Space Research Organization decided not to fund the MESO project.[308,309,310] NASA, however, decided that the favorable planetary alignment was simply too good an opportunity to miss, and in 1968, despite it being a period of reduced funding for space activity, devised the MVM (Mariner Venus–Mercury) mission and secured the requisite budgetary allocation. On the basis that savings could be made not just by using the Atlas–Centaur with a solid-fuel 'kick stage', but also by launching a single spacecraft instead of the usual pair, the mission was cost-capped at $98 million. Since the launch window was fairly brief – running from 16 October to 21 November 1973 – it would probably have been impracticable to send a backup probe in the event that the first was lost. The design of the mission was almost complete when, at a planning workshop at JPL in February 1970, Giuseppe Colombo of Padua University (a co-discoverer of the two-thirds spin–orbit resonance of Mercury) asked whether NASA had investigated the trajectory of the probe after its Mercury encounter. He pointed out that the flyby would deflect the spacecraft into a solar orbit with a period of about 176 days, which was twice the period of the planet, and suggested that with

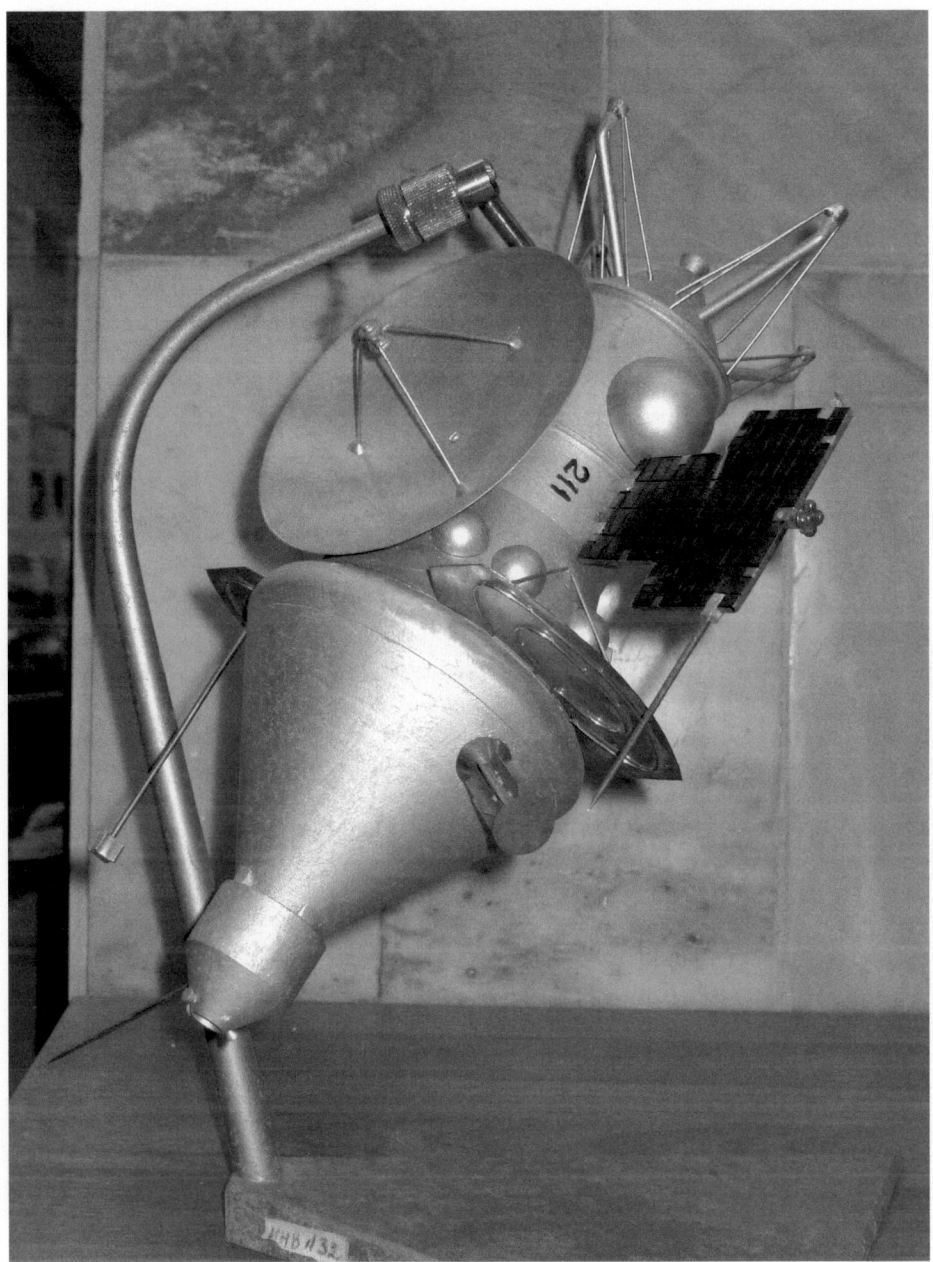

This is said to be a promotional model of a possible configuration for a Soviet probe for Mercury under study at TsNIIMash that would be powered by a combination of solar panels and RTGs, but other sources describe it as a Jupiter probe. (Courtesy of the TsNIIMash museum)

conservative propellant management and careful planning of the times and flyby ranges at both Venus and Mercury it should be possible to make the spacecraft return to Mercury at 176-day intervals. The only disadvantage of such a plan was that the illumination would be the same each time, but on the basis that this would double (perhaps triple) the scientific productivity of the mission, NASA readily accepted the suggestion and drew up a revised flight plan.[311]

The contract to build the MVM spacecraft was given to Boeing on 17 June 1971. As part of the cost-saving strategy, it was decided to reuse technology from previous missions, in particular from Mariner 9.[312] The octagonal magnesium bus was of the same dimensions as on Mariner 9, and had a mass of 18 kg. It contained the tank for 29 kg of hydrazine and the 222-N multi-start engine for in-flight maneuvering. The engine nozzle projected from the underside, which was protected by a fiberglass and aluminized Teflon 'umbrella' that was to be deployed in space to limit solar heating of the main structure. There were two solar panel wings, each 269 cm long and 97.5 cm wide, together capable of providing 820 W at perihelion, where insolation would have four times the power as in the vicinity of Earth. Each panel could be rotated on its own in order to vary its exposure to the Sun and thereby regulate its temperature. A tubular low-gain antenna would be used during maneuvers, otherwise the 137-cm-diameter parabolic double-frequency high-gain antenna would gimbal to hold Earth in its beam. An omnidirectional antenna would be able to receive commands at any stage of the flight. Mounted on the upper side was the two-degree-of-freedom scan platform with the scientific instruments, which included an imaging system based on that of Mariner 9 and an ultraviolet spectrometer that would be used throughout in order to (1) make astronomical observations of the deep sky, (2) analyze the upper atmosphere of Venus during the flyby, and (3) determine whether Mercury had an envelope of its own. The imaging system had four cameras: two had a focal length of 1,500 mm for a narrow 0.36 × 0.48-degree field of view, and two had a focal length of 62 mm for a wider field of view. There was an eight-position carousel that included color filters. The design of the single 700 × 832-pixel vidicon sensor enabled a pair of cameras to be serviced: when a wide view was required, a mirror would block the light from the narrow-angle optics and direct the light from the wider optics onto the vidicon.[313,314] Although it had a digital tape recorder, the probe was also to transmit imagery in real time using an upgraded communications system capable of operating at 117,600 bps, which was sufficient to send as many as 1,000 images during each fast Mercury flyby. Other instruments included a magnetometer on a 6.1-meter-long deployable boom. In fact, this instrument was in two parts: one at the tip of the boom and the other mounted half-way along to enable the magnetic effects of the spacecraft to be measured and compensated for in calibrating the real sensor. The scientific suite was completed by a charged-particle detector, a plasma analyzer, an infrared radiometer to measure the temperature of Mercury's surface, and an ultraviolet spectrometer fixed in position to point at the Sun and scan a planet's limb on entering and exiting its shadow.[315,316] Planetary atmospheres and ionospheres would be studied by radio occultations in which the probe would receive a radio carrier sent by the Deep Space Network and retransmit this at a slightly offset frequency until contact was

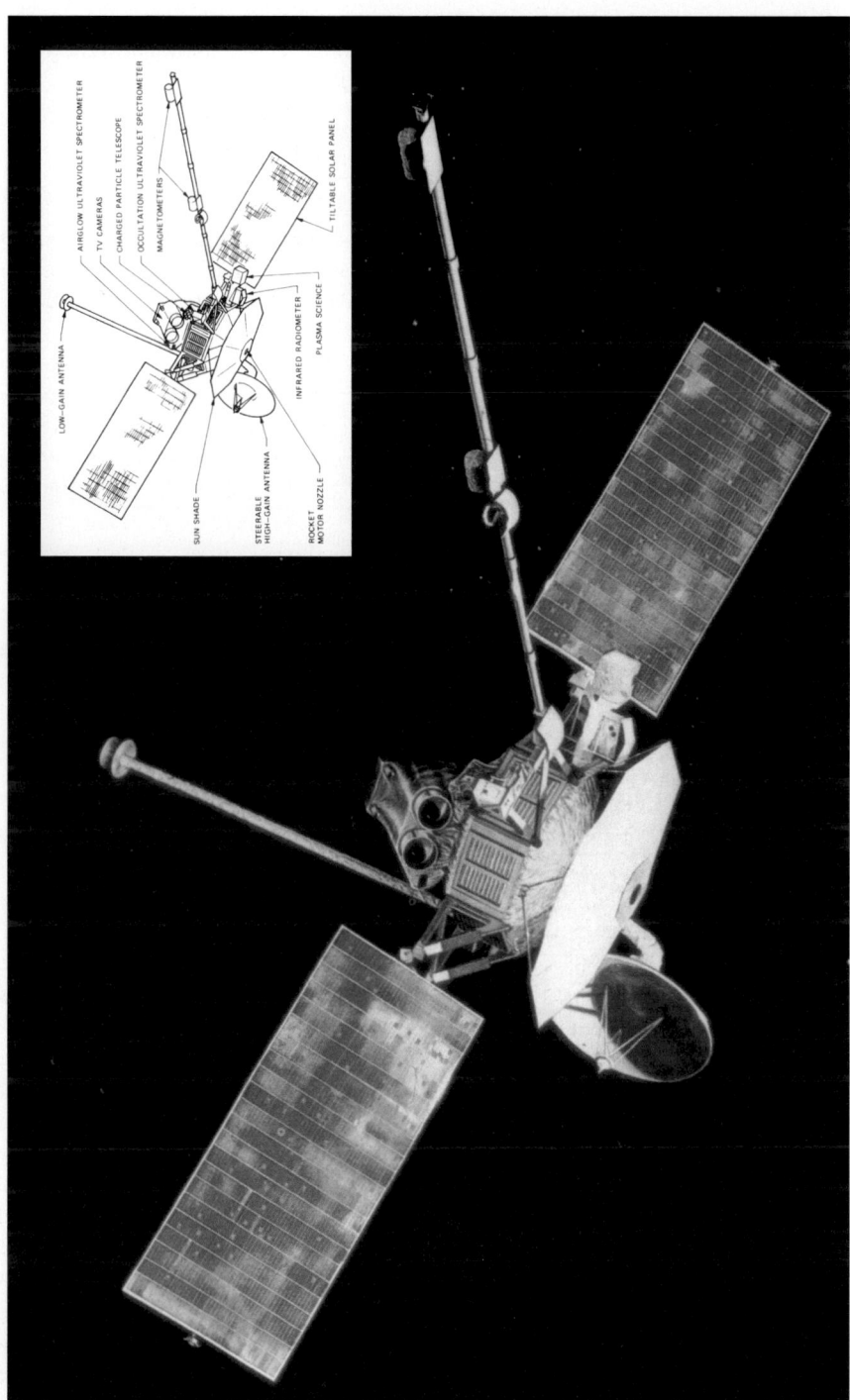

A depiction of Mariner 10 in space, with its major systems explained. (Courtesy of NASA/JPL/Caltech)

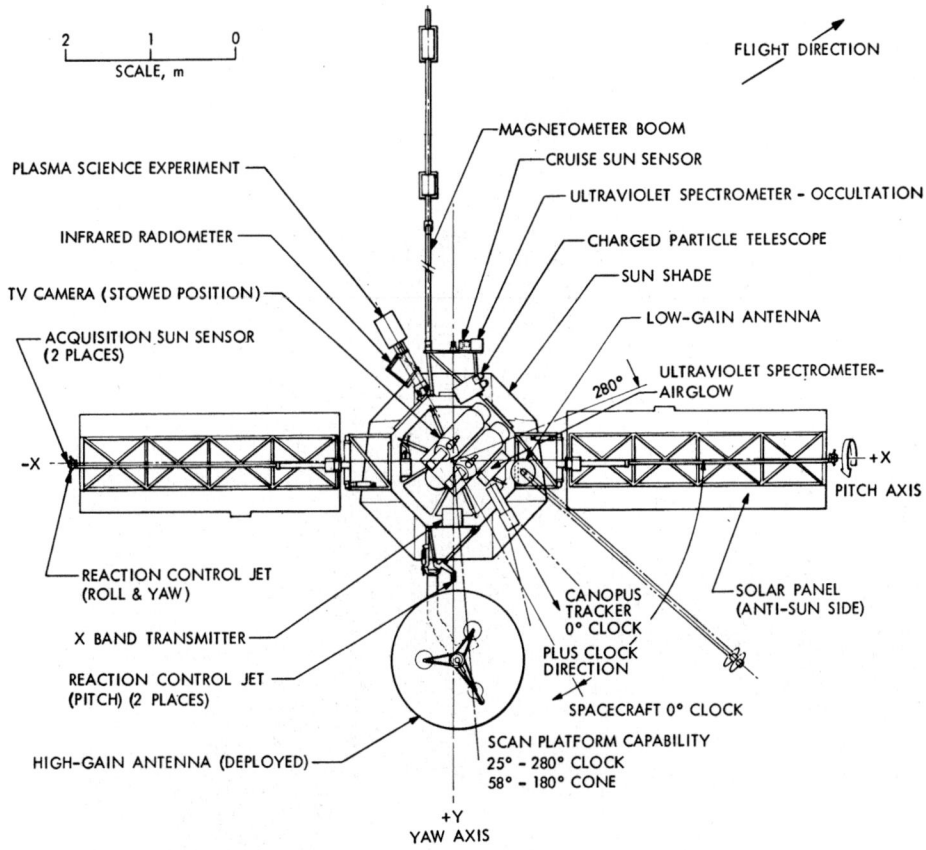

A line drawing of the Mariner 10 spacecraft. (Courtesy of NASA/JPL/Caltech)

lost upon crossing the limb. The tracking data would yield a measurement of the mass and gravity field of Mercury.[317,318,319] Overall, the instruments accounted for 78 kg of the spacecraft's launch mass of 502.9 kg.

A modified Atlas–Centaur lifted off at 18:45 UTC on 2 November 1973, within a 90-minute launch window. When the Centaur entered parking orbit, this was the first time that this stage had employed this technique to dispatch a deep-space probe, and soon thereafter it executed the 'escape' burn to insert Mariner 10 into a heliocentric orbit ranging between 0.70 and 1.11 AU inclined at 2.6 degrees to the ecliptic. The jettisoned stage then vented its remaining propellant to impart a small additional thrust to deflect its trajectory and ensure that it neither interfered with the spacecraft nor hit Venus.[320] As another new operational regime, Mariner 10 was to calibrate its instruments early, in near-Earth space. The particle detectors were activated 3 hours after launch, and the ultraviolet spectrometer 4 hours later, followed immediately by the imaging system. In order to test the cameras, two mosaics of the Earth were taken 16 hours 13 minutes after launch from a range of 200,000 km, showing cloud

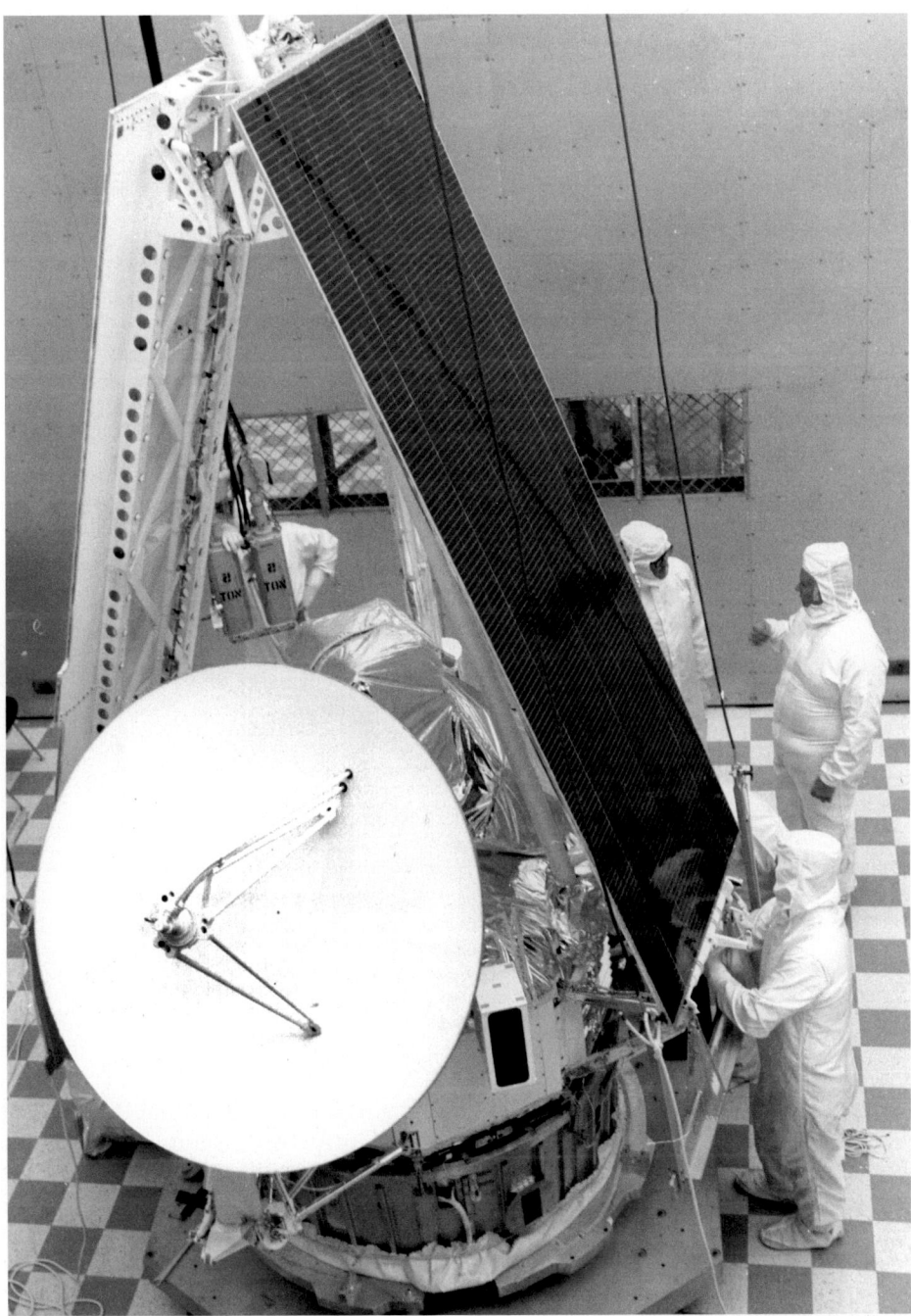

Mariner 10 is prepared for launch. The dark slit in the foreground is the Canopus star sensor.

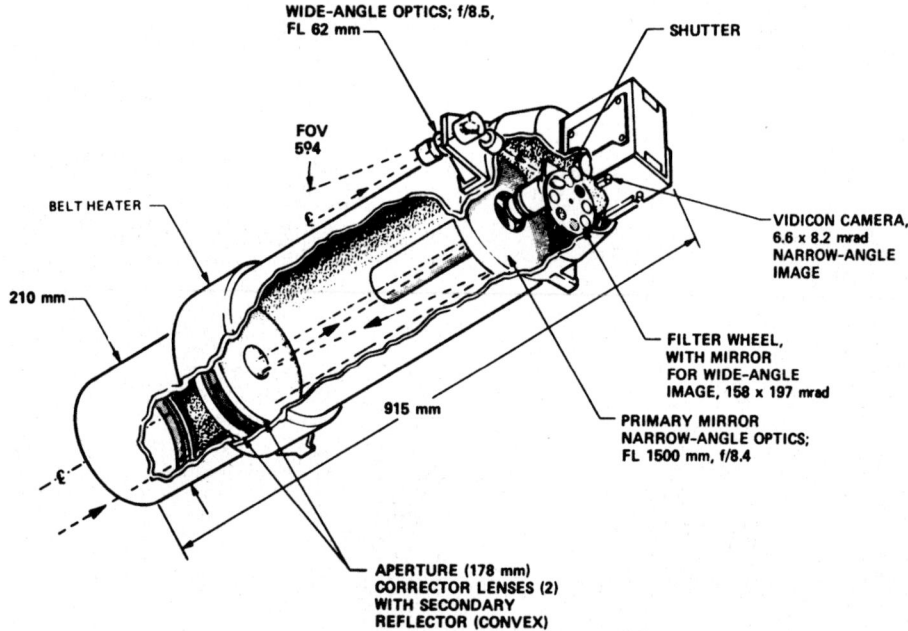

The optics of the Mariner 10 camera. The filter wheel included a mirror position to block the light from the narrow-angle optics and redirect that from the wide-angle optics to the vidicon sensor, thereby enabling one sensor to serve both systems.

formations over the Gulf of Mexico and the Pacific Ocean. The cameras were then slewed to take six mosaics of the Moon from a range of 110,000 km, providing an overhead perspective of regions in the northern hemisphere near the pole that had never been imaged before, or had been imaged only at a low oblique angle, thereby providing as a bonus a welcome contribution to lunar exploration. The mission then encountered the first of what would become a large number of technical difficulties. First, when the plasma detector was activated and reported no trace of the solar wind it was suspected that the door of the sensor had failed to open fully. Next, the heater system, which was designed to keep the vidicons warm while the imaging system was not operating, failed when it was activated; this meant that the cameras would have to remain powered for the entire flight so as to ensure that they did not freeze. As the focus of the optical systems was sensitive to temperature, test images were taken of star fields, including the Pleiades cluster. Despite these technical issues, the mission was off to an excellent start. On 13 November the spacecraft made a course correction of 7.78 m/s to place its flyby of Venus within the narrow spatial and temporal window required for the gravity-assist to Mercury. A second smaller correction on 21 January 1974 further refined the encounter by 1,384 km and 2 minutes.

Mariner 10 soon began to collect interesting scientific data. During several routine scans of the sky using the ultraviolet spectrometer, it provided proof of a decreasing

This striking image of a cloud-covered Earth was taken by Mariner 10 soon after it was safely on its way, as a test of its camera.

solar wind flux with increasing heliocentric latitude. These observations also yielded data on the interstellar medium, and detected several stars.[321,322] After observing the Gum Nebula, it turned its attention to comet C/1973E1 Kohoutek. Discovered by the Czech astronomer Lubŏs Kohoutek on 7 March 1973, this was a long-period type of comet that rarely ventured into the inner solar system, and so was widely expected to be the 'comet of the century', but when it made its perihelion passage at Christmas it failed to live up to this billing. Nevertheless, it was one of the first such objects for which satellites and deep-space probes were in position to provide good scientific data. In particular, between 11 and 24 January 1974 (i.e. during the interval of 2 to 4 weeks after the comet reached perihelion on 26 December) Mariner 10 took spectra of its tail and nuclear region in the extreme-ultraviolet and Lyman-alpha hydrogen wavelengths from a range of some 100 million km. The Lyman-alpha data provided intensity maps of the 30-million-km-diameter cloud of hydrogen that surrounded the comet, together with estimates of the rate of gaseous hydrogen

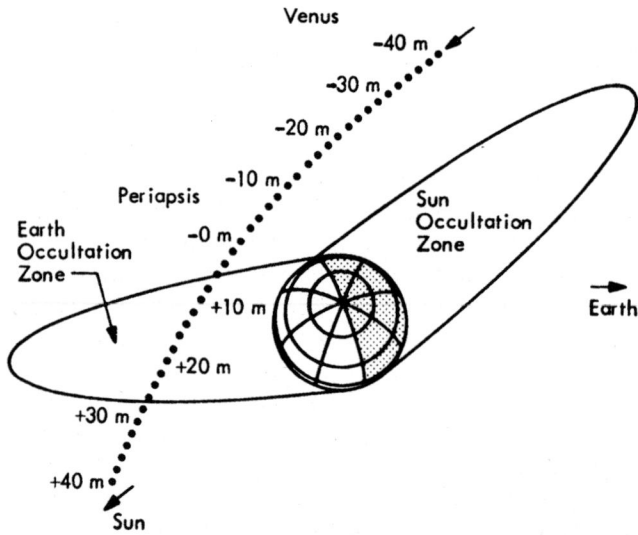

The geometry of Mariner 10's flyby of Venus as viewed from the perspective of the north ecliptic pole, with dots marking its position at 2-minute intervals. Note that the trajectory avoided penetrating the shadow of planet (marked Sun Occultation Zone). (Courtesy of NASA/JPL/Caltech)

production and how this decreased as the comet withdrew from the Sun. Attempts to take pictures were foiled by the comet's unexpected faintness.[323]

Immediately after this series of observations, Mariner 10 suffered another major mishap. While calibrating its gyroscopes on 28 January, one failed and sent spurious signals to the attitude control system which, thinking the spacecraft to be unstable, fired thrusters in an attempt at recovery, thus using 0.6 kg of compressed nitrogen, which was 16 per cent of its total reserve. In order not to jeopardize the forthcoming Venus encounter, it was decided to switch attitude control from the gyroscope-based system to the system that used positional sightings of the Sun and Canopus. It was later discovered that oscillations of the 6-meter-long magnetometer boom (whose weightless characteristics could not be fully determined during ground testing) were inducing instabilities in the attitude control system. A mysterious malfunction in the primary power distribution system left the spacecraft with very tight margins for the remainder of the flight. A problem also developed in the high-gain antenna feed that limited the signal power.[324]

Although the flyby trajectory was designed to lead to a Mercury encounter, and so was not optimized for Venus science, it was possible to make further observations of the manner in which the planet interacted with the solar wind and to sound the upper atmosphere by limb-crossing radio-occultations. Furthermore, some instruments had been especially equipped to study Venus – for example, the cameras had ultraviolet filters to image structure in the visible cloud layer. However, the range of sensitivity of the infrared radiometer had been selected for the predicted temperature at the surface of Mercury, and a solar occultation was not possible.[325] The encounter sequence started on 28 January, as the instruments were activated one by one and calibrated. At 16:50 UTC on 5 February, while 8,000 km above the night-side, the picture-taking started. The first picture showed the thin bright line of the day/night terminator cusp against the sky and dark north polar area. While Venera 2 may have been the first mission to take close-up pictures of Venus, this probe had not been able to transmit its results to Earth. The pictures from Mariner 10 were therefore eagerly anticipated. At 17:01 Mariner 10 made its closest approach to the planet at a range of 5,768 km, and 6 minutes later it passed behind the night-side limb virtually at the equator. In fact, the experiment exploited the steerable high-gain antenna to maintain Earth in the beam for as long as possible by slewing it as the Venusian atmosphere refracted the incoming radio waves, and, as expected, the signal from the spacecraft continued for several minutes after the geometrical occultation. The first frequency was lost in the altitude range 52–53 km, while the second remained detectable until 45 km.[326] The probe reappeared over the day-side at a mid-southern latitude. While it was out of contact, the spacecraft stored its imagery and other data on tape. The flyby reduced the spacecraft's heliocentric speed by 4.41 km/s to 32.3 km/s, deflecting it into an orbit ranging between 0.387 and 0.839 AU with a period of 176 days, as required to produce a series of encounters with Mercury.[327,328] The Centaur stage is estimated to have made a 45,000-km flyby at around the same time.

The Venus encounter provided good scientific results. Tracking data taken over a period of 10 days centered on the encounter provided a third estimate of the mass of the planet, thereby refining the accuracy of this quantity.[329] A total of 4,165 pictures were able to be taken over an 8-day period, covering two full rotations of the upper atmosphere. The closest images had a cloud-top resolution of about 120 meters, but

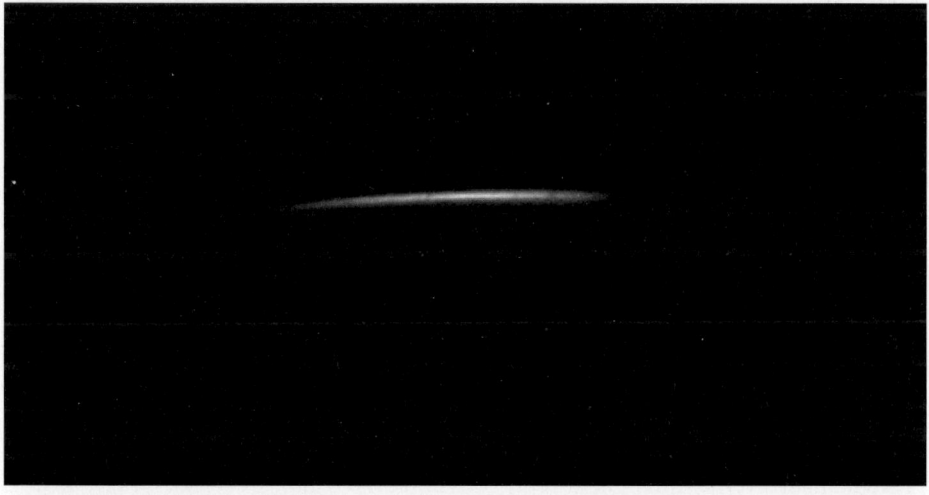

This is the first image of Venus ever returned by a spacecraft flying in the vicinity of the planet. It was taken viewing the night-side, and the streak is the bright polar cusp.

by the end of the encounter sequence on 13 February the resolution had declined to 130 km (nevertheless, this was still twice as good as had been achieved from Earth). The pictures were taken in eight imaging sessions, mostly with ultraviolet filters that telescopic studies had shown to be suitable, and were later made into a time-lapse movie. Unfortunately the post-encounter imaging plan had to be curtailed in order to preserve the life of the vidicons – vital for the Mercury flybys – lest they had been damaged by the failure of their heating system early in the mission.[330] The images of the atmosphere gave the first documented example of a meridional air flow known as a 'Hadley cell' – a circulation system in which warm air ascends near the equator and migrates at high altitude toward each pole, whereupon it cools, descends and returns at low altitude. This simple model of dispersing solar heat from the equator had been proposed for Earth's atmosphere in the eighteenth century, but the Earth's rapid rotation complicates the situation. Venus, however, rotates much more slowly, and the lower tilt of its rotation axis does not impose major seasonal differences, and a Hadley cell is possible. However, although the planet rotates extremely slowly, the upper atmosphere circles the planet in 4 days, resulting in wind speeds of 100 m/s at an altitude of 60 km, and this super-rotation 'winds up' the air stream that is moving poleward to create the sideways-Y pattern with a vortex at each pole that had been faintly discerned telescopically, and was now strikingly evident. Furthermore, at the subsolar point, and in the area immediately downwind of it, there were many small 'cellules', suggesting that the insolation was inducing convective currents that were interacting with the super-rotating flow to create bow-wave-like structures. Overall, however, the atmosphere was remarkably regular, with no traces of eddies, cyclones or other large-scale structures.[331,332,333] By placing Mariner 10 down-sun of Venus for more than 10 days during the approach, the trajectory had particularly assisted in the study of how the planet interacts with the solar wind, and the magnetometer had noted distortions and fluctuations of the magnetic field for 6 days prior to the flyby. The plasma instrument similarly benefited from the unique trajectory to explore the plasma tail downwind of the planet. In addition, the instrument was able to study the ionospheric bow shock on the sunward side. The ultraviolet spectrometer observed hydrogen and helium in the upper atmosphere, but not deuterium, indicating that the hydrogen did not originate from the impact of comets (which are enriched in deuterium) but must either be being accreted from the solar wind or be due to the dissociation of water molecules. A significant quantity of atomic oxygen was detected (more, in fact, than had been found by Mariner 9 at Mars) which further suggested the presence of water vapor at some level in the atmosphere. The infrared radiometer produced one temperature profile in the 20 minutes prior to closest approach, scanning across both the day- and night-sides, but no temperature difference was measured, thereby confirming the measurement by Mariner 2 more than a decade earlier.[334,335]

As Mariner 10 ventured closer to the Sun, it increasingly tilted its solar panels to prevent them overheating. On several occasions the Canopus sensor briefly lost its star, almost certainly because the heat was causing the vehicle to shed particulates that were drifting across the sensor's field of view and creating a distraction. To preclude a further waste of nitrogen, the gyroscopes were turned off and the

An ultraviolet view of Venus by Mariner 10 showing the Y-shaped circulation, polar rings and bow-like waves emanating from the subsolar point. The dark spots near the terminator are artifacts caused by dust blemishes on the camera.

differential tilt of the solar panels was used to generate a solar radiation pressure torque that was sufficient to counter the natural drift of the spacecraft.[336] The third course correction was made on 16 March, the 17.8 m/s change in velocity shifting the flyby from the day-side to the night-side of Mercury to create Sun and Earth occultations. This burn had been delayed until the engine was properly aligned for firing, to avoid the spacecraft using further nitrogen to actively adopt that attitude. The closest point of approach to Mercury would be 200 km nearer than optimum, but because it would produce a second encounter no further corrections were made.[337,338,339] The next day, all experiments except the cameras were turned on.

The far-encounter imaging of Mercury was started on 23 March, 6.5 days out, and lasted until 17 hours before the flyby. The first images were taken from a distance of 5.31 million km, and revealed the presence of several bright spots on the sunlit side. In addition to several images transmitted in real time, 216 were stored on

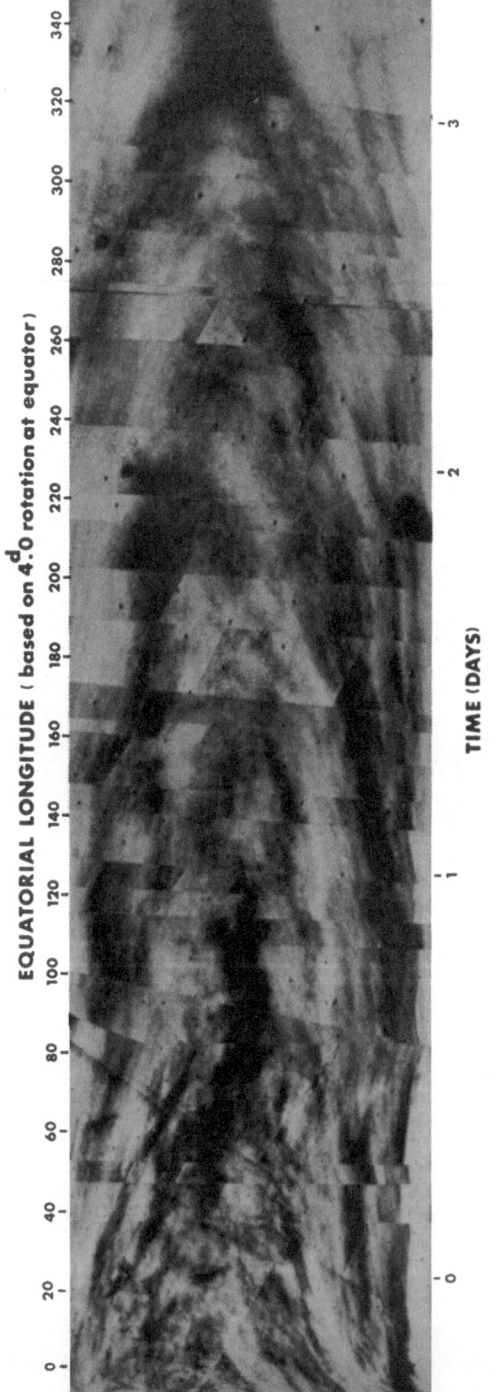

A projection of Venus at equatorial and intermediate latitudes, as the subsolar point advances from left to right with the rotation of the atmosphere over a 4-day period. (Reprinted with permission from Murray, B.C., et al., "Venus: Atmospheric Motion and Structure from Mariner 10 Pictures", Science, 183, 1974, 1307–1315. Copyright 1974 AAAS)

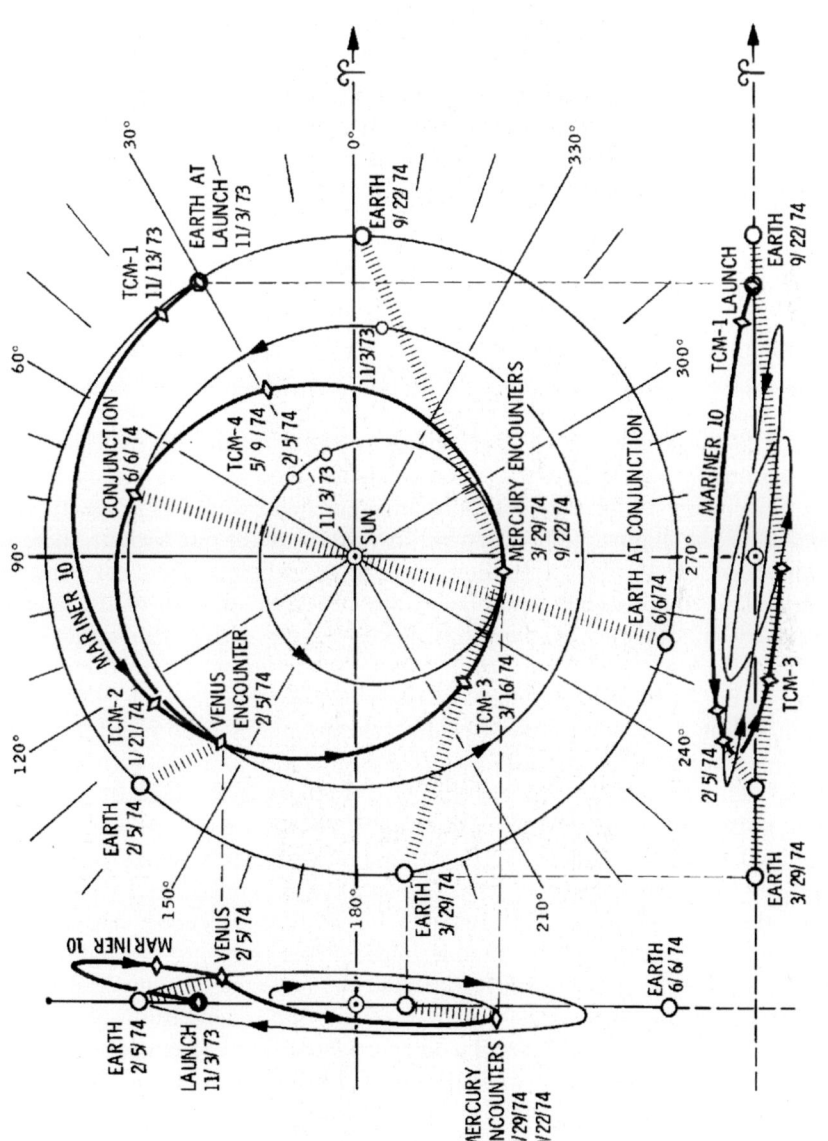

The heliocentric trajectory of Mariner 10.

tape. They showed a mottled surface of light and dark spots that were suggestive of craters. The first crater to be individually resolved was a dark spot on a large bright patch. It was named in honor of Gerard P. Kuiper, a pioneer of both terrestrial and space-based planetary astronomy, principal investigator of NASA's Ranger lunar missions and a member of the Mariner 10 television team, who died shortly after the spacecraft was launched.[340,341] On 28 March the imaging rate was increased to one frame every 42 seconds; in effect, it was 'live' television. The closest point of approach occurred at 20:47 UTC on 29 March at an altitude of 703 km. The spacecraft crossed behind the limb above the equator on the night-side and reappeared over the day-side, but there was no evidence in the radio-occultation data of an atmosphere or an ionosphere at a sensitivity of 1 Pa and 100 electrons/cm^3 respectively. During the main phase of the encounter, 612 images were transmitted to Earth in real time and 35 full frames were taped. No images were taken during the 30-minute period centered on the time of closest approach because this occurred while crossing the night-side. The celestial mechanics experiment greatly improved knowledge of the planet's mass, in addition to its detailed shape and oblateness, showing its mean density to be slightly less than that of Earth and greater than that of Venus. The most unexpected discovery was made by the magnetometer and plasma experiment: although it had been predicted that Mercury would have no magnetic field owing to its slow rotation, and that there would be a cavity in the solar wind over the night-time hemisphere (as in the case of our Moon), there was clear evidence of a bow shock resulting from the deflection of the solar wind by a magnetic field. On approaching the planet, the measured strength of the magnetic field increased smoothly to a value that was in excess of five times that of the interplanetary field, which implied a field strength at the surface of 1/150th of that at the surface of Earth. Notwithstanding the negative atmospheric results of the radio occultation, the ultraviolet spectrometer detected a tenuous helium envelope. This was probably derived from the radioactive decay of uranium or thorium, and its presence provided insight into the composition of the surface, which the spacecraft was not equipped to analyze directly. Other noble gases were also detected – argon, neon and possibly xenon. Taken together, these measurements gave an upper limit for the pressure at the surface of 10^{-7} Pa. The temperatures measured by the infrared radiometer were 300–430°C on the sunlit side, but by 'mid afternoon' had decreased to 190°C. As the instrument scanned the terminator and over onto the night-side the temperature fell below –120°C, which indicated the presence of a surficial layer of thermally insulating fine dust, as is the case on the Moon. Fluctuations of up to 2°C were observed, possibly indicating rocky outcrops free of dust. The temperature was –170°C at local midnight, and it was estimated that just prior to dawn it would have fallen to as low as –180°C.[342] Although the imagery showed the planet – or at least the portion that was illuminated – to be almost entirely covered with impact basins, craters, scarps and ridges, and rather reminiscent of the Moon, it also had its own peculiar features. For example, there were high scarps extending for hundreds of kilometers. Near the terminator that was imaged on the inbound leg was a complex terrain of hills, valleys and depressions, with the rims of many craters disrupted, and on the antipodal terminator was a 1,300-km-diameter impact basin with a number of concentric rings of cliffs and ridges. Since this feature was located at the position of one of the two 'hot spots' on the equator that face the Sun at perihelion – and thus receive more heat than the rest of the planet – it was later named the Caloris basin. Imaging continued for 13 days after the flyby, in order to secure sufficient data to attempt to gauge the diameter of the planet.[343]

An amusing incident occurred during this Mercury flyby. On both the incoming and outgoing legs Mariner 10 searched for satellites, and 2 days prior to the flyby the ultraviolet spectrometer detected a bright source over the night-side of the planet "that had no right to be there", and another was seen when withdrawing 3 days later. The premise that ultraviolet radiation of such a short wavelength from a star should be completely absorbed by neutral hydrogen in the interstellar medium prompted the conclusion that the source had to be very close, and when the appropriate calculation indicated that the object was moving relative to Mercury at a speed of 4 km/s, the team called an 'instant science' press conference to report the discovery of a small moon. Soon after, however, trajectory experts proved that the object seen during the outbound leg was the star 31 Crateris which, despite its name implying that it lies in the constellation of Crater, actually lies in Corvus. This fortuitous discovery proved that extreme-ultraviolet radiation from peculiar classes of stars is indeed detectable, and thus opened up another astronomical window in the electromagnetic spectrum. The object seen on the inbound leg was never identified.[344]

As Mariner 10 departed, it suffered a series of minor and major mishaps. First, a problem with the thermal control system increased the spacecraft's temperature and power consumption. Then the tape recorder began to operate erratically (it failed totally several weeks later), one of the transmitters showed signs of degradation, and a problem robbed the telemetry system of half of its engineering channels. Finally, it was decided to switch off the plasma instrument between encounters because it was nearing the end of its useful life. The most serious issue from the viewpoint of the scientific program was the tape recorder, but it was estimated that a combination of full-resolution and reduced-resolution real-time imaging would provide a completely satisfactory second encounter.[345] There was a vigorous debate over the trajectory for the return visit, and this had to be determined before the requisite set-up maneuver

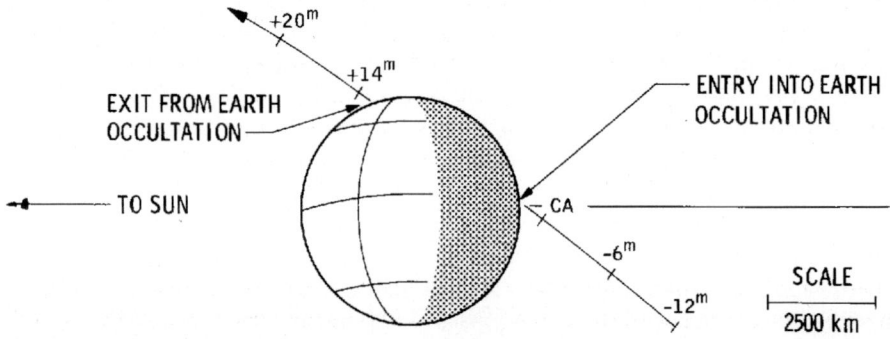

The geometry of Mariner 10's first encounter with Mercury as viewed from Earth.

could be executed. The imaging team wanted a flyby over the day-side to image the south polar regions and link up the imaging coverage taken during the incoming and outgoing legs of the first encounter, but those with particles and fields instruments wished to pass over the night-side to collect more data on the surprisingly strong magnetic field. The issue was that a night-side pass would not improve the imagery coverage, while a day-side encounter would provide little insight into the magnetic field. In the end, it was decided to undertake imaging during the second encounter and devote the third (if there should be one) to particles and fields. Part of the rationale was that by the time of the third encounter, the degradation of the vidicons would probably be so severe as to preclude imaging. The fourth course correction of the mission was executed in two parts on 9 and 10 May; the first changing the spacecraft's speed by 50 m/s in order to move the point of closest approach from 800,000 to 283,000 km, and the second of 27.6 m/s to reduce it to 45,000 km. While in transit, Mariner 10 searched for celestial ultraviolet sources.[346,347] It was in solar conjunction from mid-May to early July, reaching a minimum apparent separation of six solar radii on 6 June. In addition to searching for relativistic effects, the double-frequency radio apparatus was, for the first time, used to probe the corona and solar wind at very high latitudes.[348,349] A 3.32-m/s course correction on 2 July moved the point of closest approach for the second encounter to Mercurian latitude 40°S in order to draw in the range of the third, which would otherwise have taken place at a distance of at least 400,000 km. This had to be done before the necessary maneuver exceeded the engine's remaining correction capability.[350] The Deep Space Network introduced a new tracking technique for the second encounter, in which the signals received by a cluster of antennas were to be electronically combined to generate a stronger signal than that from any single antenna.[351]

Far-encounter imaging for the second visit to Mercury began on 17 September by calibrating the cameras using Jupiter. The trajectory requirements placed the flyby at 20:59 UTC on 21 September 1974 at the relatively high altitude of 48,069 km above the day-side. Although this altitude impaired the surface resolution, which was 1–1.5 km at best, it was sufficiently high to provide overlap between narrow-angle frames. The imagery of the southern polar regions extended the coverage of the illuminated hemisphere from 50 to 75 per cent, and cartographically and geologically linked the areas photographed during the incoming and outgoing legs of the first encounter. An outgoing sequence was taken on 22 September, and the cameras were turned off. A novel navigation strategy was implemented during the second encounter in which in excess of 100 long-exposure images were taken to measure the position of Mercury relative to the stars from the vantage point of the spacecraft, proving a technique that would be used extensively on the Voyager missions to the outer solar system. Owing to the large flyby range and unfavorable geometry, only the ultraviolet spectrometer was able to make follow-up observations of the tenuous helium envelope.[352,353]

On 30 October 1974 and 13 February 1975 Mariner 10 refined its course for the third encounter. As this was to be a particles and fields flyby it had to make a low pass over the night-side of the planet. On 7 March an additional maneuver was made to open the range by 160 km in order to preclude an accidental impact! The final trajectory would provide a flyby at an altitude of 327 km and a relative speed of 11

A long-range view of Mercury by Mariner 10 that showed the planet to resemble the Moon by being heavily cratered and streaked by rays of bright ejecta.

km/s.[354,355] One hundred hours before the encounter, the spacecraft again lost track of Canopus, but this time the normal acquisition method could not be used since it would induce oscillations in the magnetometer boom that would, in turn, prompt the attitude control system to expend the little nitrogen that remained. A gyroscope-only sequence was prepared that would make use of the Canopus sensor only in its initial phase. An attempt to reacquire Canopus was made on 12 March, but failed. At this point, an emergency request was made for the Deep Space Network's Goldstone and Canberra antennas, which were at that time allocated to tracking the US–German Helios 1 solar probe through its first perihelion. Mariner 10's attitude was eventually stabilized, and the cameras (whose vidicons, contrary to expectation, were still functional) were turned on at 10:08 UTC on 15 March, 36 hours prior to the flyby, which was at 22:39 UTC on 16 March at a latitude of about 70°N.[356] Unfortunately, there was now a problem on Earth: only the Canberra antenna had a line of sight at the time of the spacecraft's closest approach to Mercury, and as its low-noise feed had a coolant leak it could only receive real-time imagery at a much lower rate than planned. As the tape recorder had long since failed, this restriction meant that only the central quarter of each frame was able to be returned. Imaging was stopped 13

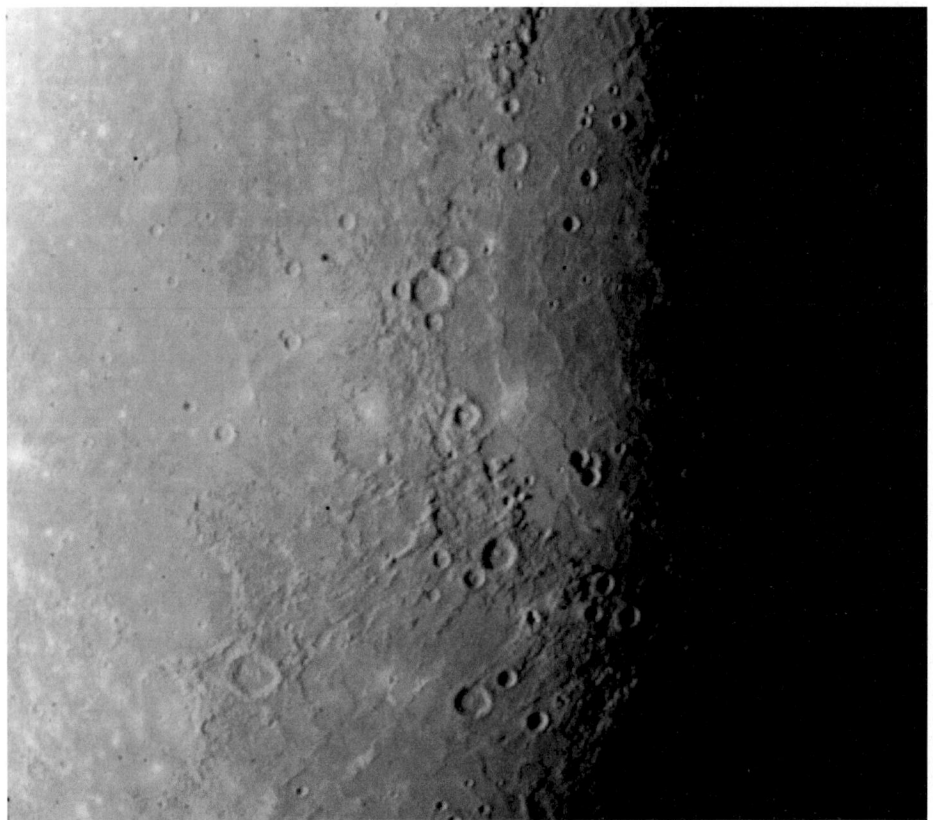

A view of Mercury by Mariner 10 showing the Caloris basin on the terminator.

minutes prior to closest approach, and resumed 5 minutes after the flyby, as soon as the spacecraft crossed over the sunlit side.[357] About 300 quarter-frames were taken for high-resolution coverage of selected areas. Ultraviolet data were also collected, but the infrared radiometer could not collect new data because it was on a fixed position on the spacecraft and on this occasion was facing away from the planet. The magnetometer proved that Mercury does indeed have a magnetic field, whose dipole axis appears to be tilted 7 degrees from the axis of rotation, and the plasma experiment added further confirmation by detecting the spacecraft's crossing of the magnetopause and bow shock at almost precisely the locations predicted after the first encounter.[358]

Mariner 10 returned in excess of 2,700 pictures during its three flybys, covering a total of about 45 per cent of the planet's surface at resolutions ranging from 4,000 to 100 meters. Mercury was shown to be broadly similar to the Moon in many respects (as had been inferred from earlier studies) but the spacecraft's images also revealed that the planet possessed peculiar features of its own. The craters differed from those on the Moon in that the rough terrain that marked where the material ejected as an

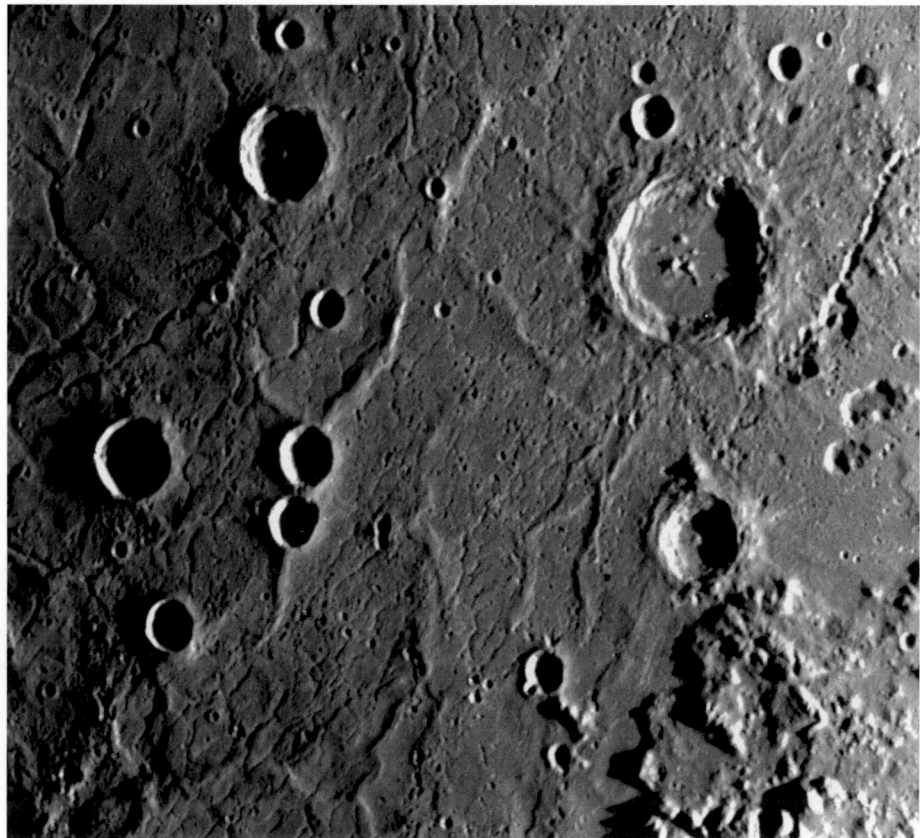

A part of the floor of the Caloris basin, showing it to be a ridged and fractured lava flow, somewhat similar to the lunar maria. The largest crater in this image is 60 km across.

impact excavated a crater fell back to the surface, was closer to the crater's rim. This was because the gravitational field of the planet is more than twice as strong as that of the Moon, and any given impact can propel ejecta only half as far. Counts of the numbers of craters of different sizes suggested that (as was thought to be the case for the Moon and Mars) the Late Heavy Bombardment battered Mercury soon after it accreted from the solar nebula. A morphological unit called the intercrater plains, which in total accounted for 45 per cent of the imaged surface, was itself heavily cratered, indicating that it dated back to the earliest days of the planet's history. The basins and some ancient craters were partially obscured by smooth plains (about 15 per cent of the imaged surface) that were not so cratered, indicating that they were relatively young. A flood-basalt origin has been suggested for both types of plain, meaning that they formed when large volumes of low-viscosity lava was erupted from widely distributed deep fissures to create large smooth sheets. The Mercurian plains may therefore be the equivalent of the maria on the Moon. Some of the most interesting features were the 'lobate scarps', so-named for their sinuous tracks on the

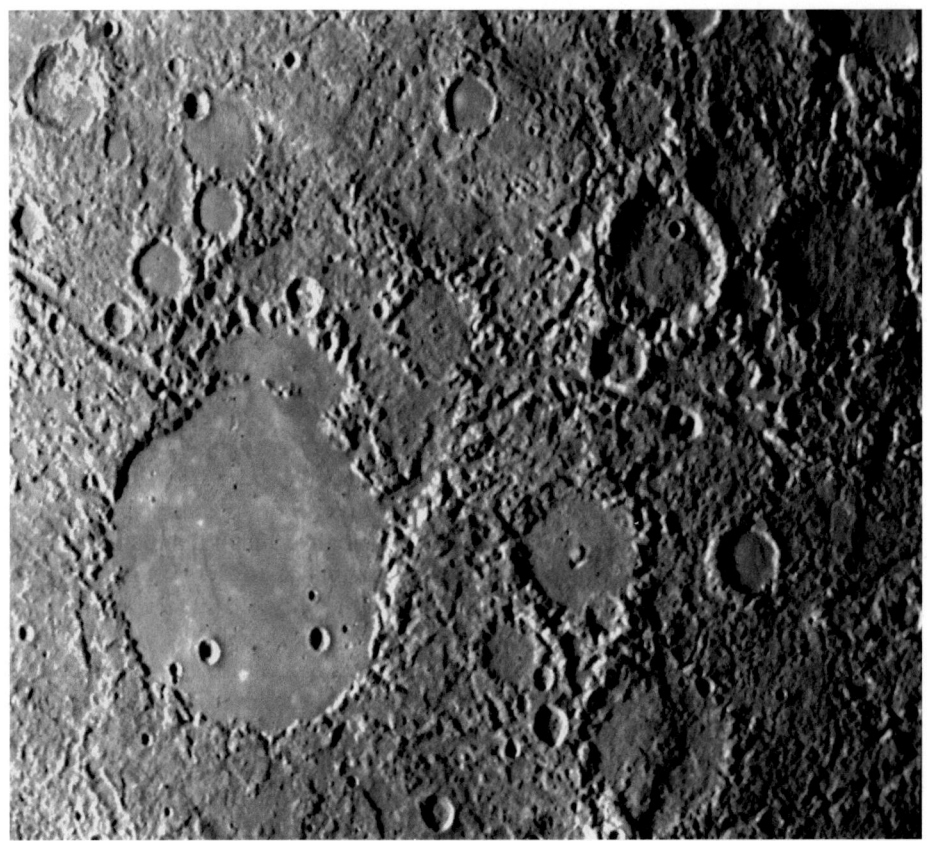

The concentration of shock waves antipodal to the Caloris basin is believed to have disrupted the surface and made this hilly terrain. The smooth floors of some craters were presumably formed later. The image is about 550 km on a side.

landscape. They ranged from 20 to 500 km in length and from several hundred to several thousand meters in height. Discovery Scarp ran for almost 400 km, and in places was 2 km tall. The fact that lobate scarps transected all types of terrain apart from some of the youngest of the smooth plains suggested that they were relatively youthful. A geometrical analysis showed that the scarps formed when compressional stresses cracked the crust and slid one block up a shallow inclination and across the adjacent surface – a situation known as a reverse fault. According to one theory, this occurred when the planet's rate of spin was slowed to its present value. That is, if it had rotated more rapidly earlier in its history, then it was most likely oblate, and in slowing it would have become more spheroidal, with the resulting stress creating the scarps. However, scarps formed by this mechanism would tend to have a latitudinal distribution that is not in evidence. By another theory, the scarps were formed as the planet cooled and contracted after an early heating phase in which it expanded. Interestingly, a complete solidification of the planet would have caused a much

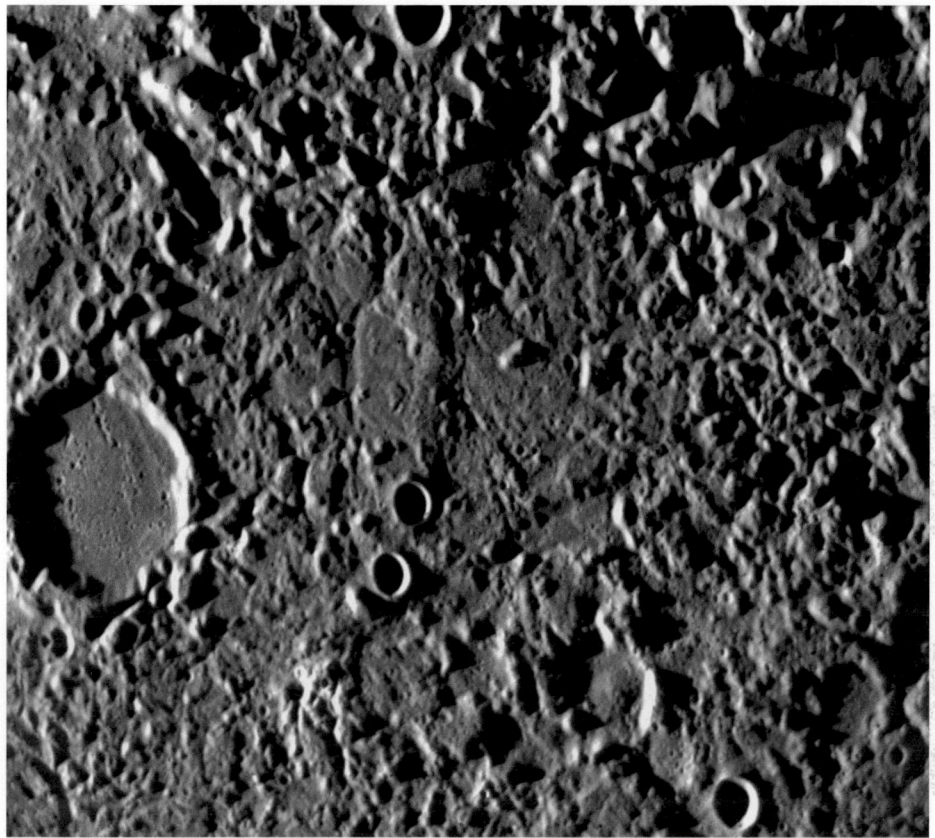

A closer look at part of the previous picture. The large crater (visible in previous image at the lower-right) is 31 km across.

larger decrease of the radius than is observed, which opened the possibility that the core may be still molten, at least in part.

The most impressive feature imaged by Mariner 10 is the Caloris basin, marked by a series of concentric rings, hills, terraces, ridges and peripheral smooth plains, in all some 2,500 km in diameter. Unfortunately, owing to the resonance between the orbits of the planet and the spacecraft, more than half of the basin was in the shadow during each encounter. As mentioned above, a complex hilly terrain covering almost 400,000 km^2 was discovered inbound to the first encounter that was later revealed to be antipodal to the Caloris basin. The hills are 5 to 10 km wide and up to 2 km high, and are separated by valleys more than 120 km long and 5 km wide. This terrain was almost certainly caused by the impact that formed the Caloris basin, because such a large impact would have generated very powerful seismic waves that, after traveling directly through the core and around the crust at the surface, would have converged and been focused at the opposite side of the planet, inducing dramatic cracking and crustal movements in excess of 1 km that left a hilly terrain. Similar terrains occur

An oblique view of the intercrater plains of Mercury.

on the Moon antipodal to the largest basins. A recent recalibration of Mariner 10's images provided false-color data that could be analyzed to infer mineral abundances and iron content, and prove a volcanic origin for the intercrater plains. Moreover, the true extent of the ejecta blankets of some of the larger craters was revealed.[359]

After the mission, the International Astronomical Union published a nomenclature system for naming the features discovered by Mariner 10. Because the god Hermes (i.e. Mercury) was the protector of arts in ancient Greece, craters were named after painters, writers and composers; plains were named after gods of other cultures and mythologies that had similar attributes to Hermes; valleys were named after radio telescopes; and scarps were named after ships associated with the exploration of the Earth. There are two exceptions, one being the crater which was named in honor of the planetary scientist Gerard P. Kuiper, and the other the small 1.5-km-diameter crater Hun Kal ('twenty' in Mayan) which was so named because the prime meridian that the astronomers defined was in darkness during all the encounters, and this small crater, which lies within half a degree of the 20-degree line, was utilized as the primary cartographic reference. Basic planetary data was also

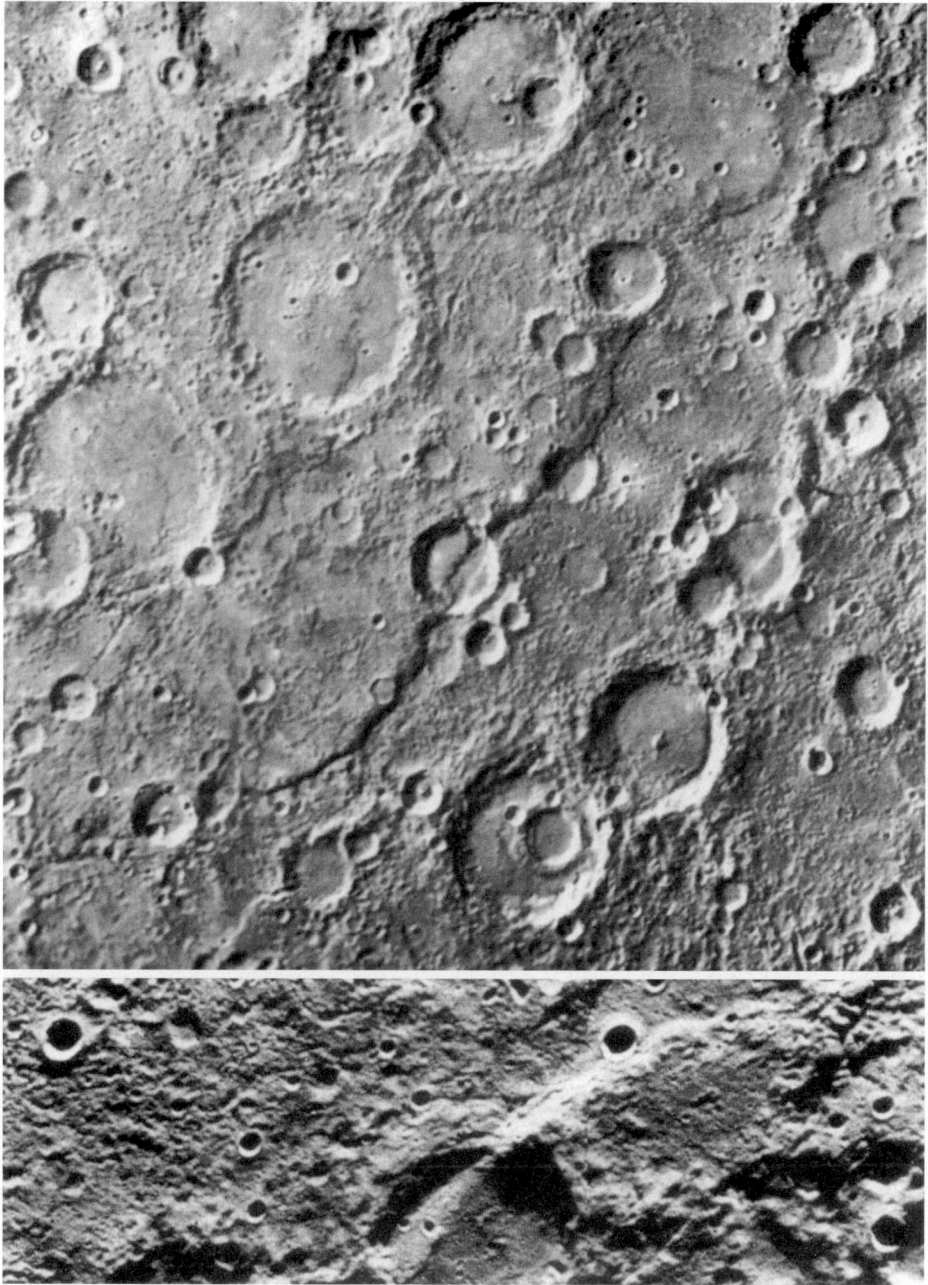

The Discovery scarp (top) is over 400 km in length and in some sections is 2 km high. A section of the 130-km Vostok scarp (bottom) transects Guido d'Arezzo, a crater that is some 80 km in diameter. These are believed to be compression faults.

refined as a result of the mission: the axial period was accurately measured at 58.646 days, and the direction of the rotational axis was established to be almost perpendicular to the orbital plane. The most unexpected discovery of course was the magnetic field which, although weaker than that of Earth, was sufficient to generate a sizable magnetosphere. Although the spacecraft did not explore it in detail, models show that the solar wind should have drawn the magnetosphere 'downwind' to create a magnetotail. A magnetic field had not been expected because it had been thought that to create a dipole a planet must have a molten core and rotate rapidly. It has therefore been suggested that this is a remanent field – a synthesis of fossilized magnetization of surface rocks.[360,361,362]

At approximately 11:25 UTC on 24 March 1975, some 8 days after its third flyby, Mariner 10 finally exhausted its supply of attitude control gas. As it slowly began to tumble, the structure that was not protected by the sunshade was exposed to the Sun and became heated. A command was sent to switch off the transmitter, and at 12:21 the signal ceased.[363] Despite its long series of problems, Mariner 10 was one of the most successful planetary missions. Unfortunately, more than three decades would pass before another mission was launched to resume the exploration of mysterious Mercury.

HOT AND HOTTER

At the end of 1965 NASA and the US State Department introduced an Advanced Cooperation Project, which was to increase the level of technological and scientific cooperation with the western European countries. In particular, it was proposed that Europe, with NASA's assistance, would build either a solar or a Jupiter probe that NASA would launch, with the Deep Space Network providing communications. The Italians and Germans were enthusiastic, the British foresaw financial difficulties in providing support, and the French were reluctant. In particular, there was concern that the unstated objective of the project was an American aspiration to divert the attention of European space engineers and scientists away from strategic areas such as launch vehicles and communication satellites, and thereby stave off long-term competition. In the fall of 1966, in view of the mixed reactions, the grand scheme was turned into a bilateral project with a single European partner, West Germany, and in June 1967 a Joint Definition Group was established. A preliminary study showed that a Jupiter mission would be beyond German technological capabilities and so interest switched to the solar probe, which was to have either a close perihelion or a high inclination to the ecliptic. The Helios mission (as the project was later named) was to study the Sun, solar phenomena, and the interplanetary medium close to the Sun. Germany was to supply two spacecraft, which Messerschmitt–Bölkov–Blohm would make, and build a ground facility to operate and control them and distribute data. NASA was to provide the launch vehicles, tracking, early operational support and train German personnel. The scientific instruments would be developed jointly. In June 1969 the memorandum of understanding was signed between NASA and the Federal Ministry for Research and Technology (Bundesminister für Forschung und

The 'spool-shaped' Helios 1 about to be enclosed in the aerodynamic shroud of the Titan IIIE–Centaur launch vehicle. The probe was covered with mirrors in order to reject as much solar heat as possible

Technologie), stipulating that Germany would provide $180 million of the project's total cost of $260 million.[364,365,366,367,368]

The Helios spacecraft was to enter a solar orbit ranging between its starting point at Earth and 0.30 AU with a period of about 180 days, and its nominal mission was to last three orbits during which no fewer than seven conjunctions would permit the corona to be investigated using radio waves. The objectives were: to measure the magnetic field and the parameters of the solar wind close to the Sun; to investigate discontinuities and shock phenomena of the wind; to study plasma oscillations; to measure the fluxes of solar and galactic cosmic rays with decreasing distances from the Sun; and to study the flux, dynamics and composition of the cosmic dust. It had a total of 10 instruments. A zodiacal light detector consisted of three telescopes with 4-cm-diameter optics illuminating photometers, oriented to monitor ecliptic latitudes in the range 16–90 degrees. Two slightly different micrometeoroid analyzers would measure the mass and velocity of interplanetary dust particles. For larger particles, a mass spectrometer would also measure the composition. The first analyzer was to scan near the plane of the ecliptic, and the second would scan a cone that included the ecliptic pole. A plasma analyzer would record the arrival direction and energy spectra of charged particles. Three magnetometers were carried: German researchers built search-coil and flux-gate magnetometers, and a triaxial flux-gate magnetometer was a joint project of NASA's Goddard Space Flight Center and various universities and institutes in Italy and jointly sponsored by NASA and the Italian CNR (Centro Nazionale Ricerche; National Research Center) to follow up a fruitful cooperation that began in 1962 and had involved the Pioneer solar probes. A plasma and radio wave experiment was to use a pair of dipole antennas with a total span of 32 meters to detect solar wind plasma radio noise. The scientific suite was completed by two cosmic-ray telescopes, including an X-ray monitor and a low-energy cosmic-ray detector. It was decided not to carry optical instruments to observe the Sun because they would yield only slightly better resolution than was possible from a satellite in Earth orbit and would suffer much higher thermal loads.[369,370,371,372,373,374,375,376,377]

A variety of spacecraft designs were considered, including 3-axis-stabilized, spin-stabilized, and hybrid solutions involving varying numbers of moving mechanisms, but it was decided that Helios should be spin-stabilized and have an antenna that counter-rotated at exactly the spin rate of the body in order to maintain the beam pointing toward Earth.[378] The spacecraft had a 16-sided 'spool-shaped' body that consisted of a short stubby cylindrical section containing electronics, equipment, instruments and a spherical tank of nitrogen for the attitude control system, with the addition of a conical skirt top and bottom. The body was 2 meters tall, with a maximum diameter of 2.7 meters. As the solar heat at perihelion would be 11 times more intense than in the vicinity of Earth and over twice that inflicted on Mariner 10 at perihelion, thermal control for Helios was a particularly critical issue, and it was addressed by placing mirrors of solar-panel glass with silvered backs for high reflectivity among solar cells on the outer faces of the skirts. It was determined that a 50:50 mix of solar cells and mirrors would ensure that the maximum temperature of the solar cells could be limited to 150°C, and still generate the requisite power. Thermal control of the equipment racks was

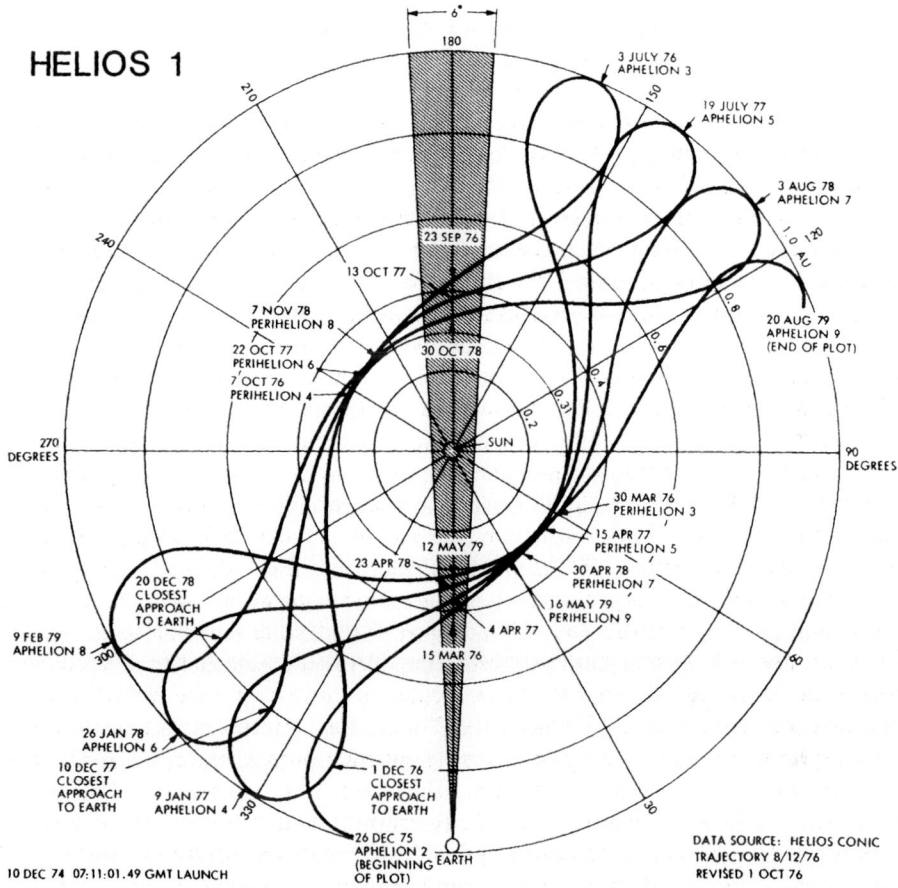

The trajectory of Helios 1 relative to Earth from December 1975 to August 1979, with the shaded section showing the portions of the orbit in which solar conjunctions occurred. (Reprinted from '10 Jahre Helios/10 Years Helios', Copyright: DLR)

ensured by louvers that would maintain an almost constant temperature throughout the orbit. The boom for the 'despun' cylindrical high-gain antenna and medium-gain antenna of the 20-W communication system projected axially from the top, and increased the height to 3.79 meters. The boom also had an omnidirectional antenna and the search-coil magnetometer. The conical adapter to mate the spacecraft to the launch vehicle was contained within the lower skirt. Hinged on the cylindrical section of the body were a pair of booms for the two flux-gate magnetometers and the two dipole antennas of the plasma radio experiment. The spacecraft would travel in the plane of the ecliptic, and for stability it would spin at 60 rpm with its axis aligned perpendicular to its orbit. Attitude determination would be by Sun sensors, and attitude control by nitrogen gas thrusters.[379] It would return data at rates varying between 8 and 4,096 bps, with the highest rate being used at perihelion when it was taking the most measurements, and would be served both by the 64-meter antennas of the Deep Space Network and also by the 100-meter Effelsberg radio telescope in Germany.

Although formally a US–German project, companies in Belgium, the Netherlands, Britain and France participated, and the Italian and Australian scientists involved in some of the instruments and studies gained experience in deep-space missions. Also, since the early thermal, vibration and other essential tests were conducted at ESTEC (European Space Technology Center) in the Netherlands, this gave ESRO invaluable experience.[380,381] JPL simulated the thermal loads for a perihelion of 0.30 AU, which was intended for the first mission, and Germany, with a view to mounting a second mission, simulated infrared loads at 0.25 AU.

The only major change to the original mission plan was made in 1970. It had been intended to launch Helios on an Atlas–Centaur. However, as the Viking missions to Mars and the Grand Tour of the outer solar system would require the new Titan IIIE–Centaur, NASA proposed assigning Helios to an early vehicle, more or less as a test of that configuration prior to trusting it with an expensive Viking, and it was decided to accept the risks in return for increasing the mass of the Helios spacecraft from 254 to 370 kg.[382,383] The first flight of the new vehicle on 11 February 1974, with instrumentation to measure the dynamic loads that it would place on a payload, ended in failure when the Centaur did not start. If the Helios launch later in the year was not successful, then NASA would face the dilemma of whether to risk Viking in 1975 or skip that launch window.

Happily, Helios 1 was launched without incident on 10 December 1974, and after spending 22 minutes in parking orbit the Centaur and the solid upper stage fired in turn to put the spacecraft into an orbit ranging between 0.307 and 0.985 AU with a period of 190 days. After it had released the spacecraft, the Centaur fired its engine twice more in an engineering study of storing cryogenic propellants in space for an extended time, and these maneuvers inserted it into a solar orbit just outside the Earth's orbit.[384,385] As Helios 1 configured itself for space, it suffered several glitches. First, one of the dipole antennas failed to deploy, limiting the ability of the plasma radio experiment to receive low-frequency waves. When the high-gain antenna was turned on it was found to interfere with the particle analyzer and radio wave receiver. This meant that the high-gain antenna would have to be run at a lower transmission power than intended and, in turn, that the 26-meter antennas on Earth allocated to the mission would be inadequate, and the burden of communicating with the spacecraft would fall to the larger antennas. Additional issues were the thermal control of the magnetometer, and interference between several instruments and the data-handling system. On 18 February 1975, Helios 1 crossed between the Earth and the Sun, and achieved its first perihelion passage on 15 March (the day before Mariner 10's final flyby of Mercury) at a relative speed of 66 km/s. At a heliocentric distance of 46 million km, the temperature on many of the spacecraft's components rose to 100°C, and the solar cells sizzled at 127°C, but it survived unscathed and the high-gain antenna problem actually rectified itself. For the 20-day period from 27 April to 15 May there was a hiatus in tracking while the spacecraft was close to the Sun in the sky. On 17 June it reached its first

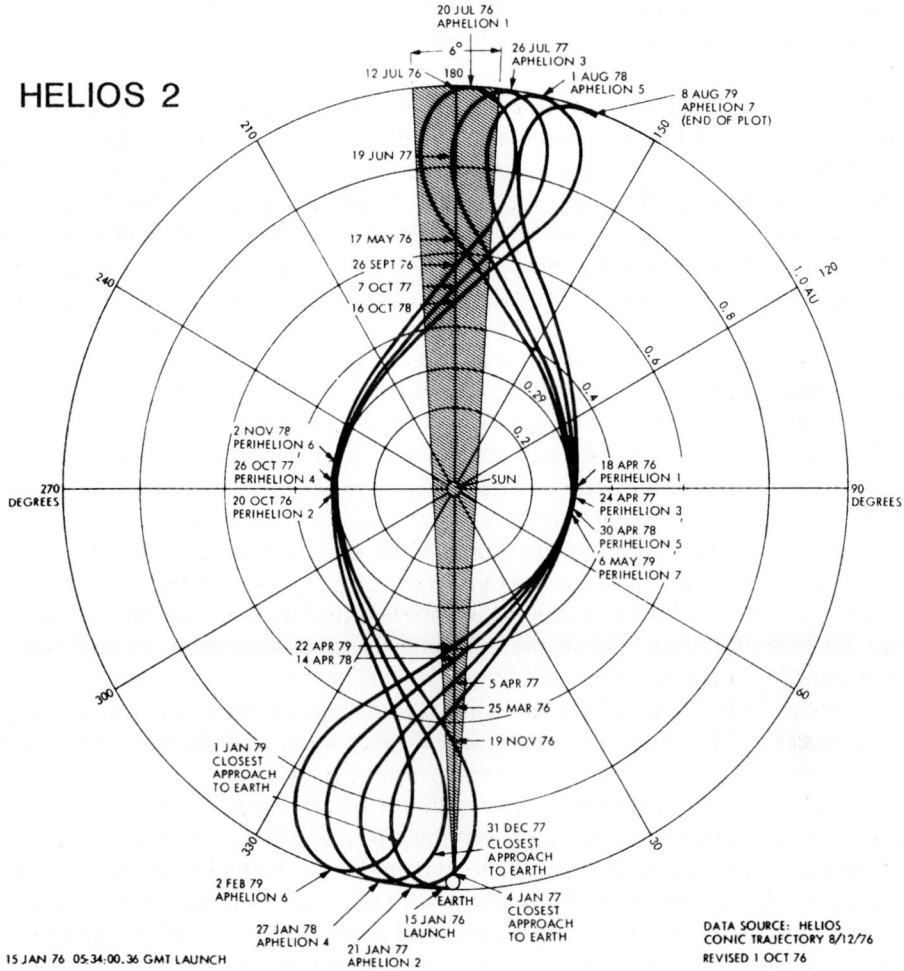

HELIOS 2

20 JUL 76
APHELION 1

26 JUL 77
APHELION 3

12 JUL 76   180

1 AUG 78
APHELION 5

8 AUG 79
APHELION 7
(END OF PLOT)

19 JUN 77

17 MAY 76

26 SEPT 76

7 OCT 77

16 OCT 78

2 NOV 78
PERIHELION 6

26 OCT 77
PERIHELION 4

270
DEGREES

20 OCT 76
PERIHELION 2

SUN

18 APR 76
PERIHELION 1

24 APR 77
PERIHELION 3

30 APR 78
PERIHELION 5

6 MAY 79
PERIHELION 7

90
DEGREES

22 APR 79
14 APR 78

5 APR 77

25 MAR 76

19 NOV 76

1 JAN 79
CLOSEST
APPROACH
TO EARTH

31 DEC 77
CLOSEST
APPROACH
TO EARTH

2 FEB 79
APHELION 6

EARTH

4 JAN 77
CLOSEST
APPROACH
TO EARTH

27 JAN 78
APHELION 4

15 JAN 76
LAUNCH

DATA SOURCE: HELIOS
CONIC TRAJECTORY 8/12/76
REVISED 1 OCT 76

21 JAN 77
APHELION 2

15 JAN 76 05:34:00.36 GMT LAUNCH

The trajectory of Helios 2 relative to Earth from launch to August 1979, with the shaded section showing the portions of the orbit in which solar conjunctions occurred. (Reprinted from '10 Jahre Helios/10 Years Helios' – Copyright: DLR)

aphelion. Although this occurred near 1 AU, it was 300 million km from Earth. There was a second, shorter solar occultation from 29 August to 1 September, followed by the second perihelion on 21 September, at which time the temperature of the solar cells reached 132°C and the thermal and radiation stresses caused some degradation of materials.[386,387,388]

Meanwhile, Helios 2 was being prepared. Based on the experience of the initial months of the first flight, the attitude control thrusters had been improved, and the high-gain antenna feed and the deployment system of the dipole antennas were modified. In addition, the X-ray detectors were improved to enable them to detect gamma rays of an energy that was triggering the Vela military satellites that the US Department of Defense had placed into Earth orbit to 'police' nuclear explosions. These bursts of gamma rays were apparently an astronomical phenomenon, and the satellites were triangulating their sources.[389] Finally, because it had been decided to send the second spacecraft even closer to the Sun, the thermal shielding was reinforced to withstand the 15 per cent increase in solar heat. It had been hoped to launch on 8 December 1975 in order to put Helios 2 into an orbit with a similar orientation to that of its predecessor, but the launch of Viking 2 in September had damaged the pad and the mission had to be postponed. There was some urgency, since in the summer of 1976 the Deep Space Network would be heavily committed to the Viking landings, at which time it would not be able to track Helios 2 through its perihelion passage. It was decided that if the mission could not be launched before the end of February 1976, the spacecraft would be placed into clean-room storage until the Viking commitment eased.[390,391,392] In the event, Helios 2 lifted off on 15 January 1976 and was injected in an orbit ranging between 0.280 and 0.995 AU with a period of 186 days. After it had dispatched the spacecraft, the Centaur, as previously, undertook engineering tests by making an unprecedented five burns over a 5-hour interval that left it in a near-Earth solar orbit.[393] This time the spacecraft's post-launch operations were controlled from the German Space Operation Center near Munich. Helios 2 aligned its spin axis toward the south pole of the ecliptic, the opposite sense to Helios 1, to enable instruments such as the meteoroid detectors to cover the entire celestial sphere. Although Helios 2 set off in better condition than Helios 1, a component in the transmission system that had failed on its predecessor malfunctioned after only 3 months. Helios 2 made its first perihelion passage at a relative speed of 70 km/s (the highest heliocentric speed yet attained by a spacecraft) on 17 April, the closer approach meaning that it endured temperatures 20°C higher than its sister, with the result that its spin thrusters overheated.

While the results of these missions did not capture the public imagination like the contemporary planetary probes, excellent data was gained on the environment close to the Sun at both the minimum point in the 11-year cycle of solar activity in 1976 and at the maximum in the early 1980s, in particular providing insights into the mechanism for the generation and acceleration of the solar wind, the interplanetary medium and cosmic rays. Although the filter wheel motors of Helios 1 jammed some 2 years into the mission, the zodiacal light detectors were able to determine the characteristics of interplanetary dust, including its spatial distribution, color and polarization, at solar distances ranging from 0.1 to 1 AU. At one point, Helios 2 developed a gas leak due to the tank overheating, and the entire nitrogen supply had to be vented. The science team used the venting to realign the spacecraft's spin axis perpendicular to the symmetry plane of the interplanetary dust, and the subsequent observations proved that dust is affected more by gravitation than electromagnetic forces, its symmetry plane being aligned closer to the mean plane of the orbits of the planets than the equatorial plane of the solar rotation. Sensors on Helios 1 detected interplanetary dust close to the Sun, thereby showing that solar heat does not deplete dust in as close as 0.09 AU.

The Helios missions made an interesting contribution to cometary astronomy

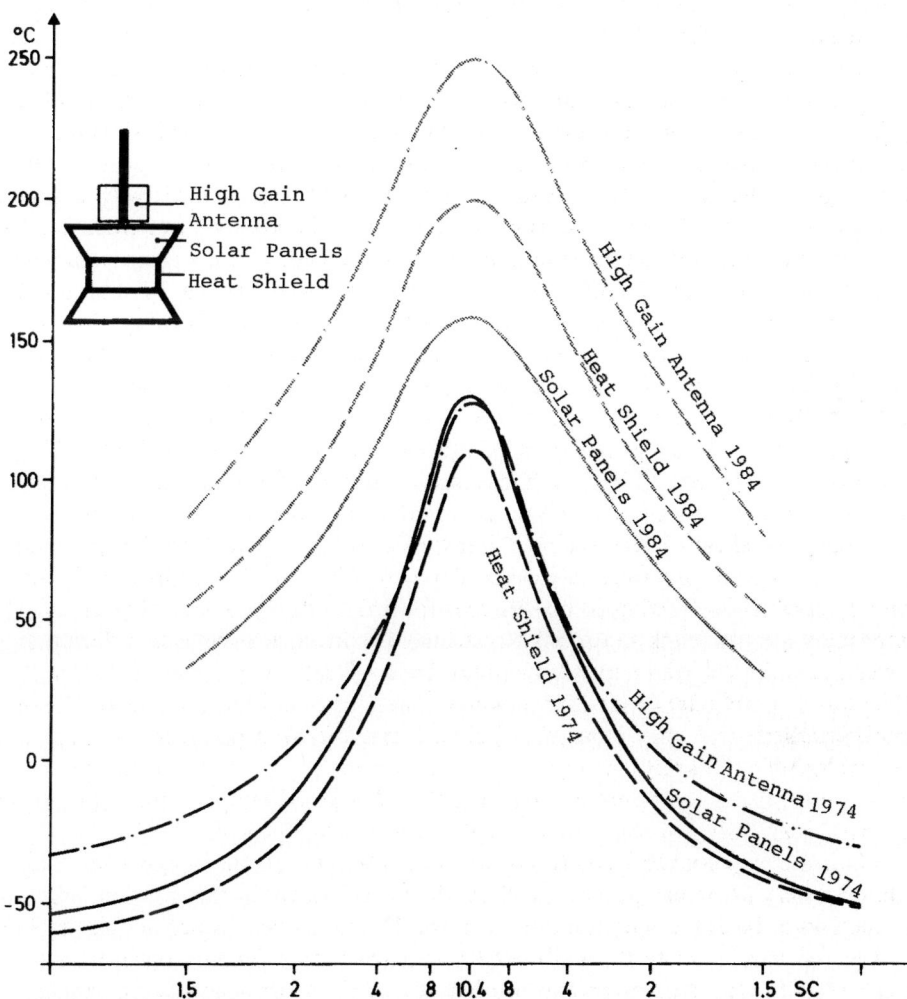

A graph of the temperatures recorded over one orbit by Helios 1 in 1975 and another in 1984. The quantity in abscissa is the solar constant, for which 1 occurs at 1 AU and 10.4 corresponds to the spacecraft's perihelion. (Adapted from '10 Jahre Helios/10 Years Helios' – Copyright: DLR)

by observing the dust and ion tails of the great comet C/1975V1 West as it withdrew from the Sun in 1976, and of comet C/1978H1 Meier in November 1978. Moreover, when Helios 2 flew 0.15 AU in 'front' of comet C/1979Y1 Bradfield in February 1980, it noted a disturbance in the solar wind that later caused a disconnection event in the comet's tail. Solar coronal structures were evident by their brightness. There were unsuccessful attempts to identify meteoroid streams against the background of dust. Helios 1's longevity meant that its zodiacal light monitoring spanned an entire cycle of solar activity.[394,395,396,397,398,399,400,401,402] The meteoroid detectors recorded hundreds of impacts, with the rates being 10 times greater near the Sun than in the vicinity of Earth. The ecliptic and polar sensors detected different populations of particles. The ecliptic sensors mostly detected particles orbiting the Sun that were being overtaken by the spacecraft, while the polar sensor also detected particles overtaking the probe and even some coming from the Sun's direction. Their velocities showed that many of the particles were not bound to the Sun, either because they had been accelerated in some manner or because they were interstellar in origin and were simply passing through the solar system. The composition of the particles resembled either stony or iron-rich meteorites.[403,404]

The plasma analyzer showed that the high-speed solar wind is correlated with the presence of coronal holes on the Sun, and also provided the first detection of singly ionized helium in the solar wind which, as it cannot exist in the corona, must have come from the cooler chromosphere. Serendipitously, the orbital geometry of Helios 1 during the period of high solar activity in 1980 was such that, from the viewpoint of Earth, the spacecraft seemed to hover beside the Sun's eastern and western limbs. Therefore, the coronal mass ejections and flares seen on the limb by the US military satellite P78-1 Solwind would often 'ram' Helios 1 several hours later and enable direct sampling to supplement the visible-light observations that were made in the vicinity of Earth.[405] The results of the Helios magnetometers were coordinated with the observations by the IMP (Interplanetary Monitoring Platform) satellites that were orbiting Earth, and with Pioneer 10 and 11 and Voyager 1 and 2 in deep space, in order to measure the direction of the magnetic field at various distances from the Sun. Measurements of the polarity of the field, either inwards or outwards relative to the Sun, made over many solar rotations, revealed 'magnetic clouds' correlated with coronal mass ejections.[406,407,408] The plasma and radio wave experiment was used to detect radio bursts and shock waves associated with solar flares, mostly around the period of the solar maximum. The two spacecraft triangulated the sources of radio noise, but one of the antennas on Helios 1 failed to deploy, drastically reducing the number of events the instrument could observe. To compensate, triangulation was occasionally done in collaboration with spacecraft such as RAE 2 in lunar orbit and ISEE 3 in near-Earth space.[409,410,411,412] The cosmic-ray detectors studied how the Sun and the interplanetary medium influenced the propagation of galactic and solar cosmic rays. Moreover, the Helios spacecraft worked with others carrying similar instruments to study how cosmic rays varied with time and distance from the Sun. In particular, cosmic-ray detectors on Pioneer 10, which traveled from 12 to 23 AU between 1977 and 1980, showed a good correlation with data collected by Helios 1 over the same interval, thereby providing an accurate determination of the gradient of cosmic rays over a large range of distances from the Sun.[413,414] The gamma-ray burst detector on Helios 2 spotted no fewer than 18 events during its first 3 years of operation, several of whose source directions were able to be triangulated in concert with satellites orbiting the Earth.[415] Owing to the low inclination of their orbits with respect to the ecliptic, signals from the Helios spacecraft could be used to investigate the inner corona during solar occultations. This could be done in two ways: (1) the spacecraft sent a signal to Earth, with the transmission traversing the corona once;

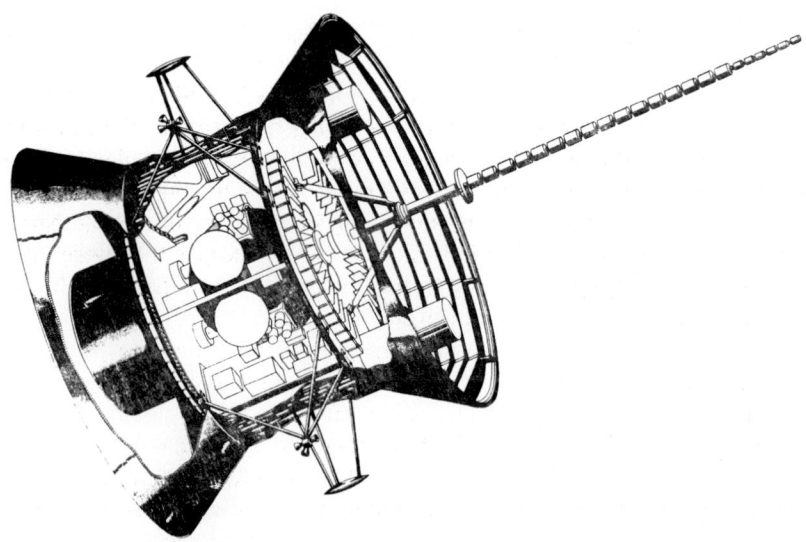

The SOREL (Solar Orbiting Relativity Experiment) probe was an ESRO proposal to reuse the Helios bus for a heliocentric spacecraft that would test gravity and relativity theories. The sphere at the center would contain a test mass whose displacement with respect to the body of the spacecraft was to be carefully monitored in order to reveal accelerations of the order of less than one-trillionth of Earth's gravity. (Courtesy of ESA)

(2) a signal was sent to the spacecraft and relayed back, providing two traverses of the corona. Such signals were able to 'sound' to a distance as small as 1.7 solar radii, to study density fluctuations in waves traveling out from the Sun. Sometimes, the telemetry link of the Helios spacecraft traveled through flares and coronal mass ejections from the point of view of the Solwind satellite, which yielded data on the electron density and velocity within these structures. Unfortunately, the transponder on Helios 1 for this experiment failed after 18 months. After the Helios spacecraft had achieved their baseline missions, the Deep Space Network was often assigned to higher priority missions, but when facilities were available they were tracked during solar conjunctions.[416,417,418,419]

On 3 March 1980 Helios 2 lost its downlink transmitter. Despite efforts to restore communications, no further intelligible data was received, and on 7 January 1981 a command was sent to shut down the spacecraft in order to preclude possible future radio interference, thus terminating the mission. Helios 1 was still healthy, but was being tracked by smaller antennas at low bit rates. By the start of its 14th orbit, the degraded solar cells were able to power the data-handling system and the transmitter together only for several months close to perihelion. In 1984 the temperature of the solar cells reached 183°C and sensors on the antenna mast reached their top limit of 250°C.[420,421] After 1984, both primary and backup receivers failed and the high-gain antenna lost Earth. After the last telemetry was received on 10 February 1986, it was predicted that the spacecraft would soon automatically shut down its transmitter due to lack of power.[422,423] Helios 2 still holds the record for the spacecraft to have made the closest (and fastest) perihelion passage. Various possible future missions were studied for the Helios configuration, including a Helios C flyby of comet Encke and a mission for ESRO named SOREL (Solar Orbiting Relativity Experiment) in which a spacecraft would be inserted into a 6-month orbit and be tracked very accurately to determine the value of some poorly known relativistic parameters and to discriminate between alternative theories of gravitation. It was decided, however, that the mission would require the development of too many expensive new technologies.[424]

SNOWBALLS WILL WAIT

The early 1970s saw great advances in cometary science. In January 1970 a comet was observed for the first time from above the Earth's atmosphere. The comet was C/1969T1 Tago–Sato–Kosaka. NASA's 2nd Orbiting Astronomical Observatory revealed that it was surrounded by an extremely tenuous hydrogen cloud that was larger in diameter than the Sun. A similar phenomenon was observed 3 months later around C/1969Y1 Bennett. These clouds were due to the dissociation of water molecules by sunlight, thereby proving that water was a key ingredient of these objects. Interest in comets was also stimulated by C/1973E1 Kohoutek, which was studied by the crew of Skylab. In April 1970 NASA sponsored a conference at the University of Arizona on comets and cometary missions, and later that summer the National Academy of Sciences held a conference to promote the role of electrically propelled vehicles in planetary exploration, and proposed a cometary mission as an excellent example. NASA briefly considered intercepting Kohoutek, but identified more realistic opportunities for the second half of the decade.

In 1971 JPL completed an internal study of the possibility of sending a spacecraft based on its Mariner to intercept comet 6P/d'Arrest when it came within 0.18 AU of Earth in April 1976, but NASA decided that the data from such a flyby would not justify the cost. In late 1971, in recognizing that an expensive Mariner-style mission would have little chance of being funded, the Goddard Space Flight Center proposed adapting the Interplanetary Monitoring Platform bus that had been used for several of the satellites of the Explorer series, by installing a course-correction engine and a despun high-gain antenna. After launch in November 1976, the Cometary Explorer would intercept 26P/Grigg–Skjellerup in April 1977 when the comet would present a particularly favorable opportunity for an encounter by being close to its perihelion within 0.2 AU of the Earth. During the approach, the comet's interactions with the solar wind, its dust environment and its chemistry would be characterized. Imaging the nucleus was not a primary objective. Astrodynamics expert Robert W. Farquhar pointed out that it would be possible to place the Cometary Explorer in an orbit that would return to Earth after 1 year, and if the probe were to be fitted with a solid-fuel rocket, it would, on its return, be able to insert itself into a high-apogee orbit in order to study unexplored parts of the magnetosphere. The mission was submitted to NASA's Comet and Asteroid Science Working Group, which enthusiastically

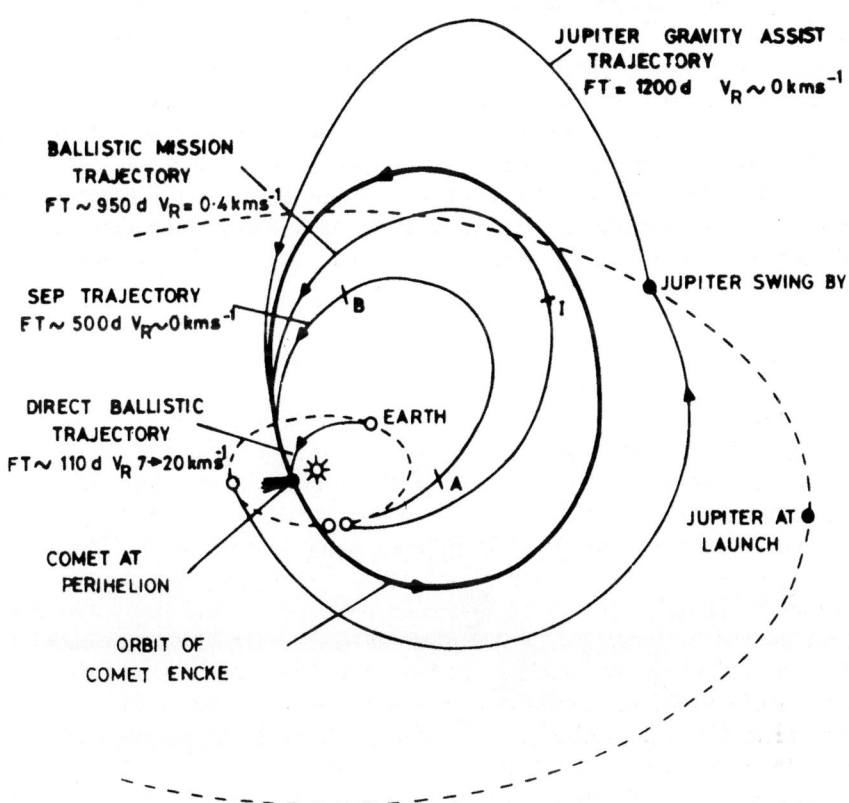

Alternative trajectories for the investigation of comet Encke, including a fast flyby ballistic trajectory, a rendezvous ballistic trajectory, a Jupiter-assisted trajectory and a solar-electric propulsion (SEP) trajectory. (Courtesy of David W. Hughes)

endorsed it, and although preliminary work was funded, this was halted in February 1973 owing to budgetary constraints. This was not the end, however. The same month Farquhar found that the Earth-orbit-insertion rocket could be used instead to send the spacecraft to a second cometary flyby, this time with 21P/Giacobini–Zinner in 1979, and the agency agreed to reconsider the mission as a double encounter.

Another proposal was to adapt a Pioneer Jupiter-class probe, but because both this and the Cometary Explorer were to be spin-stabilized it was difficult to install an imaging system into such a spacecraft, and hence imaging was not an objective of these proposals. JPL, which was the master of imaging from stabilized platforms, proposed a mission to 2P/Encke in 1980 in which a solar-electric propulsion system would progressively shape the spacecraft's orbit in order to reduce the relative speed of the encounter to less than 4 km/s, and provide a slow flyby that would facilitate a prolonged period of study. This would be a precursor for more ambitious missions, such as an Encke 'formation flight' in 1984 and a Halley interception in 1986. The probe would have six off-the-shelf instruments, ranging from the narrow-angle camera of Mariner 9 to plasma sensors of the type carried to the Moon by the Apollo missions. JPL wanted to make use of the ion thrusters that had been successfully demonstrated by the Lewis Research Center's SERT 2 (Space Electric Rocket Test) satellite. One issue was the extent to which these would affect instruments such as plasma detectors and magnetometers, and whether a build-up of efflux would degrade the performance of solar panels, optics and antennas. But in October 1973 NASA, once again citing budgetary constraints, reaffirmed its antipathy to cometary missions.[425,426,427,428,429]

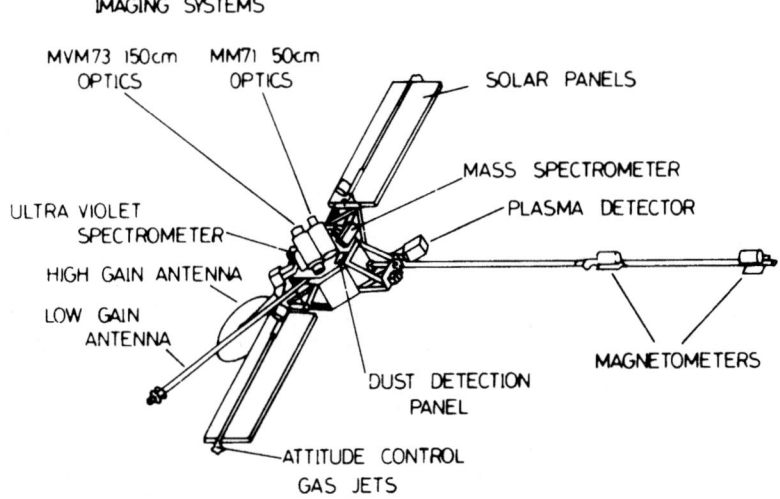

A proposal for a probe based on Mariner 10 to investigate comet Encke. (Courtesy of David W. Hughes)

Cometary probes were also studied in Europe. Messerschmitt–Bölkov–Blohm in Germany proposed to fit a Helios C spacecraft with a course-correction engine and cameras, and have it fly within 100 km of the nucleus of 2P/Encke in 1980. It would not only continue the observations of its two Helios sisters during the period of solar maximum, but would also investigate the cometary environment. Also, since the orbital period of a Helios-type mission would be approximately one-sixth that of Encke's orbit, this raised the prospect of sending the spacecraft to encounter the comet again at its next perihelion in 1984. The proposal was submitted to ESRO in 1974, but was rejected owing to its cost. The option of a German-only mission was discussed, but the German minister of scientific research judged its endorsement "unlikely".[430,431,432,433,434]

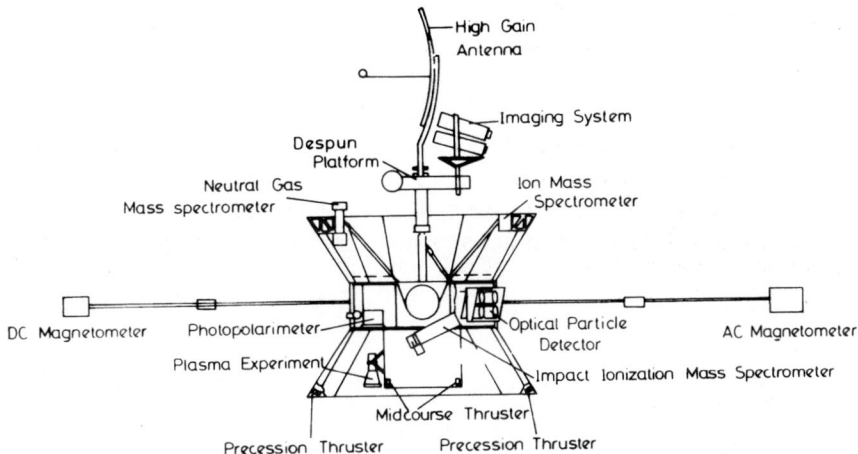

The proposal to have Helios C investigate comet Encke. (Courtesy of David W. Hughes)

POSTCARDS FROM HELL

After skipping the 1973 launch window for Venus, the Soviet Union introduced its third-generation spacecraft in 1975. It was based on the M-71 and M-73 probes, was designated 4V-1, and was to be launched by the four-stage Proton rocket. At its base was a 2.35-meter-diameter toroidal tank that surrounded the KTDU-425A engine. Above the engine was a 110-cm-diameter pressurized cylinder (narrower than the 180-cm-diameter original, and 1 meter shorter) that housed the second tank and miscellaneous systems. The engine used UDMH and nitrogen tetroxide with a throttleable thrust in the range 9.86 to 18.89 kN, and there was sufficient propellant for a total of 560 seconds accumulated over as many as seven firings. When deployed, a pair of solar panels, each 1.25 × 2.1 meters, took the total span of the bus to 6.7 meters. Set between the panels were thermal control radiators and pipes to circulate coolant fluid to both the bus and a lander, if present. The high-gain parabolic antenna was slightly smaller than on the Mars probes. The signal from the lander during atmospheric entry, and on the surface of the planet, was to be received by helical antennas. The total height of the bus was 2.8 meters. The first missions were to release a lander and then enter orbit around Venus. The lander was contained in a 240-cm-diameter spherical capsule mounted on top of the bus. The 1-meter-diameter spherical lander had a titanium pressure hull, a layer of plastic insulator to impede the inward passage of heat, and an outer titanium shell. The internal temperature was further controlled by a lithium nitrate trihydrate thermal buffer that would melt at 30°C, having absorbed a large quantity of heat. On top was a cylindrical omnidirectional antenna, 80 cm in diameter and 40 cm tall, that also enclosed the parachute system, comprising a 2.8-meter-diameter

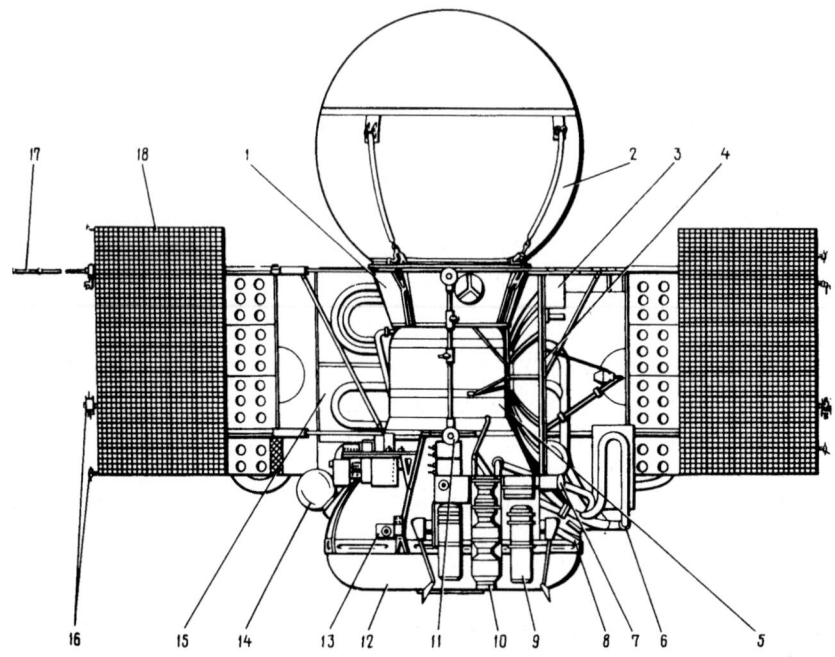

A contemporary Soviet drawing of the configuration of the Venera spacecraft for 1975 using the same bus as the Mars missions of 1971 and 1973: 1, orbiter bus; 2, descent module; 3, scientific instruments; 4, high-gain antenna; 5, cylindrical propellant tank; 6, thermal control pipes; 7, Earth sensor; 8, scientific instruments; 9, Canopus sensor; 10, Sun sensor; 11, omnidirectional antenna; 12, instrument pressurized container; 13, scientific instruments; 14, attitude control gas tank; 15, main thermal control radiator; 16, attitude control gas jets; 17, magnetometer; 18, solar panels.

drogue and a trio of 4.3-meter-diameter main parachutes. Projecting from the base of the omnidirectional antenna was a circular 210-cm-diameter 'drag plate' that was designed to retard and stabilize the probe in its atmospheric descent, and also serve as an antenna reflector. An annular landing platform was affixed to the base of the hull by shock-absorbing struts. A pair of 'chimney pipes' that emerged from the hull circulated coolant to the lander during the interplanetary cruise. The new lander was tested extensively: its safe extraction from the heat shield was tested by airdrops, and the aerodynamics of the innovative stabilization system were verified using wind tunnels.[435,436,437,438] The 'dry' mass of the 4V-1 bus was 2,300 kg. Enclosed in its entry capsule, the 660-kg lander weighed 1,560 kg. The launch mass of a 4V-1 with a lander was approximately 5,000 kg.

For photometry, polarimetry, spectrophotometry and infrared radiometry of the planet's atmosphere, the bus operated an infrared spectrometer, infrared radiometer, ultraviolet photometer and two photopolarimeters based on the VPM-73 instrument used at Mars. With the exception of one of the photopolarimeters, the optical axes of all these instruments were boresighted. Scanning would start when the limb of the planet entered the field of view during the approach to periapsis and end about 30 minutes later when the opposite limb exited from the field of view. At a periapsis of 5,000 km, the instruments would all yield a resolution of between 50 and 100 km. The bus also had an ultraviolet photometer to simultaneously measure the intensity and temperature of the upper atmosphere, and the ratio of deuterium to hydrogen. In addition, a spectrometer would search for emission lines that might indicate a night-side airglow. The bus would monitor the interactions between the solar wind and the planet with a 3-axis magnetometer, a plasma spectrometer, an ion-energy spectrometer and an electron analyzer.[439] There was also a double-frequency radio-occultation system, and some spacecraft would have a linear 'push broom' imaging system (similar to that on the M-73 probes, but with violet and ultraviolet filters for the atmosphere of Venus) and a French-built ultraviolet imaging spectrometer.[440]

The lander's payload was no less impressive. During the descent it would use five hot-wire thermometers and six barometers to profile the pressure and temperature, plus a capillary inlet tube exposed to the air flow to collect samples to enable a mass spectrometer to profile the composition of the atmosphere.[441] The MNV-75 package was to use two nephelometers to study the characteristics of the atmosphere during the descent, in order to determine its transparency and the size and optical properties of cloud particles.[442] But the big new innovation, of course, was the imaging system. Two 10-mm-thick quartz viewports, set 180 degrees apart just under the drag plate, were protected during the descent by thick metallic covers that were to be jettisoned at landing. The camera was of the same type as that of the Mars landers – that is, it had a photometer and a mirror and was to produce 517 scan lines. Its nominal field of view was 40 × 180 degrees, with a gray-scale target in the field of view to provide for photometric calibration. However, as the vantage point was only 0.9 meter above the ground and the viewing line of sight was inclined at 50 degrees from the lander's vertical axis, the ground immediately ahead would occupy most of the image and the horizon would be evident only at the edges of the panorama. Although Venera 8 had reported that some sunlight filtered through the atmosphere, floodlights were affixed to the lander's struts to increase the illumination of the ground.[443] In the seemingly never ending competition with the Americans, the Soviets hoped to obtain images of the surface of Venus in late 1975, before the first Viking landed on Mars, in order to be the first nation to return pictures from the surface of another planet. A simplified gamma-ray spectrometer would detect potassium, uranium and thorium in order to characterize the surface composition.[444] The ground density was to be measured by a densitometer in which a cesium-137 gamma-ray source on the tip of a short boom would be lowered following landing, and three scintillation sensors under the lander would count the gamma rays scattered by the ground. To ensure that the detectors collected only backscattered gamma rays, a shield mounted inside the landing ring blocked their line of sight to the source. A similar device on the Luna 13 lander had measured the density of the lunar surface. The point at which the gamma-ray source was deposited would be within sight of one of the cameras, both to provide a view of the area sampled and to observe any effects of the 2-kg spring-loaded instrument impacting the ground, to enable it to serve also as a rudimentary penetrometer.[445,446]

A model of the Venera 9 lander. The two tubes on the right were used to cool the hermetically sealed electronics compartment during the interplanetary cruise. Note also the floodlights on the landing gear struts. (Courtesy of Olivier Sanguy/Espace Magazine)

The lander's instrument suite was completed by a pair of anemometers mounted on the upper surface of the drag plate, a solar flux-meter, a photometer to measure the relative abundances of water vapor and carbon dioxide, and accelerometers to record the dynamics of the descent and impact.[447,448,449]

Two 4V-1 spacecraft were built for the 1975 window, with Venera 9 departing on 8 June and Venera 10 on 14 June. The launch mass of Venera 9 was 4,936 kg, but Venera 10 was slightly heavier at 5,033 kg, in part because its bus was heavier at 2,314 kg instead of 2,283 kg but also because it carried 1,159 kg of fuel instead of

1,093 kg owing to the fact that it would require a slightly longer orbit-insertion burn. Both were placed into transfer orbits ranging between 1.02 and 0.72 AU, and each made two course corrections (on 16 June and 15 October, and on 21 June and 18 October respectively) to put them on course for the day-side of Venus.[450]

As Venera 9 approached its target, its thermal control system chilled the lander to –10°C and the batteries were charged. On 20 October the bus released the lander and made a deflection maneuver in order to pass over the night-side on a trajectory that would take it within 1,500 km of the surface. At the time of closest approach, on 22 October, it made a long burn to reduce its heliocentric speed by 922.7 m/s to enter an orbit with an apoapsis of 111,700 km, thus becoming the first artificial satellite of Venus. Meanwhile, the lander followed a different approach to that of its predecessors, diving into the atmosphere at 03:58 UTC while traveling at a speed of 10.7 km/s on a path depressed 20.5 degrees below the local horizontal. It had no line of sight to Earth to report its data, but as the bus gained altitude over the day-side it acquired the lander's signal. The entry probe measured the atmospheric deceleration twice per second as it descended from 100 km to about 63 km, at which point, now traveling at 250 m/s, the upper heat shield was released and the parachutes were deployed. When the lower heat shield was discarded, the lander's thermometers, barometers and nephelometers started to work. After the parachutes had been released, the lander fell the rest of the way to the surface, which it struck with a vertical speed of 7–8 m/s at 05:13 UTC, 75 minutes after entry. It settled on a slope of 15 to 30 degrees at a point estimated as 31.01°N, 291.64°E, on Beta Regio, 2.1 km above the mean planetary radius.[451,452] The photometer detected a cloud of dust raised by the impact. However, one of the camera covers failed to eject, and only a 180-degree section of the planned full panorama could be returned. In a break from traditional secrecy, Soviet scientists revealed that two cameras were carried, but that data from the second was not "as good as" that of the first.[453] The illumination was much brighter than expected, at

ВЕНЕРА–9   ОБРАБОТАННОЕ ИЗОБРАЖЕНИЕ

A 180-degree panoramic view of the surface of Venus as seen by Venera 9. Both it and Venera 10 returned camera data interlaced with other telemetry that appeared as vertical bands of 'noise' in the raw pictures, but because the probes functioned for long enough to start a second panorama it was possible to remove these bands by interpolation where overlapping scans were available, and to do so 'cosmetically' in other cases. The T-shaped object at right is the gamma-ray densitometer. (Courtesy of Donald P. Mitchell and Yuri Gektin)

15,000 lux (or, as Arnold Selivanov put it, "as bright as Moscow on a cloudy day in June") rendering the floodlights redundant. Although the surface part of the mission had been expected to last 30 minutes, the lander was still transmitting when the bus passed below its horizon 53 minutes after touchdown.

Venera 10 made a similar approach, first releasing its lander on 23 October and then successfully making the 976.5 m/s braking burn on 25 October to enter an orbit ranging between 1,500 and 114,000 km. The capsule penetrated the atmosphere at 04:02 UTC on 25 October, and at 05:17 UTC the lander settled on a gently rolling plain at 15.42°N, 291.51°E, also on Beta Regio, about 2,200 km from Venera 9, at a point 1.5 km above the mean radius.[454,455] Remarkably, one of the lens covers again jammed! The lander was still transmitting when the bus passed below its horizon 65 minutes after touchdown.

Good temperature and pressure profiles were obtained by both landers, showing the temperature on the day-side at altitudes between 50 and 63 km to be 30°C higher than on the night-side. The surface temperatures were 455°C (Venera 9) and 464°C (Venera 10) at pressures of 85,000 and 91,000 hPa. Good data on the composition of the atmosphere was provided by the mass spectrometers, enabling upper limits to be specified for ammonia and argon. Unfortunately, the instrument was not capable of detecting molecules containing sulfur, such as sulfuric acid. The nephelometers collected data on the optical characteristics of the atmosphere and clouds at altitudes from 62 to 18 km. The principal cloud layer was found to extend down to 49 km, with a minimum horizontal visibility of 0.7 to 1.5 km occurring at altitudes between 53 and 51 km. The instruments also measured the sizes of the liquid droplets in the clouds.[456] On the surface, the gamma-ray spectrometers measured the percentages of potassium, uranium and thorium, and the rocks proved to be very different from those at the Venera 8 site. The rocks at the two new sites were both similar to types of terrestrial basalts. Wind speeds measured 1 meter above the surface were 0.5 m/s (Venera 9) and 1 m/s (Venera 10), with this speed being found to be constant up to an altitude of at least 30–40 meters, indicating that the atmosphere at the surface was essentially stagnant. The densitometers, which fortunately were visible in the half-panoramas returned by both landers, measured a density at the Venera 10 landing site that would indicate a relatively young soil that had not been subjected to significant erosion, but the results for Venera 9 have not been published.[457,458,459,460] The most striking results from the landers, of course, were the pictures. Although, as noted, each lander was able to use only one of its cameras, the prolonged surface activity allowed them to repeat the transmission, enabling gaps in the first scan to be recovered. The slope on which Venera 9 landed was revealed to be littered with sharp-edged slabby rocks that were dark in color, with a darker soil in the spaces between. There were few smaller fragments of rock. The densitometer might have cracked a rock as it slammed down. Venera 10 set down on a plain of large scattered rounded outcrops with linearities and fractures, separated by a darker soil similar to that at the Venera 9 site. A local darkening may have been caused by dust in the dense atmosphere that billowed at touchdown and had yet to settle when the panorama was taken 13 minutes later.[461,462,463,464]

After relaying the transmissions from their landers, both buses refined their

ВЕНЕРА-10   ОБРАБОТАННОЕ ИЗОБРАЖЕНИЕ

A 180-degree panoramic view of the surface of Venus as seen by Venera 10. Note that this site appears flatter and less rocky than the Venera 9 site. The bright object seen at the center of the picture is the discarded camera cover. (Courtesy of Donald P. Mitchell and Yuri Gektin)

orbits. Venera 9 maneuvered first into an orbit ranging between 1,300 and 112,200 km, and then into the target 1,547 × 112,141-km orbit inclined at 34.15 degrees with a period of 48 hours 18 minutes. Venera 10 settled into a 1,651 × 113,923-km orbit inclined at 29.10 degrees with a period of 49 hours 23 minutes.[465] At each periapsis passage the infrared spectrometers collected spectra in order to study how the concentration of carbon dioxide varied with factors such as latitude and distance from the subsolar point; the ultraviolet photometers noted the brightness profile to determine the speed and structure of the topmost layer of cloud; and the infrared radiometers studied the structure of the clouds in much finer detail than had been achieved by the Mariner flybys. Some features seen by an infrared radiometer were correlated with those observed by an ultraviolet photometer, which established that there were 1,500-km-wide convective cells.[466] The photopolarimeters provided data on the sizes of the particles in the clouds. The linear 'push broom' imaging system on Venera 9 made 17 scans from 26 October to 25 December, but few were published.[467] As had been noted by the landers of 1969, Venera 9 found hints of lightning and thunderstorms. In fact, on its very first periapsis passage, one of the spectrometers detected a faint airglow on the night-side, and 4 days later it detected optical flashes, possibly due to cloud-to-cloud lightning discharges.[468] The plasma instruments gave an accurate measurement of the position of the planetary bow shock, and of its thickness. Moreover, the characteristics of the plasma on the night-side of the planet were also studied. The magnetometer placed an upper limit to the planet's magnetic moment at 1/4000th that of Earth.[469,470] A total of 50 radio occultations were monitored between October and December 1975, and in March 1976. The probes always disappeared on the night-side and re-emerged on the day-side. The data yielded very good profiles of the temperature and pressure over a wide range of solar angles. In addition, these events provided profiles of the electron density over both the day and night hemispheres. Bistatic radar studies were made in which a spacecraft aimed a radio beam at the planet's surface and the reflection was recorded by a radio telescope in the Soviet Union and then analyzed to infer the dielectric characteristics of the surface, its density and degree of roughness. The various surface textures were classified as ranging from solid to loose rock. The overall reflectivity provided insights into the large-scale topography. However, this technique could only be used within certain geometric constraints, and was only able to map a small number of regions, each of which spanned only a few square degrees. Although some mountain ranges were able to be located by their high reflectivity, on average the Venusian surface appeared to be smoother than the maria of the Moon. The end of Venera 9's mission was announced on 22 March 1976, but from April to June, while passing through solar conjunction, Venera 10 was used to collect data on radio propagation in the corona down to an angular distance from the Sun of just 0.6 degree.[471]

A rare image taken by the Venera 9 orbiter of Venusian clouds in the ultraviolet; few images taken by this spacecraft have been released, and Venera 10 reportedly did not carry a camera.

LANDING IN UTOPIA

In September 1967, soon after the cancellation of the Voyager Mars landers, NASA called a meeting involving all of its field centers that had been working on Voyager, and concluded that the cancellation had been due to cost estimates spiraling out of control, not to a lack of interest in a Mars landing mission. A few months later, with the programs to send automated spacecraft to the Moon winding down, and with no planetary exploration missions funded after 1969, the agency invited proposals for new projects for the 1970s. This was the second time that NASA had overhauled its planetary program – the first time being when it realized in the early 1960s that the Mariner A and B missions would not be able to fly on the planned schedule owing to the slow pace of development of the Centaur stage. On the other hand, this shake-up

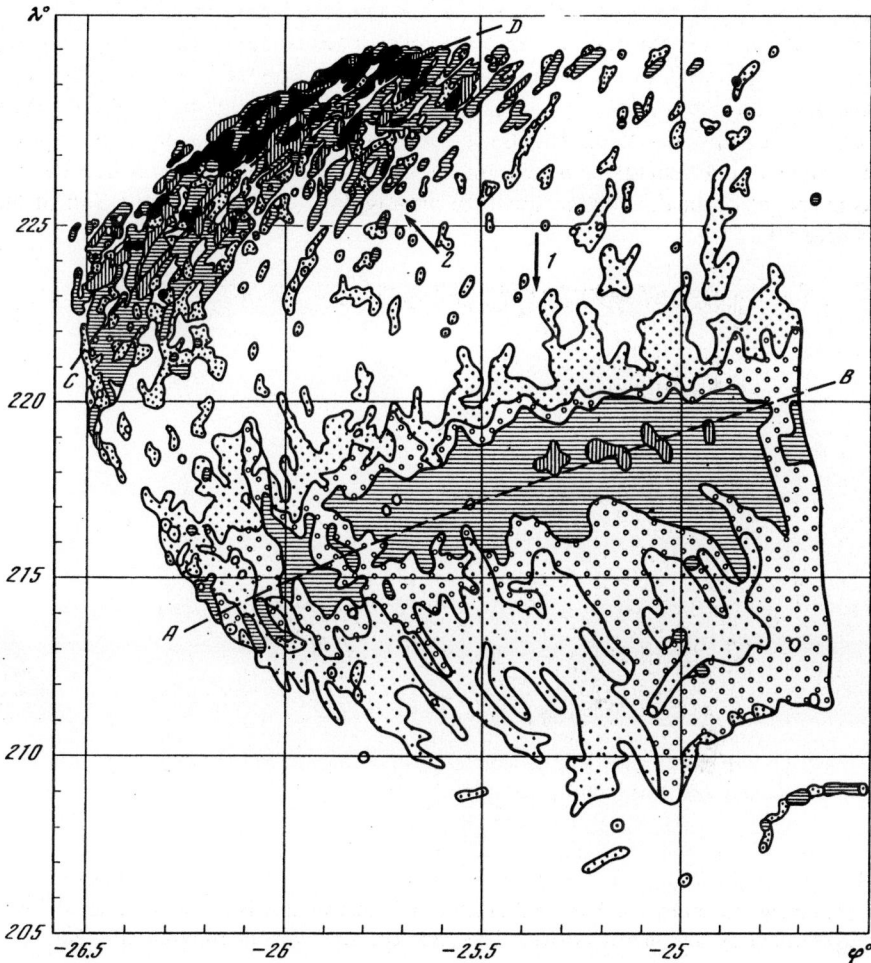

A radar map of part of the Venusian surface provided by Veneras 9 and 10 by using the bistatic radar method. (Reprinted from Kolosov, M.A., Savich, N.A., Yakovlev, O.I., "Spacecraft Radiophysical Ivestigations of the Sun and Planets". In Kotelnikov, V.A. (ed.), "Problems of Modern Radio Engineering and Electronics, Moscow", Nauka, 1985, 64–102)

provided an opportunity to formulate a stronger program, but by this point JPL had a rival. By 1967 NASA's Langley Research Center had accomplished the program to survey the candidate landing sites for the Apollo missions. At the same time, being the agency's leader in atmospheric entry vehicle dynamics, Langley made a study of a small hard-lander to make atmospheric measurements as it descended to the surface of Mars. The plan envisaged exploiting the favorable launch window in 1971 with a flyby craft to relay the probe's transmission to Earth. The expertise gained from this

An artist's impression of a Viking spacecraft approaching Mars. The lander is in the biological isolation shell mounted on top.

study had made Langley the obvious choice to develop a heat shield for the Voyager Mars lander.

The outcome of the shake-up was that NASA told JPL to develop Mariner Mars orbiters for launch in 1971, and Langley to develop an integrated orbiter/lander mission to follow in 1973 (although with the orbiter most likely being supplied by JPL) on the strict understanding that the program must cost much less than had been projected for Voyager. In recognition of the fact that the Atlas–Centaur would not be able to launch a combined orbiter/lander, the Lewis Research Center was directed to oversee the integration of the spacecraft with the Titan III vehicle. The 1971 mission was to map Mars to allow the selection of sites for landers. At the end of 1968, the orbiter/lander project was named Viking, to distinguish it from Voyager. By the end of 1969 its cost had soared from an optimistic $400 million to at least $750 million, and since at this time NASA's budget was shrinking in the wake of the first Apollo lunar landing, NASA made three drastic decisions that were designed to reduce its budgetary requirements in the years to come: (1) no more Saturn V launch vehicles would be built; (2) there would be longer intervals between the remaining Apollo missions; and (3) Viking would be postponed to 1975. The fact that the 1975 launch window for Mars was less favorable than that of 1973 meant that Viking would need both a more powerful launcher and a longer interplanetary cruise, which would increase the time the propellant for the Mars orbit-insertion maneuver would be required to remain under pressurization in the tanks. Although it was not apparent at the time, the 2-year delay proved to be a blessing, because the project suffered such extensive development problems that, in any case, it would not have been ready for launch in 1973!

Instead of being released as the orbiter approached the planet to make a hard landing and undertake a 3-day surface mission in the Soviet style, the Viking lander was to be released from orbit, perform a soft landing and operate for 90 days. This profile increased the complexity of the spacecraft and required a switch to the new Titan IIIE–Centaur, but retaining the lander until after orbit insertion would enable realistic weight margins to be assigned to the two craft, and enable the orbiter to inspect a number of potential landing sites in advance of a commitment.[472,473] Despite the cost-saving measures, Viking still became the most expensive planetary mission ever undertaken. At the time of launch, its total cost was estimated at about $1 billion – or in excess of $3 billion at current rates.

Although JPL set out to make the Viking orbiter a straightforward derivative of its Mariner orbiter, the mass of the 'piggyback' lander greatly increased the propellant required for the orbit-insertion burn. The increased power requirements necessitated the addition of a second segment to each of the existing solar panels to double their output, and the high-gain antenna had to be enlarged to perform the data-relay tasks. Nevertheless, the Mariner heritage was evident in the octagonal

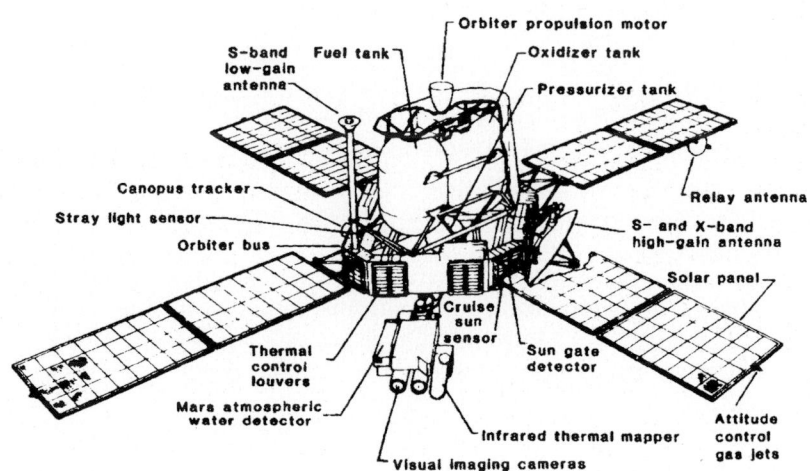

The configuration of the Viking orbiter after the lander has been deployed and the associated carriage structures jettisoned. Note that it is depicted without the thermal blankets. (Reprinted with permission from Hutchings, E. Jr. "The Autonomous Vikings", Science, 219, 1983, 803–808. Copyright 1983 AAAS)

bus, which was 2.4 meters wide and 45.7 cm deep, had alternating 140- and 56-cm-wide faces, and 16 internal bays. On the short faces were four pairs of 123 × 157-cm solar panels with a total collecting area of 15 m² for 1,400 W near Earth, diminishing to 620 W at Mars. The propulsion module was attached to one end of the bus by four attachment points. The single 1,323-N engine drew monomethyl hydrazine and nitrogen tetroxide from a pair of 140-cm-long, 91-cm-diameter tanks having a total capacity of 1,600 kg. Two redundant sets of six nitrogen thrusters at the tips of the solar panels were to provide attitude control. In addition to the 147-cm-diameter high-gain antenna, there was a low-gain antenna at the tip of a boom, and an aerial beneath one of the solar panels to communicate with the lander both during its descent and on the surface. A tape recorder could store up to 55 images for return to Earth at up to 4,000 bps. Only three instruments were mounted on the dedicated scan platform. The imaging system had two identical catadioptric Cassegrain telescopes with focal lengths of 475 mm and 3-color filter wheels, operating as vidicon cameras. It was to image the surface of the planet at periapsis with a resolution approaching 35 meters per pixel to provide the means by which to evaluate the suitability of the nominal candidate landing sites; to search for alternatives, if necessary; and, once the lander was on the surface, to monitor the site in order to assist in characterizing phenomena seen at ground level. When not supporting the lander, the cameras were to investigate geological formations and monitor seasonal variations. Several times during the development of the orbiter, it was proposed that the imaging system be deleted in order to make mass available to the lander. There was also a suggestion that it be replaced by the less expensive system developed for the 1971 mission, even though that would have offered only marginal improvements in our knowledge of Mars. An infrared spectrometer was to seek water vapor in the lowest part of the atmosphere. An infrared radiometer with seven detectors was to scan the temperature of the atmosphere, reveal surface composition differences, and analyze the composition of the polar caps. With the solar panels fully deployed, the orbiter had a span of 9.7 meters. At launch, its 'wet' mass (without the lander) was 2,328 kg. Both orbiters were delivered to the launch site in February 1975 for integration with their lander and the upper stage of the launch vehicle. JPL had even managed to trim the cost from $124 million to $103 million.[474,475,476,477]

In keeping with Langley's modus operandi, the hardware for the lander was to be built by a contractor. Several companies suggested a variety of designs with three or four legs, and in the case of McDonnell Douglas a disk that would be very stable on the surface. Designs issues included whether to use solar panels or RTGs for power, and whether to employ two panoramic cameras. It was ultimately decided to use the three-legged arrangement that had proved stable on the Surveyor lunar landers built by JPL and also in the early tests of the Apollo Lunar Module conducted at Langley (although, in fact, the actual LM had four legs). The contract was given to Martin Marietta of Denver for a three-legged lander powered by RTGs for a long life.[478] Most of the electronics and scientific instruments would be contained in a six-sided aluminum–titanium bus frame that was 1.5 meters wide and 46 cm tall, with sides alternating between 56 and 109 cm. On each of the short sides was a leg comprising a

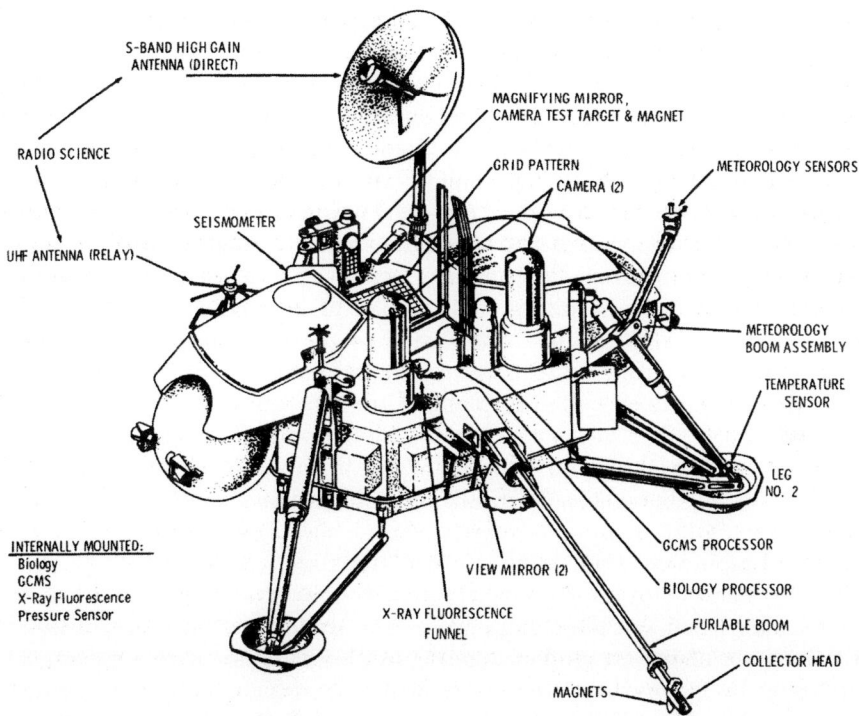

To this day the Viking lander remains the most complex spacecraft to have landed on another planet.

primary strut connected to an A frame, with a crushable aluminum-honeycomb shock absorber and 30.5-cm-diameter foot pad. In their deployed configuration, the 1.3-meter-tall legs gave the base of the bus a clearance of 22 cm. On each of the long sides of the bus there was a cluster of rockets to make the powered descent. These drew from a common supply of 85 kg of hydrazine. In order to minimize exhaust erosion and contamination of the surface, each of these clusters had 18 small thrusters and could be throttled between 276 and 2,840 N. On two of the long sides of the bus there were spherical propellant tanks that gave the craft a total length of about 3 meters. On each tank were mounted four thrusters, each providing a thrust of 39 N, to provide roll control during the descent. Alongside the tanks were mounted a pair of RTGs of the type used on the outer solar system Pioneer missions, giving up to 70 W of power and heat, and when additional power was needed their output could be augmented by rechargeable batteries. The propellant tanks and the RTGs were mounted under two squat covers of roughly pyramidal shape. On the remaining long side (dubbed the 'front balcony') were the housing and actuators for the robotic arm that was to retrieve soil samples for analysis, and a hinged mast that carried a meteorology package comprising hot-wire anemometers, thermometers and a pressure sensor.

The robotic arm – which was actually a collapsible boom made of two welded omega-section steel ribbons – was stowed very compactly in a small box, but when it was extended it became a rigid tube up to 3 meters in length. It had four degrees of freedom, and could be moved in azimuth and elevation to reach any position within a 120-degree-wide area of 9 m^2 in 'front' of the lander. The 'end effector' of the arm was a complex scoop that could rotate through 180 degrees in order to unload its soil sample into a hopper for distribution to the scientific instruments. The arm was an instrument in its own right, capable of measuring the strength of the surface, the temperature, and the magnetic properties of the soil and aeolian dust. However, it was not autonomous; it had to be guided by a controller on Earth in a step-by-step manner, and owing to the lengthy time delay between issuing commands from Earth and receiving pictures to confirm that the commands had been executed as planned, its operation was necessarily slow and complex.[479,480] There was an omnidirectional antenna to receive commands from Earth, and the lander could communicate directly with Earth at data rates of 1,000, 500 or 250 bps via a 76-cm-diameter high-gain antenna on a gimbaled mount or use a UHF antenna to relay via the orbiter at 4 or 16 kbs. Engineering and 'house-keeping' data would be returned in real time, while scientific data, and in particular bandwidth-hungry imagery, was to be stored on tape and transmitted via the orbiter when an opportunity arose. The lander had two miniaturized (for that time) state-of-the-art guidance and control computers, each weighing 114.6 kg and built by Honeywell. When their memories, which comprised a matrix of 0.05-mm-diameter wires, proved to be much more difficult to make than expected, and the development of the computers fell behind schedule, a minimum-capacity traditional graphite-core design was ordered as a backup – but not used. A two-part 3.5-meter-diameter lenticular aeroshell with a front angle of 140 degrees would protect the lander during atmospheric entry. The exterior of the aluminum cocoon was coated with an ablative material to act as a heat shield, while on the interior were two spherical tanks for 85 kg of hydrazine for four three-engine clusters for attitude control after release by the orbiter, for the deorbit burn and for attitude control during atmospheric entry.

The lander's suite was capable of testing for the presence of life in the soil and of performing general studies of the surface environment. Also, four instruments on the front of the heat shield were to directly sample the atmosphere during entry. At an altitude of 250 km, a retarding potential analyzer would study the reaction between the solar wind and the outer atmosphere. The gas composition would be measured down to 100 km by a mass spectrometer. As the capsule penetrated more deeply, data from a recovery temperature and a stagnation pressure sensor would facilitate profiling atmospheric temperature and pressure. An accelerometer would record the deceleration during most of the descent. After the heat shield was jettisoned, sensors on the lander would monitor temperatures and pressures. The instruments for use on the surface accounted for 91 kg of the lander's 576-kg 'dry' mass.

Spaced 80 cm apart on the lander's front balcony were two facsimile cameras for stereoscopic imaging from a height of 1.3 meters above the surface. As in the case of the Soviet lunar and planetary landers, a photosensor viewed the scene utilizing a rotating and nodding mirror, but different photosensors could be selected to provide high-resolution black-and-white, near-infrared, or red, green and blue pictures that could be combined for a color view. Because the image was built up by horizontally rotating the cylindrical unit containing the mirror, an image was 512 lines deep and as wide as the desired azimuth scan. This way of imaging was very slow, however, and required about 30 minutes to make a full-color panorama. In addition to facing outward, the cameras could view across the top deck where there were a number of targets, including color-calibration chips, mirrors, a magnifying mirror and a painted grid to enable the accumulation of aeolian dust to be monitored. As with most of the lander's instruments, the development of the imaging system ran over budget – in this case costing $27.3 million instead of the projected $9.8 million. An inductance 3-axis seismometer of a type developed by JPL for the Ranger lunar missions (but not used) was to sense 'Marsquakes', and it was hoped that if both landers reached the surface safely their instruments would enable the epicenters of such activity to be triangulated.

The physical properties of the soil were to be studied using the robotic arm, while magnets on the top deck within sight of the cameras and on the sampling scoop were to determine its magnetic properties, with brushes to clean the magnets to enable the observations to be repeated. The inorganic chemistry of the soil was to be analyzed by an X-ray fluorescence spectrometer, whose input was a funnel set on the top deck between the cameras. Once the arm had collected a sample from the uppermost several centimetres of the soil, the contents of the scoop could simply be tipped into the funnel, where a screen would pass rock fragments up to 1.3 cm in size, or more finely grained material could be sieved through 2-mm holes in the scoop. The analysis chamber could accept up to 25 cm^3 of material. The sample would be irradiated by two radioactive sources and the excited fluorescence analyzed to determine the elemental composition. The sample could be dumped into a cavity in order to allow a repetition.[481] Meanwhile, a Gas Chromatograph Mass Spectrometer (GCMS) would first analyze the composition of the atmosphere at ground level and then analyze a soil sample for organic material of a wide range of molecular weights, in addition to some inorganic compounds such as nitrates and nitrites. It would take two measurements by heating the sample first to 200°C and then to 500°C, each time letting the vaporized material pass through a gas chromatographic column to separate the molecules of different weights, and identify them. It was one of the most complex instruments to develop, and difficult to build within the given dimensions and mass. As a result, the initial cost estimate of $17.8 million proved unrealistic, and when the cost reached $41 million management of its development was transferred from JPL directly to the Viking Project Office and a simpler version was built.

The sophisticated biology laboratory was packed into a volume of only 0.03 m^3. It exploited the work done in the early 1960s for the Automated Biological Laboratory intended for the Voyager lander. The arm was to deposit a sample into an inlet that would distribute measured amounts to the individual instruments for incubation and processing. Whereas the GCMS would look for organic material, the instruments in the biology laboratory would seek evidence of biochemical

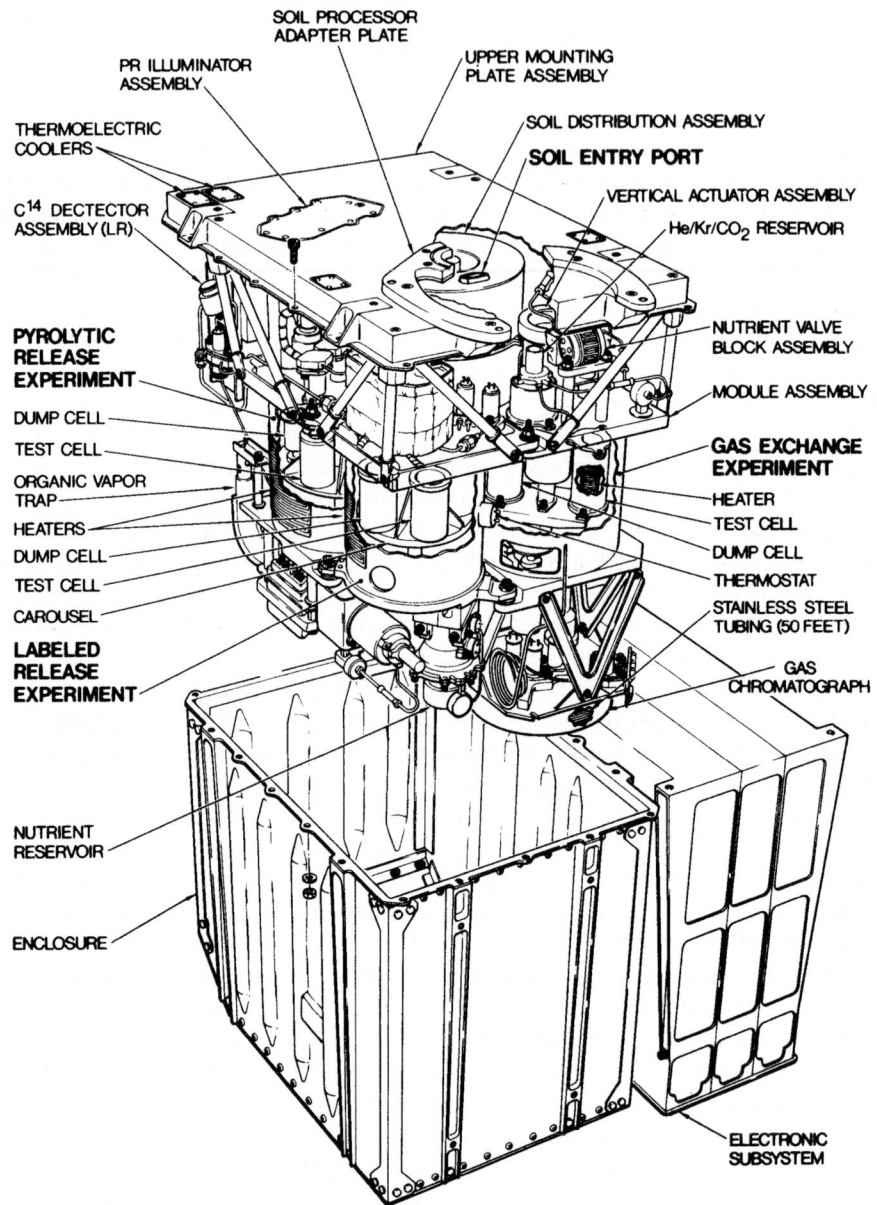

Although the Viking biology package weighed 15 kg and measured 28.75 × 33.00 × 26.34 cm with a volume of just 0.025 m^3, TRW managed to pack in three complex instruments designed to test for life on Mars.

activity. When NASA asked the Space Science Board of the National Academy of Sciences to assist in developing a strategy to determine whether there is life on Mars, Stanford University hosted a summer study in 1964 to investigate the issues. In March 1965 the draft report, *Biology and the Exploration of Mars*, said, "Given all the evidence presently available, we believe it entirely reasonable that Mars is inhabited with living organisms, and that life independently originated there." However, whereas if there were plants there would certainly be microbes, it was possible that there were *only* microbes, and thus any test for life should be aimed at microbial life. The report also said, "We have reconciled ourselves to the fact that early missions should assume an Earth-like carbon–water type of biochemistry as the most likely basis of any Martian life." It was realized that proving the presence of life would not be straightforward. Given the way in which cells function, one strategy was to seek evidence of cellular reproduction, but as this is a discontinuous process whose rate varies from species to species, and even for a single species in different conditions, employing that strategy as a test would be very difficult in the context of an exotic environment. As an *ongoing* process that can be measured in several ways (for example, by changes in acidity, or the evolution of gases) metabolism is more readily testable and also more likely to produce a definitive result. The report urged a *multifaceted* test because, "no single criterion is fully satisfactory, especially in the interpretation of negative results".

The Labeled Release (LR) experiment was an advance on the Gulliver instrument. A diluted solution of radioactively labeled nutrients like formic acid and glycine was to be introduced to a sealed sample that would be tested for the release of metabolic radioactive carbon dioxide. The Gas Exchange (GE) experiment was to measure the production or uptake of gases in the presence of water, both with and without added nutrients. If a response was obtained that suggested biological activity, the sample could then be heat-sterilized and the test repeated to confirm that the response really had been due to metabolic action. The Pyrolytic Release (PR) instrument differed in that it was to investigate the synthesis of organic matter by photosynthetic or other chemical processes, rather than its decomposition. Its sample would be incubated in an illuminated cell in which the ambient air was enriched with radioactive carbon monoxide and dioxide, evacuated, and then heated to 700°C to test for radioactive carbon resulting from biological synthesis.[482] The biology experiments had common electronics and gas reservoirs, in addition to a thermal control system to maintain the incubation temperature constant against the diurnal variation of the external surface temperature. Initially, the biology laboratory was to have had a fourth instrument, in the form of a water-turbidity experiment in the style of the Wolf Trap developed for Voyager, but this was deleted in the face of mass and budgetary pressures. Even so, the laboratory cost almost five times the original estimate of $59.5 million. In order not to jeopardize the integrity of the biology tests, and to avoid contaminating Mars with terrestrial organisms, the lander and its aeroshell were sterilized in a controlled nitrogen atmosphere at high temperature for 40 hours. To maintain the aeroshell in this state, it was enclosed by a two-piece pressurized fiberglass bioshell, 3.7 meters in diameter and 1.9 meters deep, that remained sealed until the spacecraft was on its way to Mars.[483,484,485]

In testing the lander's entry system, NASA used the preliminary (later found to be incorrect) Soviet report that the Mars 6 lander had determined the atmosphere to be 30 per cent argon. Picking up on the Voyager PEPP flights, in the summer of 1972 the Viking parachute system was subjected to full-scale balloon tests at White Sands in New Mexico.[486,487,488,489,490,491] The lander's components were then tested to prove their ability to withstand the thermal and pressure conditions that they were likely to encounter on the surface of Mars. Notwithstanding the determined efforts to control its budget, the lander was the part of the spacecraft that suffered the heaviest cost overruns, increasing from an expected $360 million to an actual $545 million.[492]

Additional tests involved the Titan IIIE–Centaur, which NASA wished to become the standard heavyweight launch vehicle for planetary missions until the introduction of the Space Shuttle. The inaugural flight on 11 February 1974 was to prove the flightworthiness of the vehicle on a trajectory similar to that required for Viking. Its payload was a small scientific satellite and a Viking Dynamic Simulator to determine the loads to which the spacecraft would be subjected. The Titan functioned as expected and the long shroud that enclosed the entire Centaur upper stage in addition to the payload was successfully released, but the Centaur failed to ignite and fell into the Atlantic. However, the perfect launch of the second Titan IIIE–Centaur on 10 December 1974, carrying the Helios 1 solar probe, cleared the way for the vehicle to be used to launch the Vikings.[493]

For the first time, the planners of a NASA planetary mission faced a problem that had previously taxed their counterparts in the lunar program: the selection of landing sites. While the Soviet, lacking high-resolution orbital imagery from earlier missions, had had no option but to target their landers at 'regions' visible from Earth, NASA was able to employ the Mariner 9 imagery to select specific sites that would satisfy both the scientific requirements and the engineering constraints in terms of accessibility, temperature, elevation, absence of obstacles, etc. When it was decided to release the lander from orbit instead of prior to the orbit-insertion maneuver, it became possible to have the orbiter inspect the candidate sites at high resolution, and then to search for alternatives if the planned sites were to prove to be unsuitable. Meanwhile, the instrument to survey water vapor would assess the suitability of each site from the point of view of its potential for life. Although some thought was given to sending one lander poleward of 65°N to a site recently exposed by the retreating polar cap, all 35 preliminary candidates were within several tens of degrees of the equator. Terrestrial radar was able to provide a sense of the roughness of the surface on a scale that was smaller than the best imagery, but geological interpretation of radar reflections was still at such an early stage that there was little correlation possible between the form of the reflection and the type of terrain evident in the imagery. The final choice of primary and backup sites was a complex process, with an official announcement being delayed for months. Because it would be summer in the northern hemisphere when the mission occurred, the sites that were selected were all north of the equator in order to increase the likelihood of finding water-driven processes. The first lander would be sent to Chryse Planitia, near where several 'outflow channels' appeared to have deposited a blanket of sediment on a low-lying plain, with Tritonis Lacus as a backup; and the second would be sent to the Cydonia plain, with the shallow shield of Alba Patera as a backup. If all went well, NASA hoped to make the first landing on 4 July 1976 – the Bicentennial of the United States of America.[494]

The hardware for the launch vehicles began to arrive at Cape Canaveral in late 1974. The first lander was delivered in January 1975, with the orbiters following in February. After the individual craft had been thoroughly tested, they were mated and tested again. In June the landers and aeroshells were cleaned, sterilized, inspected to confirm that the high-temperature treatment had not caused any damage, and finally sealed. The Viking A spacecraft was mated to its launch vehicle on 28 July. As the countdown on 11 August reached T–15 minutes, a problem was noted with a valve in the thrust-control system of the Titan, and this had to be replaced. Before the new countdown could be initiated, it was realized that the orbiter had continued to draw power from its batteries, and almost exhausted them. As it was a lengthy process to replace the batteries, NASA opted to swap Viking A

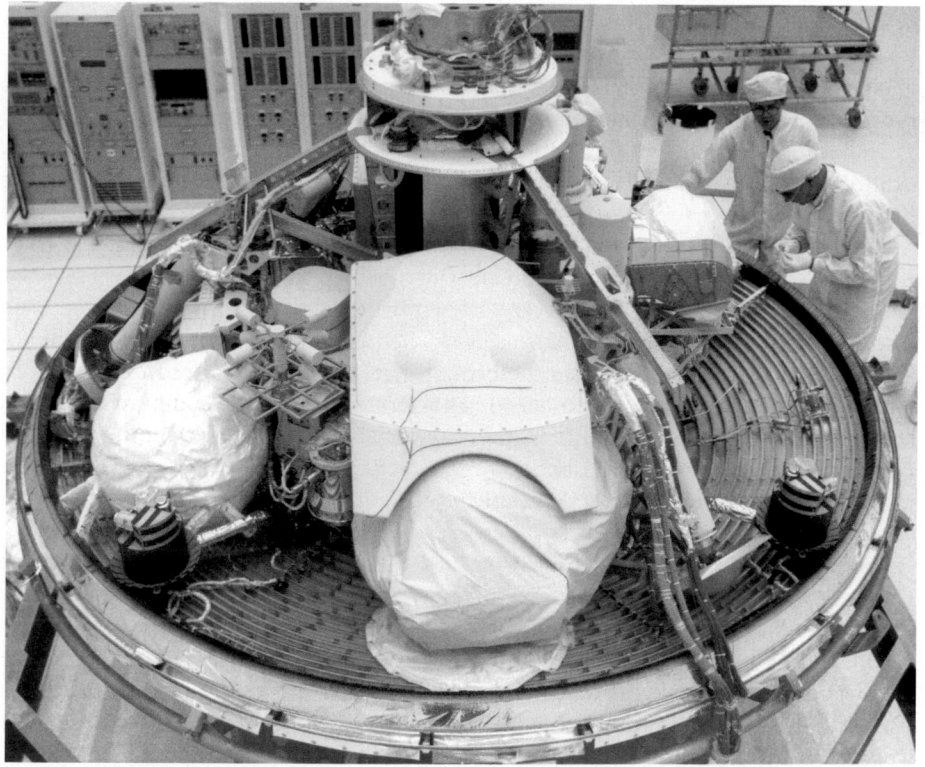

Preparing the Viking 2 lander. It is mounted inside the forward heat shield and the technicians are standing directly in 'front' of the two cameras (the white cylinders mounted on top of the craft).

for Viking B and proceed. This time the countdown was nominal, and Viking B lifted off on 20 August 1975. After a brief cruise in parking orbit, the Centaur fired to escape and send Viking 1 (as the craft now became) toward Mars. The solar panels were deployed with no difficulty, and the Canopus sensor located its star on its first attempt. The base of the bioshield was jettisoned. The transfer orbit ranged between 1.003 and 1.672 AU and would sweep through more than 180 degrees to reach Mars in 10 to 11 months, producing a slow approach that would minimize the propellant requirement for the orbit-insertion maneuver. Soon after Viking A was mated with the second Titan IIIE–Centaur and delivered to the pad, a problem developed in the craft's communications system, and it had once again to be demated. It was finally launched as Viking 2 on 9 September, a few minutes before an approaching thunderstorm would have forced a postponement. A fire ignited by one of the solid-propellant boosters caused major damage to the pad and mobile electronic and instrumentation facilities, and the repairs would delay the launch of Helios 2 by a month.[495,496,497] Although telemetry from Viking 2 was lost for 6 minutes during the ascent, which prompted some concern on the ground, it was safely injected into a transfer orbit ranging between 1.006 and 1.669 AU. In both cases the initial trajectory was designed to yield a wide miss-distance (279,000 km for Viking 2) to ensure that the unsterilized Centaur would not contaminate Mars by an inadvertent impact. On 27 August Viking 1 made a course correction to reduce the encounter range, and on 19 September Viking 2 did likewise.

Having no instruments to monitor conditions in interplanetary space, the orbiters were less active than their predecessors in transit. The landers were powered down, apart from a few basic systems and the heaters that were to prevent the reactants for the biology laboratory and propellant from freezing, although at regular intervals the chromatograph was purged of gas and recalibrated and the data recorder was tested. There were a few problems, with Viking 1 experiencing arcing events due to solar wind storms, and when the battery charger for the Viking 2 lander showed signs of distress the backup had to be brought into operation. In late March, Viking 1 took a number of calibration pictures of distant Jupiter. Soon thereafter it was found that one of the three ovens of its gas chromatograph had malfunctioned, and would probably no longer be usable. When 40 days out from Mars, the camera platform was unlatched and distant images of the planet were taken to assist in navigating the approach. As the range reduced, color picture sequences and infrared data were taken as the planet turned on its axis. Finer navigation fixes were obtained in the final days of the approach by imaging Deimos against the background of stars. On 10 June Viking 1 was to make a second course correction. In preparation, on 7 June the valve was opened to let helium pressurize the propellant tanks, but telemetry indicated that the tanks continued to rise in pressure after the regulator threshold had been achieved.[498] The obvious solution was to close a valve that would cut off the flow in order to prevent the pressure in the tanks from building up during the 12 days remaining to the orbit-insertion maneuver, but, once closed, that pressurant feed line would no longer be available, leaving the system with only one remaining line which, if it were to fail to open for the orbit-insertion burn, would result in the spacecraft sailing past the planet. As the pressure could not be allowed to rise and rupture the tanks, it was decided to leave the helium feed open, and try to control the rate of increase by splitting the final course correction into two burns on 10 and 15 June, the dates and changes in velocity being calculated to ensure that when the orbit-insertion maneuver was executed the pressure would not be high enough to rupture the tanks. However, this revision of the approach to the planet would have the effect of delaying orbit insertion by 6 hours, which would in turn displace the longitude of periapsis away from the primary landing site. It was decided to enter a higher than planned initial orbit with a period of 42.6 hours that would position the first periapsis over that site 49.2 hours later (at the time which, if the insertion had been as planned, would have been the second periapsis) whereupon an additional burn would be made to enter the intended orbit.[499] In order to preclude the same thing befalling Viking 2, which was a month behind, it was instructed to delay pressurizing its propulsion system until 12 hours prior to orbit insertion.

On 19 June, Viking 1 fired its engine for 38 minutes to slow its speed relative to Mars by 1.2 km/s, and became the fifth spacecraft to enter orbit around the planet. The valve was then closed to isolate the line between the helium tank and the faulty pressure regulator. Although the maneuver had consumed 1,063 kg of propellant the tanks contained sufficient residual pressure to make orbit trim burns in 'blow down' mode. At its first periapsis, on 21 June, it made a 132-second 'shaping' burn to enter an orbit ranging between 1,500 and 32,800 km with a period of 24 hours 39 minutes – one local day. The only major consequence of the revised orbit insertion was a delay in taking the first images of the primary landing site, which had been expected as soon as the first periapsis. When the images were received on 22 June the technicians and scientists who had spent 4 years debating the merits of the various potential sites received a nasty shock. Although the prime landing site had appeared smooth in the Mariner 9 images, the new orbiter's high-resolution camera revealed a profusion of small craters, which was bad news since impacts produce blankets of rocky ejecta. After further imagery showed craters, channels and cliff-edged mesas all across the site, it was decided on 27 June to slip the landing beyond 4 July. But the tension was still on, because Viking 2 was closing in on the planet. A tweak of Viking 1's orbit enabled it to inspect a site 250 km northwest of the original target, since the terrain appeared to become smoother in that direction, but this was also overly rough. Attention then switched to a site 580 km further west, where there were fewer fresh-looking craters. Selecting a site was not simply a matter of choosing a point, as the uncertainties of the atmospheric entry made the target an elliptical 'footprint' that extended 120 km in the direction of travel and 25 km to either side of that track. If the lander was aimed at the center of the ellipse, it was expected to have a 99 per cent chance of reaching the surface within this boundary. Each candidate site was probed by terrestrial radar, which was a powerful tool for gaining insight into the roughness at the scale that would matter in attempting a landing. Finally, on 12 July, after 20 days of heated debate, the scientists agreed on a site that represented a compromise between the roughness that could be seen in images and the fine-scale detail inferred from radar reflections. On 16 July, the spacecraft adjusted its orbit in readiness for the landing attempt.[500] At 08:32 on 20 July explosive bolts severed the connections to the aeroshell, which was allowed to drift for 7 minutes to ensure that it was at a safe distance from the orbiter when it began to use its low-thrust pitch and yaw control engines to make the 23-minute maneuver that would slow it by 160 m/s and set it on course to enter the atmosphere within the required window. Several hours later, the orbiter jettisoned the remainder of the bioshell and its supports.

The lander was accelerated by gravity as it fell, and penetrated the atmosphere at 4.4 km/s. The descent profile was more complex than that of any previous planetary lander, particularly the 'eggs' that the Soviets had dropped into the atmosphere. The retarding potential analyzer had been activated 3 minutes after the deorbit burn and now began to make measurements, and the cap that sealed the aperture of the upper atmosphere mass spectrometer was released. These instruments took measurements down to an altitude of 100 km, while the heat shield rose to a temperature in excess of 1,500°C. Once the deceleration reached 0.05 g, the thermometer and barometer in the heat shield and the descent radar were activated. The retarding potential analyzer found an abundance of ionized molecular oxygen in the upper atmosphere, possibly from the dissociation of the water vapor that was escaping from the planet. The mass spectrometer found the atmosphere to consist mostly of carbon dioxide, with some nitrogen, argon, oxygen and other gases. The ratio of nitrogen-15 to nitrogen-14 was greater than in the Earth's atmosphere, suggesting that the planet's atmosphere was denser in the past. This was consistent with the widespread erosion of the surface by flowing liquid. The low concentration of argon (1.5 per cent, rather than the almost 30 per cent inferred from the Mars 6 data) suggested that some 10 times the amount of carbon dioxide that is currently in the thin atmosphere is sequestered somewhere in the surface, as well as many meters of water permafrost.[501,502] After reaching a maximum deceleration of 8.4 g at an altitude of 27 km – mitigated somewhat by the fact that it was pursuing a 'lifting' entry profile – the lander began a long deep glide, as lift balanced some of its weight. On reaching an altitude of 5.9 km, still traveling at 400 m/s, the computer ordered, in rapid succession, the firing of a mortar to unfurl the 16.2-meter-diameter parachute, the release of the lower part of the aeroshell, the activation of the terminal descent radar, and the deployment of the landing legs. At an altitude of 1.5 km and a rate of descent of 54 m/s, it discarded the parachute and the upper aeroshell. The lander was to make the remainder of the descent using its rocket engines, first to null its horizontal velocity and then to reduce its vertical rate on a profile calculated to enable it to land at an acceptable speed. At 610 meters, the radar was switched into its low-altitude mode. As soon as one of the sensors in the landing legs indicated contact with the surface, the engines were shut off. The lander had touched down at 2.44 m/s, within 3 degrees of upright, on Chryse Planitia at 22.48°N, 47.94°W, some 28 km off target, in the mid-afternoon. It was 11:53:06 UTC on 20 July. The touchdown signal took 19 minutes to reach Earth. The final task of the navigation system before it was switched off, was to determine the orientation of the lander to enable it to locate Earth in the sky.[503,504]

The timing was such that the orbiter would pass below the lander's horizon a few minutes after touchdown and not rise again for 19 hours. The lander was therefore to return two pictures immediately, one of the foreground and the other looking out to the horizon. Some 25 seconds after landing, one of the cameras began to scan the ground alongside one of the foot pads, and the orbiter relayed this historic image to Earth, where it was assembled on screens, column by column, left to right, over a period of 20 minutes. This viewpoint had been selected to provide the maximum amount of information in the event that the lander should fail during its first Martian night. It covered 60 degrees in azimuth at high resolution, and it showed a number of angular rocks sitting on top of fine dust and fragmental debris resembling gravel that had been disturbed by rocket exhaust – in fact, dark banding on the left-hand side of the frame was evidently billowing dust that had yet to settle when the camera started its scan. At the right of the frame, dust lay on the concave foot pad. The camera then took a lower resolution 300-degree panorama. In addition to many of the structures of the spacecraft (the meteorology boom, the stowed sampler, various antennas, one of the RTG covers, the calibration aids, etc), the image showed a gently rolling surface that was reminiscent of the rock-littered desert of the American southwest. There were patches of fine dust and a small field of dunes. There were depressions downwind of some of the rocks, consistent with orbital observations of wind patterns. Most of the rocks were angular, suggesting that they were ejecta recently excavated by impacts. Some exhibited granular, vesicular or even layered textures. In some places there was what might have been exposed bedrock. Where the horizon was not obscured by hills, it was about 3 km distant, and structures were seen on or near the horizon that looked as if they were the rims of far off craters, while nearby shallow depressions filled with dust appeared to have been excavated by secondary impacts. Having transmitted its pictures, the lander closed down for the night. The orbiter would be able to relay for the lander on almost every pass, and once the lander had completed the process of locating Earth it would be able to communicate directly.

On its second Martian day (Sol 2) the lander repeated the pan in color, revealing the entire landscape to be dominated by the rusty red hue that is so characteristic of Mars telescopically. Initially, the image-processing laboratory combined the red, green and blue frames to produce the dark blue–black sky that the thin atmosphere had been expected to yield, but after they had been recalibrated the sky was found to be pinkish-orange, evidently because most of the sky brightness is due to suspended dust particles scattering sunlight. The color of the surface material was probably due to limonite – a form of oxidized iron that, on Earth, forms in the presence of water, but in the case of Mars may have formed in the absence of water by exposure to the ultraviolet light that is not blocked by the atmosphere. The initial pictures had been taken by camera 2, which served as the 'right eye' of the lander. The 'left eye' was activated on Sol 3. The first panorama from camera 1 showed a part of the landscape that had not been seen earlier, and revealed a 3-meter-wide 1-meter-tall fragmented boulder among a group of smaller rocks about 8 meters away. If the lander had come down there, it would probably have been wrecked. Initially the boulder was called Big Bertha, but when this was criticized as sexist it was renamed Big Joe. To its left, and 300 meters away, there was the rim of a small crater. The imagery was carefully studied for macroscopic signs of life: patches of vegetation or moss-like organisms, organic-looking features, or any other anomalous objects, but nothing was found. In fact, camera tests had demonstrated that the slow-scan image acquisition technique

The Viking 1 lander's first picture (top) showing one of its foot pads resting on the surface of Mars. The scan was assembled left to right, and the dark stripe at the start is probably due to dust raised during the landing. The first panorama by camera 1 on Sol 3 (below) shows hills, rocks and sandy dunes. The 3-meter boulder 'Big Joe' is between the meteorology mast and the RTG cover.

Some Viking 1 lander pictures. Top left: one of the foot pads buried in sand. Top right: a view of the work site after several days of trenching by the robotic arm. Bottom line: a sample on the inlet to the X-ray fluorescence instrument; the fully extended arm; the top deck covered with dust, showing the antenna actuator, two calibration targets (the small circle on top of the target at the left is one of the dust magnets) and the circular mirror; the arm nudges a rock; a 'stare mode' image of Phobos in eclipse, as marked by the dip in brightness in the center of the sequence.

would have had difficulty forming a coherent image of even a sedate animal such as a tortoise, a chameleon or a snake; the best that could be hoped was to note evidence of its passage. When a close examination of the imagery from the first week failed to find any surface changes, the camera was operated in a mode whereby it maintained a given azimuth and repeatedly scanned a vertical line, to seek variations attributable to wind.[505,506] Years later, the landing site was pinpointed by matching the features seen from the ground with those in high-resolution overhead imagery. However, this position was called into question in the late 1990s, and it was only in the early 2000s that the correct site was determined and the dusty lander was finally spotted sitting on the surface.[507,508]

Meanwhile, the other instruments had also been returning data. The initial runs of the gas chromatograph analyzed the composition of the atmosphere, finding it to be 95 per cent carbon dioxide, 2–3 per cent nitrogen, 1–2 per cent argon and less than 0.5 per cent oxygen. In 1983 some meteorites on Earth were found to contain trapped gas with the same composition and isotopic ratios, and it was realized that these came from

Mars. They were called the Shergottite–Nakhlite–Chassignite (SNC) class for the areas in which the first ones were recovered – Shergotty in India (in 1865), Nakhla in Egypt (in 1911) and Chassigny in France (in 1815). Most of the members of this class are significantly younger than the age of the solar system and also of the other 'mundane' meteorites, and their analysis has helped to unravel many details of the geological history of Mars. In a sense, the SNC meteorites have compensated for a long-overdue Mars Sample Return mission which, when this is finally undertaken, will probably be the single most important planetary mission.[509,510] The meteorology instruments on the Viking 1 lander recorded an average pressure of just 7.7 hPa, and a temperature varying from –86°C at 5 o'clock in the morning to –31°C at 3 o'clock in the afternoon. There was a late-afternoon breeze and variable nocturnal winds, but these were far too weak to have produced either the dust deposits seen by the lander on the rocks or the 'tails' downwind of craters seen from orbit, which were probably emplaced by the occasional dust storms.[511] The only major failure of the lander's mission was when the seismometer refused to uncage and, being almost unusable for its intended purpose, was used to measure the extent to which the lander was shaken by wind gusts. On 22 July the robotic arm was commanded to extend 30 cm, rotate to jettison the protective cover from its scoop, and then return to its stowed position. It extended and released the cover, but jammed attempting to retract. The fault was soon realized to be the locking pin that had latched the cover, and it was decided that the arm had not extended far enough to enable the pin to fall out. As operations were planned in three-day uplink cycles, instructions to extend the arm to 35 cm and then shake the pin loose were added to the sequence that was uplinked to the lander on 25 July, and the pictures taken in support of this effort showed the pin on the ground. Images of the discarded sampler cover and locking pin were analyzed to estimate the hardness of the ground.[512]

On 28 July, several days later than planned owing to the problem with the arm, the first samples were lifted and delivered to the X-ray spectrometer and the biology laboratory. Feeding the gas chromatograph proved much more difficult. The first sample did not trigger the 'level full' switch, possibly because the largest fragments could not pass through the sieve on the funnel. A second attempt on 3 August was aborted by the control system before the arm had fully retracted. Control over the arm was regained several days later. Meanwhile, a 'dry test' of the chromatograph revealed that some dirt had indeed made its way into the instrument on the first attempt. In addition to collecting more samples in the second half of August, the arm excavated a trench to study the physical properties of the soil.[513] On 20 August (Sol 30), the lander was instructed to take a series of spectacular images showing the setting Sun. The X-ray spectrometer took until Sol 31 to analyze its first sample, and then received two separate samples that barely half-filled its chamber. With an abundance of sulfur, but little aluminum or other trace elements, the Martian samples were different to any known terrestrial or lunar material. Almost 13 per cent of it proved to be iron, confirming the presence of oxides such as limonite.[514]

With Viking 1 safely on the surface of Mars, the engineers turned their attention to its twin, which was rapidly approaching. The prime landing site had been selected for being near the southernmost limit of the north polar cap in winter, and because it was about 6 km below the planet's mean surface level, the rationale being that any humidity would be concentrated in low-lying terrain where the pressure was greater. An inspection of this site in Cydonia by the Viking 1 orbiter had shown that while there were locations that appeared to be safe, none was larger than one-tenth of the area required for the landing ellipse. An image taken on the 35th orbit in search of a more suitable site in the same neighborhood gained public notoriety. It showed a group of mesas and low hills, one of which was whimsically pointed out at a press conference to resemble a human face. Independent analysts soon reported finding pyramids and ruined cities nearby. While serious scientists discounted any artificial origin for such structures, self-proclaimed 'UFO-experts' and 'conspiracy theorists' insisted that in order to prevent worldwide hysteria NASA was hiding proof of the existence of an alien civilization!

On 7 August, Viking 2 executed a 40-minute burn to slow down by 1,100 m/s and enter an orbit ranging between 1,502 and 35,728 km inclined at 55.6 degrees to the equator. On its 4th revolution it took calibration images and joined in the search for a landing site, examining Alba Patera. Although this site initially looked promising, later images showed it to be covered by ancient lava flows, and too rough. Starting on 17 August it switched to a promising-looking site on Utopia Planitia where the ejecta from the 100-km-diameter crater Mie was covered by sand dunes, possibly to a depth sufficient to hide hazards. The site was too far north to be inspected by radar for rocks, but the thermal data from orbital infrared scans was encouraging. On 21 August it was decided to attempt a landing at a point about 200 km to the southwest of Mie.[515] Viking 2 fired its engine twice, first to enable it to pass over the site, and then to reduce its apoapsis by 3,000 km in order to establish the required timing. Despite some minor glitches to the lander's descent radar and gyroscopes, at 20:19 on 3 September this was released. Just 26 seconds later, however, the orbiter's attitude control system lost power and as the vehicle started to drift unstabilized the high-gain antenna lost Earth-lock. Fortunately, the link via the low-gain antenna that did not need such precise alignment remained, enabling the engineers to monitor the lander's progress, and when at 22:37:50 UTC the telemetry showed that the orbiter had switched its tape recorder to a high data-rate to receive the lander's post-landing status report, they knew the descent was successful. It was on target on Utopia Planitia at 47.97°N, 225.71°W, where it was mid-morning. The two landing sites were 6,460 km and almost exactly 180 degrees apart in longitude in order that only one lander would be in direct line of sight to Earth at any time. The lander took its preliminary pictures and transmitted them to the orbiter, which stored them. After studying the telemetry from the orbiter, the engineers managed to reinstate its attitude control system, and once the high-gain antenna was realigned it relayed the lander's pictures – some 9 hours later than planned. As a result of the anomaly, one of two redundant gyroscopic platforms had been disabled. Suspecting that the power supply may have been disrupted by the firing of the pyrotechnic charges to separate the lander, it was decided to postpone the jettisoning of the remainder of the bioshell and its support structure, even though this would slightly impede the operation of the scan platform.[516,517]

Tilting at 8 degrees, Viking 2 had come to rest with its 'balcony' (on which the cameras were located) facing approximately north. The first image showed that one of the foot pads was resting on two small rocks, that the rocks were much more

The Viking 2 lander's first picture (top). Note the vesicular appearance of the rocks. An early panorama (below). The tilt of the lander makes the horizon appear curved. At the extreme right of the sequence, the azimuthal drive of the camera has stopped and the elevation drive is rescanning the same line.

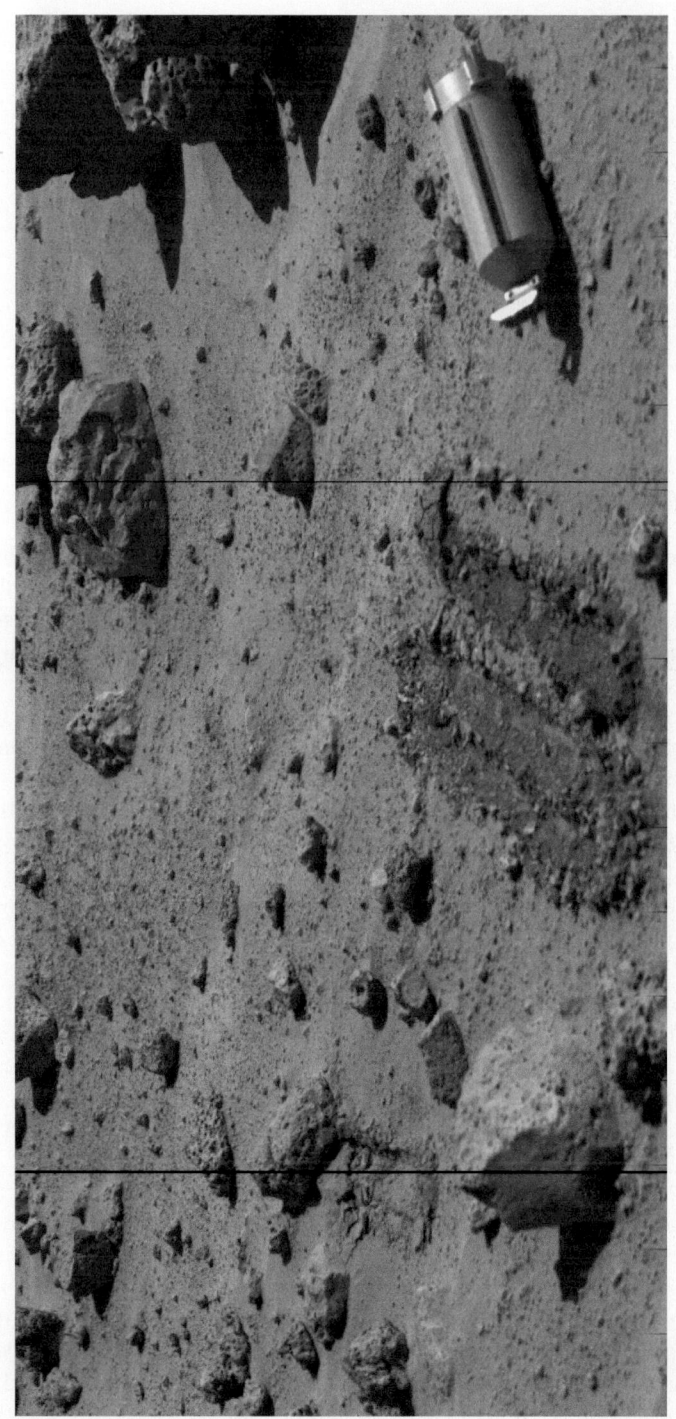

The site sampled by the Viking 2 lander's arm. The cylindrical object at right is the arm's discarded cover.

Pictures by the Viking 2 lander. Left to right: the rear of the 'head' of the robotic arm showing the two small circular magnets; sunrise; a September 1977 morning picture showing frost on the ground; a view looking aft across the vehicle's upper deck; Phobos as a bright dot in the night sky.

pitted by vesicles than those at Chryse, that there was less dust between the rocks, and that the rocket exhaust had exposed a harder crust that resembled an evaporation terrain known as 'duricrust' or 'caliche'. Although at first sight the panoramic view looked similar to Chryse, it was actually very different terrain, being completely flat, with the horizon almost devoid of topography – the only horizon feature was a low bright plateau that corresponded to one of the blankets of ejecta deposited by Mie. Moreover, and most alarmingly in view of the reason for its selection, the site was almost saturated with boulders of a wide range of sizes. Intriguingly, there were a number of polygonal troughs about 1 meter wide that were free of rocks and had small drifts of dust on their floors. Theories for their formation ranged from fluvial features to cracks created by groundwater undergoing freezing and thawing cycles, which could be expected to be more common at this latitude. In contrast to the Viking 1 site, there were no signs of exposed bedrock or large deposits of dust – in fact, there was no evidence of the dune fields that had been inferred from overhead imagery and had made the site seem so attractive.[518,519] Due to the absence of topographical references for triangulation, it was difficult to identify the site in the imagery from the Viking orbiters, and it was not found until very high resolution pictures were taken almost 30 years later, whereupon it was determined that Viking 2 had landed several kilometers west–southwest of a small pedestal crater named Goldstone.[520]

When it was realized that Viking 2 was tilted because it was resting on a boulder, there was concern that this might have damaged the belly. If the thermal insulation had been breached, the electronics would probably fail during the first Martian night as cold seeped into the instrument compartment. But it had been lucky, and when the orbiter passed overhead the next day the lander established contact. When the entry and descent data was finally relayed to Earth, it confirmed the data from Viking 1, although the atmosphere had been significantly colder because Viking 2 had entered over the morning side of the planet. The engineering data revealed that less than half a second before the vehicle had made physical contact with the surface, its radar had locked onto a false target and ordered the engines to increase throttle. This not only slowed the landing to a gentle 2 m/s, but also issued a blast that blew away dust and caused the increased erosion observed in the first image.[521] The diurnal temperatures ranged between –32°C and –82°C, and the light breezes were gusty after sunrise.[522] The atmospheric pressure at both sites was found to be steadily falling, probably due to the progressive condensation of carbon dioxide over the winter pole.

Viking 2 was able to uncage its seismometer, but the instrument's location on the top deck proved to have been a poor choice, since it made it extremely sensitive to (among other things) vibrations caused by the wind and the mechanical operations of the tape recorder, robotic arm, high-gain antenna actuator and camera motor. On Sol 80 it did detect a weak natural quake, but without a second station it was impossible to triangulate the epicenter of the event.[523] The arm was configured without incident, and on 12 September it supplied soil to the biology laboratory and pebbles to the X-ray spectrometer. Despite this promising start, the arm suffered a faulty limit-switch that impaired some operations. On 25 September it set out to collect the first sample for the gas chromatograph. As the test was for organic compounds, it was decided to use the hoe of the scoop to break through a patch of duricrust to sample material that might have been protected from the sterilizing solar ultraviolet. When taking further samples in early October, the arm nudged aside small rocks to sample the protected material just beneath. On 8 October, Viking 2 took sunrise pictures to complement the sunset by Viking 1. The lander then suffered its first major problem, losing the capability to use one of two transmitters in direct communication with Earth. New samples from under small rocks were obtained in late October, with an improved procedure reducing the time between moving the rock and taking the sample to less than half an hour – the issue being the time that the newly exposed soil was subjected to sunlight.[524] On the night of Sol 48 (23 October), the cameras were turned for the first time to the sky, to take a series of pictures of Phobos through different filters. The satellite's reflectivity, spectrum and composition indicated it to be similar to the carbonaceous chondrite meteorites, and hence ancient, rather than being a chip off a thermally differentiated body. This contrasted with the contemporary theory of how the solar system formed, which postulated that such material originated in the colder realms further from the Sun. It seemed that early in its history, Mars had captured Phobos, and probably also Deimos.[525] The magnetic properties investigation, which used images of the magnets located on the deck of the lander and on the scoop of the arm, showed a substantial fraction of the dust to be strongly magnetized. The properties of the soils – inferred from how the foot pads displaced the material, the force that the arms had to apply to dig into it, and the visual appearance of the excavated trenches – showed that the two landing sites were similar, although the Utopia site had more small pebbles.[526] However, it was also apparent that there were differences in the dust at the two sites, indicating that despite the global dust storms the dust had not been uniformly mixed.[527]

The most eagerly anticipated results were, of course, those from the instruments of the biology laboratory that were to determine whether there was microbial life on Mars. In the case of Viking 1, the problem with the arm prevented the samples from

A picture of the Viking 2 lander on the surface of Mars taken in 2006 by the Mars Reconnaissance Orbiter.

being collected until 28 July. The Pyrolytic Release experiment was the first to start. After 5 days of incubation, the atmosphere was extracted from the chamber and the temperature of the sample was raised to 625°C to vaporize (or pyrolyze) any organic material that microbes may have synthesized. Counts would measure the extent to which the sample had taken up the radioactive carbon, and the strong positive signal caused a sensation. A second sample gave much weaker results, as did as the third taken from the same spot near local noon. Two further samples were analyzed in an effort to reproduce the initial strong response, but gave only weakly positive results, as indeed did the final sample that was incubated for 139 sols. For Viking 2, it was decided to incubate the first sample in a dark chamber (it was the only sample of all the tests to be processed in this manner) and the weakly positive result implied that light was not required to drive the reaction. The second and third samples (the latter collected from beneath a rock) gave the only definitely negative results. A hardware problem invalidated the results of the fourth test, and precluded further testing.[528]

Although not the first to start, the Gas Exchange experiment, which added carbon dioxide, krypton and nutrients to its sample, was the first to return results, yielding a spectrum after only 2 hours of incubation. However, this was just to provide a point of reference prior to the experiment proper. In the 7-day 'humid' phase, the nutrient solution was injected into the chamber to humidify the sample without wetting it. In the following 'wet' phase, the sample would be soaked and incubated for 7 months. As the humid phase began, the carbon dioxide unexpectedly increased by a factor of five, and the oxygen level was 15 times more than could be accounted for. One day later, the spectrum was unchanged apart from the oxygen, which had increased by a further 30 per cent. Something in the sample had been producing oxygen, but it was not yet possible to say whether this was due to microbes or was simply an inorganic chemical reaction. In contrast, the wet phase was an anticlimax, with the release of carbon dioxide gradually slowing and the oxygen concentration diminishing. A second sample was 'heat sterilized' to act as a 'control' and humidified, and this too released oxygen. Three processing cycles were made at Utopia, including one with a sample collected from under a rock. Less carbon dioxide was produced at Utopia than at Chryse, with the sample from under a rock producing the least.

The Labeled Release experiment on Viking 1 was given its first sample on Sol 8, and it injected the nutrient two sols later. Surprisingly, the sample promptly released radioactive gas, as if microbes were feasting on the nutrients, but after 10 hours the counts leveled off. A second injection of nutrient was made seven sols later; this not only failed to prompt a second positive reaction, the counts actually fell somewhat. The biological nature of the reaction was tested using the second part of the initial sample (which had been preserved in the apparatus) by heating it to 160°C prior to the first injection on Sol 29, but this produced no response. Finally, the third cycle, starting on Sol 39, was a 50-sol incubation involving three injections of nutrient, with results very similar to the first test. Viking 2 started incubations on Sols 11, 34 and 54. In the second test the sample was heated to just 50°C in 'cold sterilization', the rationale being that any microbes present would be killed by a temperature much higher than that to which they were accustomed, but a chemical agent would not be affected unless it was very volatile. In this case, the magnitude of the reaction was reduced to half. The third sample, taken from beneath a rock, gave similar results.[529] The fact that similar outcomes were obtained from sites on opposite sides of Mars suggested that the reaction was probably ubiquitous.

When the biology instruments ran out of their supplies of nutrients and the helium that was used to purge the chambers, they were turned off: in the case of Viking 2 on 28 May 1977 after 259 days, and Viking 1 on 30 May after 308 days.

In reviewing the overall results, most scientists felt that the cause of the reactions was chemical, not biological. It was reasoned that if the soil was rich in peroxides – probably the result of reactions stimulated by long-term exposure to solar ultraviolet – these would decompose on being exposed to water in the presence of iron, which was abundant in the soil. This explained the initial strongly positive Labeled Release response, why the reaction did not recur upon the second injection of nutrient (the peroxides had been fully decomposed), and the negative results from the 'sterilized' soil (the peroxides were very sensitive to heat). Only the weakly positive signal from the Pyrolytic Release instrument remained to be explained, and this reaction did not behave in the manner expected of biological activity: in particular occurring after a 3-hour heating cycle at 175°C. A theory involving microbial life had to explain how organic matter survived in the highly oxidizing environment suggested by the other experiments.[530] It has even been proposed that the oxidizer may have come from microbes through the use of a mix of hydrogen peroxide and water, instead of just water, as their intracellular fluid. This characteristic would not only make organisms well-suited to the low-temperature environment of Mars, but would also make them vulnerable to exposure to liquid water – in which case the 'wet' experiments would have been lethal![531] It was the gas chromatograph that drove the proverbial nail into the coffin of Martian life. It was capable of detecting organic compounds that comprised more than 10–100 parts per billion of the soil. In fact, in tests it detected organics in Apollo lunar dust that probably derived from carbonaceous meteorites. On Sol 17, Viking 1 heated its first sample to 200°C for 30 seconds. It detected no organics, but this was as expected because that temperature was too low to break up complex compounds. Six sols later, the sample was heated to 500°C and, to the surprise of the scientists, a large amount of water was released, but no organics other than traces of the cleaning solvents known to be present in the instrument. Two samples were analyzed by each lander, with the same results. The lack of carbon-based materials presented a serious obstacle to any theory that interpreted the results of the biological laboratory as due to microbes! Even if Mars had proved to be sterile, organic material of meteoritic origin had been expected, and its absence was a clear sign that the soil was hostile to organics.[532,533,534,535] The post-Viking scientific consensus was that there is no life on the *surface* of Mars. This did not exclude niches underground, where the sterilizing agent produced by ultraviolet would be absent, and where water might be preserved as a permafrost of ice grains. The only plausible strategy for a future search of life would be to use an orbiter to locate subsurface ice and then to send down a lander with a drill to collect samples from a depth of several meters.

Despite the consensus, some scientists – in particular, Gilbert Levin, the leader of the Labeled Release team – insisted that the Vikings had indeed found life. Part of their case was that no instrument had really detected the putative oxidizing compounds; these were an ad-hoc assumption to explain the results. As for the failure of the chromatograph to detect carbon, they noted that while this instrument did not detect organics in supposedly sterile soil from Antarctic, a positive response was obtained when such a sample was incubated in the more sensitive Labeled Release instrument. Furthermore, samples that had been appropriately enriched with hydrogen peroxide for testing did not match the results obtained from Mars.[536,537] As noted by Klaus Biemann of the Massachusetts Institute of Technology, leader of the gas chromatograph team, there would need to be at least 1 million microbes in the sample for their organics to be found at the instrument's sensitivity of 10–100 parts per billion. A typical temperate sample of terrestrial soil can contain hundreds of millions of bacteria per cubic centimeter. If only the living cells were present for analysis, then 1 million bacteria would have been far too few to be detected by the instrument. In terrestrial soil, the amount of dead organic matter often outweighs the living material by a factor of 10,000, and if Martian microbes were the same they would probably have been able to be detected by the organic wastes and dead cells they produced. However, if they were adapted to the local environment and recycled their wastes, and if solar ultraviolet destroyed the dead cells, microbes that were not able to be detected may well have been present to produce the reactions reported by the biology laboratory. More recent tests have shown that gas chromatography may not be appropriate to the search for organics on Mars, as at high temperatures the iron-rich soil would oxidize the organic compounds to carbon oxide, so diminishing the scope for detecting life. It is possible this process falsified the chromatograph results and that microbes did indeed cause the results of the other experiments.[538] To sum up, Gilbert Levin opined: "The accretion of evidence has been more compatible with biology than with chemistry – each new test result has made it more difficult to come up with a chemical explanation, but each new result has continued to allow for biology." All other things being equal, he noted that if a terrestrial sample had given the observed results, "we would unhesitatingly have described [it] as biological".[539] Vance Oyama, principal investigator for the Gas Exchange experiment, was skeptical: "There was no *need* to invoke biological processes."[540] Norman Horowitz, leading the Pyrolytic Release experiment, agreed, but admitted that it was "impossible to prove that any of the reactions ... were *not* biological in origin".[541] Harold Klein, in overall charge of the biology investigation, later recommended that the assumption that Martian microbes would be similar to terrestrial ones should be dismissed, and that scientists ought to consider whether the Viking data suggested any clues as to "whether there might be some less obvious kind of life on Mars".[542]

On 30 September 1976, having delivered its lander, the Viking 2 orbiter increased the angle of its orbit to 75 degrees to gain a better view of the north polar region, where the summer season officially started in July. Meanwhile, the Viking 1 orbiter adjusted its orbit to relay for the second lander. Its own lander, now without an orbital relay station, entered a period of reduced operations in which it returned mostly routine meteorological data directly to Earth. During October alone, Viking 2 took 700 high-resolution images of the north polar region, documenting the layered terrain nearest the pole, the surrounding fields of dark sand dunes, and the plain that underlies both of these. The layered terrain was similar to that seen at the south pole, and appeared to consist of layers of dust mixed with ice and other volatiles that had accumulated over the years, masking the older terrain so extensively that only a few ancient craters could be recognized. Studying the layers, scientists realized that they might preserve a history of how the climate had changed over hundreds of thousands of years.[543] This raised the enticing prospect of one day sending a lander to dig into this terrain to 'read' this story. The water detector found abundant water vapor over the poles. In fact, the data showed that the residual north polar cap (i.e. the part that does not melt in summer) is water ice, perhaps several kilometers thick. However, if all the vapor were to be condensed evenly around the planet, it would form a film of water only one-tenth of a millimeter thick. Later observations of the south polar cap revealed it to be mostly carbon dioxide, although it was not possible to rule out the existence of a small cap of water ice beneath.[544]

The Viking orbital imagery was much better than that from Mariner 9, in part due to camera improvements, but also because the air was clearer. Although Mariner 9 had waited for the global dust storm to abate before it began mapping, there had still been a lot of dust in the atmosphere. Moreover, while Mariner 9 had shown very few small craters, leading geologists to believe that few existed, the Vikings established that there were thousands of such craters. In addition, water-cut features occurred at a much finer scale than previously recognized, revealing the drainage systems to be more complex than had been believed. Although Valles Marineris had tributary canyons, the new imagery showed that the dominant factor in its formation was not erosion but steeply inclined crustal faulting and subsidence, making it a rift valley or

A remarkable teardrop-shaped 'island' some 40 km in length seen by the Viking 1 orbiter in the southern Chryse region during the spacecraft's fourth orbit.

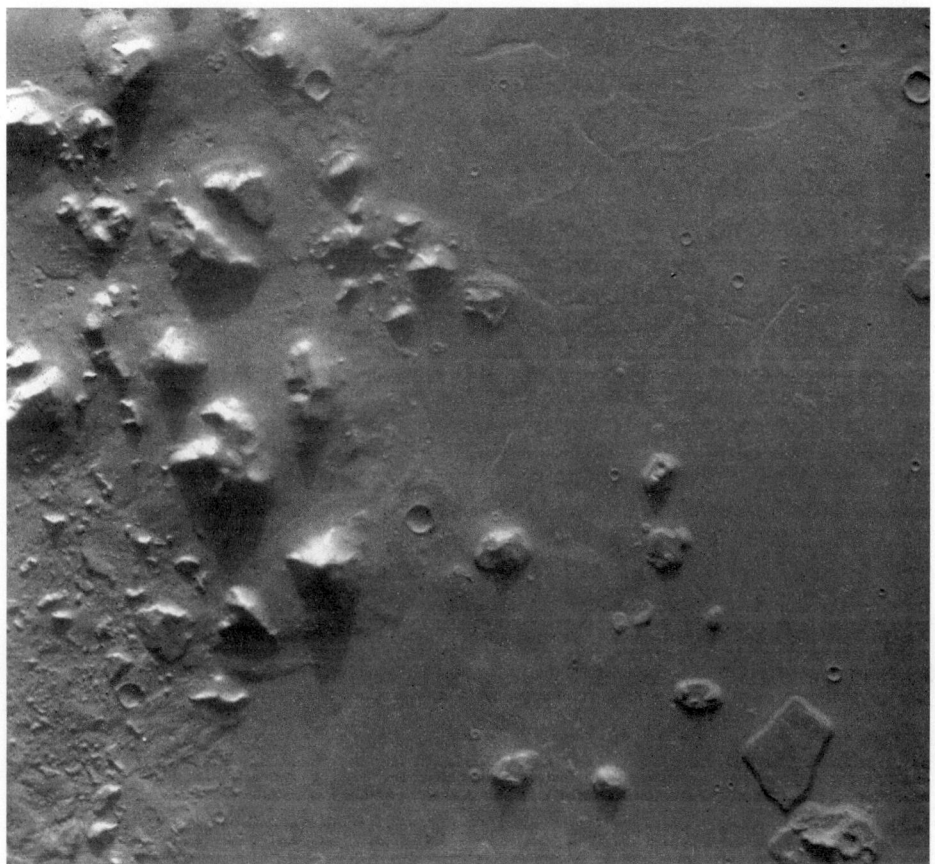

This image of Cydonia taken by the Viking 1 orbiter on 26 July 1976 is the most infamous of the entire NASA planetary program, since it contains a mound whose appearance resembles the face of a sphinx.

graben; and the many constituent canyons proved to be poorly linked to one another, with floors that did not exhibit slopes consistent with water action. Different types of channel (which, of course, had nothing to do with the 'canali' drawn by Schiaparelli and Lowell) were mapped. Those with well-developed networks of tributaries were most common on the ancient terrain of the southern hemisphere and equatorial zone, while outflow channels with widths of many tens of kilometers originated primarily in the patches of chaotic terrain north of Valles Marineris, and drained northward to debouch on the low-lying Chryse Planitia. Within these channels were terraces and distinctive teardrop-shaped islands. All of these features, and in particular the well-developed southern hemisphere channels, would imply a warmer climate early in the Martian history. A wide variety of meteorological phenomena were studied, ranging from clouds and hazes to layers of suspended dust. The 'W-shaped' high-altitude clouds that were long familiar to

An oblique view of the summit of Olympus Mons poking above a sea of clouds taken by the Viking 1 orbiter on 31 July 1976. Bright clouds in this locale had been known since at least the times of Schiaparelli, who called the area 'Nix Olympica' (Snow of Olympus) for that very reason.

This Viking 1 orbiter oblique view of the Argyre basin and the atmospheric layers above is one of the most beautiful pictures taken by the Viking orbiters.

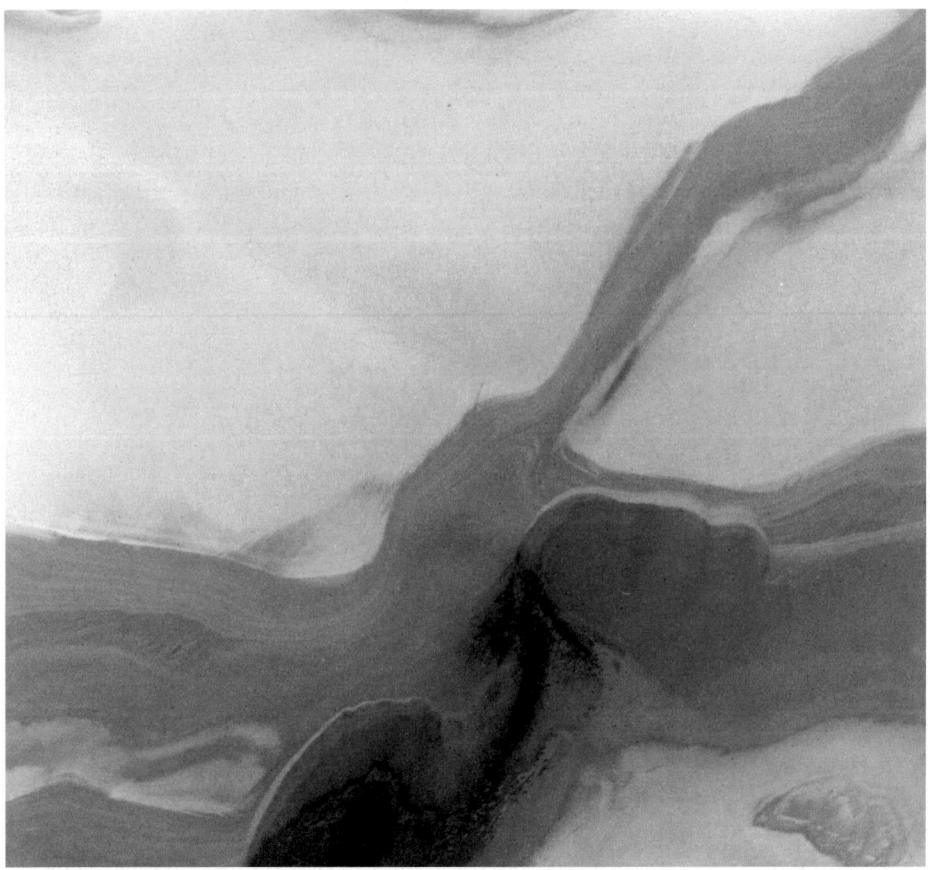

A high-resolution close up of the retreating northern polar cap taken by the Viking 2 orbiter. It shows extensive layered terrain and cliffs, as well as a variety of erosion features.

telescopic observers, and which Mariner 9 established to be associated with the volcanoes in the Tharsis region, were closely monitored. Characteristic repetitive 'wave clouds' were discovered to form where the prevailing wind swept over terrain relief such as crater rims. The new imagery and thermal mapper data enabled models of the planet's current meteorology and climate to be developed.[545,546,547] The Viking pictures were carefully compared with those taken by Mariner 9 to search for changes, but the only differences were dark or bright windblown dust streaks that had either appeared or disappeared. A subject of particular interest was the morphology of the volcanoes, and their lava flows. The summit caldera of Mons Olympus was imaged at a resolution of 18 meters per pixel, and the overall height of the edifice was revised to 27 km, making it the tallest known mountain in the solar system.

Between 11 November and 13 December 1976, Mars was on the opposite side of the Sun from Earth, and an experiment was performed to determine, with the utmost accuracy, the round-trip travel times of radiowaves in the vicinity of the Sun. The

On 28 September 1978 the Viking 1 orbiter imaged a dust storm sweeping over the Viking 1 landing site.

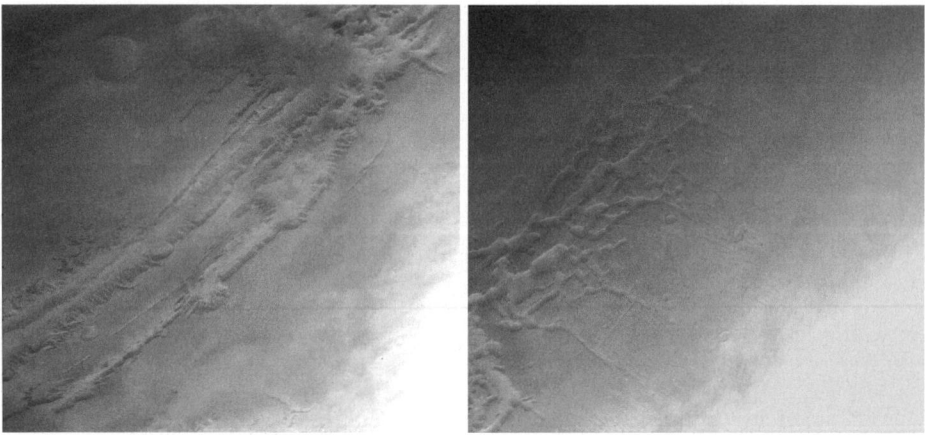

Two portions of the Valles Marineris imaged by the Viking 1 orbiter, showing the parallel canyons Tithonium and Jus Chasma (left) and the complex fretted terrain of Noctis Labyrinthus (right).

results matched the prediction of General Relativity to within 0.1 per cent.[548] During solar conjunction the landers autonomously continued the long incubation of their biology samples. On 16 November NASA said that the mission had accomplished its primary scientific objectives, but hoped that the four vehicles would continue to work for at least another 18 months in order to monitor seasonal variations during a full local year, both on a global basis from orbit and in detail at the landing sites. It was also possible to pursue secondary imaging objectives. In particular, there would be encounters with Phobos and Deimos when the orbits of the orbiters and moons intersected, and the mission planners calculated the optimum times to make burns to set up a series of close flybys. Viking 1 had an opportunity with Phobos in February 1977, and Viking 2 for Deimos in October. By coincidence, 1977 would also mark the 100th anniversary of Asaph Hall's discovery of the Martian moons. Viking 1 made a series of three maneuvers in January and early February to establish an orbit that would repeatedly encounter Phobos in the last two weeks of February. Of a total of 17 daily encounters, the 8th (on 20 February) took the spacecraft to within 80 km of the moon's surface. In an unprecedented operational effort, measurements of the position of Phobos and estimates of its mass from radio tracking were used in almost real time to ensure that the scan platform was correctly pointed on later flybys. The success of this 'adaptive' technique was such that a total of 125 useful pictures were obtained. The objectives of this Phobos campaign were to estimate the mass and volume of the moon, to obtain imagery of part of its surface at a resolution of 10 to 20 meters per pixel, and to obtain high-resolution infrared coverage. About 80 per cent of the moon's surface was mapped at a resolution of 30 meters, and its mass and volume (and hence density) were determined with sufficient accuracy to impose constraints on its composition, confirming the lander's evidence that it is similar to carbonaceous chondrite meteorites. As already shown by Mariner 9, the surface of Phobos is dominated by sharp craters and a network of linear grooves hundreds of meters wide.[549] On 11 March Viking 1 lowered its periapsis to 300 km, at which altitude it would be able to image objects on Mars as small as 20 meters. It took additional pictures of Phobos during a 300 km flyby in May, then executed a small burn on 15 May to ensure that it did not venture too close to the moon.

After concluding their final incubations, the landers dug trenches and built small piles of dust in order to determine information such as grain size. The meteorology package followed the changing seasons, noting that, after hitting a seasonal low, the atmospheric pressure increased as the southern polar cap began to thaw and carbon dioxide entered the atmosphere. With the onset of winter in the northern hemisphere in September 1977, the Viking 2 lander saw morning frosts. Interestingly, even after the temperature reached the point of evaporation of carbon dioxide, the frost remained, indicating that it was made of water ice or a mix of water and carbon dioxide ice, but it was not possible to determine because the chemical analysis instruments were no longer operational. Early in the southern spring, and again after the southern summer solstice, the orbiters monitored two dust storms which, while global in extent, were not as intense as the storm of 1971. They were documented from their development in the Thaumasia–Solis Planum region until the atmosphere was clear again, 2 or 3 months later. From the point of view of the landers, the sky darkened, the wind speed rose to several tens of meters per second, dust was deposited on their decks, and their diurnal temperature range narrowed dramatically.

On 23 June 1977 the Viking 2 orbiter captured a striking sequence of images that showed Phobos against the Tharsis region. In addition to their aesthetic value, these pictures confirmed the determination of the shape and volume of the moon, because even its darkened hemisphere was clearly outlined against the much brighter surface of the planet.[550] When the shadow of Phobos transited the Viking 1 landing site on 24 September (Sol 419), the lander and orbiter made simultaneous observations of this 'eclipse' in order to pinpoint the lander's position. Of course, tiny Phobos never totally blocks out the Sun; it produces only a partial eclipse. And because it orbits so close to Mars, the rapidly moving shadow would pass by in as little as 20 seconds. The lander therefore monitored the event in 'staring mode', repeatedly scanning the same vertical line to record how the shadow's passage caused a brief decrease in the brightness of the sky.[551,552] In early October 1977, Viking 2 maneuvered into an orbit that would take it within 1,000 km of Deimos at 5-day intervals, starting on 5 October. On 15 October the spacecraft passed just 26 km over the surface of the tiny moon, and its high-resolution images provided an explanation for the apparent dearth of craters (in relation to Phobos, at least, which is pocked by craters large and small). There is no crater of comparable proportion to Stickney; the largest craters are only 3 km, and the surface is covered by a thick layer of regolith that blankets all but the largest of its topographic features. In fact, being in a wider orbit around Mars than Phobos, Deimos, though smaller, is better able to retain the material that is 'ground up' by meteoroid impacts. The surface is littered with boulders. There are no grooves such as disfigure Phobos. These observations enabled the Mariner 9 measurements of Deimos's orbit and dimensions to be refined.[553] On 23 October

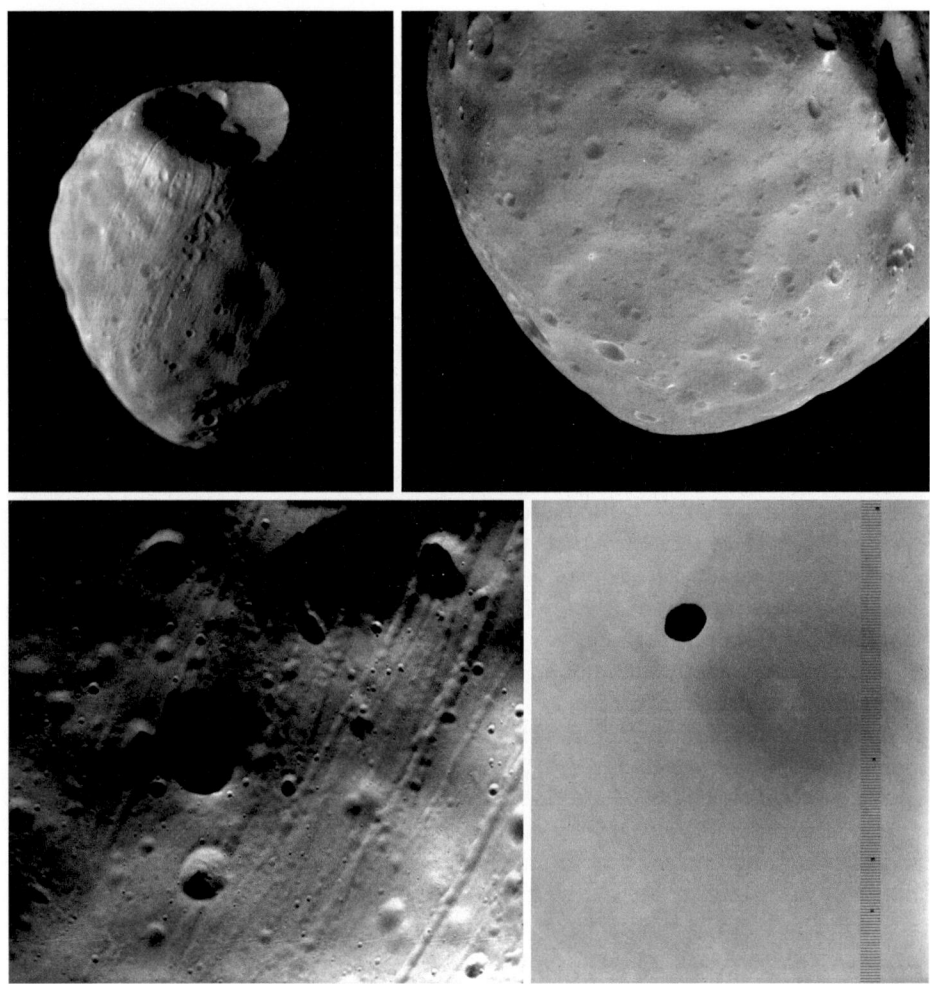

Pictures of Phobos taken by the Viking orbiters. The image at upper left shows the large crater Stickney and the grooves that appear to radiate away from it, and these are shown in higher resolution in the picture at lower left. The image at lower right shows the silhouette of Phobos passing in front of Mars with Ascraeus Mons in the background.

Viking 2 lowered its periapsis to 300 km. Tracking data acquired over the next 6 months measured the spacecraft's radial accelerations with respect to the planet, not only facilitating a detailed map of the gravitational field around the whole range of longitudes and ranging between latitudes 30°S and 65°N, but also identifying mass concentrations at the locations of each of the large volcanoes – the largest being associated with Olympus Mons. There were also gravity 'lows', in particular associated with Valles Marineris.[554] Other activities during this low-periapsis phase of the mission were bistatic radar observations (as had been done by the Soviets using spacecraft orbiting Venus) in which the reflection of the radio

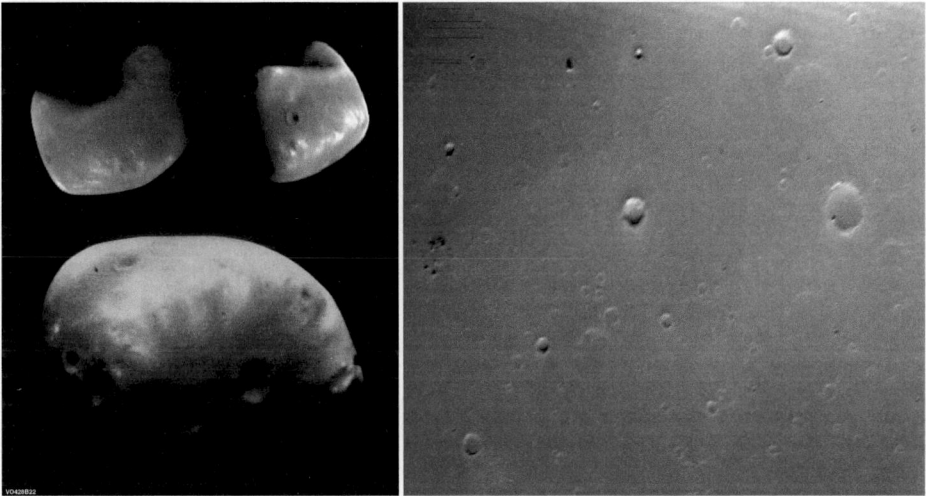

Pictures of Deimos taken by the Viking 2 orbiter. The smoothness of the surface is particularly evident in the high resolution picture at right.

waves issued by the orbiter were received on Earth. The target of this study was the Hellas basin, and the objective was to measure its roughness and surface slopes.[555]

As the operation teams working on Viking dwindled from 1,000 people during the summer of 1976 to fewer than 100 for the extended mission, and subsequently to 25, the engineers at JPL devised ways to teach the computers of the orbiters to monitor some of their own telemetry and automatically deal with some of the tasks that had previously been left to controllers on Earth. Viking 2 which had suffered a series of gas leaks in its attitude control system in February and March 1978, was the first to be updated. The usual action when a leak was detected was to briefly fire the thrusters to dislodge the particles that were presumed to be clogging the valves. Because the orbiters were no longer able to be tracked on a full-time basis, several hours could elapse between the detection of the leak and taking remedial action. Auto-diagnosis software offered the prospect of immediate intervention which, by reducing the waste of the precious gas, would lengthen the life of the probe. Although this reduced the leaks to one-third of the previous rate, Viking 2 exhausted its nitrogen on 25 July 1978, which terminated its mission. For most of its time, the orbiter had been lugging around the adapter and the remaining part of the bioshield of its lander, having discarded this 'dead weight' only on 3 March 1978. Other life-extending software for the Viking Continuation Mission that started in June 1978, and the Orbiter Completion Mission in late 1979, was designed to protect the surviving orbiter against stray light from Mars, its moons, or spacecraft structures prompting the Canopus sensor to lose its attitude reference. The software was also to enable the spacecraft to shut down its instruments and other hardware in the event of anomalous battery discharges. The Deep Space Network support for it was greatly reduced to make resources available for the Voyager Jupiter encounters. In 1980 it returned to full operation in order to cover some gaps in the high-resolution coverage of the planet, to image potential sites for future landers (including that which would be selected for Mars Pathfinder in 1997) and to conduct radio occultations. By the summer of 1980, having spent 4 years orbiting the planet, its attitude control gas was running so low that it was evident that the mission was nearly over. One further software update was made to enable the spacecraft to shut off its transmitter once it lost attitude control, which occurred on 6 August on the 1,489th orbit. Projecting the orbit forward indicated that Viking 1 will enter the atmosphere in about 2019.[556,557] Meanwhile, the buffer batteries on the Viking 2 lander had lost power, resulting in the shutdown of its cameras and remaining scientific instruments. The problem was noted on 31 January 1980, and starting in mid-March unintelligible transmissions were received. Even if Viking 2 had survived this problem, it would not have lasted much longer, as by the end of 1978 it had lost both the primary and secondary transmitters feeding the high-gain antenna and was reliant on an orbital relay station to return data to Earth. The mission, therefore, would last only as long as there was an operational orbiter. On 12 April 1980 it was commanded to shut down.

The final phase of the project was the Viking Monitor Mission, involving only the Viking 1 lander. On Sol 921 it was reprogrammed to make weekly weather reports and take a sequence of five images every 37 days in order to monitor short-term and seasonal variations in the hope that it would be able to return data until 1994, which was the latest possible date that could be specified in the antenna-pointing software. It monitored the erosion of the five piles of dirt that had been built by the robotic arm, and the deposition of dust and wind effects. Dust was seen to settle after global dust storms in the fall and winter of the second Martian year. A sequence of images of nearby Big Joe in the third winter showed how a small dust storm dramatically lowered the transparency of the atmosphere. Unfortunately, by then the anemometer had partially failed, and was giving only pressure data. By Sol 2,230, when images taken during the fourth autumn showed that another dust storm was in progress, the lander was ailing, with three of its four batteries not recharging correctly. Although a new procedure to solve the charging problem was uplinked on 19 November 1982, there was no confirmation of its receipt. When further attempts to contact the lander failed, the software update was re-examined and found to have been loaded into the part of the memory reserved for the vital antenna-pointing software! Although the high-gain antenna was no longer pointing at Earth, efforts to make contact continued for the next 3 months, but to no avail. Unfortunately, the low-gain antenna, designed to receive commands, had failed early in the mission. In March 1983 support for the Viking Monitor Mission was terminated, and the end of one of the most successful of NASA's planetary missions was announced on 21 May. The Viking 1 lander had operated on the surface of Mars for 2,245 sols, or 3.3 local years.[558,559,560]

Overall, the Viking orbiters returned a total of 52,663 pictures and mapped 97 per cent of Mars at a resolution of 30 meters, in addition to relaying some 4,500 images from the landers and providing a wealth of other data. Thomas A. Mutch, the leader

Three pictures of 'Big Joe' taken by the Viking 1 lander on (left to right) Sols 1,705, 1,742 and 1,853. At the time of the center image the atmosphere was laden with dust from a storm.

of the lander imaging team, died trekking in the Himalayas in 1980. The Viking 1 lander was dedicated the Mutch Memorial Station, and a commemorative plaque was made for a future astronaut to affix to the vehicle.[561,562] Following his death in 2000, the Viking 2 lander was dedicated to Gerald A. Soffen, the chief scientist of the Viking project.

PIGEONS, ROVERS, SNIFFERS

Immediately after Viking 1 successfully landed on Mars, JPL presented its vision for the future of solar system exploration, involving a series of amazing 'Purple Pigeon' missions in the 1980s that would capture the imagination of the public, in contrast to dull scientific 'Gray Mouse' missions. This plan included a spacecraft that would be propelled by the pressure of solar radiation to encounter comet Halley in 1986; an orbiter to map Venus by radar; a mission that would enter orbit around Saturn with a Titan atmosphere-and-lander probe; an 'asteroid tour'; a spacecraft that would enter orbit around Jupiter with a hard lander for Ganymede, the largest Jovian moon; and a robotic base to resume the exploration of the Moon. Of these, only the Venus radar mapper and the Saturn orbiter would actually fly; the latter reaching its target almost 30 years after its conception, as the Cassini–Huygens mission. For Mars, the Purple Pigeon would be launched in either 1981 or 1984 with two landers, each with a pair of remotely controlled rovers that would drive up to 1,500 km over a 2-year period. The 200–250-kg rovers would carry a camera, a robotic arm and instruments for geochemical and biological analyses, and their explorations would be coordinated.[563,564]

In fact, proposals to send rovers to follow up on the Viking landers were made in the early 1970s. At that time, it appeared that the most obvious option would be to build a third orbiter/lander using the existing backup systems, revised as appropriate for the new mission, with a launch date possibly as early as 1979. The payload of the 'Viking 3' lander would include new biology instruments designed to characterize any potential life responses reported by its predecessors. When it was recognized that Viking had sufficient growth potential to carry a rover, proposals made in 1973 included (1) a small tethered rover with a range of 100 meters to set up experiments outside the area affected by the landing jets, to perform stereoscopic imaging of distant targets, and to collect samples for analysis, or (2) a 100-kg vehicle to be carried 'piggyback' on the lander on a foldable truss that would be deployed after landing, and, being fully autonomous, would travel several kilometers. In addition to carrying instruments to analyze samples, its camera would enable it to 'look over the horizon' of the stationary lander – which would greatly assist in characterizing the landscape. Possible rover configurations were studied by Martin Marietta, with help from Messerschmitt–Bölkov–Blohm of Germany, which assembled a team of 30 to work on the project. Despite suggestions that the mission might become a joint US–European project, it seems that no official contacts were made between NASA and ESRO.[565,566,567] Interest in a rover was reinvigorated by the fact that the Vikings were immobile. An unusual proposal was offered by Martin

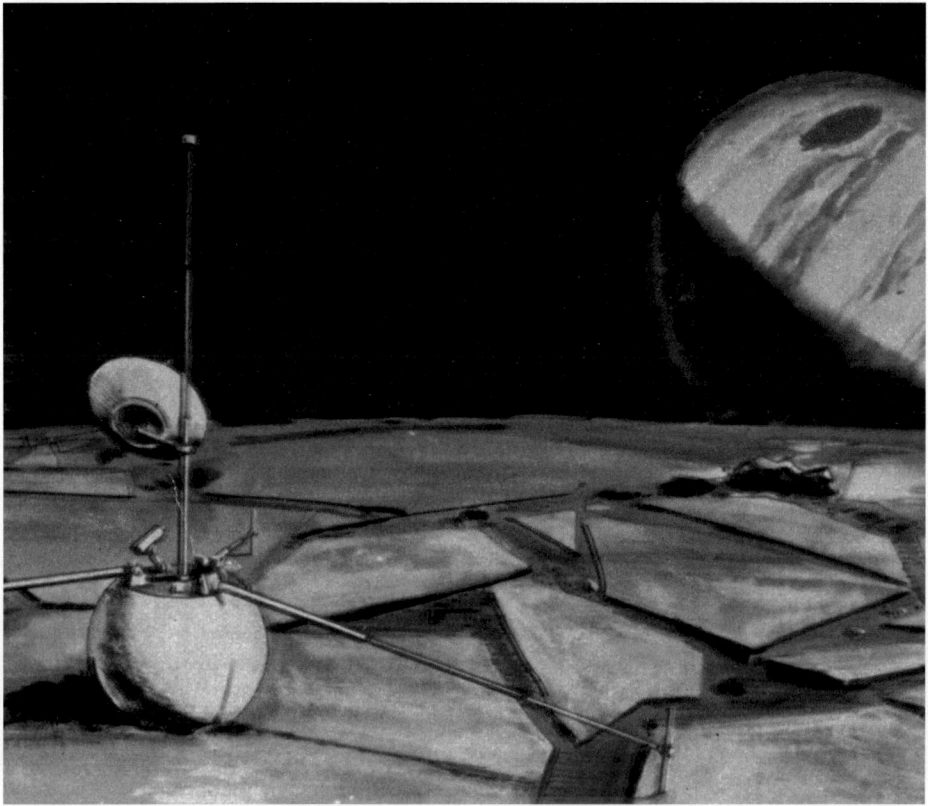

The 'Purple Pigeon' Ganymede lander.

Marietta for the 1981 window, in which the foot pads of the backup Viking lander would be substituted by twin independently swiveled motorized tracks that would provide a top speed of 150 meters per hour, the ability to climb a 20-degree slope, and an overall range of about 50 km; during which traverse it would make frequent stops to conduct experiments. The principal advantage of this option over a smaller autonomous rover was that the entire payload suite would be available at each area of study.[568]

Larger autonomous rovers with masses of up to 400 kg were considered for the 1980s. Alternatives included a desk-sized rover mounted on motorized tracks, and a six-wheeled rover with an articulated body. The RTG-powered vehicles would have autonomous navigation, detect and avoid obstacles, recognize scientifically interesting sites, and undertake preliminary analyses without intervention from Earth.[569,570,571] There were also studies of simpler options. In particular, the discovery that the rarefied Martian atmosphere could yield strong winds suggested the amazingly simple possibility of the Mars Ball, which would be an inflatable beach-ball-like rover that would simply roll across the surface, either blown by the wind like a tumbleweed, or propelled by an onboard system to provide some degree of control over its direction of travel. It would pause to investigate interesting sites simply by deflating. The wind-blown versions, several meters in diameter, could carry up to 30 kg of payload, located at the center of mass, that could possibly include a sample scoop. A French invention, Mars Balls were studied at JPL, and in 1985 by students of the University of Arizona. A probe to deliver a number of Mars Balls was proposed by France to ESA in 1979 as a cooperative mission with NASA but not funded.[572,573,574,575]

A completely different form of autonomous vehicle could survey Mars from the air. In 1975 Dale Reed of NASA's Dryden Flight Research Center started to work on a project of high-altitude air sampling in a context unrelated to that of planetary exploration. A small drone dubbed the 'Mini-Sniffer' was to take air samples while flying at altitudes in excess of 30 km in the wake of an SR-71 spyplane traveling at Mach 3, in order to assess the possible threat to the stratosphere posed by large supersonic airliners. The drone's wing was tailored for the extremely rarefied air at that altitude (or, in engineering parlance, for low Reynolds numbers). To obviate the problem of installing a large compressor to enable an air-breathing engine to operate in very thin air, it was equipped with an 'anaerobic' hydrazine engine that powered a piston engine by using the gas released by the decomposition of hydrazine over a catalyst bead. To address the problem that after several hours of continuous use the catalyst would start to release particles that could clog valves, and reduce the engine efficiency, Reed sought assistance from JPL, which had a history of using hydrazine decomposition systems on its spacecraft; where the potential of the Mini-Sniffer as a post-Viking flyer for Mars was immediately recognized. In fact, the density of the air at an altitude of 30 km on Earth is comparable to that near the surface of Mars, and in one-third gravity the airplane would not need to fly so rapidly to generate lift, and would therefore require less power. It was calculated that a Mini-Sniffer would have an endurance of 40 hours on Mars, and could fly 8,000 km.[576] As the design of the Mini-Sniffer would have caused a reduction in performance owing to aerodynamic penalties of lower lift and larger drag, JPL funded Developmental Sciences Inc. to design a new airframe. This 300-kg 'Astroplane' had a high-aspect-ratio wing especially designed for flight on Mars. It had six hinge points at 3 meter intervals to enable it to be stowed inside a Viking aeroshell and unfold to a span of 21 meters. The fuselage comprised the engine, which could be either electric or hydrazine-powered, a tractor propeller, fuel tanks, compartments for electronics and a scientific payload of 40–100 kg that would include a 12-cm-resolution camera mounted in a transparent bubble under the fuselage for a hemispherical field of view. Also mounted on the fuselage were two Viking landing-engine pods to ensure a soft landing, and possibly several vertical take-offs. The aircraft would be able to drop seismometers, meteorological stations, and even mini-rovers along its route. The mission called for an 'air force' of a dozen aircraft to be launched on three carrier probes in a single 1-month launch window using the Space Shuttle (which was optimistically expected to fly weekly turnrounds). The carriers would enter 24-hour orbits around Mars with their periapses at 500 km, and would deploy aircraft singly, as directed. On entering the atmosphere, the aeroshell would open and as it descended by parachute the aircraft would unfold, deploying the 6.35-meter-long fuselage and then its wing, at which point it would start its engine, release the parachute, and start its cruise. The mission would be carried out mostly at an altitude of about 750 meters, although heights up to 7.5 km would be attainable, and would last between 7 and 31 hours (depending on the engine, payload mass, and number of stops) during which it would travel up to 10,000 km, making magnetic, gravity and geochemical surveys, analyzing the atmosphere, taking high-resolution images, and seeking evidence of subsurface water, geothermal fields and active volcanism. For navigation, it would use an inertial system, a radar altimeter and a terrain-avoidance system. Communications would be entirely through the carrier orbiters. After the flight, the airplane would land and operate on the surface until it expired.[577] To test the landing maneuver, NASA modified a sailplane to enable its tailplane to 'pop up' at low altitude and cause the wing to enter a deep stall, causing the plane to drop to the ground at an almost vertical angle.[578] However, the agency eventually decided to concentrate on other priorities, and the Mars airplane concept was shelved.

Low-cost Mars orbiters were also studied that would exploit the Pioneer bus that was under development for missions to Venus. Two baseline Pioneer Mars missions were outlined. One was to dispatch penetrators to perform surface and subsurface studies at several locations, and the other was to perform orbital observations of the atmosphere and surface processes across the planet. Spear-shaped penetrator probes were designed to impact at speed and embed themselves deep into the ground. This technology had been developed by the US military to deploy networks of sensitive seismometers to stealthily monitor troop movements and clandestine underground nuclear tests. The penetrators for Mars would exploit research undertaken by the US Atomic Energy Commission's Sandia Laboratories to deliver nuclear warheads at high speed so as to bury them deep underground in order to make them effective against 'hard targets' such as bunkers, and at the same time (or so it was believed) minimize the fallout and contamination. Since the 1960s, Sandia had experimented with thousands of 'terradynamic' vehicles of various sizes and masses, arriving at speeds of almost 1 km/s, and by the mid-1970s the technology was deemed to be sufficiently mature to be applied to planetary exploration. At impact, the penetrator was to split in two, with the aft body that held the antenna and transmission system remaining on the surface while the instrumented and tethered forebody buried itself to a depth of several meters. Each penetrator would have an accelerometer to record the dynamics of the impact, a seismometer, and instruments to analyze the soil composition. All the hardware, including the electronics, RTG power supply, and instruments, would be designed to withstand an impact deceleration of 1,800 g. A small penetrator was even devised that could be released by a Viking lander during its descent, with instruments to test for near-subsurface water. In 1975 the Space Science Board of the National Academy of Sciences recommended penetrators as the highest priority post-Viking Mars mission. One important advantage of penetrators was that they could be targeted at high-altitude terrains where there was insufficient atmospheric braking for a conventional lander – such as the flanks of the Tharsis volcanoes. The 'Pioneer Mars Orbiter' jointly designed by the Ames Research Center, Hughes and Sandia, was to have six canisters, each of which was to fire a penetrator into the intriguing layered terrains at the poles.[579,580,581] The Pioneer Venus Orbiter would need only simple modifications to perform the aerometry mission. It would enter an elliptical orbit that had its periapsis as deep in the atmosphere as possible in order to measure the composition and heat balance of the ionosphere and upper atmosphere, and the interactions with the solar wind. When at this low periapsis, a gamma-ray spectrometer would also collect data on the elemental abundances of the surface.[582] Other study options included sending a Pioneer bus carrying several small hard landers, and an orbiter with instruments to search for water deposits.[583]

Although Mars Sample Return would be the ultimate challenge for an automated mission, the sample that would be returned to Earth might be as little as a few grams. Two Earth-return strategies were studied, one involving a single spacecraft and the other involving two spacecraft. In both cases, the vehicle that lifted off from Mars with the sample would enter orbit around the planet. In the single-spacecraft case, the vehicle would wait several hundred days for the Earth window to open and then set off for home, but this strategy would require landing a very large mass on Mars, possibly as much as five times that of Viking. In the other case, the craft with the sample would rendezvous with another vehicle that would execute the 'escape burn'. This strategy would greatly reduce the mass that had to be landed on Mars, but, owing to the long time delay, would require a reliable system for rendezvous and docking without any intervention from Earth. There was some flexibility in the two-spacecraft strategy: if it were to prove impossible to launch the second spacecraft during the same window as the sampler, it could be sent later; indeed it could be sent later in order to reduce the peak funding, although the longer the sampler was required to function awaiting the rendezvous, the greater would be the operating costs. Once on Earth, the samples would be subjected to state-of-the-art laboratory analyses to identify the chemistry, mineralogy and absolute ages of the material, and to gain insights into its potential organic history.[584,585] To protect Earth from contamination by infectious Martian life forms, there was also a feasibility study of an Orbital Quarantine Facility to receive the sample and assess its biological risk. But the development of a space station would greatly complicate an already complex mission, as it would require fitting the return craft with an engine to slow down to enter orbit around Earth, as opposed simply to releasing a capsule to make a high-speed atmospheric entry.[586]

However, for two main reasons all these proposals fell by the wayside. First, the protracted development of the Space Shuttle obliged NASA to divert funds from its scientific program for nearly a decade. And, in any case, the failure of the Vikings to find life on Mars undermined public interest in the planet as a place to explore.

Scenarios for post-Viking Mars missions. Top: a naive depiction of a Viking lander being assembled in Earth orbit by Space Shuttles; a JPL drawing of a small probe similar to Ames' proposed Pioneer Mars Orbiter carrying six penetrators, and with a penetrator embedded in the polar terrain. Middle: 'Viking 3' fitted with tracks for mobility; the model of a large JPL-designed tracked rover. (Courtesy of NASA/JPL/ Caltech) Bottom: the high-altitude Mini-Sniffer III drone that was to become the basis for the projected Mars Airplane; a Viking-based Mars Sample Return spacecraft.

THE VENUSIAN FLEET

The 1978 window for Venus saw no fewer than four launches, two American and two Soviet, involving a grand total of 10 probes.

Soon after Venera 4 and Mariner 5 visited the planet, the Goddard Space Flight Center, eager to enter the field of planetary exploration, began to work on a low-cost mission with an atmospheric capsule for Venus, but NASA headquarters transferred the project to Ames because it desired Goddard to specialize in other fields of space science. Moreover, Ames had started work on planetary entry capsules in the 1960s, and in June 1971 it had flown the PAET (Planetary Atmosphere Experiment Test) in which a prototype had re-entered the Earth's atmosphere from a suborbital flight in order to test how an atmospheric capsule for a planetary mission would collect scientific data.[587] In keeping with its tradition of solar and planetary probes, Ames named the new project Pioneer Venus. The initial suggestion was to launch two identical buses in the 1976–1977 window, each with four atmospheric probes to determine the composition of the clouds, and the composition and structure of the atmosphere from high altitude down to the surface. An orbiter would follow in 1978 to determine the structure of the upper atmosphere and ionosphere, its interaction with the solar wind, the planetary-scale characteristics of the atmosphere, and the planet's gravitational field. In addition the orbiter would assist in triangulating the directions of the celestial gamma-ray bursts.[588,589,590]

At about the same time, planetary scientists in Europe started to work on a Venus Orbiter proposal. After ESRO had rejected the MESO proposal for a mission to Mercury simply due to the cost, it was realized that the US Delta launch vehicle could send a spacecraft to Venus carrying about 80 kg of payload relatively cheaply – indeed, Ames was designing its missions to use this rocket for that reason. ESRO issued preliminary study contracts to German and Italian companies, who proposed simple spin-stabilized craft based on the Helios probes and the SIRIO experimental communication satellite, respectively. The expectation was for a feasibility study to be awarded in 1972, and a funding decision in 1973. In April 1972, however, NASA and ESRO convened a meeting on cooperation in space science projects, and one of the topics was European participation in Pioneer Venus. Under the expected work breakdown NASA would provide the spacecraft bus and deliver it to ESRO, which would install mission systems, integrate the scientific instruments (provided by both US and European scientists), carry out qualification and acceptance tests and return the spacecraft to NASA for launch and in-flight operations. The prospect of sharing costs would make funding easier to gain for each community. But in February 1973, just as ESRO was about to approve funding, NASA warned that it would not be able to guarantee its participation until Congress approved the Pioneer Venus mission – a decision on which was not imminent – and with the project in limbo ESRO diverted its funds elsewhere.[591]

One of NASA's difficulties in 'selling' the Pioneer Venus mission was that it was unlikely to yield results that would spark public interest. Nevertheless, it was finally approved in August 1974. By then, a review had rejected the idea of dispatching the atmospheric probes and the orbiter in separate launch windows, and the plan was to

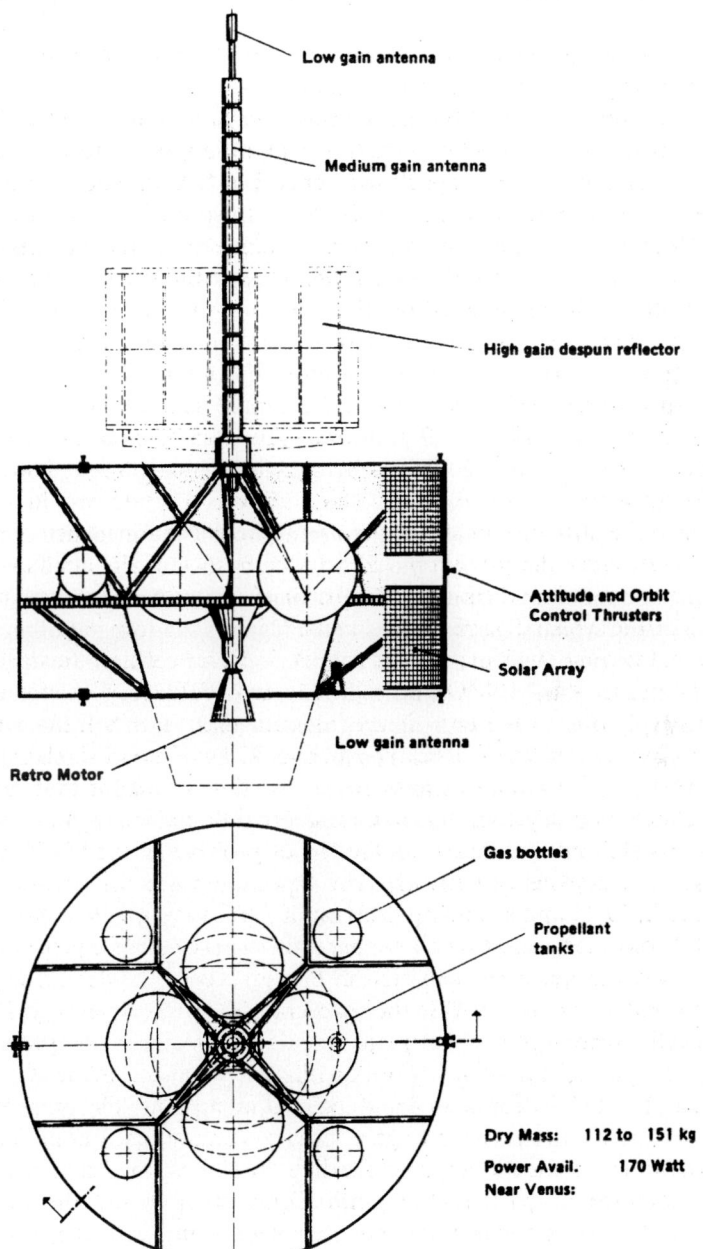

A line drawing of the Venus Orbiter proposed to ESRO by the German firm MBB. (Courtesy of ESA)

send an orbiter and a single bus carrying probes over an interval of a few months in 1978. Hughes won the contract to supply the spin-stabilized spacecraft.[592] The cost was reduced by maximizing the commonality between the two spacecraft (up to 78 per cent of the bus systems and structures were the same) and by using components from earlier missions whenever possible, and although the Atlas–Centaur was eventually selected instead of the Delta, this facilitated a relaxation of the mass limitations which, in turn, allowed less expensive technologies to be employed. Furthermore, no prototype or spare spacecraft were built.[593] The overall project was cost-capped at a relatively cheap $200 million. Both spacecraft used the same 2.53-meter-diameter, 1.22-meter-high drum-shaped bus derived from a successful Hughes communications satellite.

In the case of the orbiter, a Thiokol TE-M-604 solid-fuel rocket giving a thrust of 17.8 kN was inserted into the thrust tube of the bus to undertake the orbit-insertion burn – the first use of such an engine for this purpose. Attitude determination was by star and Sun sensors, with attitude control by seven 6.5-N thrusters which drew on a pair of tanks holding a total of 32 kg of hydrazine, including a reserve of more than 6 kg for a possible extended mission. These thrusters would also perform course corrections during the interplanetary cruise and orbit maintenance around Venus. An equipment shelf connected to the thrust tube by 24 struts held all the electronics, most of the scientific instruments, a data storage system, and rechargeable batteries for the periods in which the orbiter was in the planet's shadow – at other times, up to 312 W would be provided by a 6.4-m^2 cylinder of solar cells on the drum. Mounted on the top deck of the drum was the star mapper, a foldable 4.78-meter-long boom for the magnetometer, a 2.99-meter-tall despun mast with a 1.09-meter-diameter parabolic high-gain antenna, a backup high-gain antenna and an omnidirectional antenna. A second omnidirectional antenna was mounted under the bus. The communication system could transmit at either 10 or 20 W as required. The launch

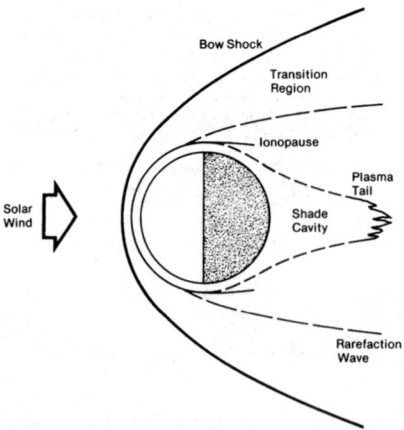

The model of the Venusian environment and its interactions with the solar wind used in planning the Pioneer Venus mission.

mass was 553 kg.[594,595,596] It carried no fewer than 12 scientific instruments. A photopolarimeter would use a 3.7-cm-diameter telescope to collect polarimetric data on the Venusian cloud cover and take spin–scan images in the same manner as the Pioneer missions sent to Jupiter and Saturn. A neutral and ion mass spectrometer would measure the populations of neutral and charged particles in the altitude range 500 to 150 km. An eight-channel infrared radiometer would profile the temperatures from 150 to 60 km. An ultraviolet spectrometer would measure sunlight reflected by the clouds, and study the 'ashen light' phenomenon. At 9.7 kg, the final instrument was the most power-hungry and bandwidth-hungry in the suite. It was a 20-W radar system designed to provide an insight into the nature of the surface beneath the clouds.

The Pioneer Venus Orbiter being prepared for launch at the Kennedy Space Center's Spacecraft Assembly and Incapsulation Facility.

The idea of installing a radar on an orbiter to map the surface of Venus was not new; NASA had awarded study contracts as early as 1959, and JPL had considered a radar system appropriate as a follow-on to the Mariner A flyby. The technique was shown to be practicable in 1964. Given the technical constraints of the bus, the radar system and the orbit, a surface resolution of 15 to 20 km was expected and would improve at lower periapsis.[597,598] On 21 December 1964 the US military launched its first top-secret radar reconnaissance satellite, codenamed Quill. This operated for several days but, although a success, the data was not deemed to be of sufficient use – especially against such targets as moving ships and vessels – to justify flying further missions.[599,600] Throughout the 1960s and 1970s, terrestrial radars had explored parts of the surface of Venus at resolutions of up to 10 km, revealing a series of radio-bright regions, the elevated terrain of Maxwell Montes, possible volcanic structures like Theia Mons and Rhea Mons, and some circular structures that looked as if they could be impact craters.[601] The Soviet orbiters had also performed limited experiments in bistatic radar mapping of the surface. In addition, ESRO had envisaged a radar mapper on its Venus Orbiter.

The radar for the Pioneer Venus Orbiter was also built by Hughes. It was to use a small 38-cm-diameter antenna to make two kinds of measurements: in its altimetric mode it would measure the instantaneous height of the orbiter above the surface; and in its mapping mode, in which it would compensate for the complex motions of the spacecraft, it would inspect a 7×23-km oval footprint directly below. When its elliptical orbit brought the spacecraft within 4,700 km of the planet, the radar would activate in altimetric mode, and on closing in to 500 km it would switch over to mapping mode. Because the vehicle was to spin at 5 rpm, the radar would operate for approximately 1 second per 12-second rotation.[602] The instruments also included a magnetometer, a solar plasma analyzer, a charged-particle analyzer, an electric field detector, an electron temperature probe, and two photomultipliers with which to detect celestial gamma-ray bursts. As usual, celestial mechanics experiments would be conducted, as well as measurements of aerodynamic braking at periapsis.[603]

The 881-kg multiprobe bus was similar, but without the orbit-insertion engine, the data storage system and high-gain antenna, having instead a pair of omnidirectional antennas, one mounted on the top deck and the other underneath, together with a medium-gain horn antenna. The reduced solar cell area of 5.22 m² provided 214 W. Four atmospheric capsules and their retention and separation systems were mounted on the top deck. The capsules consisted of a large probe, mounted on the spin axis of the bus, and three identical much smaller probes set at 120-degree intervals around the periphery of the deck.[604,605,606]

The 316.5-kg 'large' probe comprised a 79-cm-diameter spherical shell of forged titanium, a conical aluminum and carbon phenolic heat shield 142 cm wide that had a front cone angle of 90 degrees, and a nomex aft cover with a teflon window that was transparent to radio wavelengths to transmit its data. Inside the 1-cm-thick shell, which was pressurized by nitrogen at 2,000 hPa to withstand the external pressure of the planet's lower atmosphere, was a 2.5-cm layer of kapton for thermal control and beryllium supports for the scientific instruments, batteries and 10-W transmitter. There were apertures in the shell to accommodate 8 thermally controlled industrial

The Pioneer Venus Multiprobe being prepared for launch. The conical heat shields of the large probe and two of the three small probes are evident.

sapphire instrument windows, and a window cut from a South African brownish diamond of a quality that was unsuitable for use in jewelry. Two other inlets were to feed atmospheric gases to instruments. On the outside of the shell were a series of vanes in a configuration designed to spin the probe during its descent at about 1 rpm, drag plates and fixtures for the parachute. The 3.6-meter-diameter main parachute was first tested in the absence of cross winds by being dropped inside the Vehicle Assembly Building at the Kennedy Space Center, one of the world's tallest closed structures. The entire parachute system was then tested at speeds up to Mach 0.8 by dropping test vehicles from an F-4 Phantom jet. Finally, the full-scale system was verified by a series of drops from high-altitude balloons.[607,608] There were seven instruments. At three points during the descent, a gas chromatograph based on that of the Vikings was to analyze an atmospheric sample. In addition, a neutral mass spectrometer was to monitor the atmosphere continuously from an altitude of 67 km down to the surface. In its control system, an Intel 4004 marked the first use of a single-chip microprocessor by NASA on an instrument for a planetary mission. An atmospheric structure instrument using hot-wire thermometers immersed in the air stream, pressure strain gauge sensors and various accelerometers was to produce temperature, pressure and density profiles. A US/French nephelometer would determine the vertical structure of the atmosphere, to prove the presence or absence of particles and layering. The thermal flux, cloud opacity and concentration of water vapor were to be measured by a six-channel infrared radiometer. The sizes and densities of the cloud particles would be measured at various heights by a laser spectrometer capable of detecting particles in the size range 0.005 to 0.5 mm. The amount of solar radiation able to penetrate the atmosphere to different depths was to be measured by a radiometer in order to determine the extent to which the very high surface temperature was due to the greenhouse effect.[609] Each 93-kg 'small' probe comprised a 46-cm-diameter spherical shell of titanium pressurized with xenon, a conical heat shield 76 cm wide, and an aluminum aft cover. Because there was no parachute, the heat shield and aft cover were not to be released during the descent. The apertures for the three instruments were installed on protrusions from the upper hemisphere of the shell, and had spring-loaded doors. The atmospheric structure instrument and nephelometer were identical to those on the large probe. There was also a radiometer to profile the net heat flux with altitude. It had two small diamond windows, and aerodynamic vanes were to ensure that the freely falling probe rotated to give the instrument an all-round view.[610,611]

All the probes were designed to survive a temperature of 495°C and an external pressure of 115,000 hPa. As a late addition that required substantial modifications, all the probes had temperature sensors buried in their heat shields for a heating rate experiment. The large probe was designed for a 400-g deceleration on entering the atmosphere, and the small ones for a 565-g deceleration, but none was specifically designed for contact with the surface.[612] The multiprobe bus had its own neutral and ion mass spectrometer to measure the composition of the upper atmosphere before it disintegrated on entry, this data being required to complement that from the probes at lower altitudes.[613]

To test the parachute system of the Pioneer Venus large probe, models were released inside one of the largest enclosed structures on Earth, the Vehicle Assembly Building at the Kennedy Space Center.

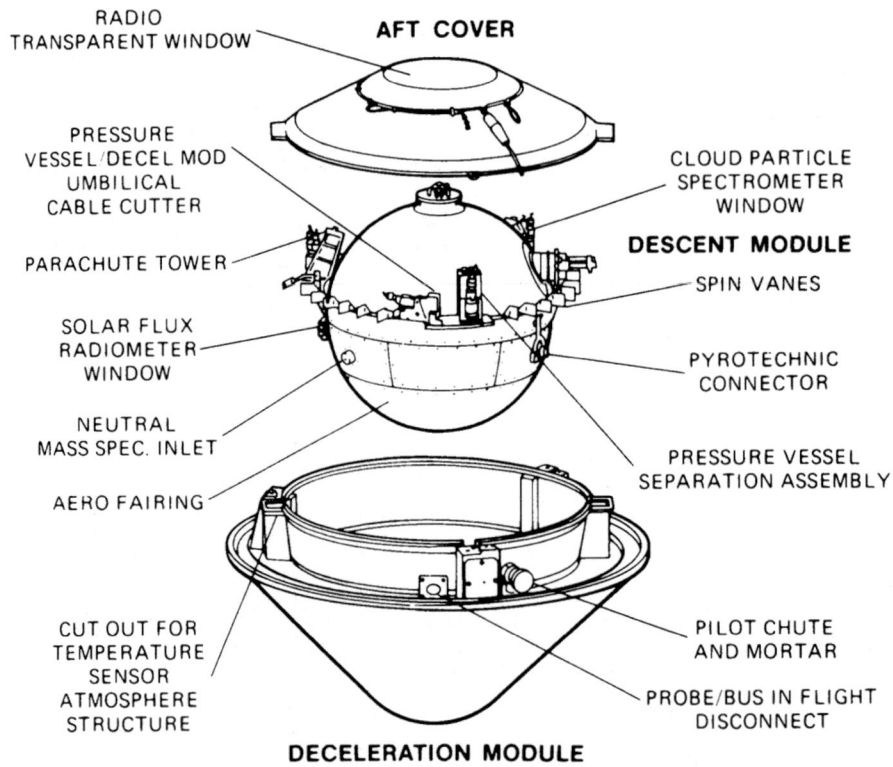

RADIO TRANSPARENT WINDOW

AFT COVER

PRESSURE VESSEL/DECEL MOD UMBILICAL CABLE CUTTER

CLOUD PARTICLE SPECTROMETER WINDOW

PARACHUTE TOWER

DESCENT MODULE

SPIN VANES

SOLAR FLUX RADIOMETER WINDOW

PYROTECHNIC CONNECTOR

NEUTRAL MASS SPEC. INLET

AERO FAIRING

PRESSURE VESSEL SEPARATION ASSEMBLY

CUT OUT FOR TEMPERATURE SENSOR ATMOSPHERE STRUCTURE

PILOT CHUTE AND MORTAR

PROBE/BUS IN FLIGHT DISCONNECT

DECELERATION MODULE

The Pioneer Venus large probe.

The Soviets had skipped the 1977 window for Venus to gain time to develop new scientific instruments within the tight budget available for planetary research. The new mission was very similar to Veneras 9 and 10, except that the bus was not to enter orbit. After releasing its lander, the spacecraft was to make a flyby,[614] for two reasons: first, since the 1978 window was more expensive in energy terms than that of 1975, the scientific payload would have to have been reduced in order to carry the additional propellant that would be required for the Proton's transfer orbit burn; and, second – and more importantly – a distant flyby would enable the bus to be in radio contact with its lander for longer than if it were to enter orbit, thereby enabling more data to be returned from the surface (contact with the most recent landers had been lost not when they succumbed to the environment, but when their orbital relay had passed below the local horizon). As usual, two spacecraft were prepared. As the bus was no longer expected to return data pertaining to Venus, its scientific load was drastically reduced, and mostly to study deep space. The suite included a variety of plasma and charged-particle detectors, counters and spectrometers, a magnetometer, and a French extreme-ultraviolet spectrophotometer. However, two new instruments were added: the French–Soviet SIGNE-2MS (also known by its Russian name of

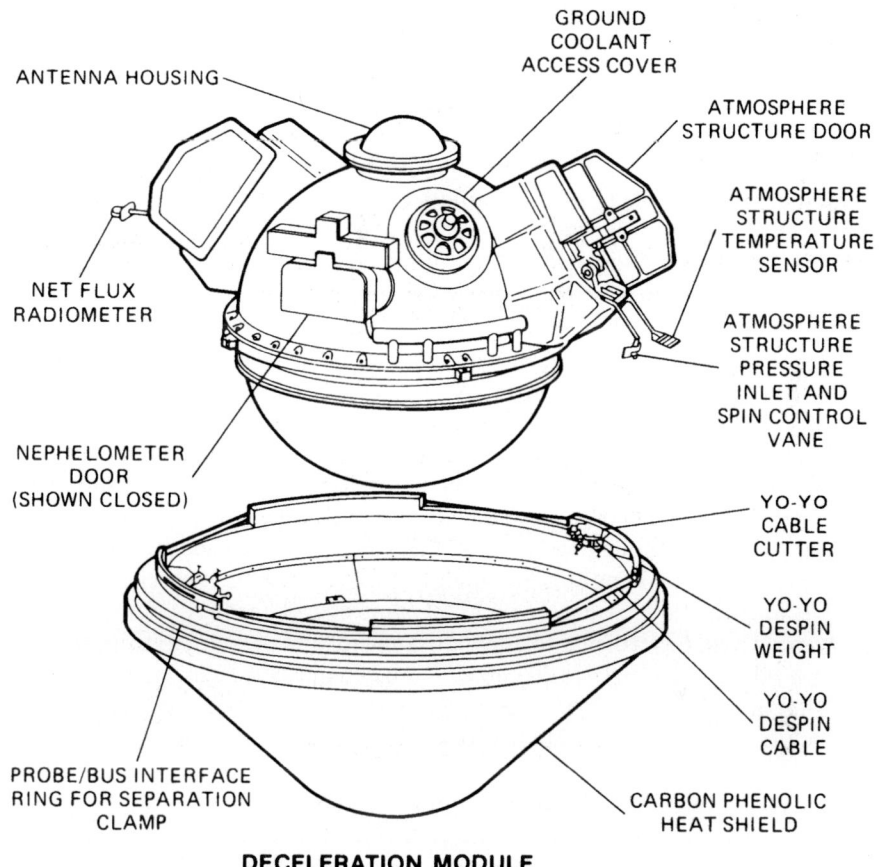

GROUND
COOLANT
ACCESS COVER

ANTENNA HOUSING

ATMOSPHERE
STRUCTURE DOOR

ATMOSPHERE
STRUCTURE
TEMPERATURE
SENSOR

NET FLUX
RADIOMETER

ATMOSPHERE
STRUCTURE
PRESSURE
INLET AND
SPIN CONTROL
VANE

NEPHELOMETER
DOOR
(SHOWN CLOSED)

YO-YO
CABLE
CUTTER

YO-YO
DESPIN
WEIGHT

YO-YO
DESPIN
CABLE

PROBE/BUS INTERFACE
RING FOR SEPARATION
CLAMP

CARBON PHENOLIC
HEAT SHIELD

DECELERATION MODULE

One of the Pioneer Venus small probes.

Sneg; 'snow'), and the Soviet Konus ('cone'), which used six scintillation devices. These instruments were to be used in concert with similar ones on Soviet and US probes and satellites (including the Pioneer Venus Orbiter) to triangulate the sources of celestial gamma-ray bursts to within an accuracy of about 5 arcminutes, and to study the time structure and evolution of individual bursts. Two SIGNE detectors were carried on each spacecraft in order to cover the entire sky.[615]

As Veneras 9 and 10 had functioned so well, the design of the landers had been retained, but the main parachute system had been cut from a triple to a single canopy to speed the descent and accommodate new instruments. As usual, temperature and pressure sensors were carried, together with the nephelometer and anemometer. Two color cameras had an improved resolution of 3.8 mm (as opposed to about 5.5 mm previously) at a range of 1.2 meters. The lens caps had been redesigned to preclude a repeat of the problem that had jammed one cap on each of Venera 9 and Venera 10. This time, one camera was to make a full 180-degree scan using a clear filter, then reverse and scan 60 degrees with color filters, before finishing in the clear. The other camera was to scan the entire 180-degree panorama through color filters. There were color calibration chips within the field of view of each camera.[616] Also in the field of view was the 60-cm-long PrOP-V penetrometer built by VNII Transmash. This 2.1-kg instrument was to be lowered using a single-fold boom, and had a rotating wedge to measure the physical and mechanical properties of the soil (with two dials to indicate the angle through which it rotated and the time taken to reach a given depth) and a sensor with which to measure the electrical resistivity.[617,618] Three instruments were to investigate the atmosphere. The 'Sigma' gas chromatograph was to detect sulfur compounds, the noble gases, carbon monoxide, etc. A completely new 9.5-kg mass spectrometer was to determine the isotopic ratios of the main components and the abundances of minor constituents, particularly the noble gases. There was also a 10-kg atmospheric X-ray fluorescence analyzer.[619,620,621,622] The photometer that had been carried by Veneras 9 and 10 was replaced by an integrated optical spectrometer and photometer.[623] The gamma-ray spectrometer had been replaced by a more complex instrument to perform soil analyses, using a 26-kg external drill drawing 90 W of power built by I.V. Barmin to collect a sample from the upper surface. Although the hollow drill auger was designed to function in the surface environment for only a few minutes, this would be long enough for it to penetrate to a depth of 3 cm in rocks of almost any hardness. Then a series of diaphragms would be retracted to allow the atmospheric pressure to force the sample through a tube to the analysis chamber.[624] The chamber was enclosed in a double-walled titanium container with a 'heat sink' material in between that was capable of withstanding a pressure of 100,000 hPa, in order not to compromise the main pressure hull. Once loaded, the chamber would be sealed and drawn down to very low pressure (relative to ambient conditions), and then the sample would be irradiated by two radiation sources and its X-ray fluorescence detected by discharge counters.[625] The final instrument was Groza ('thunderstorm'). This consisted of an external acoustic sensor and a 25-cm-diameter loop antenna to detect low-frequency radio emissions from lightning. Consideration was also given to adding a descent camera to take pictures of the surface from an altitude of 35–40 km, after the probe had broken through the cloud base, but it was decided that owing to the optical characteristics of the dense atmosphere it would be impractical to image the surface from above about 1 km, and even then the quality would be poor.[626] As a result of the simplified mission, the two spacecraft for the 1978 window for Venus were some 500 kg lighter than their predecessors, at 4,450 and 4,461 kg.[627,628]

The first spacecraft to reach the pad was the Pioneer Venus Orbiter (also known as Pioneer 12 and Pioneer Venus 1). It lifted off on 20 May 1978, on the first day of the 21-day window. The Centaur placed it into a solar orbit ranging between 0.7 and 1.3 AU. It was to fly an interplanetary transfer that subtended a wide heliocentric angle in order to make a slow approach to Venus, and thereby minimize the orbit-insertion burn. Three and one-half hours after launch, the magnetometer boom was deployed, and the following day the high-gain antenna was despun for the first time. Meanwhile, the photopolarimeter had been tested by imaging the crescent of Earth. A single course correction was performed on 1 June. The Pioneer Venus Multiprobe

The PrOP-V instrument mounted on the Venera 11 and 12 landers to measure the strength of the Venusian surface. (Courtesy of VNII-TransMash)

(also known as Pioneer 13 and Pioneer Venus 2) followed on 8 August 1978, on the second day of its 27-day window, one day late owing to a problem with the launch vehicle. It was placed into a solar orbit ranging between 0.7 and 1.11 AU on a faster transfer than its partner. A course correction on 16 August reduced the 14,000-km flyby range to the required collision course.[629]

The two Soviet probes were faultlessly launched by a slightly modified version of the deep-space Proton on 9 and 14 September 1978, just before the window closed. Venera 11 entered a slower solar orbit ranging between 0.69 and 1.01 AU, and made its first course correction on 16 September. Venera 12 pursued a faster orbit ranging between 0.67 and 1.01 AU and corrected its trajectory on 21 September.[630,631]

As it had been the first spacecraft of the fleet to set off, the Pioneer Venus Orbiter was the first to arrive. On 2 December it adopted the appropriate attitude for orbit insertion and was spun up in order to even out any asymmetry in the thrust from the solid rocket motor. On 4 December it crossed the planetary bow wave at a range of some 10,000 km from the planet, and at 15:51 UTC it disappeared behind the limb as viewed from Earth. It fired its orbit-insertion engine 7 minutes later, for a duration of 30 seconds. Owing to doubts about the effective operation of the solid-propellant engine after several months in space, a backup orbit-insertion command was sent at 16:14, as the spacecraft emerged from occultation. The early tracking indicated that the period of its orbit was 23 hours 11 minutes, which was 42 minutes less than intended, and at 378 km the periapsis was 15 km lower than intended, but the other parameters were as planned – in particular the 105-degree inclination and 66,900-km apoapsis.[632] It was the third spacecraft from Earth to enter orbit around Venus, and NASA's first. Several hours later, the spin rate was cut to 5 rpm. On approaching the second apoapsis, the first ultraviolet images of the crescent of Venus were taken by the photopolarimeter. Of the other instruments, only the radar failed to return usable data, as it had been activating at the wrong spin angle and had scanned the sky instead of the planet (although this problem was later rectified).[633] Meanwhile, the spacecraft lowered the periapsis of its orbit to the operating altitude of 160 km.

Having flown a faster route, the Pioneer Venus Multiprobe was not far behind. On 16 November it released the large probe on a trajectory that would deliver it near the equator on the day-side, and 4 days later the small probes were released at a precise spin-rate and spin-angle to ensure that they spread out like pellets from a shotgun on trajectories that gave rise to the designations North, Day and Night Probe. Finally, on 9 December, just a few hours from the planet, the bus made its own targeting maneuver. Each probe remained silent until 22 minutes before the computed time of entry, whereupon an internal timer was to activate it. The large probe was the first to arrive, issuing its first telemetry at 18:29 UTC on 9 December, with the small probes doing so over the next 10 minutes. At 18:45 UTC the large probe slammed into the atmosphere at 11.6 km/s, experiencing a peak deceleration of 280 g that slowed it to 200 m/s in just 38 seconds. It then jettisoned its aft cover and initiated the parachute deployment sequence, which concluded with the release of the spent heat shield and the activation of the scientific instruments at an altitude

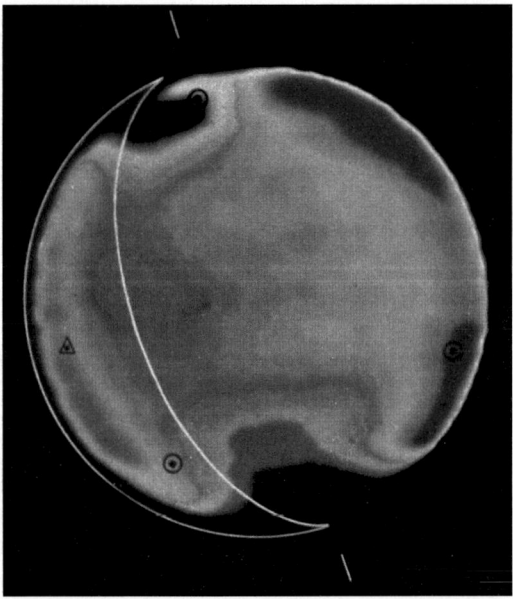

A ground-based infrared map of Venus showing the entry sites of the Pioneer Venus small probes (circles) and of the large probe (triangle). The illumination phase of the planet at the time of their arrival is also depicted.

of 64 km. Eighteen minutes after entry, the parachute was released and the spherical shell entered free-fall at an altitude of 45 km. At 19:40 UTC, after a descent lasting 54 minutes 21 seconds, the transmission ceased as the probe made contact with the surface of Navka Planitia at 4°N, 304°E, at a speed of less than 9 m/s. Five minutes prior to its predicted time of atmospheric entry, each small probe unwound two diametrically opposed 2.4-meter-long cables in order (through the conservation of angular momentum) to reduce the spin rate of 48 rpm imparted at the time of their deployment for stability in transit to 15 rpm for entry, and then the cables were cut. These probes endured even greater decelerations, and then opened their instrument doors at altitudes of about 70 km and started to transmit. They took between 53 and 56 minutes to reach the surface. The North Probe came down in the highlands of Ishtar Terra at 60°N, 4°E. The Night Probe continued to transmit for 2 seconds after the 2 m/s impact at 27°S, 56°E. After touching down at 32°S, 318°E, the Day Probe reported for no less than 67 minutes 37 seconds until its internal temperature increased and its radio amplifier failed. Of course, unlike the Soviet landers, the Pioneer Venus probes were not equipped with surface science experiments. When the Day Probe was halfway through its impromptu surface mission, the bus slammed into the atmosphere and returned data on conditions in that environment for 63 seconds prior to disintegrating at an altitude of 110 km, with its debris falling near 41°S, 284°E. During this time, the orbiter had been well positioned to view the entry sites.[634,635,636]

The bus spectrometers reported the composition of the upper atmosphere from an altitude of 700 km down to 150 km, finding hydrogen, helium, atomic and molecular oxygen, nitrogen, carbon dioxide and atomic carbon.[637] Below an altitude of 50 km, the temperature profiles from the four probes showed differences of no more than a few degrees, despite their entry sites being separated by several thousand kilometers. The atmosphere appeared to be isothermal at high altitude, with a sharp inversion in the altitude range 60–70 km, possibly corresponding to the horizontal shear zone of the 4-day super-rotating upper atmospheric wind. Measurements in the cloud were complicated by chemical reactions between the thermometer and the sulfuric acid in the droplets. All four probes lost temperature data in the altitude range 12–14 km, at the same time as several other instruments had anomalies and the nephelometer reported an increase in the ambient light; the cause of this failure has never been definitively resolved, but it has been suggested that a global thunderstorm may have been in progress. The measured surface pressures ranged from 86,200 to 94,500 hPa, and the extrapolated temperatures were 448–459°C.[638] The nephelometers identified three cloud layers of differing stratification and particle sizes, confirming the report from the Venera probes of a cloud deck at an altitude of about 50 km. The particle-size spectrometer found both sulfuric acid and free sulfur clouds, and a haze layer that extended from immediately below the cloud deck down to an altitude of 30 km. The nephelometer on the Day Probe also noted the dust that was raised by its own impact settling over a span of several minutes.[639,640] Cloud layers were also detected by the radiometers, which provided accurate readings of the solar energy reaching the surface.[641] The gas chromatograph measured the abundances of carbon dioxide, molecular nitrogen and oxygen, sulfur dioxide, argon, neon and water vapor, mostly confirming the Venera data. In fact, the known 98 per cent abundance of carbon dioxide was insufficient to explain surface temperature, and the detection of water and sulfur dioxide completed the 'runaway greenhouse' model of the atmosphere.[642] The mass spectrometer drew in a sample of gas every kilometer down to the surface, but in passing through a cloud it ingested a droplet that, until it evaporated, partially blocked the inlet. A particularly interesting finding was a much greater ratio of argon-36 to argon-40 than is present in the Earth's atmosphere. Argon-40 is formed by the radioactive decay of potassium-40 in rock. Argon-36 cannot have a similar radiogenic origin, and hence is believed to be a relic of the primordial solar nebula from which the planets condensed. The low ratio of argon-36 to argon-40 on Earth is thought to be evidence that our atmosphere is considerably younger than the planet itself (and, in fact, the current hypothesis is that the primordial atmosphere of the Earth should have been destroyed in the giant impact that gave rise to the Moon). The Pioneer Venus data therefore hinted that the Venusian atmosphere has had a quite different history.[643] Finally, as in the case of the Venera probes, the Pioneer entry probes were tracked in order to measure wind speeds. A 200 m/s wind was observed in the middle cloud layer, reducing to 50 m/s at the cloud base, and to just 1 m/s near the surface.[644,645,646,647]

Of the two Soviet probes, Venera 12, which had followed a faster trajectory, was the first to arrive. It corrected its trajectory for the second time on 14 December and released its lander on 19 December, then the bus made the deflection maneuver for a 35,000 km flyby. On 21 December the lander entered the atmosphere at 11.2 km/s, the parachute was opened at an altitude of 62 km and shed at 40 km. Unfortunately, excessive vibrations during the descent broke the seal on the hermetic tube that was to transfer the surface sample to the analysis chamber. At 03:30 UTC it touched down on Phoebe Regio at 7°S, 294°E, where the Sun was 70 degrees above the horizon. It returned data from the surface for a record 110 minutes but, with the lens caps of both cameras failing to eject and the sampling system rendered unusable, the results fell far short of expectations. Although the engineers were unable to determine why the lens caps had jammed, one theory suggested that the heaters for the gas-ejection system had failed; but another possibility was that the caps had become sealed on by an unexpected thermal expansion. Venera 11 corrected its trajectory on 17 December, released its lander on 23 December and set up its flyby. The lander entered the atmosphere early on 25 December and landed at 03:24 UTC at 14°S, 299°E, some 800 km south of Venera 12. It returned data for 95 minutes, the link being cut when the bus passed below the horizon. Frustratingly, Venera 11 suffered precisely the same problems as its partner – failing to return images and to collect a sample.[648,649,650] On the positive side, good data was returned by most of the other instruments on both probes. The mass spectrometers drew in a total of 22 gas samples, yielding a total of 176 spectra. The isotopes of carbon and the noble gases were measured, and water, sulfur and possibly chlorine were detected. The unexpected argon isotopic ratio that was discovered by the Pioneer probes was confirmed: the measured ratio was close to unity, whereas on Earth argon-40 is 300 times as abundant as argon-36. The gas chromatograph was calibrated at altitudes in the range 65–54 km, and then a total of eight analyses were made, identifying carbon dioxide, nitrogen, argon and carbon monoxide, in addition to traces of sulfur compounds. It was only possible to put an upper limit on the presence of oxygen, as the experiment to measure its abundance proved to have been contaminated. Overall, the results were in good agreement with those from the Pioneer probes. The X-ray fluorescence apparatus on Venera 12 was active as the probe descended through the altitude range 60–45 km, and established that chlorine and sulfur compounds were the main constituents of the aerosols in this region. The fact that no results of the experiment on Venera 11 have been published may indicate that it failed. The spectrophotometer began to operate at an altitude of about 65 km, and reported that only 3–6 per cent of sunlight reached the ground, even though the Sun was within 20 to 25 degrees of the zenith at the Venera 11 site. It also noted the absorption by carbon dioxide and water droplets in the clouds, and the emergence from the cloud base at an altitude of 47–48 km. The nephelometer gave similar results to the instruments on Veneras 9 and 10. As a lander descended, its bus recorded the Doppler shift of an oscillator on the lander, and later relayed this to Earth to enable wind-speed profiles to be calculated throughout the altitude range. Venera 11 yielded ambiguous results, but Venera 12 mostly confirmed the results of previous probes, measuring wind speeds of 40 to 50 m/s in the upper reaches of the atmosphere. The ultraviolet spectrophotometers on the buses made diametrical scans of the planet's disk, and took data on the hydrogen and helium content of the upper atmosphere. The Groza experiment on Venera 11 detected pulsed electromagnetic fields, and in particular six strong bursts as the probe descended through the altitude range 17–13 km, the intensity declining as the altitude reduced further, suggesting violent thunderstorms up to 7,000 km away.[651] Venera 12 encountered much calmer atmospheric conditions. Its acoustic detector was saturated by aerodynamic noise as the probe descended, but picked up the operating sounds of the instruments, and also the subsequent surface activities. Venera 11 also noted a loud noise some 32 minutes after landing.[652,653,654,655,656,657,658,659]

The planet's gravity deflected the paths of the Venera 11 and 12 buses, causing them to depart in solar orbits having the same perihelion distance at 0.715 AU, and aphelia at 1.116 and 1.156 AU respectively. During the cruise to Venus and after the flyby, they used their gamma-ray burst detectors to report data on this phenomenon, yielding detailed time-profiles for 143 events, from which an extensive catalog (the first of its kind) was compiled. This data was combined with that from the Prognoz 7 high-apogee Earth satellite in order to triangulate the sources on the celestial sphere. Although the directions of some bursts were able to be pin-pointed within very small error boxes, in all but one case a search of sky surveys failed to indicate the source. The exception was a burst of 5 March 1979 that was detected by no fewer than nine Soviet and American spacecraft, and the resulting very small error box contained an intriguing supernova remnant that was physically located in the Large Magellanic Cloud – although it was far from clear that this alignment was anything other than a coincidence. On the other hand, the Konus data had sufficient resolution to show the 'finger print' of the very intense gravitational and magnetic field of a compact object such as a neutron star or black hole, providing for the first time an indication of the type of object that might be responsible for gamma-ray bursts.[660,661,662,663,664,665] The French ultraviolet spectrophotometers were calibrated by observing Venus and went on to collect comprehensive data on Lyman-alpha and helium emissions in the local interstellar medium.[666] They were turned toward comet Schwassmann–Wachmann (or according to other sources comet Bradfield) but no calibration for stray sunlight could be made, and the data was impossible to interpret.[667] The last transmission dates of the buses has not been published, but it is known that the last gamma-ray burst detection by Venera 12 occurred on 5 January 1980 and the spacecraft returned spectrophotometric data until 19 March, and the last gamma-ray burst detected by Venera 11 occurred on 27 January 1980.[668,669]

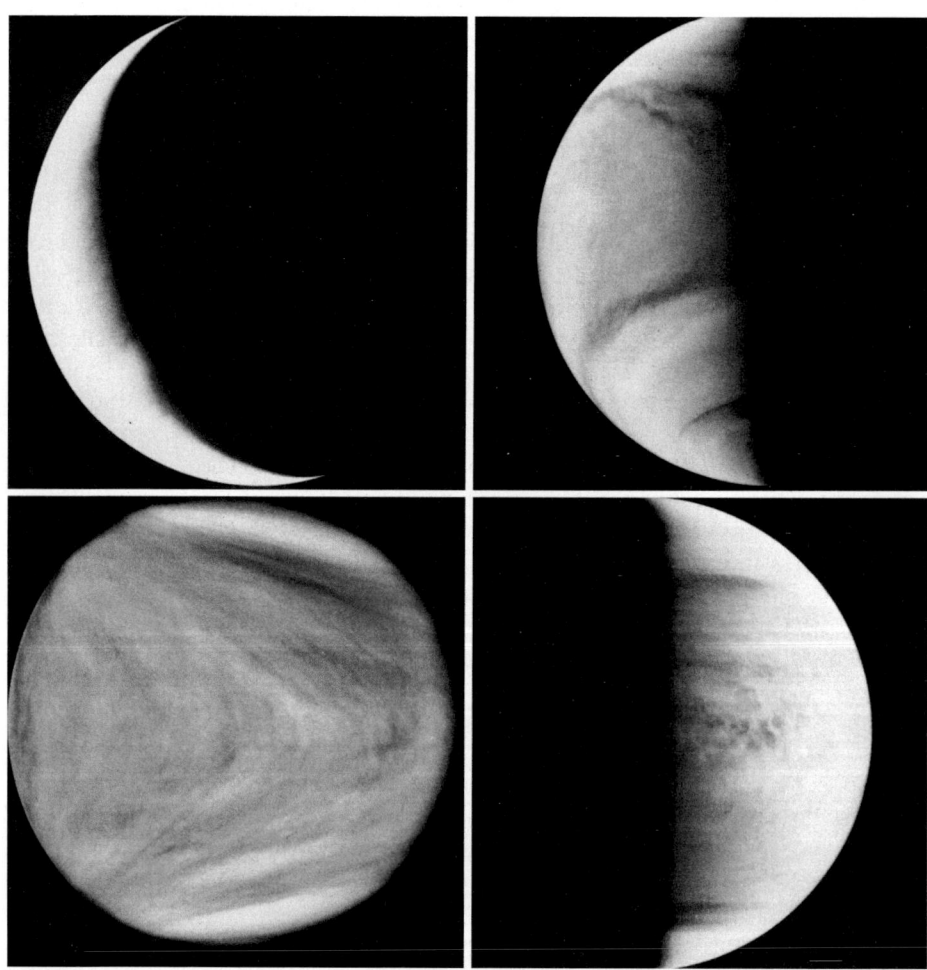

Pioneer Venus Orbiter imagery of Venus in ultraviolet light at different illumination phases.

The Pioneer Venus Orbiter detected its first gamma-ray burst on 21 May 1978; one day after launch.[670] Its orbit of Venus took it through the planetary bow shock twice per revolution and, over time, the magnetometer observed how the ionosphere interacted with the solar wind, and how the shock's geometry varied with the solar cycle, and investigated the cavity in the solar wind 'downstream' of the planet. The spacecraft observed oxygen ions escaping the planet with the solar wind after having been ionized by solar ultraviolet in the highest atmosphere, even upwind of the bow shock itself. In fact, the orbiter discovered that although the main atmospheric gas is carbon dioxide, complex chemical reactions cause the ionosphere to be made mostly of ionized oxygen. A weak ionosphere was also found on the night-side, where ions made on the day-side were able to migrate in the absence of a global magnetic field. The ultraviolet spectrometer observed the airglow phenomenon, and detected for the first time 'patchy' auroras on the night-side arising from the collision of solar high-energy particles with molecules in the upper atmosphere. One of the most important findings was the confirmation that hydrogen ions and atoms leak to space, with the implication that Venus once had abundant water in its atmosphere, if not actually on its surface, and that over the eons the water that reached the upper atmosphere had been dissociated by solar ultraviolet and by interactions with the solar wind. This inference was supported by the ratio of deuterium to hydrogen measured by the ion mass spectrometer on the orbiter for the ionosphere and by the mass spectrometer on the large probe in the lower atmosphere.[671] The ultraviolet spectrometer identified compounds in the upper atmosphere that revealed the roles of sulfur dioxide, water vapor and sulfuric acid in the 'sulfur cycle' that occurs in the clouds. An intriguing observation was a progressive decline of the concentration of sulfur dioxide over the years, possibly caused by the decline of volcanism that was active around the time of the spacecraft's arrival, or perhaps simply due to a variation in the atmospheric circulation.[672] The electric field sensor picked up signals that might have been due to lightning associated with volcanic eruptions, as dust in a roiling cloud will become charged and issue electrostatic discharges.

Although limited by operational constraints, the photopolarimeter was able to take images of the planet at differing phase angles to document not only the 4-day super-rotation of the upper atmosphere, but also changes in the morphology of the clouds over longer times. The sideways-Y pattern glimpsed by telescopic observers using ultraviolet filters, and clearly imaged by Mariner 10, remained the most significant meteorological feature, but there were week-long periods during which it was not so evident. Polarimetry gave a measurement of the size of the particles in the sulfuric acid smog that forms the clouds, and indicated the presence of a thin layer of haze above the clouds.[673] The infrared radiometer provided data on the atmosphere until it failed on 14 February 1979. It found the day-side and night-side to have very similar temperatures up to high altitudes, the day-side being slightly warmer, and confirmed the conclusion from the entry probes that the highest reaches of the atmosphere are isothermal. Moreover, the orbiter found that above the cloud tops the atmosphere is warmer at the poles than at the equator, possibly owing to a substantial layer of haze overlying everything except the poles, or perhaps to an unknown heat transportation mechanism. A 'collar' of cold clouds rising to an altitude of 75 km was observed for the first time surrounding the north pole, within which were two warm 'eyes', one on each side of the pole, often connected by an 'S' feature. These were the warmest places on the entire planet at some wavelengths, and may have been clearings in the main cloud deck.[674,675] Many radio-occultation 'soundings' were made by the double-frequency technique. The upper atmosphere proved to be sufficiently dense that the drag exerted on the craft could be measured to further characterize the air density and composition.[676]

The radar data from the first two orbits of Venus was lost, and after the 14th orbit the instrument malfunctioned, necessitating a long pause to understand and work around the problem. Data lost could not be recovered as planned during September 1979, as the Deep Space Network was then occupied by Pioneer 11's encounter with Saturn, but support was available from April 1980 until the radar was deactivated on 19 March 1981. The radar altimeter covered 93 per cent of the band from 74°N to 63°S with a vertical accuracy better than 200 meters and a surface resolution better than 150 km. Although the planet proved to be almost flat, with only 5 per cent of its surface deviating more than 2 km above or below the average radius, the instrument revealed plains and continents not previously seen by terrestrial radar. Of particular interest were an 1,800-km-diameter circular feature east of the Aphrodite continent that looked as if it might be a vast impact crater, and a long valley (since named Diana Chasma) that plunged to a depth of 4 km in parts – in fact, the lowest point on

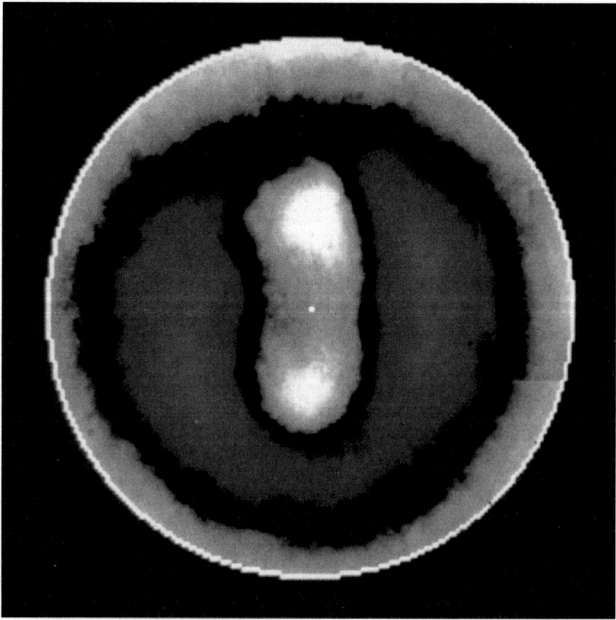

This map is an average of infrared measurements of the north pole of Venus, and it shows the mysterious bipolar vortex, the two 'eyes' of which are the warmest parts of the atmosphere at some wavelengths.

the entire planet – which, if it were to prove to be a rift valley, would be the first evidence of activity in the style of 'plate tectonics' to be identified on a planet other than Earth. The highest point on Venus was confirmed to be in Maxwell Montes, at an elevation of some 11 km above the average radius. Even at the low resolution of the data, this was clearly a complex feature with terraces, a spur ridge and radar-bright hills.[677,678,679] Owing to operational constraints, the radar mapper could return data only between 50°N and 20°S, but despite its relatively poor 20–40-km resolution at a time when terrestrial radars could achieve a resolution of 1 km in some regions, it was able to map parts of the planet beyond the reach of terrestrial radars. Although the data enabled different surface units to be identified in terms of their brightness at radar wavelengths (a parameter that depends not only on the profile of the landscape and its texture, but also on the composition of the material) little could be deduced of the actual geology. Nevertheless, a number of features in the equatorial zone were classified as volcanic coronae – a type of feature not present on Earth.[680,681,682] The maps of the gravity field inferred from how the spacecraft's orbit evolved in time suggested that the continents of Ishtar and Aphrodite might be in a state of isostatic equilibrium, a condition in which a surface elevation is supported by a force exerted by the dense upper mantle.[683]

Even before the spacecraft reached Venus, NASA had approved funding to enable the orbital mission to be extended beyond the nominal 243-day period (i.e. one local year) so as to permit more extensive scientific monitoring and to fill in any gaps that might occur in the radar altimeter's coverage. Over the first two years, the spacecraft executed in excess of 100 maneuvers and returned over 40 gigabits of data. From orbit 600 on 27 July 1980 the engines ceased to be used to maintain the periapsis, and the orbit was allowed to be perturbed by solar and planetary gravity and also by the pressure of solar radiation. The latitude of periapsis was allowed to migrate freely, reaching the equator in 1986, then continuing south. Meanwhile, the altitude of periapsis increased from its initial 160 km to 2,270 km in June 1986, before starting to fall back at a rate that would lead to atmospheric entry in 1992. As the periapsis reached its maximum altitude, the spacecraft was able to penetrate the rarely explored region just upstream of the bow shock, in order to study a variety of shock-wave precursor phenomena.[684] Permitting the orbit to drift not only saved fuel, of which only 4.5 kg remained when this new strategy was initiated, but also allowed 'noise-free' measurements of the gravity field.

While orbiting Venus, the spacecraft also made a range of celestial observations. First, it observed a series of magnetic field disturbances that occurred (in large part) when the planet was in conjunction with asteroid (2201) Oljato, which appeared to be an extinct comet that was followed by a tail of material millions of kilometers in length.[685] Then in April 1984 it made ultraviolet scans of comet Encke as that object reached perihelion, and measured the rate at which the nucleus was issuing water. In September 1985 it observed Giacobini–Zinner, just before another mission made the first in-situ observations of such an object.[686] But Encke and Giacobini–Zinner were just rehearsals for a more exciting study for which the Pioneer Venus Orbiter would again hit the scientific headlines in 1986. It had been computed that on 3 February Halley's comet would pass 0.27 AU from Venus, and at the perihelion passage of its

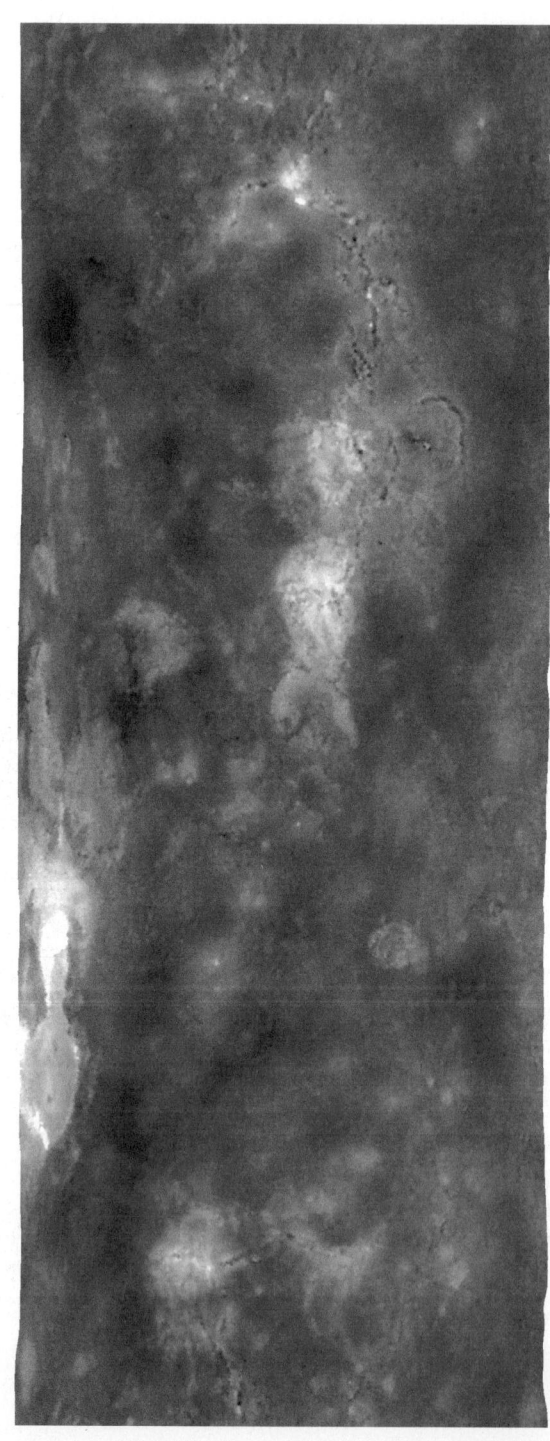

An altimetry radar map of Venus obtained by the Pioneer Venus Orbiter.

76-year orbit 6 days later it would still be well positioned in the Venusian sky, but a difficult object to see from Earth because it would be near solar conjunction. The ultraviolet spectrometer made daily observations of the comet from 28 December 1985 through 7 March 1986, with the exception of 25 days in January when the spacecraft was out of contact because the Sun blocked the line of sight to Earth. This filled a gap in the coverage of this most important of comets by other satellites and space probes. In fact, the geometry in early February was such that the ultraviolet emission could be presented as an image that showed the hydrogen coma to be 12.5 million km across. Moreover, the emissions by hydrogen, oxygen and carbon ions enabled the water production rate at perihelion to be calculated to be about 40,000 kg/s.[687,688] Other comets were observed in later years, including C/1986P1 Wilson, C/1987B1 Nishikawa–Takamizawa–Tago and C/1987U3 McNaught.[689]

As predicted, by 1991 the periapsis was in the southern latitudes and down to an altitude of 700 km. The original plan to restart the radar in order to improve the map of the southern hemisphere was apparently shelved because the Magellan spacecraft had entered orbit on 10 August 1990 with a radar capable of mapping the entire planet at a higher resolution. In September 1992 the Pioneer Venus Orbiter made the first of a series of maneuvers designed to prevent its periapsis dipping below 130 km, in order to counter atmospheric drag until late October, when the various perturbing forces would begin to raise the periapsis once again. But after the 6th maneuver it

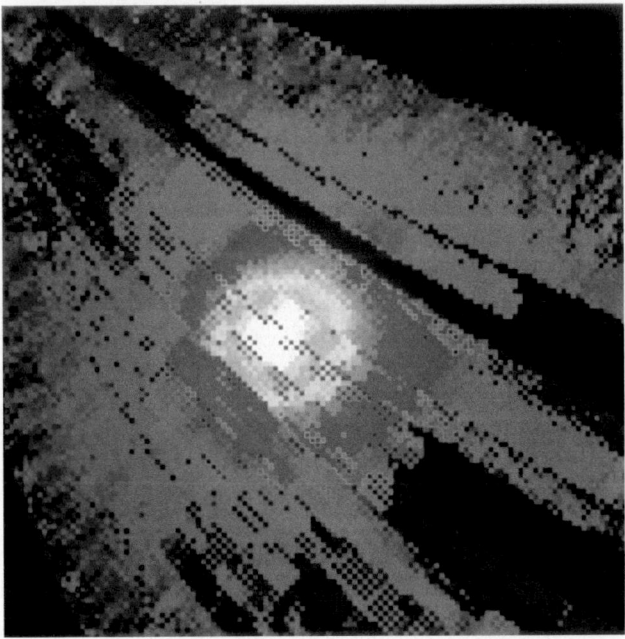

An image of comet Halley built from ultraviolet scans taken by the Pioneer Venus Orbiter. In the ultraviolet range, the usual cometary structures (coma, tails, etc.) are masked by a large and more or less spherical cloud of hydrogen.

became evident that there was so little fuel remaining that the thrusters were venting mostly the gas used to pressurize the tank. In early October the last drops of fuel were used to put the craft into a spin. The final low passes provided unique in-situ observations of the atmosphere. On 8 October, when the spacecraft ended its 5,055th orbit, it was down to 128 km and succumbed to the drag, concluding its remarkable 14-year mission.[690,691]

The results of the Soviet and American missions to Venus in 1978 showed that there was a layer of haze (a fine smog) in the altitude range 90–70 km, that the main cloud base was at 48 km, and that there was thin haze down to about 30 km, below which the atmosphere was clear. The maximum opacity was at 50 km. Because the lower atmosphere was stagnant, the base of the troposphere (the convective part of the atmosphere, which on Earth is the lowest layer) began at the base of the cloud deck and was confined to the upper atmosphere. In 1969, telescopic measurements of the index of refractivity had suggested that the condensates in the atmosphere were acid-laden water droplets, and this proved to be the case. The aerosol of sulfuric acid is the result of photochemical oxidation at altitudes exceeding 60 km. The dissociation of carbon dioxide or sulfur dioxide in the upper atmosphere yields atomic oxygen that oxidizes sulfur dioxide to sulfur trioxide, which is hydrated into sulfuric acid droplets. As these 'rain out' and fall, they are thermally disrupted on reaching the 100°C temperature at an altitude of about 49 km and yield sulfur trioxide which, upon encountering carbon monoxide, regenerates sulfur dioxide and carbon dioxide, which completes the precipitation cycle. In the lower atmosphere, there is a steep thermal profile in which the temperature decreases with increasing altitude by about 8°C/km. Therefore, although for a specific elevation the surface temperature will be uniform both from pole to pole and from the day-side to the night-side, the temperature varies over the 13-km range of elevation from the summit of the tallest mountain to the floor of the deepest depression. At no point on the surface is water stable; the water that remains is confined to the cooler upper atmosphere, and is progressively diminishing because the hydrogen atoms released by photodissociation leak to space. Liberated hydroxyl radicals combine with rising sulfur dioxide and enhance the manufacture of sulfuric acid. Any free oxygen that reaches the surface will oxidize the hot rock, removing it from the atmosphere. In retrospect, it is not surprising that astronomers in the 1950s had difficulty fathoming conditions on the planet.

THE COLOR OF VENUS

The Soviet Union did not attempt any missions to Venus in 1980, marking only the third time since 1961 that it had missed an opportunity, but it prepared two modified 4V-1M probes for the next launch window. Because it was again decided to trade an orbital mission in favor of a longer period of contact with the landers on the surface, the buses carried only a reduced instrument suite that included cosmic-ray and solar wind sensors, plus the new SIGNE-2MS gamma-ray detector. For the first time, an Austrian instrument was flown on a deep-space probe. On the occasion of a visit by the IKI director to that country in 1976, it had been agreed that a Soviet probe would fly an experiment supplied by Austrian scientists. Owing to the limited development time available, it was decided to adapt a magnetometer that was in development for the Shuttle-borne Spacelab, and to place it at the tip of a 2-meter-long boom on one of the solar panels of each of the two probes. The magnetometers were to take data on the structure of the interplanetary magnetic field and the passage of shock waves, and during the flyby they were to study how the solar wind interacted with the planet's ionosphere.[692,693] The lander was modified by adding small sawtooth wedges around its edge, spaced 5 cm from tip to tip, in order to improve its aerodynamic stability sufficiently to preclude the vibration mode that, on the most recent missions, had broken the hermetic seal of the surface-sample-retrieval system. An extensive suite of instruments increased the mass of the lander to 760 kg. The ejection mechanism for the lens caps of the color cameras was redesigned. The drill and the X-ray fluorescence soil analyzer (whose operations had previously been frustrated) were retained, as were the accelerometer, the temperature and pressure sensors, the X-ray fluorescence aerosol analyzer, the nephelometer and the PrOP-V penetrometer. The gas chromatograph had been much improved, and redesignated Sigma 2. A modified and more sensitive version of the mass spectrometer weighing 9.5 kg was included to determine the isotopes of neon, krypton and xenon, and to refine the relative abundance of argon as a check on the earlier surprising measurement.[694] The Groza 2 had new sensors to detect coronal discharges, and a 0.88-kg uniaxial seismometer to measure the vertical displacement of an electromagnet with sufficient accuracy to detect a perturbation as small as one-millionth of a centimeter.[695,696] The visual spectrophotometer had been improved to provide a profile of how well ultraviolet radiation penetrated the atmosphere.[697] The VR-3R humidity analyzer was a new instrument to measure the moisture content of the upper atmosphere, and to verify spectroscopic measurements made from Earth and by previous spacecraft.[698] In addition, the simple 'Kontrast' experiment was devised to assess the oxidizing characteristics of the atmosphere: an asbestos frame impregnated with a chemical reactant (a sort of space-age litmus paper) was placed on the landing ring. This would be protected by a metal cover to prevent reactions with the acidic clouds, but in the hotter environment below the cloud base the rivets would melt, allowing the cover to be torn away by the air flow. Color changes of the reactant following landing would be documented by one of the cameras.[699] Finally, a small solar cell panel was mounted on the landing ring to measure light reaching the surface. The initial plan was to target the probes on a plain east of the mountains of Phoebe Regio, southeast of Beta Regio, but after consultation with Harold Masursky of the US Geological Survey and a study of the radar data from the Pioneer Venus Orbiter it was decided to move one landing site nearer the mountains, to a region of rolling plains that looked as if it might be part of the ancient crust.[700,701] The 4,363-kg launch mass included the 1,645-kg descent capsule with its 760-kg lander.

On 30 October 1981 a Proton successfully dispatched Venera 13. On 4 November Venera 14 gave chase. Both were inserted into a solar orbit that ranged between 0.99 and 0.7 AU and had a period of 285 days. Venera 13 made course corrections on 10 November 1981 and 21 February 1982. On nearing Venus, the internal temperature of its lander was chilled to –10°C, and the capsule was released on 27 February. The bus performed the deflection burn for the planned 36,000 km flyby. The capsule hit the atmosphere at 2:55 UTC on 1 March. The parachute opened 90 seconds later, at an altitude of 62 km, and the instruments activated. The parachute was released at 3:05 UTC, at an altitude of about 47.5 km. At 3:57:21 the lander touched down at a vertical speed of about 7.5 m/s. It was at 7.55°S, 303.69°E, some 1.4 km above the mean radius. The Sun was 36 degrees off the zenith.[702] The surface temperature was 465°C, the pressure was 89,500 hPa, and there was a 0.57-m/s breeze blowing. The surface transmission lasted 127 minutes, almost three times the expected survival duration, during which time the two cameras, between them, took 11 full 180-degree panoramas and 10 partial pictures spanning 60 degrees in azimuth. The landing site appeared to be a gently undulating plain between a pair of low hills, beyond which could be seen the profile of a more distant hill, the flanks marked by striations and steps. In the foreground were light-hued flat slabby rocks rising several centimeters above a mixture of dust and rock fragments that occupied the gaps in between. The dust appeared to be very finely grained. Instrument readings indicated that the cloud of dust raised by the lander's impact took about 40 seconds to settle. Moreover, the comprehensive camera data not only documented the breeze blowing away the dust that had settled on the landing ring, but also variations in the illumination. There was also a hint that the lander shifted slightly some 50 minutes after touchdown, as the material it rested upon settled.[703,704,705,706,707]

ВЕНЕРА-13   ОБРАБОТКА ИППИ АН СССР И ЦДКС

The 180-degree panoramas of Venus taken by the two cameras on Venera 13. Note the rock slabs that abounded at this site. A variety of spacecraft appendages can be seen on the surface, including the PrOP-V (in the top image), the discarded camera cover (both images), the photometric reference chip, and the sawtooth pattern on the landing ring that was used to stabilize the lander during its free-fall descent. There is also a pentagon-shaped Lenin pennant on the landing ring (top image). (Courtesy of Donald P. Mitchell and Yuri Gektin)

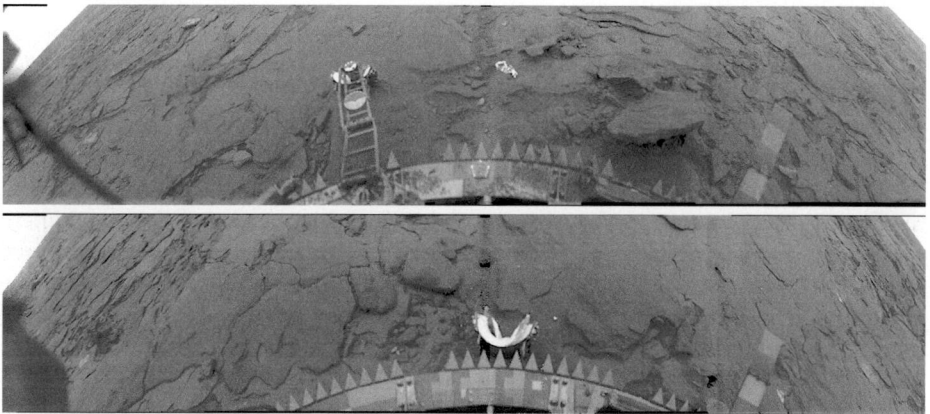

ВЕНЕРА-14   ОБРАБОТКА ИППИ АН СССР И ЦДКС

The 180-degree panoramas of Venus taken by the two cameras on Venera 14. Note that the landing site was far less dusty than that of Venera 13. On the top image, the PrOP-V swung down right onto the camera cover! (Courtesy of Donald P. Mitchell and Yuri Gektin)

Venera 14 attempted a course correction on 14 November 1981 but a malfunction prevented the burn. It was rescheduled for 23 November, and this time was made successfully. The spacecraft refined its trajectory on 25 February 1982.[708,709] After its release on 3 March, the capsule hit the atmosphere at 5:53 UTC on 5 March. The parachute opened at 5:54, and was jettisoned at 6:06 at similar altitudes to its predecessor. The lander touched down at 7:00:10 at 13.055°S, 310.19°E, about 1 km above the mean radius, some 1,000 km from Venera 13, where the Sun was 35.5 degrees from the zenith.[710] The temperature was 470°C, the pressure was 93,500 hPa, and the wind was 0.37 m/s. It took six panoramas during 57 minutes of operation on the surface which showed the site to be on a 500-meter-tall hill that was almost devoid of dust, with rocks showing stratification suggestive of overlapping lava flows, and outcrops of a somewhat lighter hue.[711,712,713,714] In fact, the rocks at both sites proved to have a distinct orange-reddish hue. The only major problem of the second landing was that one of the ejected lens caps came to rest just where the boom of the PrOP-V deployed, with the result that the penetrometer measured the strength of that material rather than the surface of the planet![715,716] The radar map made by the Magellan orbiter a decade later would show that Venera 14 had landed on the eastern slope of a 75-km-diameter volcanic corona feature. In contrast, the revised Venera 13 landing ellipse was in a geologically complex area that included two steep volcanoes and a small impact crater, which possibly accounted for the dust and pebbles.[717] Impact dynamics were studied by the Bizon-M accelerometer, which also provided data on density profiles during the descent.[718,719] The drill system was activated about 1 minute after landing, once its components had become thermally stabilized. As it was designed to operate at 500°C, the tolerances of its mechanical components were chosen to ensure that they would fit together and work properly in a thermally expanded state. Moreover, the moving parts were lubricated by molybdenum disulfide, which would serve this function up to a temperature of 1,000°C. One minute of drilling provided a sample of about 1 cm³, which was delivered to the teflon-coated analysis chamber 4 minutes after landing. The X-ray fluorescence analyzer on Venera 13 produced 38 spectra of the elemental composition of the material, and that on Venera 14 provided a further 20 spectra, with similar results.[720,721] The strength of the rocks was obtained from the data from the Bizon-M accelerometer, the telemetry from the drill and the PrOP-V experiment which, in the case of Venera 13, penetrated 31 mm into the ground. This confirmed that the rock was similar to a terrestrial volcanic porous rock such as tuff. These measurements were confirmed by the appearance of the terrain, and in particular by the size distribution of the small rocks.[722,723,724]

After a detailed analysis, the Doppler measurements of the radio transmissions from the landers as they descended through the atmosphere gave wind speeds of about 50 m/s, being faster at higher altitudes. There was significant turbulence in the cloud deck in the altitude range 60–50 km, with oscillations of up to 2.5 m/s.[725] The visual and ultraviolet spectrophotometer measured the abundance of water vapor at high altitudes. It also established that 90 per cent of solar ultraviolet is absorbed by the clouds that occur above 58 km. On Veneras 13 and 14, respectively, it measured the cloud base at 49 km and 47.5 km. The first simultaneous measurement of the water vapor content of the clouds by direct and by optical means confirmed that the atmosphere was indeed arid.[726] In fact, the humidity analyzer, which operated in the altitude range 50–46 km, measured vapor to be just 0.2 per cent by volume. The nephelometer, which operated from 62 km down to the surface, indicated some fine structure in the clouds, particularly the presence of 100-meter-thick layers just above the cloud base.[727,728,729] The aerosol analyzer of Venera 14 operated in the altitude range 63–47 km, and reported more sulfur but less chlorine than had had been noted by Venera 12. No results were released for Venera 13, possibly indicating that the instrument failed.[730,731] The Sigma 2 gas chromatograph, which was activated 4 hours prior to atmospheric entry for calibration, collected samples in the altitude range 58–49 km and made a total of nine analyses, identifying for the first time molecular hydrogen, krypton and sulfidric acid, and a hint of chloridric acid.[732] The mass spectrometer ran from 26 km down to the surface and took a total of 250 spectra. One major controversial result was that this time the ratio of argon-36 to argon-40 was found to be closer to the terrestrial value, rather than the values close to unity that were measured by previous probes, both Soviet and American.[733] The Groza 2 experiment measured the electromagnetic field at various altitudes, but did not detect any lightning and electrical discharges.[734] The microphone recorded sounds both during the descent and on the surface, where it noted the lens caps of the cameras hitting the ground, the operation of the drill and the transfer of the sample to the analysis chamber. Other noises might have been the wind blowing across the surface; certainly, the cameras indicated dust movement on the landing ring.[735] The ultra-sensitive seismometer was not activated until the other instruments had finished their surface activities; it was then interrogated in 8-second bursts. The instrument on Venera 13 did not find anything interesting, but Venera 14 detected two displacements, one some 950 seconds after landing at an amplitude not exceeding 0.0001 cm, and the other after 1,361 seconds of an even lower amplitude, but it was not possible to say whether they were seismic activity, the settling of the lander or wind gusts.[736] Finally, the 'Kontrast' experiment found the atmosphere at ground level to be more highly oxidizing than had been inferred from the oxygen concentrations measured by earlier probes.[737]

The encounters deflected the buses into solar orbits ranging between 0.715 and 1.123 AU. They collected data on solar X-ray bursts using the SIGNE instrument, and participated in a second interplanetary network for the triangulation of gamma-ray bursts that detected some 150 events, many of which were able to be accurately localized.[738,739,740] Venera 13 and Venera 14 each fired their main engines again on 10 June and 14 November respectively, to provide engineering data in a rehearsal of a maneuver that would be made by the Venus–Halley (Vega) mission scheduled for the mid 1980s.[741,742] Venera 13 returned data on the solar wind until 12 April 1983 and Venera 14 until 16 March 1983, after which they were probably shut down.[743]

'PURPLE PIGEONS' FROM THE COLD

In May 1966 a cooperative scientific program was established between NASA, the French CNES (Centre National d'Etudes Spatiales) and Argentina's CNIE (Comision Nacional de Investigaciones Espaciales) to jointly investigate the circulation of the high-altitude winds in the southern hemisphere. Accordingly, in August 1971 NASA launched the 84-kg CAS 1 (Cooperative Application Satellite) built by the French, who named it Eole after the Greek god of winds. With the satellite established in a polar orbit, Argentina released no fewer than 479 helium balloons over an interval of 3 months loaded with pressure and temperature sensors and a transponder. The task of Eole was to poll the balloons on every orbit to download their data and measure their positions to within several kilometers, to enable the direction and speed of the winds to be monitored. The Eole program was an outstanding success, with the last surviving balloons being deliberately blown up in December 1972, after which the satellite tracked buoys and ships until 1974.[744,745,746] In late 1967, Jacques Blamont, one of the scientists of the Service d'Aéronomie (aeronomy service) of the French scientific research center that worked on Eole, proposed a similar mission for Venus. The initial concept envisaged a Venera 4-like entry capsule releasing six balloons at high altitudes. Because the total mass had to be compatible with the 300-kg capacity of the Venera entry capsule, each 2.5-meter-diameter helium-filled balloon would have a mass of 50 kg, including its inflation system and a payload of 5 kg. At the same time, a second spacecraft would enter a circular orbit in order to relay the data from the balloons over as many days or weeks as possible. The project, named Eos – this being both a contraction of 'Eole' and 'Venus' and also a reference to 'Eosfòros' (the dawn-bringer), an ancient Greek name for Venus – was considered as a joint mission with the Soviet Union in the framework of an agreement in 1966 that, among other things, would put French instruments on missions sent to the Moon and missions to Mars and Venus. The proposal was for the Soviet Union to provide the vehicles and flight operations, and the French to provide the balloons and the tracking system for the orbital relay.[747] At the same time, Soviet engineers were developing an idea for a 3,000-kg entry probe that would deploy a 5-kg high-altitude aerostatic probe, a 400-kg medium-altitude probe and a 600-kg lander, but Babakin himself decided not to support it because it was beyond Soviet technical capabilities. High-altitude balloons were also proposed in the 1960s in America by Martin Marietta and Goodyear in the context of the Voyager Venus orbiter; in particular, Goodyear planned a balloon that would double as an inflatable decelerator and parachute for the entry phase.[748,749]

In 1972 the Soviets accepted the French proposal, but because they were about to abandon the Venera 4 bus in favor of the larger 4V-1 they suggested that rather than release a number of small balloons the new entry capsule should deploy one large PAS (Plavalyuschaya Aerostatnaya Stantsiya, buoyant aerostatic station). When the Soviets suggested deleting the orbital relay satellite, the French proposed either that NASA be asked whether its Pioneer Venus Orbiter could serve as a relay, which was not something that appealed to the Soviets, or that a network of radio-telescopes be established to track the balloon from Earth to within a positional accuracy of 6 km. However, because such a network would require to be global to provide continuous monitoring, and involve significant international cooperation, the Soviets decided to reinstate the orbiter. The plan envisaged the launch of two spacecraft in 1981, one

A model of the French Eole satellite that formed the basis for the Eos proposal for a French–Soviet Venus atmospheric balloon. (Courtesy of Patrick Roger-Ravily)

to enter orbit and the other with the capsule. The entry sequence would be similar to that of Venera 9, but when the large parachute deployed it would open the nacelle that held the balloon (stowed in a container 1 meter in length and 50 cm in diameter) and the helium tanks. The balloon would inflate as the stack slowly descended. With a fully inflated diameter of 9.5 meters, it would have a volume in excess of 3,500 m^3. It would be made of layers of teflon to withstand the acidic clouds, aluminum to reflect heat, polyester to make it airtight, and kevlar to give it stiffness. The fabric would be less than 0.1 mm thick, and weigh less than 50 g/m^2. Once the balloon had inflated, the parachute would be jettisoned and the empty helium tanks would be discarded to cause the balloon to ascend to its operational altitude of 55 km. This altitude represented a compromise: any higher, and the balloon would require to be strengthened to resist exploding in the low pressure; and any lower, it would need to withstand much higher temperatures – in both cases making the balloon too heavy. The 600-kg total mass included a scientific payload of 20 kg, 140 kg of nacelle, 35 kg of balloon, 15 kg of rigging, 55 kg of ballast, and some 300 kg for the parachute, inflation system and helium tanks. The balloon was to be released near the limb of the night-side of Venus, and be carried by the super-rotation wind over the day-side limb, where it would pass out of the line of sight of Earth. To ensure that it would be tracked for at least two passes of the Earth-facing hemisphere, it would need to survive for a duration of 10–20 days. This requirement would set the mass of ballast required to counter the slow gas leakage. During the entire mission, including the time when the balloon would be invisible from Earth, it would be tracked by the orbiter in a 24-hour orbit ranging between 1,000 and 60,000 km. In addition to the Eole-based tracking system that would comprise the main French contribution to its payload, this spacecraft would carry a scientific payload of about 100 kg.

The project proceeded well until CNES became involved in the reshaping of the European space program, dismantling ELDO and ESRO to create a single European Space Agency (ESA) and starting the project to develop the Ariane launch vehicle. For France this meant increasing its involvement in the new agency, abandoning its independent launch capacity – which had made it only the third country in the world to orbit a satellite using a nationally developed rocket – and downscaling its scientific programs. In October 1974, after in-flight inflation tests had been conducted, CNES announced that it would withdraw from the Eos project and by mid-1975 had ceased work on it. In the hope that the French would still contribute the balloon and its inflation system, the Soviet Union decided to proceed with the project, but by this time the launch had been delayed from 1981 to either 1983 (which would celebrate the 200th anniversary of the first flight by the Montgolfier brothers in a 'lighter-than-air' vehicle) or possibly to 1985.[750,751]

In fact, even more ambitious Venus exploration projects were under study in the Soviet Union. Immediately after the missions of Veneras 9 and 10, scientists voiced a keen interest in using follow-on landers to study Venusian seismology. This kind of research would pose two problems: either the lander would require a very long life, of a month, at least, or it would have to use explosive charges for 'active seismology', with the explosives having to be protected against accelerations at launch, entry and landing, to say nothing of the heat. Consideration was also given to using a separate Venera 4-like capsule to deliver a nuclear device to the surface in order to generate seismic waves, but international treaties prohibited placing atomic weapons in space, even for purposes of a scientific study such as this, and the idea was soon discarded. Attention switched to the possibility of developing a DZhVS (Dolgozhivushaya Veneryanskaya Stanziya; long-duration Venusian probe) that would incorporate a seismometer and operate for at least one month. It would be built like a dewar, with an external pressure hull and an internal instrument compartment, with a vacuum in between. To enable it to withstand the deceleration of the entry and landing, the inner compartment was to be suspended by shock-absorbing titanium rods. However, heat would be able to reach the inner compartment in three ways: (1) the rods would conduct heat, (2) the metal on the inner side of the external pressure hull would sublime to gas, thus undermining the vacuum and allowing heat to be transmitted across the gap by convection, and (3) the inner side of the external pressure hull would reach such a high temperature that it would transmit heat to the inner sphere by radiation. In practice, this design could not ensure a surface

The wind-driven KhM-VD rover prototype designed for use on Venus. (Courtesy of VNII-TransMash)

mission longer than about 5 days. Consideration was also given to replacing the empty interstice with a thick layer of aerogel and adding a substantial heat sink of lithium nitrate trihydrate, like the one already used on the 4V-1-type Venus landers. At that time, it became known that engineers at the Radioelectronic Institute of Minsk in Belarus had succeeded in developing electronic components to be used for geological drilling and prospecting that could withstand temperatures in excess of 250°C, whereupon it was decided to investigate how long a lander using such components would be able to work on the surface of Venus. In its final design, the DZhVS would have two compartments: an upper one for components and instruments possessing limited tolerance, and a lower pressurized compartment housing the long-duration components and instruments. It was planned that the 'traditional' instruments would characterize the site in the first hour after landing, and then the upper compartment would be shut off and the lower compartment would start its long mission powered by an RTG delivering 30 W, with its instruments providing data on seismic and volcanic activity, meteorology, wind noise, radioactivity, and the variability of the illumination in the visible and infrared. The lander would be capable of communicating directly with Earth, but a spacecraft was to enter orbit around the planet to serve as a relay station. Many components for the DZhVS were built and tested, including high-temperature electrical connectors and cables, radiation sensors, RTGs, seismometers and other instruments. Unfortunately, by the early 1980s the Lavochkin design bureau was overloaded, and the engineers working on the DZhVS were redirected to other projects, such as the Venera orbiters for radar mapping and the international Vega mission. Despite the DZhVS not having flown, technologies that were developed for it were used for Venus probes (starting with Venera 11 – in fact, the soil samplers and seismometers on these missions were developed for the DZhVS) and other aeronautical projects.[752]

Preliminary studies were also made by VNII Transmash of building some kind of rover for Venus, and some basic hardware choices (i.e. materials, motors, cabling and lubrication systems) were made. In fact the PrOP-V penetrometer developed by VNII Transmash was flown on Venus landers in order to determine the mechanical characteristics of the soil on which a rover would operate. There were also studies of how best to power a 'Venerakhod', since the surface would be too dimly lit and too hot for solar panels. The Soviet engineers considered the idea of a rover powered by two windmills and batteries. Two prototypes with masses of up to 160 kg were built and tested between 1984 and 1987 on the volcanic Kamchatka peninsula, but these studies did not reach the stage of being assigned to a particular mission.[753,754,755]

SOURCES

Citations refer to the chapter references section at the end of the book.

1 Stuhlinger-1970
2 Perminov-2002
3 Mitchell-2004f
4 Varfolomeyev-1998c
5 Perminov-2002
6 Mitchell-2004f
7 Kerzhanovich-2003
8 Lardier-1992f
9 AWST-1971a
10 Marov-1974
11 Bertaux-1976
12 Wilson-1987i
13 Perminov-1999e
14 Gatland-1972g
15 Botvinova-1973
16 Caroubalos-1974
17 Steinberg-2001
18 Epstein-1974
19 Poquérusse-2004
20 Wilson-1987i
21 Perminov-1999e
22 Gatland-1972g
23 AWST-1971c
24 Kemurdjian-1992
25 Kovtunenko-1992
26 VnIITransmash-1999
27 Ball-1999
28 Kemurdjian-1990
29 VnIITransmash-2000
30 AWST-1971c
31 Perminov-1999e
32 NASA-1971b
33 Ezell-1984m
34 Wilson-1987j
35 Gatland-1972h
36 Siddiqi-2002k
37 Kraemer-2000a
38 Koppes-1982k
39 Harvey-2007b
40 Huntress-2002
41 Dawson-2004a
42 Kraemer-2000a
43 Caroubalos-1974
44 Perminov-1999e

45 Ezell-1984n
46 Bohlin-1973
47 Kliore-1973
48 Hartmann-1974
49 Ezell-1984o
50 Wilson-1987j
51 Siddiqi-2002k
52 Veverka-1977
53 Huntress-2002
54 AWST-1971b
55 Huntress-2002
56 AWST-1971b
57 AWST-1971d
58 Perminov-1999e
59 Gatley-1974
60 Zellner-1972
61 Murray-1973
62 Pollack-1975
63 Carr-1976
64 Hartmann-1974
65 Arvidson-1974
66 Born-1975
67 Veverka-1977
68 Barth-1974
69 Barth-1974
70 Pearl-1973
71 Kliore-1973
72 Sheehan-1996b
73 Hartmann-1974
74 Dolginov-1987
75 Botvinova-1973
76 Caroubalos-1974
77 Steinberg-2001
78 Epstein-1974
79 Poquérusse-2004
80 Wilson-1987i
81 Siddiqi-2002l
82 AWST-1972a
83 Kliore-1973
84 Barth-1974
85 Ness-1979a
86 Dolginov-1987
87 Wilson-1987i
88 Siddiqi-2002l

89 Kolosov-1985
90 AWST-1972c
91 Wilson-1987j
92 ESRO-1966
93 Kraemer-2000b
94 Wolverton-2004b
95 Fimmel-1976a
96 Wilson-1987k
97 Gatland-1972i
98 Dyal-1990
99 Fimmel-1976b
100 Van Allen-1972
101 Judge-1974
102 Fimmel-1976c
103 Judge-1974
104 Gehrels-1974
105 Burgess-1982a
106 Davidson-1999c
107 Dawson-2004b
108 Lozier-2005
109 Teegarden-1973
110 Fimmel-1976f
111 Webber-1975
112 Fimmel-1976d
113 Turnill-1984b
114 Cunningham-1988a
115 Wolverton-2004c
116 Schuerman-1977
117 Soberman-1990
118 Toller-1982
119 van der Kruit-1986
120 Kinard-1974
121 Fimmel-1976e
122 Zwickl-1977
123 Wolfe-1974
124 Prakash-1975
125 Kennel-1977
126 Null-1976
127 Cruishank-1976
128 Wu-1978
129 Fimmel-1976g
130 Wu-1978
131 Chase-1974
132 Smith-1974

133 Prakash-1975
134 Simpson-1974
135 Van Allen-1974
136 Trainor-1974
137 Fillius-1974
138 Null-1976
139 Kliore-1974a
140 Fimmel-1976h
141 Cruishank-1976
142 McElroy-1975
143 Kliore-1974a
144 Ingersoll-1976a
145 Fimmel-1976i
146 Kliore-1974b
147 Chase-1974
148 Woiceseyn-1974
149 Fjeldbo-1975
150 Fountain-1974
151 Swindell-1974
152 Coffeen-1974
153 Ingersoll-1976a
154 Fimmel-1976j
155 Fountain-1978
156 Ingersoll-1976a
157 Ingersoll-1976b
158 Wolfe-1974
159 Keenan-1975
160 Mudgway-2001a
161 Dawson-2004c
162 Fimmel-1976k
163 Humes-1975
164 Wilson-1987l
165 Fimmel-1976l
166 Smith-1975
167 Mihalov-1975
168 Fimmel-1976m
169 Swindell-1975
170 Null-1976
171 Baker-1975
172 Duxbury-1975
173 Simpson-1975
174 Fillius-1975
175 Ip-1980
176 Null-1976
177 Kliore-1975
178 Baker-1975
179 Ingersoll-1975
180 Smith-1975
181 Peters-1987

182 Jopikii-1995
183 Wilson-1987l
184 Butrica-1996c
185 Fimmel-1980a
186 Wolverton-2000
187 Dyer-1980
188 Turnill-1984c
189 Covault-1979
190 Elson-1979a
191 Fimmel-1980b
192 Wolfe-1980
193 Ingersoll-1980
194 Elson-1979b
195 Fillius-1980
196 Anderson-1980
197 Van Allen-1980
198 IAUC-3483
199 Simpson-1980
200 Van Allen-1980
201 Trainor-1980
202 Fillius-1980
203 Null-1981
204 Kliore-1980
205 Smith-1980
206 Acunha-1980
207 Humes-1980
208 Elson-1979c
209 Dobbins-2002
210 Gehrels-1980
211 Judge-1980
212 Mudgway-2001b
213 AWST-1979
214 Ingersoll-1980
215 Judge-1980
216 Anderson-1980
217 Wolfe-1980
218 Helton-1975
219 Wilson-1987k
220 Flight-1990
221 Wu-1988
222 Gangopadhyay-1989
223 McKibben-1985
224 Jokipii-1995
225 Verschuur-1993
226 Kayser-1984a
227 Gazis-1995
228 Van Allen-1984
229 Landgraf-2002
230 Armstrong-1985

231 Anderson-1985
232 Armstrong-1987
233 Anderson-2001
234 Nieto-2001
235 Anderson-1998
236 Nieto-2005
237 Brooks-2005
238 Mudgway-2001c
239 Wolverton-2004d
240 Giorgini-2005
241 Cesarone-1983
242 NASA-1971c
243 Gregory-1973a
244 NASA-1974b
245 Dixon-1974
246 Perminov-2002
247 Mitchell-2004e
248 Surkov-1997a
249 Blamont-1987b
250 Marov-1974
251 Marov-1974
252 CIA-1973a
253 Surkov-1997a
254 Perminov-2002
255 Mitchell-2004e
256 Wilson-1987m
257 Siddiqi-2002m
258 AWST-1972b
259 AWST-1972d
260 Dollfus-1998a
261 Perminov-1999f
262 Wilson-1987n
263 Siddiqi-2002n
264 Lardier-1992g
265 Dollfus-1998b
266 Dollfus-2004
267 Dollfus-1983
268 Poquérusse-1978
269 Poquérusse-2004
270 Bertaux-1976
271 Perminov-1999f
272 Wilson-1987n
273 Siddiqi-2002n
274 Surkov-1997b
275 Clark-1986
276 Flight-1974a
277 Perminov-1999f
278 Wilson-1987n
279 Siddiqi-2002n

280 Surkov-1997c
281 Kolosov-1985
282 Perminov-1999f
283 Wilson-1987n
284 Siddiqi-2002n
285 Dollfus-1983
286 Dollfus-1988b
287 Santer-1985
288 Santer-1986
289 Surkov-1997c
290 Dolginov-1987
291 Poquérusse-1978
292 Poquérusse-2004
293 Steinberg-2001
294 Bertaux-1976
295 Kolosov-1985
296 Kolosov-1985
297 Flight-1974b
298 Perminov-1999f
299 Wilson-1987n
300 Siddiqi-2002n
301 Surkov-1997b
302 Perminov-1999g
303 VnIITransmash-1999
304 Kemurdjian-1992
305 Lantranov-1996
306 Sagdeev-1994a
307 Butrica-1996d
308 Zak-2004
309 Murray-1989d
310 CIA-1969
311 Kraemer-2000c
312 Anderson-1978
313 MVM-17
314 Anderson-1978
315 MVM-14
316 MVM-19b
317 Wilson-1987o
318 Siddiqi-2002o
319 Koppes-1982l
320 Dawson-2004d
321 Kumar-1978
322 Ajello-1979
323 Kumar-1979
324 MVM-15
325 Dunne-1974
326 MVM-19b
327 MVM-18
328 Howard-1974

329 Howard-1974
330 Dunne-1974
331 Murray-1974
332 Schubert-1981
333 JPL-1976a
334 MVM-19b
335 JPL-1976b
336 MVM-21
337 MVM-20
338 MVM-21
339 MVM-22
340 MVM-24
341 MVM-29
342 MVM-28
343 MVM-24
344 NewScientist-1974
345 MVM-34
346 MVM-30
347 MVM-31
348 Tyler-1981a
349 MVM-32
350 MVM-33
351 Mudgway-2001d
352 MVM-36
353 JPL-1975a
354 MVM-37
355 MVM-38
356 MVM-38
357 MVM-37
358 JPL-1975b
359 Robinson-1997
360 Strom-1979
361 Strom-1987
362 Murray-1975
363 MVM-38
364 Sebesta-2003
365 Sebesta-1997
366 Porsche-1968
367 Kraemer-2000d
368 Siddiqi-2002p
369 Porsche-1968
370 Leinert-1984
371 Leinert-1980
372 Grün-1984
373 Schwenn-1984
374 Neubauer-1984
375 Mariani-1984
376 Mariani-2005
377 Villante-1984

378 Porsche-1968
379 Unz-1970
380 Sebesta-2003
381 ESRO-1970
382 Kraemer-2000d
383 Langereux-1974
384 Dawson-2004e
385 Langereux-1974
386 Kutzer-1984
387 Winkler-1976
388 AWST-1975
389 Cline-1979a
390 AWST-1975a
391 AWST-1975b
392 AWST-1975c
393 Dawson-2004f
394 Leinert-1984
395 Leinert-1980
396 Leinert-1981
397 Leinert-1978
398 Hanner-1978
399 Jackson-1990
400 LeBorgne-1983
401 Hick-1991
402 Leinert-1989
403 Schmidt-1980
404 Grün-1984
405 Schwenn-1984
406 Neubauer-1984
407 Mariani-1984
408 Villante-1984
409 Gurnett-1984
410 Kellogg-1984
411 Kayser-1984b
412 Weber-1977
413 Kunow-1984
414 Trainor-1984
415 Cline-1979b
416 Edenhofer-1984
417 Volland-1984
418 Woo-1978
419 Porsche-1980
420 Winkler-1983
421 Kehr-1984
422 Mudgway-2001e
423 Mudgway-2001f
424 Israel-1974
425 Gregory-1973a
426 Gregory-1973b

427 Gregory-1973c
428 Farquhar-1972
429 Farquhar-1999
430 Doty-1975
431 Hughes-1977
432 Russo-2000
433 Pletschacher-1976
434 Farquhar-1972
435 Wilson-1987p
436 Siddiqi-2002q
437 Lardier-1992h
438 Lorenz-2006
439 Vaisberg-1976
440 Mitchell-2004e
441 Surkov-1997d
442 Marov-1976
443 Florensky-1977
444 Surkov-1997e
445 Surkov-1997f
446 Surkov-1977
447 Surkov-1977
448 Wilson-1987p
449 Siddiqi-2002q
450 Sagdeev-1994b
451 Basilevsky-1992
452 Garvin-1984
453 Covault-1976b
454 Basilevsky-1992
455 Garvin-1984
456 Marov-1976
457 Surkov-1997d
458 Surkov-1997e
459 Surkov-1997f
460 Surkov-1977
461 Garvin-1981
462 Florensky-1977
463 Vinogradov-1976
464 Andreeva-1976
465 Ksanfomaliti-1976
466 Ksanfomaliti-1976
467 Mitchell-2004d
468 Russell-1991
469 Gringauz-1976
470 Dolginov-1976
471 Kolosov-1985
472 Ezell-1984p
473 Kraemer-2000e
474 Ezell-1984q
475 NASA-1975
476 Wilson-1987q
477 Koppes-1982m
478 Kraemer-2000f
479 Ezell-1984r
480 Becker-2004
481 Surkov-1997g
482 Horowitz-1977
483 NASA-1975
484 Ezell-1984t
485 Wilson-1987q
486 NASA-1975
487 Ezell-1984s
488 Ezell-1984t
489 Ezell-1984t
490 Kraemer-2000g
491 Moog-1973
492 Koppes-1982m
493 Dawson-2004g
494 Ezell-1984u
495 AWST-1975a
496 AWST-1975b
497 AWST-1975c
498 Spaceflight-1976
499 Baker-1977a
500 Baker-1977b
501 McElroy-1976
502 Ezell-1984v
503 Ezell-1984t
504 Ezell-1984v
505 Mutch-1976a
506 Davidson-1999d
507 Carr-1980
508 Malin-2005
509 Bogard-1983
510 Sheehan-1996c
511 Hess-1976a
512 Shorthill-1976a
513 Baker-1977d
514 Clark-1976
515 Baker-1977b
516 Mudgway-2001g
517 Baker-1977c
518 Mutch-1976b
519 Baker-1977e
520 Malin-2005
521 Shorthill-1976b
522 Hess-1976b
523 Anderson-1976
524 Baker-1977e
525 Pollack-1978
526 Shorthill-1976b
527 Hargraves-1976
528 Horowitz-1976
529 Levin-1976
530 Horowitz-1977
531 Houtkooper-2006
532 Horowitz-1977
533 Ezell-1984w
534 Klein-1976
535 Mazur-1978
536 Adelman-1986
537 DiGregorio-2004
538 Navarro-Gonzáles-2006
539 Ezell-1984x
540 Ezell-1984x
541 Ezell-1984x
542 Ezell-1984y
543 Cutts-1976
544 Farmer-1976
545 Arvidson-1978
546 Leovy-1977
547 Spitzer-1980
548 Reasenberg-1979
549 Tolson-1978
550 Duxbury-1978a
551 Spitzer-1980
552 Mutch-1978
553 Duxbury-1978b
554 Sjorgren-1979
555 Simpson-1979
556 Hutchings-1983
557 Mudgway-2001h
558 Arvidson-1983
559 Mudgway-2001h
560 Smith-1983
561 Hutchings-1983
562 Mudgway-2001h
563 Fink-1976a
564 Peebles-1981
565 AWST-1973
566 Covault-1976a
567 Ulivi-2006
568 Fink-1976b
569 Pritchard-1974
570 Peebles-1981
571 Burke-1992
572 Peebles-1981
573 Hajos-2005

574 Hilton-1985
575 ESA-1979
576 Reed-1978
577 Taylor-1980
578 Morton-2000
579 Simmons-1977
580 Spaceflight-1977
581 Hughes-1974
582 NASA-1974b
583 Pritchard-1974
584 Pritchard-1974
585 Minear-1978
586 Adelman-1986
587 Bugos-2000
588 Kraemer-2000h
589 Wilson-1987r
590 Hughes-1978
591 Ulivi-2006
592 Kraemer-2000h
593 Hughes-1978
594 Wilson-1987r
595 Hughes-1978
596 Colin-1977a
597 Kovaly-1970
598 Butrica-1996e
599 Day-2001
600 Day-2007
601 Pettengill-1980
602 Butrica-1996f
603 Colin-1977b
604 Wilson-1987r
605 Hughes-1978
606 Colin-1977a
607 Wilson-1987r
608 Hughes-1978
609 Colin-1977a
610 Wilson-1987r
611 Hughes-1978
612 Hughes-1978
613 Colin-1977a
614 Huntress-2002
615 Niel-1976
616 Mitchell-2004d
617 Ball-1999
618 VnIITransmash-1999
619 Gel'man-1979
620 Surkov-1997h
621 Surkov-1997i
622 Istomin-1979

623 Moroz-1979
624 Barmin-1983
625 Surkov-1997j
626 Perminov-2004
627 Wilson-1987s
628 Siddiqi-2002r
629 Hughes-1978
630 Wilson-1987s
631 Wotzlaw-1998
632 Wolverton-2004e
633 Elson-1978a
634 Hughes-1978
635 Wilson-1987r
636 Colin-1979
637 von Zhan-1979
638 Seiff-1979
639 Ragent-1979
640 Knollenberg-1979
641 Tomasko-1979
642 Oyama-1979
643 Hoffman-1979
644 Counselman-1979
645 Luhmann-1994
646 Elson-1978b
647 Elson-1979a
648 Wilson-1987s
649 Siddiqi-2002r
650 Vekshin-1999
651 Russell-1991
652 SAL-1979
653 Istomin-1979
654 Gel'man-1979
655 Moroz-1979
656 Ksanfomaliti-1979
657 Kerzhanovich-1980
658 Wilson-1987s
659 Siddiqi-2002r
660 Mazets-1981
661 Mazets-1979
662 Estulin-1981
663 IAUC-3356
664 Vedrenne-1979
665 Caraveo-1995
666 Chassefière-1986
667 Blamont-1987c
668 Mazets-1981
669 Chassefière-1986
670 Evans-1979
671 Hartle-1983

672 Luhmann-1994
673 Travis-1979
674 Taylor-1979a
675 Taylor-1979b
676 Keating-1979
677 Pettengill-1979
678 Pettengill-1980
679 Butrica-1996f
680 Senske-1987
681 Senske-1989
682 Senske-1990
683 Pettengill-1980
684 Craig-1985
685 Russell-1984
686 Craig-1985
687 Stewart-1987
688 Smyth-1991
689 Flight-1987
690 Ryne-1993
691 Strangeway-1993
692 Besser-2004
693 Schmidt-1981
694 Istomin-1982
695 Ksanfomaliti-1982a
696 Ksanfomaliti-1982b
697 Moroz-1982
698 Surkov-1982a
699 Florensky-1983
700 Wilson-1987t
701 Siddiqi-2002s
702 Garvin-1984
703 Wilson-1987t
704 Siddiqi-2002s
705 Sagdeev-1982
706 Florenskii-1982
707 Selivanov-1982
708 Siddiqi-2002s
709 Verigin-1999
710 Garvin-1984
711 Wilson-1987t
712 Siddiqi-2002s
713 Sagdeev-1982
714 Florenskii-1982
715 Ball-1999
716 VnIITransmash-1999
717 Ivanov-1992
718 Ball-1999
719 VnIITransmash-1999
720 Surkov-1982c

721 Surkov-1997j
722 Surkov-1983
723 Ball-1999
724 VnIITransmash-1999
725 Kerzhanovich-1982
726 Sagdeev-1982
727 Wilson-1987t
728 Surkov-1982a
729 Moroz-1982
730 Surkov-1982b
731 Surkov-1997i
732 Mukhin-1982

733 Istomin-1982
734 Ksanfomaliti-1982a
735 Ksanfomaliti-1982c
736 Ksanfomaliti-1982b
737 Florensky-1983
738 Bogovalov-1984
739 Bogovalov-1985
740 Hack-1993
741 Wilson-1987t
742 Siddiqi-2002s
743 NSSDC-2004b
744 Langereux-1971

745 Langereux-1972
746 Espace Information-1979
747 Blamont-1987d
748 Perminov-2005
749 Gross-1966
750 Blamont-1987e
751 Carlier-1995
752 Perminov-2001
753 Kemurdjian-1992
754 Matrossov-2004
755 Matrossov-2006

3

The grandest tour

THE JOURNEY OF THREE LIFETIMES

With the first Mariner missions underway and the Voyager Mars landers apparently set to start in the late 1960s, JPL began to consider the Navigator missions that were to venture through the asteroid belt to explore the outer solar system. The design of this spacecraft was to be sufficiently flexible for planetary flybys and orbital missions (in some cases ferrying soft landers) and to perform high-resolution imaging and a multitude of other experiments. It would use the powerful (and expensive) Saturn V launch vehicle, be nuclear powered and, because it was thought that missions to the outer planets would not be feasible using conventional chemical propulsion since the time spent in the interplanetary transfer would be excessive, ion thrusters were to be used for continuous acceleration to the objective. In parallel, JPL issued contracts to both Lockheed and General Dynamics to study whether the existing or imminent technologies might facilitate a mission that would attempt to reconnoiter the asteroid belt in 1967–1975 and perform a flyby of Jupiter in 1973–1980. General Dynamics offered four RTG-powered designs, ranging from a simple 'spinner' capable of only limited scientific observations, up to a complex 3-axis-stabilized spacecraft capable of a comprehensive study.[1,2,3] Meanwhile, a discovery was made that would not only make missions to the outer solar system more manageable and less time-consuming, but would also revolutionize planetary exploration.

By 1965, the celestial navigators at JPL had realized that the trajectory of a probe making a Venus flyby would, if conditions were just right, be able to be deflected by that planet's gravity to Mercury, and they set out to determine whether this gravity-assist technique could be used to explore the outer solar system. Jupiter is the most massive planet in the system and, orbiting just beyond the asteroid belt, is uniquely capable of providing a powerful gravity-assist to destinations beyond. For centuries, it had been observed to deflect the comets that strayed too close. Elliott Cutting, the head of the advanced projects group, asked postgraduate Gary Flandro to investigate the possibility. Walter Hohmann, Gaetano Arturo Crocco, Michael

Minovitch and Krafft Ehricke had all described the theory, but Flandro's challenge was to apply it. To gain a first approximation, he used the relatively crude technique of describing a perturbed path as a series of 'patched' trajectories that alternated between having the Sun or a planet as their foci. He found that launches in the latter half of the 1970s offered many opportunities to reach the outer planets, owing to the fact that their heliocentric longitudes all fell within a fairly narrow arc. This alignment only occurred every 170 years or so, and if it were not exploited this time, the opportunity would not recur until the middle of the twenty-second century!

Three main solutions were identified. The first, for a launch in the late summer of 1977, offered flybys of Jupiter, Saturn and Pluto. In fact, this J–S–P mission would enable Pluto to be visited in 1986, long before Neptune could be reached by other trajectories. An added bonus was that at the time of the encounter Pluto would be approximately in the direction of the solar apex (the direction in which the Sun travels around the center of the galaxy) and would yield an early opportunity for the spacecraft to cross the heliopause (which, at that time, was optimistically thought to be 20 AU from the Sun) and report on conditions in interstellar space. Sharing the same launch window was the J–S–U–N solution, which offered a comprehensive series of encounters with each of the 'gas giants', concluding with Neptune in 1989. Although a complete J–S–U–N–P tour was possible, this solution offered itself only twice per millennium. The third solution involved a launch in late 1979 and offered a flyby of each of the gas giants apart from Saturn, and hence was labeled J–U–N. In

Gary Flandro. (Courtesy of Gary Flandro)

fact, several other opportunities existed, with launches running through to the mid-1980s, but the earlier cases were the best available for a medium-lift launch vehicle. There would be further opportunities after the mid-1990s, when Jupiter was once again in position, but only for single-planet encounters beyond. In comparison to a 'ballistic' trajectory, a gravity-assist would greatly shorten the flight time to the outer planets: from 30 years to just 10 for Neptune, and from 45 years to 9 for Pluto. On the other hand, as the spacecraft's heliocentric speed would be increased with every encounter in transit, it would be traveling too fast to allow orbit insertion when it reached the outer planets, thus limiting such probes to flybys. The J–S–P and J–U–N options were more attractive from an engineering point of view, since their short durations would increase the likelihood of the craft still being functional on reaching Pluto or Neptune. The J–S–U–N option had the disadvantage that the spacecraft would have to cross the plane of the planet's system of rings just outside the A ring, posing the risk of a collision with a ring particle damaging, or perhaps even destroying, the vehicle.[4,5] In mid-1965 Flandro presented his findings to Homer Joe Stewart, the chief scientist at JPL, who labeled it the Grand Tour of the Outer Solar System.[6] In fact, the term 'Grand Tour' had been coined by Crocco for his 1956 investigation of a manned Earth–Mars–Venus–Earth flight. It was also the name given to the voyages of intellectual exploration that well-educated offspring of well-to-do north-European families undertook during the eighteenth and nineteenth centuries, the most famous being that by Johann Wolfgang von Goethe in Italy

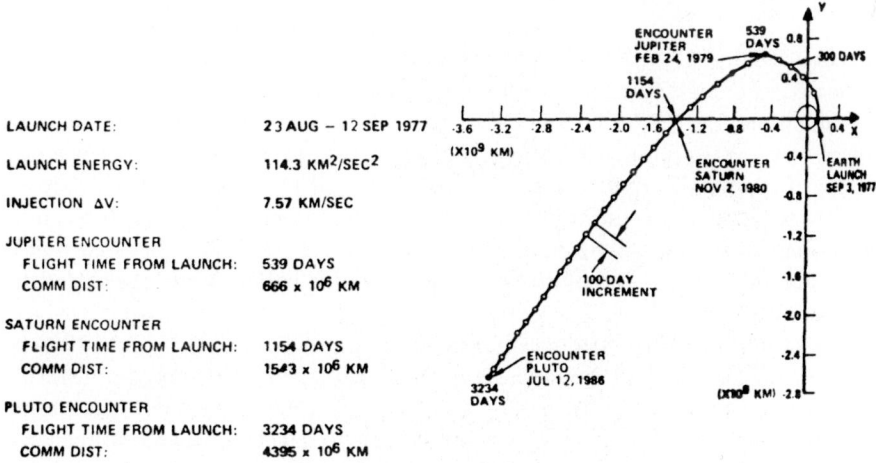

The 'Grand Tour' trajectories discovered by Gary Flandro would enable a minimum of two spacecraft launched in the late 1970s to use Jupiter's gravity to investigate all of the outer planets long before they could be reached by conventional trajectories. In this case, the J-S-P trajectory would launch in 1977 and visit Jupiter in 1979, Saturn in 1980 and Pluto in 1986. JPL intended this to be the first Grand Tour mission, but that project was canceled. However, after Pluto was deleted in favor of a close look at Titan this trajectory was actually flown by Voyager 1.

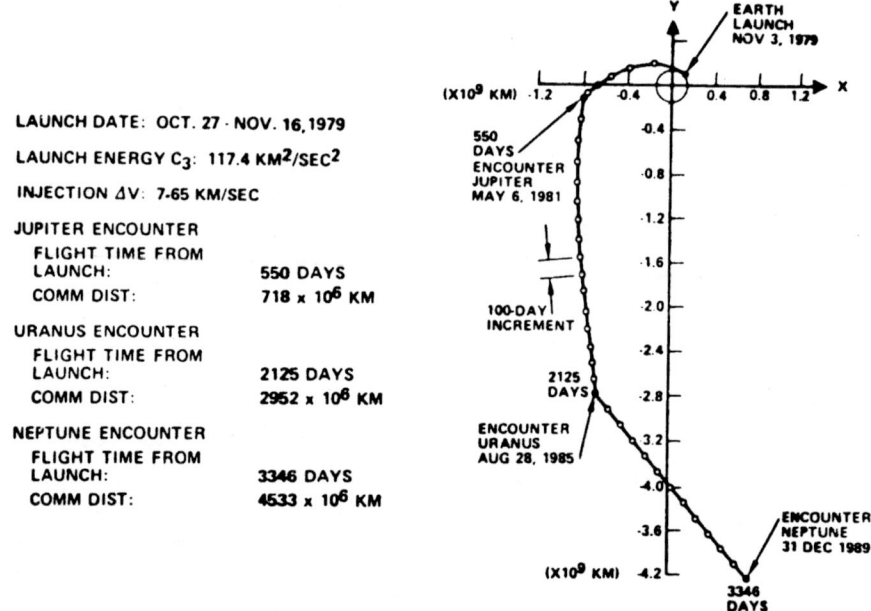

LAUNCH DATE: OCT. 27 - NOV. 16, 1979

LAUNCH ENERGY C₃: 117.4 KM²/SEC²

INJECTION ΔV: 7-65 KM/SEC

JUPITER ENCOUNTER
  FLIGHT TIME FROM
  LAUNCH: 550 DAYS
  COMM DIST: 718 x 10⁶ KM

URANUS ENCOUNTER
  FLIGHT TIME FROM
  LAUNCH: 2125 DAYS
  COMM DIST: 2952 x 10⁶ KM

NEPTUNE ENCOUNTER
  FLIGHT TIME FROM
  LAUNCH: 3346 DAYS
  COMM DIST: 4533 x 10⁶ KM

The J-U-N trajectory would launch in 1979 and visit Jupiter in 1981, Uranus in 1985 and Neptune in 1989. It was to have been flown by the second round of Grand Tour missions.

between 1786 and 1788. After publicly announcing this enticing opportunity, JPL set out to determine the feasibility of mounting such a mission.[7]

At this time, the Soviet Union was also conducting studies of robotic missions to Jupiter, but it is not known whether the opportunity for a Grand Tour was a factor. In view of the low reliability of Soviet spacecraft and the lack of a worldwide deep-space tracking network, the studies were probably academic in nature. Nevertheless, Soviet scientists could still boast in a May 1971 *New York Times* interview that they were studying the feasibility of dropping instrumented balloons into the atmospheres of the gas giants.[8,9,10]

NASA did not initiate serious planning for a Grand Tour mission until 1968, after the trajectory had been analyzed in detail. Meanwhile, with the successful Mariner 4 and Mariner 5 flybys of Mars in 1965 and Venus in 1967 respectively, the engineers at JPL felt confident that they would soon be able to develop a spacecraft capable of operating for a decade or more. Two launch vehicles were under consideration. One was to fit the US Air Force's Titan III with a Centaur upper stage and a small motor to give the probe a final boost. The other was the proposed 'Intermediate 20' version of the Saturn V, consisting of the first and third stages of the standard Saturn V plus a Centaur upper stage. While the Titan III variant would be able to place up to 700 kg into a typical Grand Tour trajectory, this increased to 4,300 kg for the Saturn V variant. A spacecraft designed to operate for the duration of a Grand Tour mission

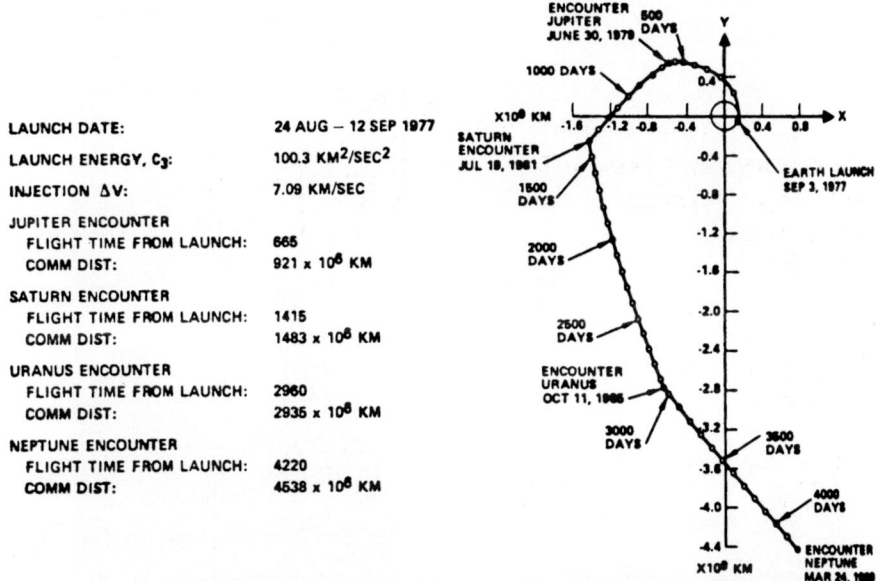

LAUNCH DATE: 24 AUG – 12 SEP 1977

LAUNCH ENERGY, C_3: 100.3 KM^2/SEC^2

INJECTION ΔV: 7.09 KM/SEC

JUPITER ENCOUNTER
  FLIGHT TIME FROM LAUNCH: 665
  COMM DIST: 921 x 10^6 KM

SATURN ENCOUNTER
  FLIGHT TIME FROM LAUNCH: 1415
  COMM DIST: 1483 x 10^6 KM

URANUS ENCOUNTER
  FLIGHT TIME FROM LAUNCH: 2960
  COMM DIST: 2935 x 10^6 KM

NEPTUNE ENCOUNTER
  FLIGHT TIME FROM LAUNCH: 4220
  COMM DIST: 4538 x 10^6 KM

The J-S-U-N trajectory would launch in 1977 and visit Jupiter in 1979, Saturn in 1981, Uranus in 1985 and Neptune in 1989. It was not chosen for the Grand Tour because it was believed at that time to require too long a mission duration, but as events were to transpire it would be the profile flown (with some revision of the dates) by Voyager 2.

would require onboard artificial intelligence and heavily redundant systems to make it resilient – in effect, defense in depth against components succumbing to the harsh environment – thereby making it much heavier than the contemporary Mariner, whose design was constrained by the Atlas rocket. The Saturn V option was attractive because one vehicle would be able to launch a pair of spacecraft. Three NASA centers submitted proposals. Ames offered a version of its spin-stabilized Pioneer Jupiter spacecraft, the contemporary form of which had an expected lifetime of 5 years, adapted to deliver an atmospheric probe to Jupiter. The Marshall Space Flight Center (MSFC), which had managed the development of all the agency's man-rated launchers, including the Saturn V and the Space Shuttle, proposed using Shuttle technology to dispatch a series of spacecraft to deliver landers to the Jovian moons Io and Ganymede, and to study Saturn's ring system. Because this would be more than four times heavier than its rivals, those who felt that it would be so costly that it would inevitably be abandoned (as had the ill-fated Voyager Mars landers) dismissed it as the 'Grandiose Tour'. In any case, because the Shuttle would not be available for the 1977 launch window, the MSFC mission would have to use one of the later windows which, for vehicles more powerful than the Titan III, would be available until January 1984. This approach would have shortened the flight time to

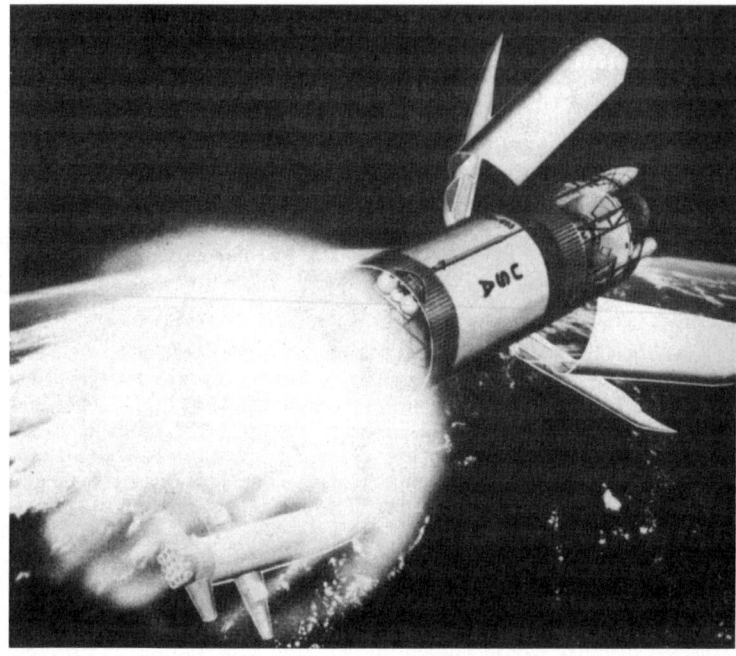

The Marshall Space Flight Center's proposal for the Grand Tour envisaged leaving Earth on a specially developed stage that would be delivered into Earth orbit by the Space Shuttle. Owing to its scale and cost this proposal was dismissed by its critics as the 'Grandiose Tour'.

Neptune to 6 years, but at the cost of an even greater flyby speed which, even if it was not intended to enter orbit, would have severely limited the time available for observations. In hindsight, it is doubtful whether the Shuttle could have attempted a flight of this complexity as early as 1983 or 1984.[11]

The winning Grand Tour proposal was submitted by JPL, for a 656-kg spacecraft named the Thermoelectric Outer Planet Spacecraft (TOPS). It consisted of a central bus with a high-gain antenna that was designed to unfurl in space like an umbrella 4.3 meters in diameter for a data rate of 90 kbps at Jupiter and, depending on the tour option, 1.1 kbs at either Neptune or Pluto. It also had two stubby stacks, each with a pair of RTGs, and the scientific instruments were set on booms that projected from the opposite side in order to shield them from the neutron radiation that would be emitted by the power units. To enable the spacecraft to make a comprehensive study of the gas giants, their retinues, and the particles and fields of interplanetary space in the outer solar system, the scientific instruments – with an overall mass in excess of 90 kg – included a magnetometer, plasma, radiation and cosmic-ray detectors, three cameras with different fields of view, ultraviolet and infrared spectrometers, photometers, an infrared radiometer and a radio-astronomy receiver. Since the spacecraft was required to operate so far from Earth, the primary innovation was to be its triply-redundant majority-voting computer known as the STAR (Self-Testing And Repairing). It would monitor an internal telemetry system and on identifying a fault would autonomously switch to a backup system. JPL built a three-rack 'bread board' prototype of the STAR in 1969 to prove the principles by which it was to operate.[12] The plan envisaged two probes being launched in 1977 for the J–S–P trajectory, and two more in 1979 for the J–U–N trajectory. Nevertheless, some scientifically interesting observations would not be possible. For example, no craft would be in the vicinity of Saturn when Earth passed through the plane of the ring system in 1979 and 1980; a well-timed radio occultation would have given a direct measure of the thickness and composition of the rings. Also, flying inside the C ring in order to perform close up observations of Saturn would not only pose a risk to the craft, but would also result in unacceptably fast flybys of future targets.[13,14,15,16,17,18]

Although by the end of the 1960s spending on the Apollo lunar program was well past its peak, NASA's budget was in steep decline. Despite this, the Viking project that would send orbiter/landers to Mars and some form of Grand Tour were accepted as the two principal projects for US planetary exploration in 1970s, each of which was expected to cost no more than $1 billion. NASA also asked the Space Science Board of the National Academy of Sciences to compare the merits of a reconnaissance of all the outer planets with a mission involving only Jupiter and Saturn, and against an intensive study of Jupiter and Saturn by orbiters. In 1969, when NASA's budget still appeared likely to be able to fund both the Grand Tour and a Jupiter orbiter, the Space Science Board recommended that both be pursued, and also suggested that a TOPS be launched in 1974 or 1975 on a test mission that would approach Jupiter on a trajectory that would result in the spacecraft being deflected out of the plane of the ecliptic and later flying over the Sun's polar regions. In March 1970 the Grand Tour gained an official, though tentative, presidential endorsement when it was mentioned by Richard Nixon in his statement about the future of the US space program, being described as the mission that would reveal "the mysterious planets of the outer solar system". But one year later the project was derailed when scientists and astronomers on the Space Science Board lobbied for the building of a Large Space Telescope (the project that would eventually become the Hubble Space Telescope), with the result that the lowest possible budget priority was recommended for the Grand Tour. Many on the Space Science Board, including its chairman, deemed the mission to be too risky since the requisite technology either did not yet exist or remained to be proven. NASA sought a reassessment. This reinstated support for the Grand Tour but, because it was becoming clear that the Grand Tour would exceed $1 billion (although only by a small margin), it also urged NASA to consider a simpler and much cheaper Mariner-class Jupiter and Saturn flyby mission instead. However, it was the Space Shuttle that finally killed the Grand Tour. The concept behind the Shuttle was that the high development costs would be fully compensated by a reduction in operating costs, and since it would be the National Space Transportation System, all existing rockets would be obsolete. But the development cost would be so great that NASA was forced to cancel other projects in order to reassign funds. By December 1971, the agency had agreed to delete both the Grand Tour and the NERVA (Nuclear Engine for Rocket Vehicle Application) engine – work on which had started in the mid-1960s and offered the prospect of slashing the flight time of a human mission to Mars. The decision was reported one month later, just a few weeks after the fanfare announcement that the development of the Shuttle had been officially approved. One factor that contributed to the loss of the Grand Tour was NASA's decision that the spacecraft would be developed 'in-house' by JPL, without the significant input (and the political leverage) of an industrial contractor.[19,20]

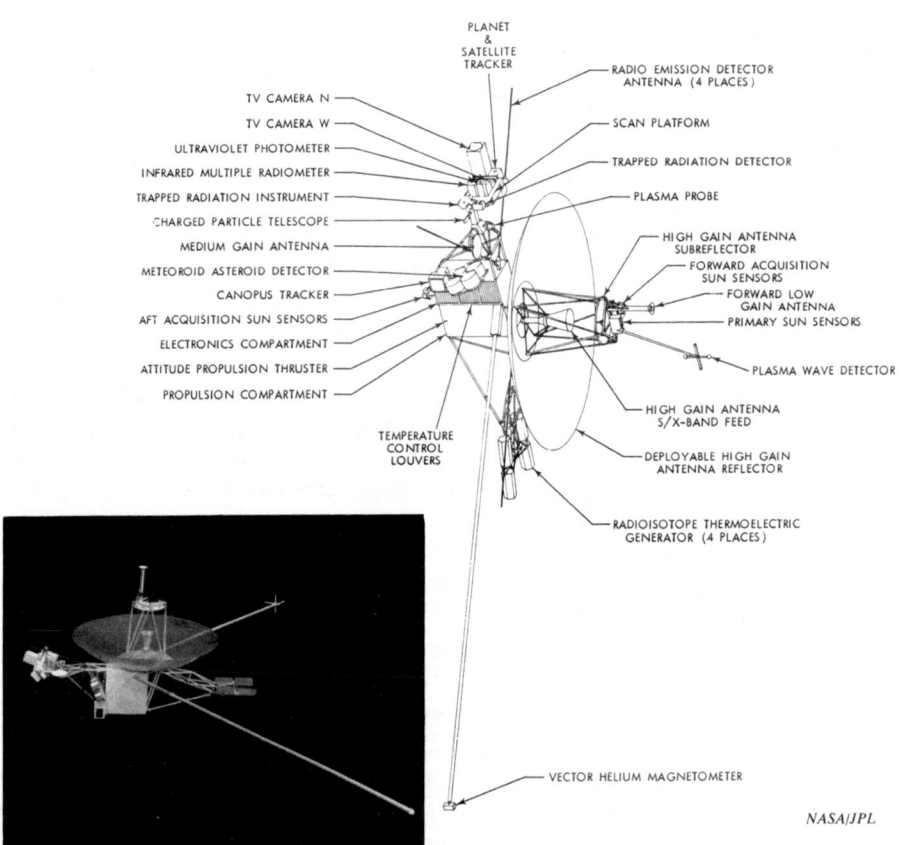

NASA/JPL

A line drawing and a rendition of the TOPS spacecraft with which JPL proposed to fly the Grand Tour missions.

GRAND TOUR REBORN

JPL rapidly recovered from the cancellation of the Grand Tour, pursuing the Space Science Board's recommendation that it develop a simpler, less ambitious and less costly mission. As its name implied, the new Mariner Jupiter–Saturn (MJS) mission was to follow up the Pioneer Jupiter flybys to further study Jupiter and Saturn and to investigate their retinues, which the Pioneers were not really equipped to study. In keeping with JPL's modus operandi, two identical spacecraft were to be launched. As with TOPS, the MJS spacecraft would have to be capable of overcoming anomalous situations autonomously, but with the launch date at least 5 years away, in 1977, the development of a suitable computer was not expected to be an issue. With the roots of a constituency having been established for the Grand Tour, JPL rapidly mobilized support for the MJS mission. On 22 February 1972, just 6 weeks after the demise of the Grand Tour, the Space Science Board endorsed its successor as being first-rate science at an affordable cost. NASA officially approved the $360 million mission on 18 May, and Congress followed up with an additional appropriation to allow work to start promptly.[21] Owing to its heritage of the past Mariners and the already designed Viking orbiters, the development of the vehicle was expected to be straightforward. In July, NASA invited the scientific community to submit proposals for instruments, and selected nine instruments from the 77 responses; a 10th instrument to study plasma waves in space was added in July 1974 because it was lightweight and would make use of the antennas that were in place for planetary radio astronomy.[22,23]

The flyby of Jupiter by Pioneer 10 in December 1973 had a significant influence on the early planning for the MJS mission, because it revealed that radiation close to the planet was thousands of times more intense than expected. Although MJS would not penetrate so deeply into the system and, as yet, only 18 months after the start of the project, no hardware had been built, this finding sent the engineers back to their drawing boards to redesign most of the electronics and install radiation shielding to augment that already planned or to be provided by the spacecraft's structures. Fortunately, the budget was able to absorb the $13 million cost of this effort. All electronic parts were designed to survive a radiation dose of twice the amount that they were expected to encounter. Also, some particularly sensitive components – such as the relatively new integrated circuits for the data management system and some of the more complex instruments – were selected only after they had been thoroughly investigated. It was not appreciated at the time, but it is now believed that the hardening of the electronics in this redesign was one of the factors that enabled the spacecraft to continue to function reliably in space for more than three decades.[24,25] Meanwhile, the celestial navigators had conducted a thorough study of the options, computing more than 10,000 possible trajectories. Various constraints were then used to identify the most operationally feasible and scientifically rewarding cases. For example, the relative positions of the Earth, the Sun and the craft during an encounter had to be favorable for communications, as it would be pointless to arrange the encounter for a period when the spacecraft was on the far side of the Sun. It was desired not only to use a gravity-assist at Jupiter to reach Saturn, but also to pass as close to as many of the satellites as possible in the two systems, and to arrange occultations by having the spacecraft pass behind the planets as viewed from Earth, in addition to passing through their shadows. In some cases, occultations would be set up for some of the satellites. The scientific requirements called for at least one spacecraft to pass close to Io, which, after the Pioneer flybys, was believed to be the most interesting Jovian satellite, and to Titan, the Saturnian moon with an atmosphere rich in organic compounds, while also not penetrating too deeply into Jupiter's radiation belts or venturing too close to Saturn's ring system.[26]

The planetary alignment that offered the Grand Tour was too enticing to be given up lightly, and the trajectory study delivered a bonus. In order to recover most of the science that had been expected from the Grand Tour, JPL urged using the J–U–N window to launch one or two Mariner Jupiter–Uranus (MJU) spacecraft to deliver an atmospheric probe to Uranus, with the option of going on to Neptune, but NASA was not receptive, essentially because it would require the agency to buy additional Titan rockets from the Air Force at a time when it expected to start testing the Space Shuttle.[27,28]

An even cheaper option was discovered in 1974. One MJS spacecraft would be launched in a window that would enable it to conduct a close inspection of both Io at Jupiter and Titan at Saturn, and if were to succeed then it would have achieved all of the project's scientific objectives. The other MJS spacecraft would be launched in the J–S–U–N window. (Both windows would be open over the interval of several weeks in the summer of 1977.) If the first spacecraft was completely successful, and if the second was still healthy upon reaching Saturn, it could go on to Uranus and, if it survived long enough, Neptune. Although NASA was appreciative of the option, it ordered JPL not to openly promote the possibility of an extended mission. It would have been possible to send the first spacecraft on the J–S–P trajectory, but given that little was known of Pluto, a close inspection of Titan was judged more significant, and since this would deflect the craft significantly north of the ecliptic it would rule

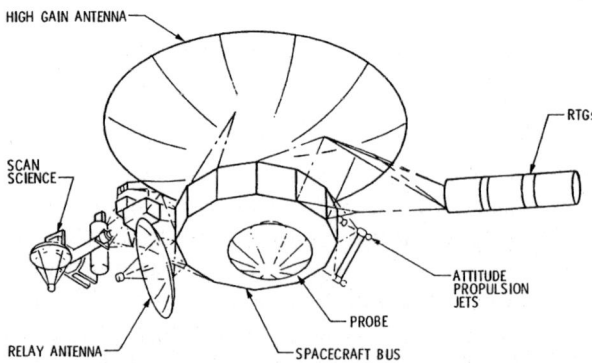

A drawing of a Mariner–Jupiter–Uranus spacecraft with an atmospheric probe for Uranus.

out further planetary encounters. The opportunity to attempt a reconnaissance of Uranus and Neptune would therefore depend on the success of the first spacecraft; if it were to fall short, then its running mate would have to attempt the Titan flyby.

Although it had been widely presumed that the spacecraft for MJS would become Mariners 11 and 12, in early 1977 it was decided to give the project a name, and its personnel were asked to make suggestions. Names such as Nomad, Pilgrim and even Planet Trek were rejected in favor of Voyager. And although the window for the J–S–U–N trajectory would open first, this was a slower route, and therefore it was decided to name the missions in the order in which they would arrive at Jupiter and Saturn, which meant launching Voyager 2 first.[29]

THE SPACECRAFT THAT COULD

The principal structural unit of the Voyager spacecraft was a 10-sided aluminum bus 188 cm by 47 cm tall. Most of the electronics and systems were contained within the bus to provide a benign thermal environment and to protect them against damage by micrometeoroids and the harsh Jovian radiation. At the center was a 71-cm-diameter spherical titanium tank with 104 kg of hydrazine for trajectory and attitude control, capable of a total change in velocity of 190 m/s. A truss on top of the bus carried the 3.66-meter-diameter aluminum honeycomb high-gain antenna for both scientific and 'housekeeping' communications in deep space. Preliminary activities in the vicinity of Earth would use an omnidirectional antenna mounted on the feed of the high-gain antenna. The transmitters could operate at 2.3 GHz in the S-Band or 8.4 GHz in the X-Band. The high data rate of 115.2 kbps at Jupiter offered by the higher frequency was the key to achieving the planned scientific yield, and although it had been tested on Mariner 10 and the Viking orbiters it was still considered experimental. Both the S-Band and X-Band 23-W transmitters were redundant (i.e. there were four in all) as were the receivers. The 500-Mbit tape recorder had sufficient capacity to store about 100 images or other data for replay to Earth. Despite the high-gain antenna being the largest yet flown on a planetary mission, the signal received from Jupiter would be of the order of 10^{-18} W/m^2. To further improve communications, the antennas and amplifiers of the Deep Space Network (DSN) were upgraded to minimize the 'noise' in their systems.[30]

Serious problems were encountered during development when some components of the X-Band system could not be obtained. The activities planned for the Jupiter encounter would be feasible using just the S-Band system, but not only would the output from Saturn be dramatically reduced, so little data would be able to be returned from further out as to make an extended mission pointless. However, with less than a year remaining before launch, the required components became available.[31] This incident also had beneficial side-effects that would only become evident during the extended mission: to reduce the amount of data to be transmitted to Earth, the data management system was given a dedicated data error compensation algorithm. On paper, this was in case a failure were to prevent the X-Band system being used during the Saturn encounter, but it was an open secret

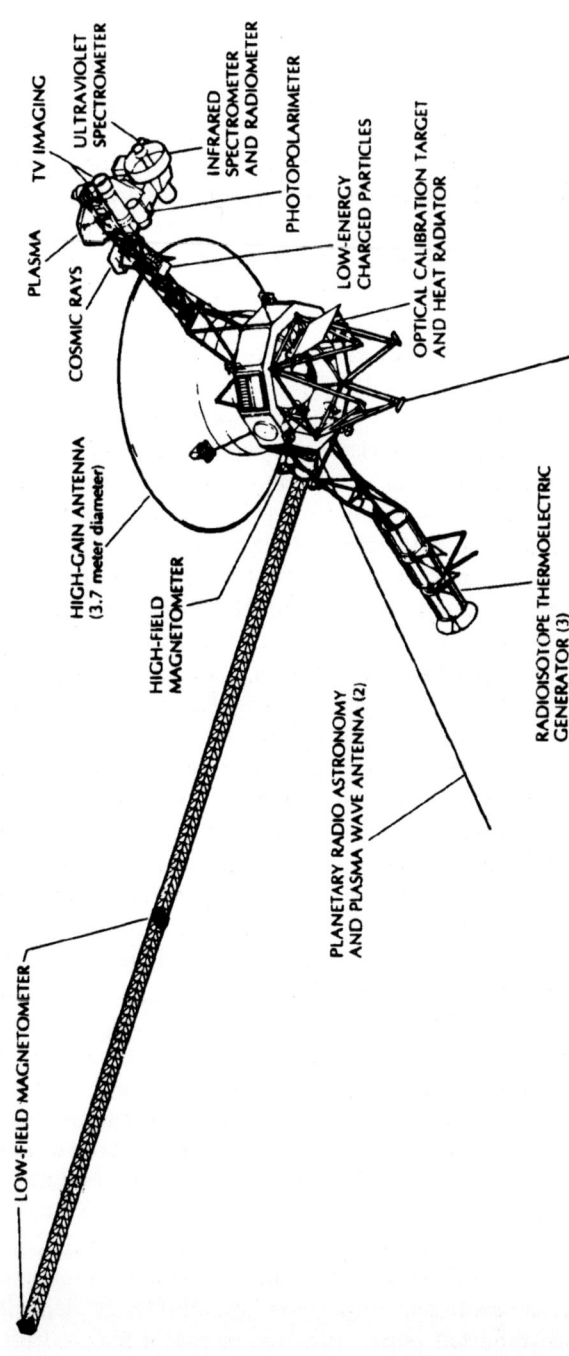

TV IMAGING

ULTRAVIOLET
SPECTROMETER

PLASMA

INFRARED
SPECTROMETER
AND RADIOMETER

COSMIC RAYS

PHOTOPOLARIMETER

LOW-ENERGY
CHARGED PARTICLES

OPTICAL CALIBRATION TARGET
AND HEAT RADIATOR

HIGH-GAIN ANTENNA
(3.7 meter diameter)

HIGH-FIELD
MAGNETOMETER

RADIOISOTOPE THERMOELECTRIC
GENERATOR (3)

PLANETARY RADIO ASTRONOMY
AND PLASMA WAVE ANTENNA (2)

LOW-FIELD MAGNETOMETER

The design of the Voyager spacecraft was dominated by the high-gain antenna that was required to maintain extremely-long-range communications with Earth.

that these functions would increase the amount of data that could be returned from an extended mission to Uranus and Neptune.

Affixed to the base of the bus was a four-legged truss to connect the spacecraft to the solid-fuel upper stage – the same motor as used for the Pioneer Jupiter missions. During the 43-second burn, the Voyager spacecraft would maintain its attitude using its own thrusters, then discard the spent stage. A rectangular radiator with a special coating to enable it to double as a scientific instrument calibration panel was added to the truss. A trio of booms projected out to the sides. One was to maintain the three 39-kg plutonium RTGs a safe distance from the sensitive electronics located inside the body. Once the spacecraft was in space, the hinged boom deployed by swinging upward 90 degrees. The RTGs would provide 450 W of power at launch, but natural decay would reduce this to less than 400 W for the Saturn encounter 5 years later. On the opposite side of the bus from the RTG boom, and also designed to swing up, was a 2.3-meter-long graphite epoxy boom that had instruments mounted along its length and a scan platform at its end for the imaging, spectrometric, radiometric and polarimetric instruments that would require to be aimed by slewing within 2 degrees of freedom. The epoxy fiberglass truss mounted near the base of the RTG boom was of a different type. It was stored in a compact form, and once in space was extended at an upward-inclined angle to a length of 13 meters to position the magnetometers away from the body of the spacecraft. Also projecting to the sides were a pair of 10-meter-long beryllium–copper aerials splayed out at an angle of 90 degrees.

Unlike the Pioneer Jupiter spinners, Voyager was to be 3-axis stabilized, with its high-gain antenna pointing at Earth. This strategy reflected the emphasis on imaging and other remote-sensing observations. However, to improve the particles and fields investigations at times during its encounters the spacecraft would execute a series of 360-degree roll maneuvers to 'sweep' the environment around it. It would determine its attitude by Sun sensors protruding through the high-gain antenna dish and by Canopus sensors. As another preparation for an extended mission, the Sun sensors were fitted with amplifiers to boost their sensitivity in the dim realm beyond Saturn. Attitude control and course corrections would be by a total of 16 thrusters, each with an output of 0.89 N, set in clusters on the upper and lower faces of the bus. In a departure from previous JPL practice, these jets used the more efficient hydrazine monopropellant rather than compressed nitrogen. The attitude control system was required to ensure that the 'pencil' beam of the high-gain antenna – the only means of communicating with Earth during most of the mission – was always properly aimed at Earth; which, given that the width of the beam was less than one-sixth of a degree, would be quite a challenge. Each spacecraft had three redundant computers, with memories ranging from 4 to 8 kb: one to decode commands received from Earth and issue them to the others; one to handle data flowing from the instruments; and the third to manage the attitude of the spacecraft and command the scan platform. All the computers were sufficiently autonomous to react to problems without input from Earth. An important feature was that about one-third of the memory of one of the computers would be able to be reprogrammed in-flight. This capability was another example of planning for an extended mission.[32,33] In fact, most of the

Preparing a Voyager spacecraft for launch. Note that the scan platform is in its most compact configuration with covers protecting its optics. (Courtesy of NASA/KSC)

technologies for the TOPS mission that were under development when the Grand Tour was canceled were exploited by the Voyagers. Two notable exceptions were the STAR computer and the 'umbrella' antenna, but at a later date JPL would use the latter on the Galileo Jupiter orbiter.[34]

As the Voyagers were to depart the solar system, they were each provided with a message for any extraterrestrial civilization that might intercept them. This message took the form of a 30-cm-diameter gold-plated copper disk with recordings of audio and images. It was enclosed in a protective jacket inscribed with symbols to indicate the location of Earth with respect to several pulsars (as on the plaques of Pioneers 10 and 11), plus instructions on how to play the disk using a cartridge and needle, both of which were provided. The message was mounted on one side of the bus, together with an ultra-pure chip of uranium-238 to enable the duration of the spacecraft's trip through space to be calculated from the ratios of the radioactive decay products – the half-life of this 'mission timer' being a little over 4.5 billion years. The audio on the disk included salutes in 55 languages covering all human history and representing an estimated 87 per cent of the population, ranging from the ancient Akkadian and Hittite to Chinese dialects and African tongues. It continued with a selection of 35 sounds of Earth, including whale song, ocean surf, laughter and kisses. There was then 90 minutes of music featuring a variety of Eastern and Western classics and ethnic music including a Bulgarian shepherdess's song, Chuck Berry's *Johnny B. Goode*, and an aria from Mozart's *Magic Flute*. The 115 images on the disk were encoded in analog form, together with instructions on how to decode them. They included views of Earth, animals, and a variety of human occupations. Finally, the disk contained messages from President Jimmy Carter and the Secretary General of the United Nations, Kurt Waldheim. Piqued by criticism that the Pioneer plaque had been pornographic, sexist or ethnocentric, Carl Sagan, who again was the driving force behind the design of the message, endeavored to include a selection of messages and images that represented as many of the world's ethnicities as possible, and in as neutral a manner as possible with regard to the human sexes, but in doing so he attracted criticism for having given the false impression of a peaceful and loving humankind, akin to a reassuring TV commercial.[35,36,37]

The 10 scientific instruments had a total mass of 115 kg. The most important was the imaging system, which was built in-house by a JPL team. The 38.2-kg package comprised a pair of vidicon cameras incorporating 800 × 800 pixel arrays. The wide-angle camera used optics with a focal length of 200 mm to provide a field of view 3 degrees wide, and the narrow-angle unit used optics with a focal length of 1,500 mm to provide a field of view 0.4 degree wide. To produce color images and to analyze atmospheres for various chemicals, both units had carousels of filters. In the wide-angle camera, the filters were clear, blue, green and orange, plus one for sodium and two for methane (one of which was optimized to image the atmosphere of Uranus); the narrow-angle unit had two clear, two green, violet, blue, orange and ultraviolet filters.[38]

An infrared spectrometer capable of discriminating about 2,000 wavelengths was to analyze the structure and composition of atmospheres, being particularly tailored for the Jovian atmosphere. It could identify hydrogen, helium, methane, ammonia,

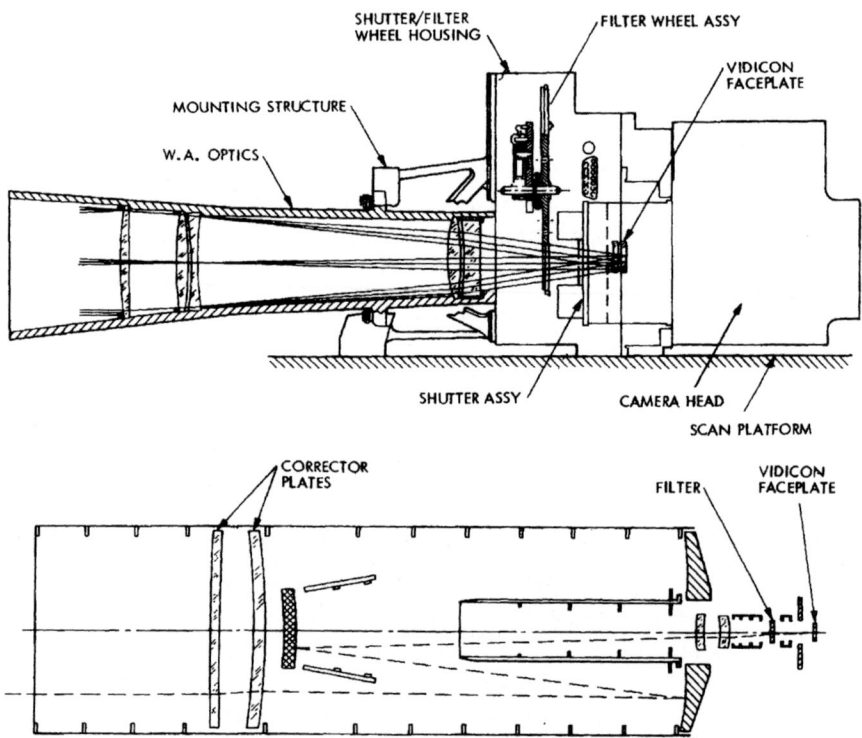

SHUTTER/FILTER
WHEEL HOUSING

FILTER WHEEL ASSY

VIDICON
FACEPLATE

MOUNTING STRUCTURE

W.A. OPTICS

SHUTTER ASSY

CAMERA HEAD

SCAN PLATFORM

CORRECTOR
PLATES

FILTER

VIDICON
FACEPLATE

The optical arrangement of Voyager's cameras: the wide-angle camera (top) and the narrow-angle camera. (Courtesy of Bradford A. Smith. Reprinted from Smith, B.A., et al., "Voyager Imaging Experiment", Space Science Reviews, 21, 1977, 103–127. With kind permission from Springer Science and Business Media)

phosphine, water, oxides of carbon, simple compounds of silicon and some organic compounds. It could also be used to study the structure of an atmosphere and to map the temperatures of the planets and their retinues. Its optical system used a 51-cm-diameter gold-plated telescope. An uprated version of the spectrometer was built at the last minute to give better performances at large distances from the Sun, such as during a Uranus flyby, but this failed during vibration tests and the original was used instead. The uprated version did not pass its flight qualification trials until 6 weeks after the mission had begun.[39] An ultraviolet spectrometer was to study the upper regions of atmospheres, and their interactions (such as auroras) with magnetospheric phenomena. The suite on the scan platform was completed by a photopolarimeter to measure the brightness and polarization of light reflected by planets, satellites and rings. Two investigations utilized the 10-meter-long aerials. One was the planetary radio astronomy package to study radio emissions from the planets at close range. It had been discovered in the mid-1950s that Jupiter was a source of radio noise, and it was already evident that this was in some way influenced by the position of Io in its

Lowering the aerodynamic shroud of the launch vehicle over Voyager 2. The circular disk on a side of the bus is the 'Sounds of Earth' recording.

orbit around the planet. The instrument that shared the use of these aerials was the plasma wave detector. The 13-meter-long boom had 3-axis magnetometers, one pair mounted midway along its length to measure high fields, and another pair at its tip to measure low fields, an arrangement that would enable scientists to correct the measurements of the low-field sensors for the residual magnetic field imparted by the metallic mass of the spacecraft. The suite was completed by three particles and fields instruments: a plasma analyzer, a low-energy charged-particle detector and a cosmic-ray detector. As usual, the spacecraft's radio signal would be used both for navigational purposes and to measure gravitational fields, and during occultations to study the atmospheres and ionospheres of the planets and their moons.[40,41]

Each Voyager spacecraft had a mass of 825 kg at launch, including fuel and other expendables, but with the fully-fueled solid rocket motor it weighed 2,066 kg. They would use the final two of the seven Titan IIIE–Centaur launch vehicles that NASA had procured.

LAUNCH AND TEETHING TROUBLES

Three spacecraft, designated VGR77-1, -2 and -3, were delivered to Cape Canaveral in 1977. Two were intended to be launched, and the third could be launched if one of the others were to become disabled, or, if appropriate, be cannibalized for spares. The window for the J–S–U–N trajectory would open on 20 August 1977, and last for 30 days. At the beginning of August testing revealed faults in two of the computers on VGR77-2, which was replaced by VGR77-3, and although this suffered a number of minor problems, including the replacement of the low-energy particle experiment after the failure of some of its components, this spacecraft was cleared for launch. In addition, with only a few minutes remaining on the clock the launcher developed a stuck valve, but this was resolved and lift off occurred at 14:29 UTC on 20 August, less than 5 minutes behind schedule. Almost immediately, the JPL engineers were alarmed. Simply stated, the launch vehicle was rolling more rapidly than the spacecraft's gyroscopes were designed to measure, or would ever be called upon to measure once in space, with the result that the output from the inertial platform was meaningless. After spending 43 minutes in parking orbit, the Centaur made the 'escape' maneuver, and shortly thereafter the solid rocket motor was fired to complete the departure sequence.[42] Voyager 2 left in a solar orbit that ranged between 1 and 6.28 AU and would deliver it into close proximity of Jupiter in July 1979.

Voyager 2's first hours further alarmed its builders. Although the various booms and aerials deployed, there was no confirmation that the boom for the scan platform had locked into place. Also, after trying to stabilize the spacecraft using the primary and secondary thrusters in turn, the computer activated the backup attitude control system. And, as commonly occurred at the start of a mission, the star sensor was distracted by the insulation debris that was floating nearby. A new software release to overcome these problems was hastily written and tested, and then uploaded to the

A Titan IIIE–Centaur lifts off with Voyager 2.

spacecraft.[43] Attempts were made to lock the loose boom into place, but to no avail. A detailed examination of pictures that were ordered to be taken of star fields found that the boom was within 0.5 degree of its deployed position, which suggested that it had swung up and locked correctly, and that the malfunction was in the sensor that was meant to indicate this fact.

Ten days out, the spacecraft fired its engines for the first time, to make a course correction. This revealed the engines to be less effective than their specification. An analysis established that the spacecraft's structures were obstructing rather more of the 'field of view' of the thrusters than was expected. When this underperformance was projected forward, it showed that the propellant would be marginal for reaching Saturn, which in turn ruled out an extended mission. However, by clever navigation and advancing the Saturn-targeting maneuver from the intended point 70 days after Jupiter to as near as possible to the moment of closest approach to that planet, it was possible to rescue the mission. Moreover, it was possible to make additional savings by revising the spacecraft's attitude to aim its star sensor at Deneb, the brightest star of Cygnus, instead of Canopus, to minimize solar radiation pressure perturbations. Despite its teething troubles, by 2 September all the instruments had been tested and found to be functional.[44]

Meanwhile, some hardware and software modifications were made to the repaired VGR77-2, which was to be launched next (in particular to strengthen the deployment springs of the boom for the scan platform) and the second launch vehicle was set up on the pad. The mission was delayed 5 days to permit checking of the modifications, and then launched on 5 September 1977. A malfunction that caused the fuel and oxidizer to mix at an incorrect ratio in the second stage of the Titan did not threaten the integrity of the launch vehicle, but the resulting underperformance at shutdown left the Centaur traveling at a lower speed than planned. After a 45-minute coast in parking orbit, the Centaur extended its burn in order to compensate, in the process consuming 550 kg of additional propellant. It was a very close call, because when it shut down upon achieving the desired speed the Centaur had just 3.4 seconds of usage remaining. If the Titan's shortfall had been marginally greater, then Voyager 1 would not have been able to reach Jupiter. Furthermore, if that particular rocket had been assigned to Voyager 2, which had a different velocity requirement, that spacecraft would not have been able to attempt the post-Saturn part of its mission.[45] Voyager 1 left Earth in a solar orbit that ranged between 1.01 and 8.99 AU and would enable it to reach Jupiter 4 months ahead of its twin. Having learned lessons from Voyager 2, the JPL engineers encountered fewer problems activating Voyager 1. On 18 September, 11.66 million km away, the spacecraft pointed its scan platform toward home and snapped the first-ever picture to include both Earth and the Moon.

With both of their spacecraft safely in transit to Jupiter, the Voyager team was in a happy mood. However, there were also causes for uncertainty and sadness. Although the approval of the Galileo Jupiter orbiter in July had staved off fears that JPL might have to forgo deep-space exploration owing to a lack of projects, the community at large faced the same dilemma since, other than Galileo and the Pioneer missions to Venus that were being managed by Ames, NASA had no further

The first picture ever taken showing the disks of Earth and the Moon together was taken on 18 September 1977 by Voyager 1. (Courtesy of NASA/JPL/Caltech)

projects approved. Moreover, the Centaur stage that had launched a whole decade of American probes seemed to have reached the end of the line because 'expendable' launch vehicles were to be phased out in favor of the Space Shuttle, which was to carry the next generation of planetary missions in its payload bay and dispatch them using a three-stage solid-fuel rocket.

As they left the inner solar system, both Voyagers continued to suffer a variety of problems. Voyager 2, for example, briefly lost attitude control when it jettisoned the cover of its infrared spectrometer, and in December the sensitivity of the instrument was greatly degraded when a bonding material crystallized in the cold of deep space and slightly warped the mirrors and optics; to overcome this the instrument's heater was switched on to evaporate the pollutant.[46,47] On 15 December, the faster-moving Voyager 1 overtook Voyager 2 at a distance of 124 million km from Earth. At that time, the two spacecraft were 17 million km apart. Serious problems began 2 months later. On 23 February 1978, when Voyager 1 was commanded to conduct a series of slews to test the articulation mechanism of its scan platform, this seized. The loss of the scan platform would be disastrous for the imaging mission. After tests using an engineering model at JPL, the spacecraft was ordered in March to put its platform through a series of slews, initially slowly and for short arcs, and then through larger angles. On 31 May the position at which the seizure had occurred was tested again, and the platform appeared to behave properly. It was concluded that the problem had been caused by small dirt particles contaminating the actuator gear, and that these had been either crushed or displaced by the subsequent test motions.[48] Just as the engineers were trying to coax the reluctant scan platform on Voyager 1 to life, Voyager 2 suffered a major trauma.

With their attention firmly focused on Voyager 1, the flight controllers at JPL had neglected to send any commands to Voyager 2, and on 5 April, with a week having elapsed since it had last received instructions from Earth, the 'command loss timer' alerted the computer. According to its instructions, Voyager 2 was to assume that its receiver had failed, and switch from the primary to the backup. However, the backup receiver suffered an almost incapacitating glitch when a capacitor in the circuitry for its frequency-tracking loop failed, preventing it from locking on to the signal from Earth, whose frequency continually varied owing to the Doppler shift prompted by a variety of factors, including the Earth's axial rotation. This was a problem that had been noticed in ground tests of the transmitter, but was not fully understood. Without this circuitry, the system would only receive a signal in a bandwidth (the width of the range of frequencies that could be detected) that was less than one-thousandth of the designed value.[49] The engineers were not too concerned, because if after 12 hours the spacecraft had not received a signal with its backup system it was to switch back to the primary, which it duly did. However, within 30 minutes of being reactivated this receiver suffered an electrical short in its power supply that blew the fuses and permanently disabled it. If the mission was to continue, and if humanity was to stand any chance of gaining a close look at Uranus and Neptune, Voyager 2 would have to rely on its 'tone deaf' backup system. Seven days after switching over to the primary receiver, and having had no instructions from Earth, the spacecraft again switched to the backup. Fortunately, the DSN had developed an oscillator for its transmitters that could be controlled by computer, and thus readily vary its frequency. It was ready with a sequence of commands and a transmission whose frequency changed to allow for the motion of the antenna as Earth rotated on its axis. On 13 April, Voyager 2 finally acknowledged a command from Earth. When it was discovered that not every command was being received successfully, it was realized that the receiver's central frequency (in the middle of its narrow bandwidth) varied with the temperature of the hardware – which varied with the orientation of the spacecraft and the distance from the Sun and nearby planets – but varied mostly with changes in the electrical loads as equipment was switched on and off. As the receiver emergency was slowly brought under control, the spacecraft was coaxed through a critical course correction on 3 May to target the Jupiter flyby. The mission would continue, but maintaining contact with the ailing spacecraft would be considerably more difficult than intended. Transmissions to Voyager 2 would have to take into account every possible cause for frequency shifts, including the relative velocity of Earth and the spacecraft, and the axial rotation of the Earth, which alone could induce a shift more than 30 times the bandwidth of the defective receiver. JPL developed a detailed thermal model of the spacecraft and its systems in order to be able to predict the temperature of the receiver to within 0.1°C. The actual temperature and electrical loads were monitored on an ongoing basis, and the central frequency was periodically checked. Nevertheless, communications were still lost from time to time.[50,51] A major milestone in resolving the issue was achieved in October, when the spacecraft's memory was filled with sequences of commands that would enable it to conduct a completely autonomous 'minimum mission' at Jupiter and Saturn in the event of its remaining receiver failing permanently. Also, following a rehearsal between 12 and 14 December 1978 that included the full set of platform motions of the Jupiter flyby sequence, the scare with Voyager 1's scan platform was declared to be over.

Meanwhile, both spacecraft had been reporting data on the solar wind and the interplanetary environment, and making infrared and ultraviolet observations of the sky. They also passed unscathed through the asteroid belt, but unlike their Pioneer predecessors they were not equipped to make any particular observation of this part of the solar system.[52]

JUPITER: RING, NEW MOONS AND VOLCANOES!

In June 1978, when it was 9 months from Jupiter, Voyager 1 started to take images and radio astronomy measurements. The pictures showed a considerable amount of detail in the atmosphere, but were still of lesser resolution than the best images taken by the terrestrial telescopes monitoring the long-term variations in the visible and infrared in support of the mission. By December, however, the spacecraft's pictures surpassed the view from Earth, and they were scrutinized to identify the atmospheric features best suited for study during the forthcoming encounter, the targets for which would have to be programmed in advance.

On 4 January 1979 Voyager 1 switched from its 'cruise phase' to its 'observatory

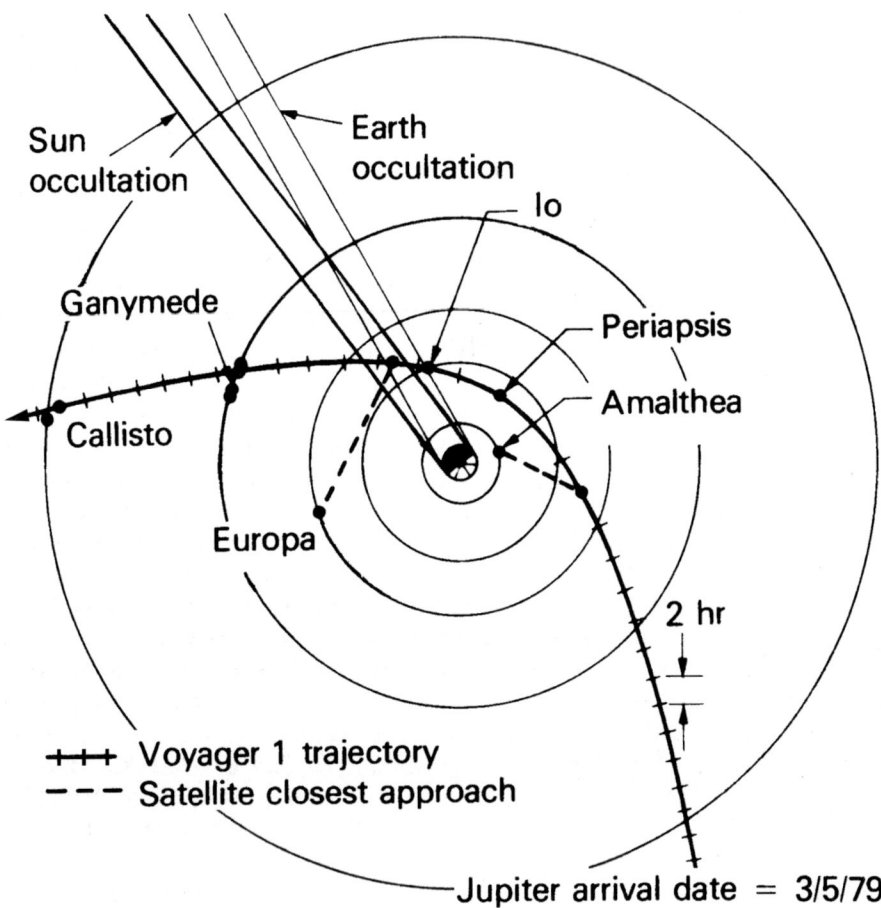

The trajectory of Voyager 1 through the inner Jovian system, as seen from above the planet's north pole.

phase', during which it would routinely monitor Jupiter as it closed in on the planet. Two days later, it began to snap a picture every 2 hours using four filters to enable a color image to be produced. The atmosphere had changed somewhat since the visits of the Pioneers a few years earlier, with the South Equatorial Belt being darker and some regions north of the equator being brighter, and was much more dynamic than expected, with striking changes occurring around the Great Red Spot. Details on the satellites were also becoming evident for the first time. On 22 January, Voyager 1 first detected streams of hot plasma escaping from the planet's magnetosphere at speeds of thousands of kilometers per second, and as it closed in over the ensuing weeks the composition of these streams began to change, with sulfur increasingly enriching the plasma of hydrogen and helium ions. Starting on 30 January, the spacecraft took a picture every 96 seconds for a duration of 100 hours for a color movie that documented 10 rotations of the planet. Another time-lapse movie that was taken that month consisted of a color picture every local day, centered on the Great Red Spot in order to study its circulation. This sequence showed many small white spots approaching the Great Red Spot from the east, circling around it once or twice at speeds of about 100 m/s, with each circuit taking between 6 and 7 days, before being sheared, with one half being swallowed by the storm and the other escaping. This sequence strikingly showed the anticyclonic motion of the Great Red Spot – an indication that it was a high-pressure structure that towered high above its surroundings. The three white ovals located south of the Great Red Spot that had first appeared in the 1930s also showed anticyclonic motion, with hints of an internal spiral flow. At least four elongated brown spots (dubbed barges) were located near the southern edge of the Northern Equatorial Belt, along the line of shear between the westward flow to the north and the eastward flow to the south. Infrared measurements indicated that these were 'warm', which was interpreted to mean that they were breaks in the cloud cover that provided a view of the deeper regions. Even before reaching the planet, Voyager 1 had established that although the atmospheric circulation appeared to be regular on the large scale – such as when seen from Earth or in the low-resolution images from the Pioneers – it was actually chaotic at scales below 1,000 km.[53,54]

This picture of Jupiter and the Great Red Spot was taken by Voyager 1 on 1 February 1979, from a distance of over 30 million km. The atmosphere is already showing its tumultuous complexity.

Although Voyager 1's Jovian encounter coincided with some of the busiest times of the Pioneer Venus Orbiter mission, the two planets were on opposite sides of the sky as seen from Earth for most of this period. But with one spacecraft rising as the other set, this placed such a heavy load on the DSN's 64-meter antennas that there was very little time available for routine maintenance, calibration and any other downtime.[55] By February, Voyager 1 was so close to Jupiter that the disk spilled over the image frame, and first 2 × 2 and then 3 × 3 mosaics were needed to fully document it. On 10 February it crossed the orbit of Sinope, a small satellite orbiting 23 million km from the planet. Three days later, it returned one of the most impressive images ever taken by a planetary mission. This portrayed the southern hemisphere of Jupiter, 20 million km away, vividly showing the Great Red Spot and, in transit, the reddish disk of Io with several indistinct bright white and yellow markings. Nearby could be seen the disk of Europa, white and virtually featureless. Meanwhile, the other instruments were making their own observations. The ultraviolet and infrared spectrometers in particular, scanned the planet's disk and the surrounding space several times per day. The particles and fields scientists were eagerly awaiting the vehicle's crossing of the bow shock, and the subsequent penetration of the magnetosphere. By mid-February, Voyager 1 began to take long-range images of the four Galilean satellites, following each as it traveled its orbit. One of the factors in selecting the encounter trajectory had been to provide fairly close flybys of each of the large satellites. The axial rotations of these moons were synchronized with their orbits. It would be possible to provide good close-up coverage of the innermost satellites, but not the more slowly orbiting outer ones, and these long-range views would fill in the gaps to some extent. From 60 until 12 days before the flyby, the spacecraft took a total of 9,300 pictures. The ultraviolet spectrometer discovered strong emissions originating from unexpectedly spectacular auroral displays at the planet's poles. Meanwhile, the plasma wave and the radio astronomy instruments made observations of charged-particle radiation and very-low-frequency radio bursts that seemed to be correlated with Io's motion.

On 26 February Voyager 1 closed to within 100 Rj of the planet, the approximate location of the bow shock during the Pioneer encounters, but there was no trace of it. Solar activity was more intense than it had been in 1973–1974, and the increased pressure of the solar wind had compressed the planet's magnetosphere. At this point, the ultraviolet spectrometer team added another piece to the puzzle that was Io. They reported that immediately after they had first aimed their instrument at the planet the previous year from a range of 1 AU they had detected a prominent torus of doubly ionized sulfur and oxygen that occupied the orbit of Io and overlapped the already discovered sodium torus. With the encounter imminent, more and more members of the press and television crews arrived at JPL to cover the event.[56] The much-anticipated crossing of the bow shock finally occurred on 28 February, at a distance of 86 Rj. As revealed by the particles and fields instruments on Voyager 2, now several months behind, the solar wind was quite gusty. With the magnetosphere rapidly expanding and contracting, the bow shock washed back and forth over Voyager 1 four times over the next 2 days, until the spacecraft was at a distance of about 56 Rj, half that at the time of the Pioneers. The fact that the magnetosphere

A spectacular picture of Jupiter, Io (hovering over the Great Red Spot) and Europa taken on 13 February from a distance of 20 million km.

was so confined prompted concern that radiation belts around the planet would be so intense as to damage the electronics and other components of the vehicle in spite of the heavy shielding and built-in safety margins.

On 3 March the magnetopause washed outward over Voyager 1 for the final time at 47 Rj and the spacecraft entered the relatively calm magnetosphere. Several hours later it crossed the orbit of Callisto, the outermost of the Galilean satellites, but the moon itself was not present; the closest approach to Callisto was to occur on the outbound crossing. So far the best imaging resolution of any of the large moons was 200 km per pixel, which was not yet sufficient to allow any significant inferences to be made about their geology; this resolution being roughly equivalent to viewing the Moon with the naked eye. However, as the spacecraft continued its approach a number of circular features became evident on Io that looked as if they might be impact craters, and there was also an intriguing heart-shaped dark feature 1,000 km in diameter that bore no resemblance to anything found in two decades of solar system exploration. The orbit of Ganymede was crossed on 4 March, but again the

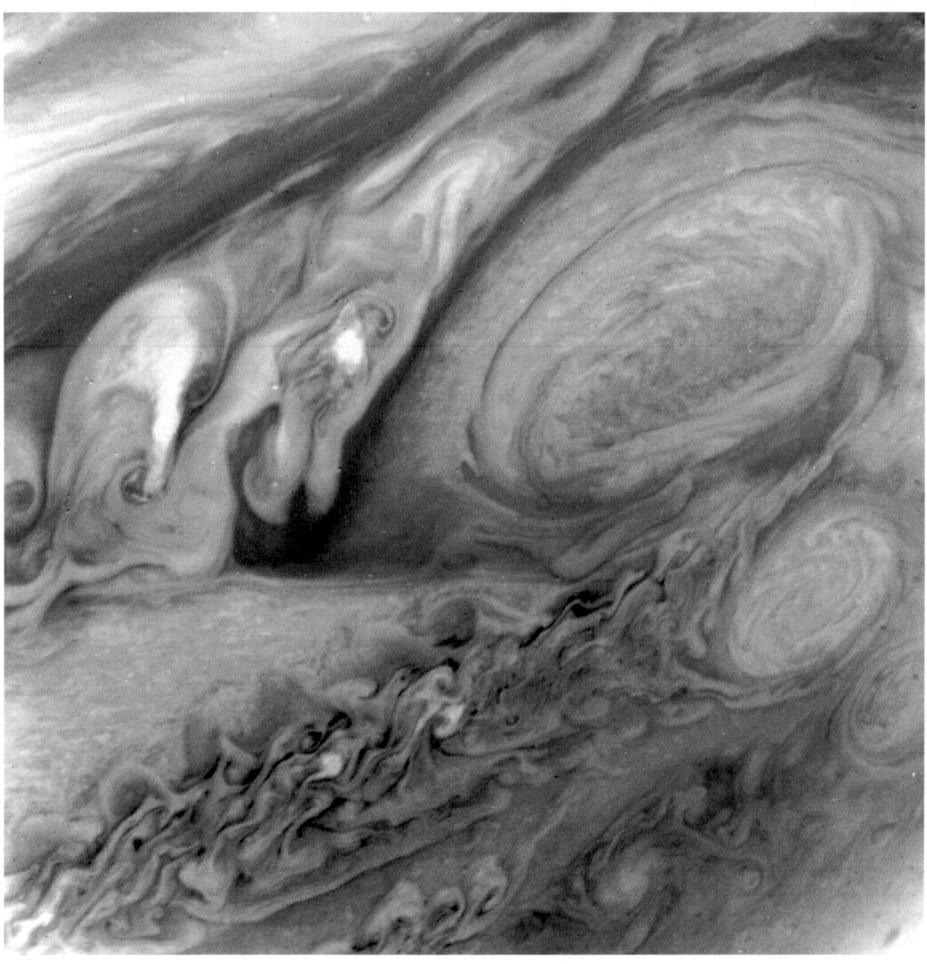

Seen through the orange filter on 1 March from a distance of 5 million km the Great Red Spot displays a wealth of details. The smallest features that can be resolved are about 100 km across. Note the turbulent region to the left of the Great Red Spot and one of the white ovals immediately to its south.

closest approach to this moon was to occur on the outbound leg. Next the spacecraft flew by Europa at a range of 2 million km, taking the best pictures of this moon of the whole encounter. Its surface was mostly white and bland, and the absence of craters at a resolution of 40 km per pixel suggested that it was relatively young. An intricate network of dark streaks that ran for thousands of kilometers appeared to criss-cross the entire moon. But these images were only a teaser, since Voyager 2's trajectory would pass much closer to Europa. Images of Amalthea, the innermost moon then known, were also obtained. Later that day, Voyager 1 returned a series of filtered images of Io from a range of 860,000 km that had a resolution of 16 km per

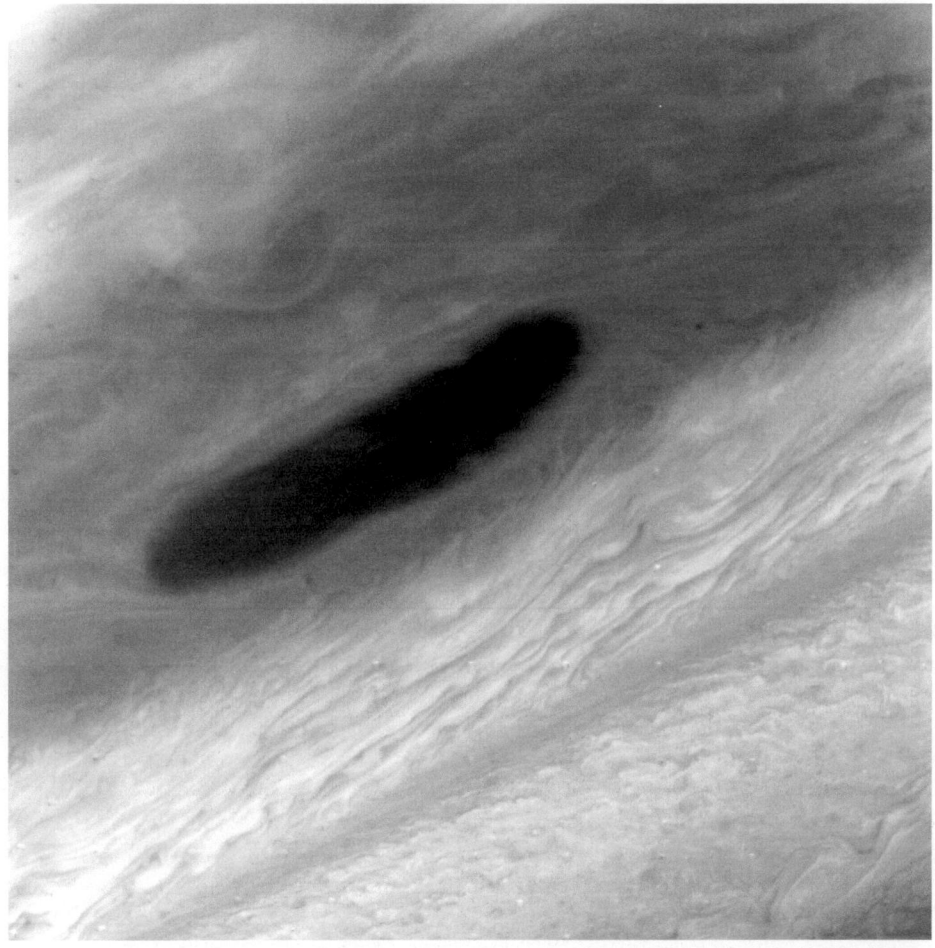

This picture taken on 2 March from a distance of 4 million km shows one of the large brown ovals in the northern equatorial belt. It is probably a hole in the upper layer of ammonia cloud that provides a view of the darker, chemically different and warmer atmosphere below.

pixel. These were assembled into a color rendition that prompted the name of the 'pizza picture', as the garish red, orange and whitish splotches made Io look remarkably like a pizza – at least by American pizza standards. It was far from evident what most of these features were. While none of the circular features was recognizable as an impact crater, there were some intriguing dark circular objects. As the spacecraft continued to close in on this enigmatic moon, each new image was eagerly scrutinized.

Voyager 1 passed through the equatorial plane of the Jovian system 16 hours 52 minutes prior to the scheduled closest approach to the planet. The possibility that there might be a ring around Jupiter was apparently first considered by the Soviet astronomer S.K. Vsekhsvyatskii, who theorized in the early 1960s that volcanoes on the Galilean moons would throw out a large amount of debris, some of which would create a family of comets bound to Jupiter, while the remainder would form a ring around it. An anomalous dip in the charged-particle counts at a particular distance from Jupiter during the Pioneer flybys had hinted that the particles were being absorbed by an unknown object – either a ring or a moon. The two Pioneer scientists responsible for this discovery suggested that as Voyager 1 passed through the equatorial plane it should aim its cameras at the region where this ring would reside, to view it edge on. In addition to the trails of many stars in the background (the field of view happened to include M44, the Beehive star cluster, in the background) the 11-minute exposure by the narrow-angle camera viewing between the limb of the planet and the orbit of Amalthea clearly showed a bright line in the equatorial plane corresponding to a ring some 30 km in thickness that extended out to an altitude of about 57,000 km above the cloud tops, or half-way to Amalthea's orbit. A wide-angle image had been taken at the same time, but the radiation had 'fogged' it completely white. It was decided to revise Voyager 2's encounter sequence to take additional pictures in order to enable the full extent of the ring to be measured. This discovery made Jupiter only the third planet known to possess a ring system, because 2 years earlier a telescopic study had found a ring around Uranus, albeit a rather unspectacular one. Ironically, it is still not clear whether the ring around Jupiter is sufficient to account for the depletion of charged particles noted by the Pioneers, and could be just a remarkable coincidence. However, while Vsekhsvyatskii had correctly predicted the existence of the ring and of volcanoes (at least in the case of Io), his actual reasoning was erroneous.[57]

Voyager 1's trajectory took it within 420,200 km of Amalthea, providing the first opportunity to inspect a body intermediate in size between the tiny moons of Mars and bodies like the Moon, Jupiter's Galilean satellites and the planet Mercury. Very little was known of Amalthea apart from its orbit. It was revealed to be an elongated object some 270 km in length and 160 km across, with a distinctly reddish hue. Even at the highest resolution of 8 km per pixel, few details were evident. However, there was a depression that might be an impact crater, and some enigmatic white patches. The infrared spectrometer found it to be slightly warmer than it should be if it were just heated by the Sun and by Jupiter, possibly indicating that it draws energy from the radiation belts.[58,59] When pictures of Jupiter taken shortly after the Amalthea sequence were scrutinized a year later, a tiny dark spot was found to correspond to a silhouette view of a hitherto unknown satellite about 80 km across. Pictures taken by chance on the previous orbit showed both the object and its shadow. Initially labeled 1979J2 by the International Astronomical Union, and later named Thebe, this moon was traveling between the orbits of Amalthea and Io.[60] Another moon was found in pictures of Jupiter taken 20 hours prior to closest approach. It was initially mistaken for Adrastea (a small moon that would be discovered by Voyager 2) but turned out to be the innermost of the retinue, traveling more than 50,000 km inside the orbit of Amalthea. Initially designated 1979J3, and later named Metis, it was about 40 km in size.[61] After crossing inside the orbit of Io at 12:05 UTC on 5 March Voyager 1

The bright diagonal streak crossing this long-exposure picture taken by Voyager 1 starting at 19:12:36 GMT on 4 March is the thin Jovian ring; this is the first picture ever to show it.

made its closest approach at a planetocentric distance of 348,890 km, snapping a picture every 48 seconds as it passed some 280,000 km above the cloud tops. The gravity-assist transformed the spacecraft's eccentric solar orbit into a hyperbolic trajectory with an eccentricity of 2.3 that would take it to Saturn.

Orbiting in the most intense part of the radiation belt, Io was 'linked' to the planet by electrical currents of 1 million amperes which had been dubbed 'flux tubes'. The mission planners had decided that Voyager 1 should fly through this structure, even though to do so would rule out a later Ganymede occultation. Although the crossing was detected by the instruments, subsequent analyses showed that the spacecraft had missed the center of the tube by a distance of 5,000 to 10,000 km.[62] Meanwhile, the spacecraft was rapidly approaching Io, catching up with it 'from behind' and taking

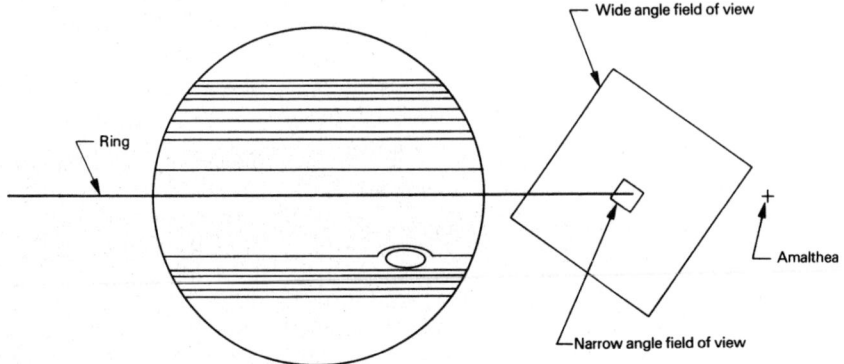

The geometry of Voyager 1's discovery image of the Jovian ring. The wide-angle frame was ruined by radiation.

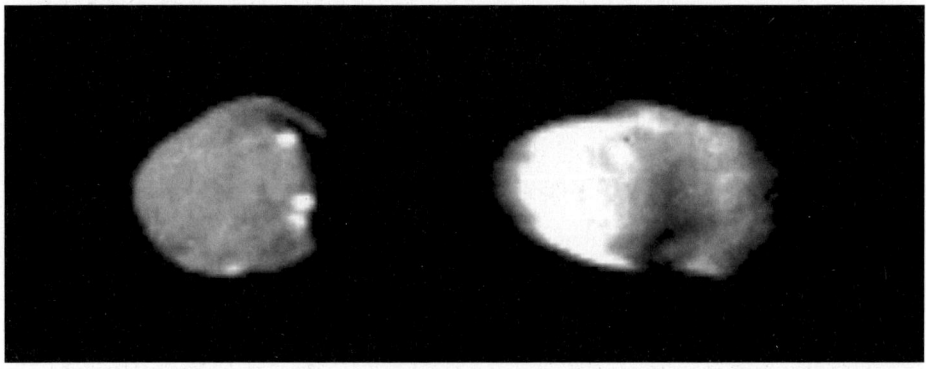

Two of the best pictures of Amalthea. The irregular body is about 270 km long and 160 km wide.

pictures of its trailing hemisphere. At 15:47 UTC the spacecraft reached its point of closest approach to Io, passing the southern polar region at an altitude of 21,000 km. This was the closest that either Voyager spacecraft would approach any moon in the Jovian system, and arranging it had been one of the scientific priorities of the entire Voyager project. Viewed in close up, Io was a complete shock. Whereas a lunar-like crater density had been expected, not a single impact crater could be discerned to the highest resolution of 600 meters, and the dark spots that had been presumed to be craters when seen at longer range were now shown to be a hitherto unknown type of feature. Some of these spots comprised a small black central dot surrounded by a patch of a somewhat lighter-hue, beyond which was a dark halo, usually fan-shaped but often butterfly- or heart-shaped. In excess of 100 such features with diameters greater than 25 km were counted. There were also large yellowish and white areas.

A beautiful picture of Io on the limb of Jupiter taken by Voyager 1 on 4 March 1979. Some intriguing dark spots were evident on the moon's surface.

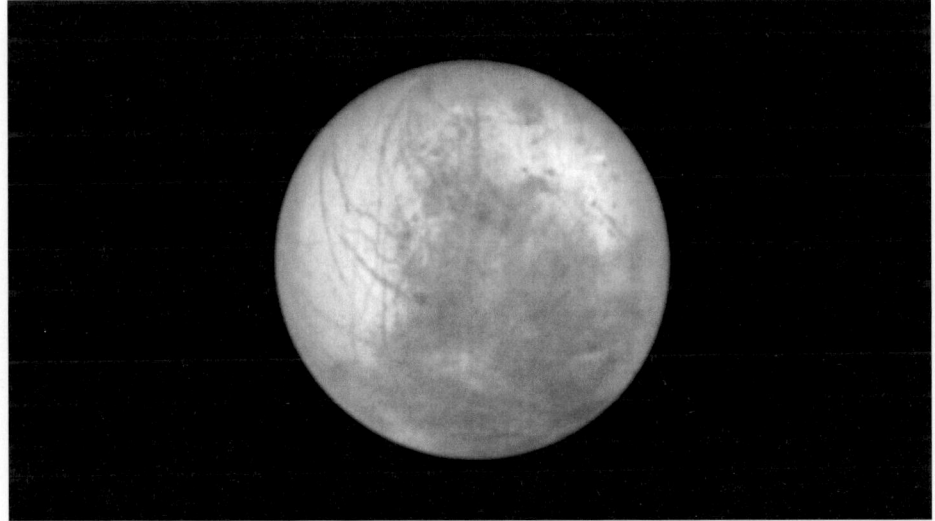

A picture taken by Voyager 1 of Europa on 4 March from a distance of 2 million km, which was about as close as the spacecraft came to this enigmatic moon. Despite the low resolution, some darks streaks are visible criss-crossing the moon's surface.

Evidently, either Io had somehow been shielded from impacts – which contradicted the reasoning that because the population of impactors would have been accelerated by Jupiter and would have made very violent impacts, Io must be the most intensely battered of the Galileans – or it had indeed suffered impacts and had been resurfaced within the last 100 million years. The presence of what appeared to be sinuous fluid flows running out from some of the dark spots in the best imagery suggested that any resurfacing was the result of volcanic activity. If so, then the dark spots were calderas similar to that of Yellowstone Park in Wyoming, although on a much larger scale. Layered terrain near the south pole looked as if it had been deposited by a succession of eruptions. Although the surface was remarkably flat in general, there were also mountains that suggested the crust had cracked and large blocks had been pushed up at an angle. However, it was not immediately evident from the imagery just when this activity had taken place.[63,64,65]

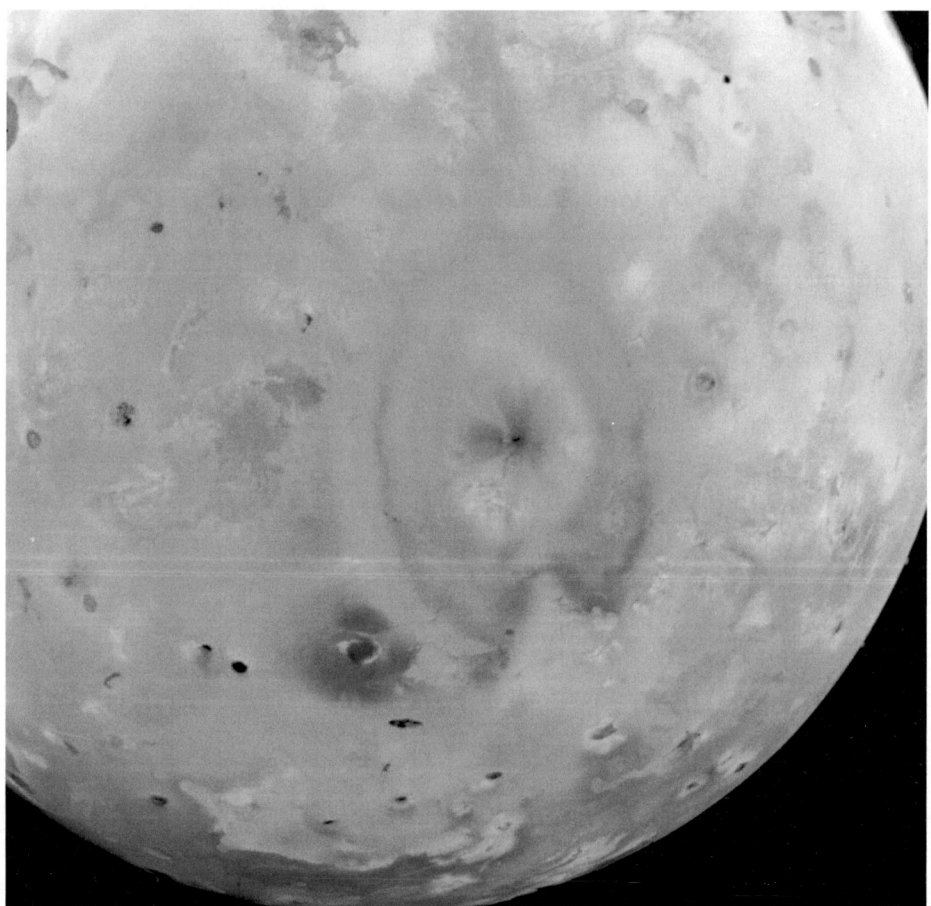

A picture of Io taken by Voyager 1 on 5 March from a range of about 400,000 km. The large heart-shaped feature puzzled scientists for several days, until it became evident that it is 'fallout' from a tall plume issued by the volcano Pele at its center.

Less than 30 minutes after Io, Voyager 1 passed behind the limb of Jupiter as seen from Earth. The spacecraft had switched its transmitter to a strong carrier tone, and it slewed around at a rate calculated to maintain Earth in the beam of its high-gain antenna as the signal was refracted by the Jovian atmosphere, with the manner in which this occurred yielding data on the ionosphere and upper atmosphere. Some 53 minutes into the 126-minute radio occultation, it entered the planet's shadow and the instruments scanned the limb to measure the selective absorption of sunlight in order to provide further information on the composition of the atmosphere and to derive a temperature profile independently of the radio-occultation experiment, with the data being recorded on tape for subsequent replay.

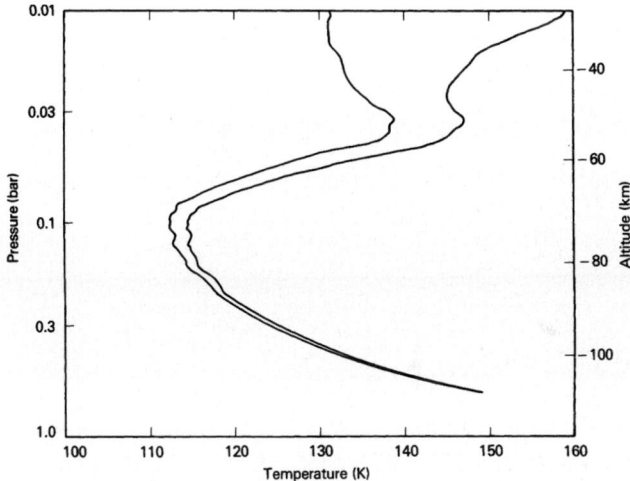

Monitoring the radio signal from Voyager 1 as it passed behind and re-emerged from Jupiter's disk as viewed from Earth provided temperature and pressure profiles of the Jovian atmosphere. The two profiles represent different 'boundary conditions' of the same data.

While the spacecraft was over the night-side of Jupiter the infrared spectrometer measured atmospheric temperatures, and the cameras took long-exposure images that would reveal a lot of lightning activity illuminating the clouds and the presence of a ghostly double aurora near the north pole. There were even bright trails created by meteors burning up in the atmosphere at speeds of 60 km/s. The radio-astronomy instrument detected bursts of radio from lightning discharges from storms deeper in the atmosphere. In view of an experiment in the 1950s by Stanley Miller, a student of Harold Urey, in which prebiotic molecules were synthesized simply by subjecting a mixture of methane, water, hydrogen and ammonia to repeated electric discharges, the presence of lightning fueled speculation that a form of life may have developed to float in Jupiter's atmosphere.[66,67,68]

The night-side of Jupiter imaged by Voyager 1 while in the planet's shadow. The arc on the left is the glow of the Jovian aurora, while intense flashes of lightning illuminate the clouds at right.

While out of contact with Earth, Voyager 1 passed within 733,760 km of Europa, but the view was of the night-side of the moon and no images were taken. Shortly after the radio occultation ended, the spacecraft exited the planet's shadow. After the Pioneers had produced questionable atmospheric profiles, the Voyager mission team had decided to use a dual-frequency system, a greater power, and an oscillator that was better protected from radiation. When Voyager 1 was occulted, it was traveling almost perpendicular to the limb at 12°S, at a longitude where the Sun was about to set, and it emerged at 1°N shortly before sunrise. As hoped, the data provided good temperature profiles. It also provided profiles of the concentration of electrons in the ionosphere at sunset and sunrise that were very different from those of the Pioneers, possibly owing to the change in solar activity over the intervening years. Both prior to and after the occultation by Jupiter, the spacecraft's line of sight to Earth passed through the torus that occupies Io's orbit, and the manner in which this affected the radio signal provided a measurement of the density of the plasma inside the torus. In theory, during the low-latitude egress

A medium-resolution picture of Ganymede showing dark polygonal terrain, linear grooves and bright crater ejecta.

from behind the planet the line of sight should have also crossed the Jovian ring but there was no evidence of a 'dip' in the radio signal to indicate this.[69,70] The hours that followed the occultations were devoted to the task of downloading the stored data.

Early on 6 March, Voyager 1 passed within 114,710 km of Ganymede, the largest member of Jupiter's retinue. Many of its craters were surrounded by white splotches of material that the impacts had excavated from the mostly icy surface. There were also a number of white streaks that were suggestive of fault lines where the crust had been displaced horizontally. On the larger scale the surface appeared to comprise two types of terrain: there were dark patches, often polygonal in shape and evidently ancient, and there were brighter grooved areas that seemed to span a

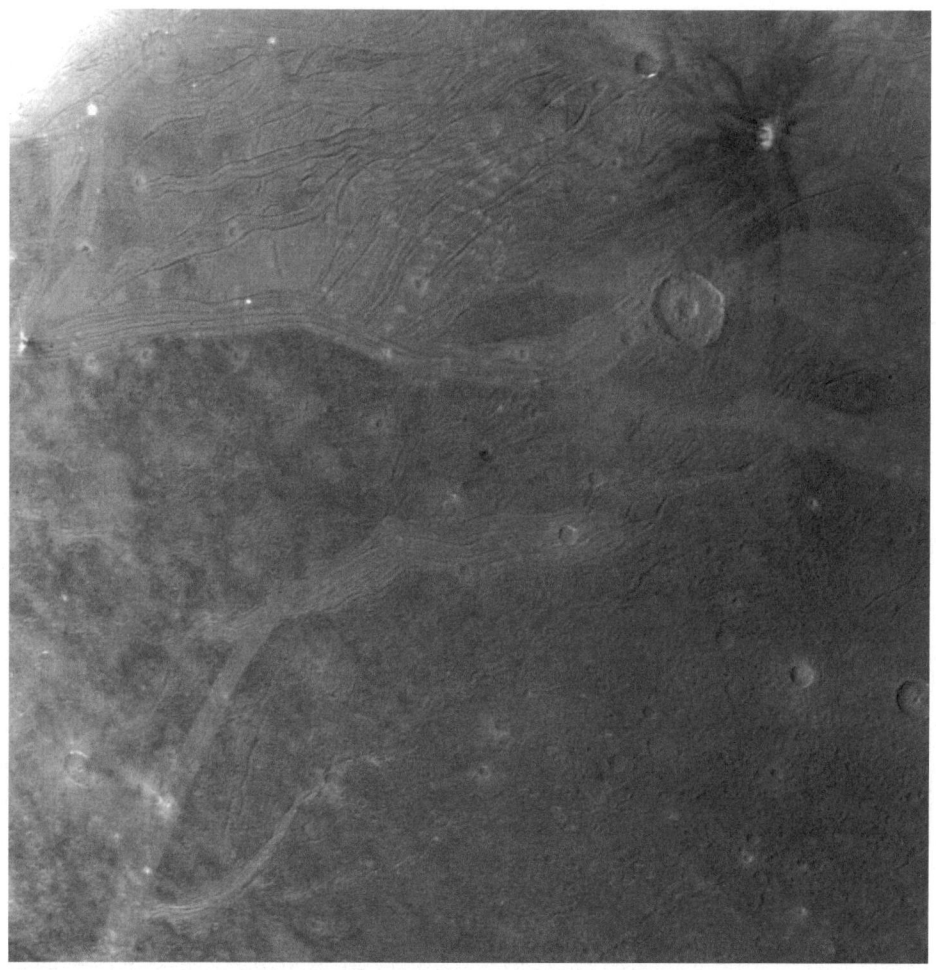

A closer look at the grooved terrain on Ganymede.

variety of ages. The bright areas (dubbed sulci) were made up of parallel ridges 10 to 15 km wide and standing about 1,000 meters tall that seemed to mark the highest elevations on the entire surface.[71,72] Near the time of closest approach, the ultraviolet spectrometer observed for several minutes as Ganymede occulted the ultraviolet-bright star kappa Centauri, and the data put an upper limit on the surface pressure of any envelope that the moon might possess.[73]

Thirteen hours after Ganymede, Voyager 1 passed within 126,400 km of Callisto, revealing it to be particularly heavily cratered, including a remarkable impact basin (later named Valhalla) that comprised a central bright area some 600 km in diameter and a series of concentric rings extending out 1,500 km from the center, like ripples. For some time this was believed to be the largest such feature in the entire solar system, but subsequent research would transfer this status to the Aitken basin in the southern region of the far side of our own Moon. A remarkable fact about the rings of Valhalla was that they did not show any relief – in fact, there was very little relief on the moon. The icy material had evidently been sufficiently mobile to reduce elevations and fill in depressions. The solid-state flow had reduced craters to smooth circular spots known as palimpsests. A similar process was also evidently at work on Ganymede, although the activity there had broken and displaced the crust.[74,75] The ultraviolet spectrometer sought emissions by atomic species excited by sunlight, but found no measurable atmosphere around Callisto.[76]

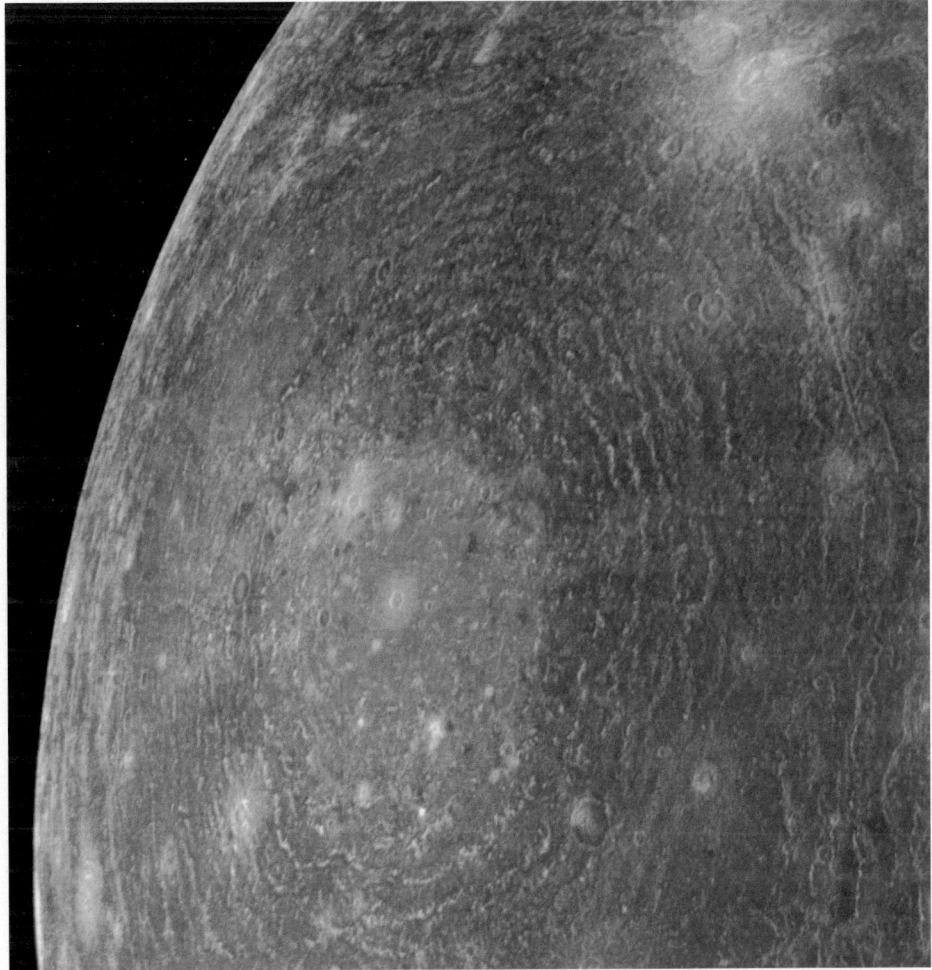

On 6 March Voyager 1 took this picture of the Valhalla impact structure on Callisto from a distance of 350,000 km. The actual basin (the light area at the center of the 'bull's-eye') is 600 km in diameter. The concentric rings extend out to a radius of 1,500 km.

By 8 March, the encounter was winding down and the press and the science teams were both beginning to depart JPL, the latter with data that would require years to fully analyze and appreciate. However, it was on that day that the greatest discovery of the encounter, and very possibly of the entire project, originated. Voyager 1 was commanded to turn toward Io, now 4.5 million km away, and take a long-exposure in which the moon would be an overexposed crescent set against the background of stars. The object of the exercise was to measure the moon's position with respect to as many stars as possible, to enable the navigators to pinpoint the current position of the spacecraft in space in order to calculate precisely the trajectory that it had flown through the Jovian system, and hence determine the accuracy of the gravity-assist for Saturn. The picture was studied by navigation engineer Linda Morabito, who had to identify as many stars as possible. A 'bump' that projected out from Io's limb gave every appearance of being a view of a distant satellite that was partly masked by the disk of Io, but she knew that there could not have been a moon in that position. Intrigued, over the ensuing hours and days she consulted other members of her team, and finally a member of the imaging team, who announced that it was an umbrella-shaped plume of gas and dust from a volcano located on the limb. For the first time, an active volcano had been discovered on another solar system body! Furthermore, with this insight, a bright spot located on the terminator in the image was interpreted as another volcanic plume rising high enough to catch the Sun.

Over the next few days, scientists analyzed the earlier images of Io, and located a total of eight active volcanoes. The International Astronomical Union later decided to name Io's volcanoes after mythological associations with fire and eruptions, and to draw on a variety of cultures for inspiration. On the 8 March picture, the volcano on the limb was named Pele after the Hawaiian goddess of fire, and the one on the terminator was named Loki after the Norse blacksmith god. Further analysis showed that the Pele vent was sited at the center of the 1,000-km-wide heart-shaped feature that had so puzzled scientists in the earlier pictures, which was now realized to be a deposit of material that had been erupted and pursued a ballistic arc to fall far from the vent.[77,78] One volcano was also detected by chance by the infrared spectrometer, when it noted a temperature that was some 150°C higher than the surrounding terrain. Correlating the infrared data with imagery showed the 'hot spot' to be a dark fissure immediately north of Loki's caldera. Initially, there was some consternation because the measured temperature was also almost 100°C too low for liquid sulfur. But the measured temperature was the average of the sensor's field of view and, because the surface was generally frozen, the site of activity might well have been much hotter – perhaps even hot enough for the volcanic fluid to be silicate. The infrared data showed there to be many smaller 'hot spots' that could be small volcanoes, cooling flows of material, or sites of subterranean activity.[79]

The discovery that there were volcanoes on Io provided the final piece of a jigsaw that had been building up over the years. Specifically, it explained the source of the sulfur and oxygen in Jovian space: the volcanic plumes were rich in sulfur dioxide. In fact, although most of the material erupted by Io falls back to the surface, a small

This *un*remarkable image taken at 13:28:25 UTC on 9 March 1979 triggered the most important Voyager discovery in the Jovian system. Taken for navigational purposes to measure the position of Io against the background of stars (some of which are evident in the picture) it also recorded two volcanic plumes on the satellite: an umbrella shaped one on the limb and another catching the Sun high above the terminator. The inset identifies the sites of these plumes on a close up image of Io taken several days earlier as being the volcanoes that were later named Pele and Loki.

part (possibly less than 0.1 per cent) escapes and, on becoming electrically charged by the particles circulating in the radiation belts, is 'picked up' by Jupiter's magnetic field and joins the torus of material that occupies Io's orbit.[80] Sulfur might also be responsible for the yellow hue of the moon's surface. Indeed, sulfur dioxide – a gas commonly emitted by terrestrial volcanoes and thanks to spectroscopic observations already known to be on Io – might be responsible for the white deposits. Sulfur and salts such as sulfates, nitrates, carbonates and various other volcanic products might

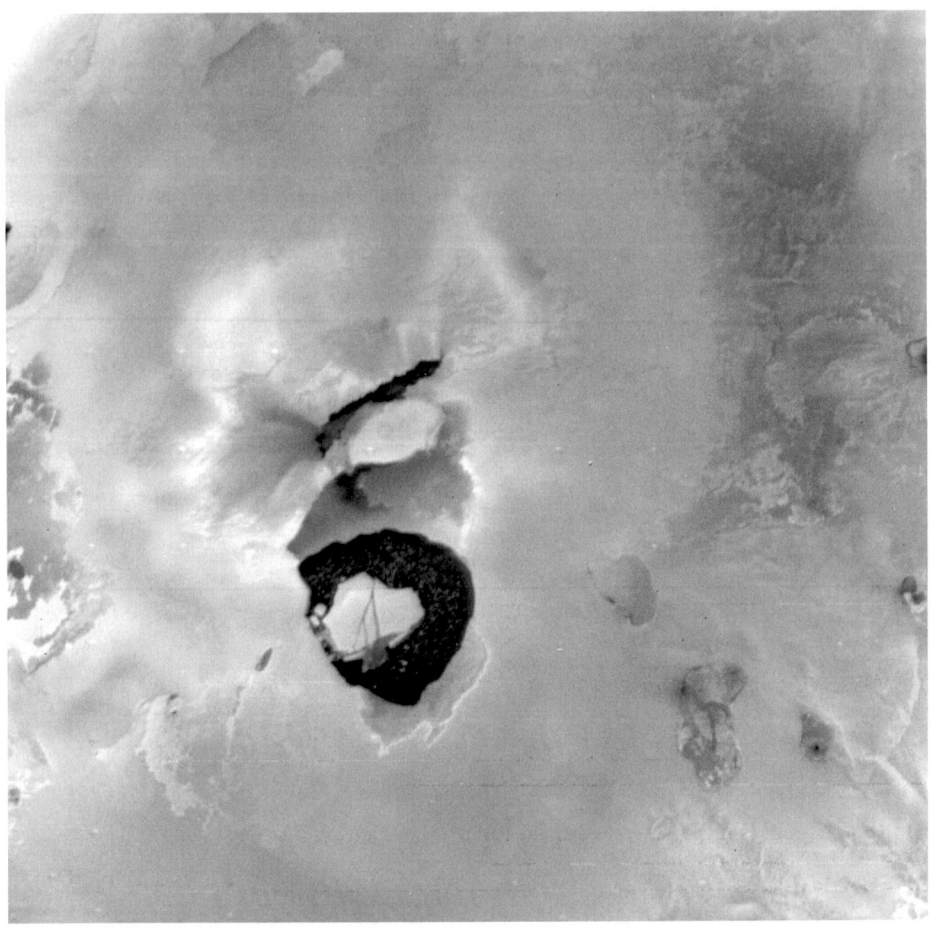

A close up of the lava lake in the caldera of Loki taken by Voyager 1. Remarkably, there are 'icebergs' afloat. The source of the plume was located in the dark streak to the north, which also displays a butterfly pattern of fallout.

cause some of the features found in Io's spectrum. Intriguingly, blue–white patches in calderas that were evident in some pictures but not in pictures taken hours earlier, might be fresh clouds of sulfur dioxide 'snow'.[81,82] Remarkably, just 1 week before the 8 March navigation picture was taken, the journal *Science* had published a paper in which three researchers posited that gravitational tides caused by Jupiter and the other Galilean satellites that maintained the ellipticity of Io's orbit might deform the moon sufficiently to melt its interior, and they speculated that Voyager 1 might find widespread volcanism. It is also possible that the currents that flow in the flux tubes that connect Io to Jupiter contribute to heating the moon's interior. The active vents were releasing a tremendous amount of heat to space – indeed, the overall heat flow was much greater than in the case of Earth. An analysis that correlated the data from

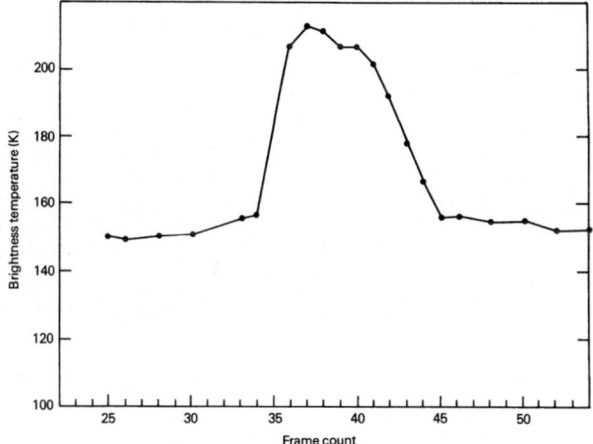

Voyager 1's infrared spectrometer and radiometer scanned across the Loki caldera, recording the variation in temperature.

other instruments with the imagery concluded that if the current activity was typical, then the volcanoes could deposit a blanket of material 10 to 100 meters thick over the entire moon every 1 million years, which explained why there were no surviving impact craters.[83,84]

The scientific results from Voyager 1's Jovian flyby had been amazing, with most instruments returning an unprecedented amount of data on a multitude of bodies that often revealed unexpected characteristics. It had been an impressive follow-up to the Pioneer reconnaissance. In a sense, too, it had been like exploring a different solar system, since the bodies that make up the Jovian system do not resemble any of the terrestrial planets, nor each other. A total of about 18,000 pictures were returned of Jupiter, its satellites and the ring. Measurements by the ultraviolet spectrometer revealed that Io's torus extends almost 6 Rj from the planet and also 1 Rj above and below the planet's magnetic equator, in which it resides, as distinct from the plane of the moon's orbit, which matches the rotational equator. The prodigious energy required to maintain the plasma in the torus at 100,000°C was supplied by the magnetic field. Of course, the density of material in the torus was so low that a vehicle could pass through it unscathed, as Voyager 1 did twice. The ultraviolet spectrometer was also able to measure the temperature of the highest reaches of the Jovian atmosphere, and to identify some of the hydrocarbons present there. Indeed, Jupiter was found to be a strong emitter in the ultraviolet, both due to sunlight exciting molecules in the upper atmosphere and because the high-energy particles that reached the poles by traveling along the magnetic field lines stimulated auroral emissions. Interestingly, the auroras seemed to occur at the latitude at which the magnetic field lines that threaded the Io torus made contact with the atmosphere. Correlating ultraviolet observations by the Pioneer missions and Voyager 1, and those made by a series of sounding rockets and astronomical satellites during the

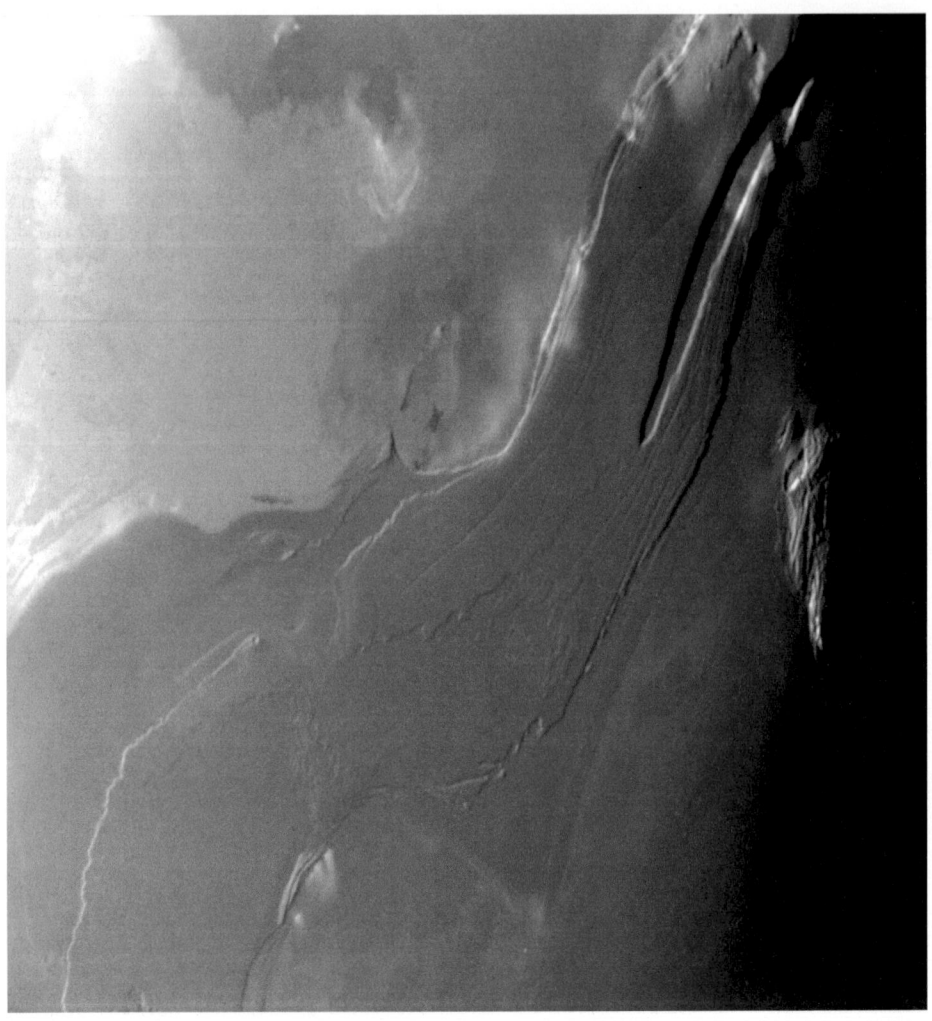

As Voyager 1 made its closest approach to Io it gained high-resolution imagery of irregular mesas and tilted massifs in the south polar region.

intervening years, it seemed that the appearance of Jupiter had changed: in 1973 the disk of the planet was much dimmer, as indeed was the plasma torus, and no auroral activity was detected.[85] The infrared spectrometer took more than 50,000 spectra. Its results for the Jovian atmosphere showed evidence of hydrogen, methane, ethane, water vapor, ammonia and a variety of compounds of carbon and phosphorus, but it was not possible to identify the 'chromofores' that were responsible for the colors of the clouds; nevertheless, it was believed that these chemicals included a variety of sulfur compounds. Significantly, the abundance of helium was confirmed to be close to the solar value. Working in its radiometer mode the instrument provided good

Linda Morabito displays the picture of Io on which she discovered the plumes from active volcanoes. (Courtesy of NASA/JPL/Caltech)

vertical profiles of the temperature in the atmosphere, as well as scans of the stratosphere and troposphere. Particular attention was paid to the Great Red Spot, which was shown to be a sort of stratospheric 'cold spot', consistent with it being a high-pressure structure that towered above the ammonia clouds.[86] Owing to the use of thrusters to control the spacecraft's attitude, which introduced relatively large trajectory disturbances, no celestial mechanics measurements were to be collected at Jupiter by either Voyager spacecraft, although they were scheduled for Saturn.[87] The Pioneers had already provided excellent data of this type.

Voyager 1 had suffered a number of radiation-induced problems, but these were manageable. The main clock and the computers had lost mutual synchronization and the cameras had fired their shutters 40 seconds before the specified time. This sometimes occurred while the scan platform was still slewing to the target, with the result that many of the pictures of Io and Ganymede were blurred, and hence useless. The filter carousel of the photopolarimeter failed 6 hours prior to closest approach, resulting in the loss of data on Jupiter's clouds and the particles within them. The instrument had three wheels: one wheel contained a series of color filters, a second was drilled with holes of various sizes to control the angular width of the field of view, and the third had polarizing filters in various orientations. It was concluded that excessive use of the wheels prior to the encounter with Jupiter had prompted them to start to function erratically, and the radiation had dealt the final blow. In addition, several hours of data had been lost without the possibility of recovery owing to rain at the various DSN stations, which attenuated the X-Band data link. Otherwise, the spacecraft departed Jupiter in a healthier state than either of its predecessors. As with approaching Jupiter, leaving it involved repeated crossings of the magnetopause and the bow shock as gusts in the solar wind caused the magnetosphere to fluctuate. In fact, the spacecraft crossed the bow shock no less than six times at distances ranging from 199 to 258 Rj before finally leaving the Jovian environment behind on 20 March. On 13 April the encounter sequence was deemed to be complete, thereby concluding a 98-day period of almost continuous observations. There was little respite for the engineers and scientists, however, for even as Voyager 1 left the stage, its sister probe was heading in.[88]

THE RETURN TO JUPITER: LIFE, PERHAPS?

Voyager 2's Jovian encounter was to be similar to that of its sister, but in reverse: whereas Voyager 1 had made close passes by the outer Galilean satellites on its way out of the system, Voyager 2 would visit them on the way in. This strategy derived from the fact that the rotation of each satellite was synchronized with its orbit, and catching them on opposite sides of their orbits would enable the pair of spacecraft to map most of the surface of each moon at a useful resolution. And since Europa had previously been poorly presented, a close flyby had been arranged this time. Overall, however, Voyager 2's trajectory was not optimized for Jovian science but rather to keep the J–S–U–N option open. But there had been last-minute changes. Voyager 1 had paid particular attention to Io, and although this moon had not featured highly in the plan for Voyager 2, the discovery of volcanoes prompted the addition of several long-range observations to search for active volcanoes, and a 'volcano watch' when the spacecraft was near to the moon. Also added were spectrometry and imagery of Jupiter's night-hemisphere and auroras, and a series of pictures to be taken of the ring in order to determine its size. Several engineering improvements were made. As Voyager 2's trajectory would not pass inside the orbit of Europa, the spacecraft was not expected to receive as much radiation as its predecessor. Nevertheless, it was told to resynchronize its main clock and computer timers every hour in order to minimize any radiation-induced time-reference drift. Other than the known problems affecting its command receiver that would inevitably reduce the scientific output, the only major hardware issue was that the photopolarimeter wheel had stuck, with the result that only color polarimetry could be taken.[89]

Just 6 weeks after Voyager 1 made its final observation, Voyager 2 began its own, and was soon yielding results: for example, whereas Voyager 1 had not encountered streams of plasma until it had closed to 600 Rj, Voyager 2 began to detected them at 800 Rj. The particles and fields instruments sought evidence of interactions between the solar wind and the Jovian magnetosphere. A color picture was taken of Jupiter at each turn in order to make a time-lapse movie with the field of view centered on the Great Red Spot. On completing this movie on 27 May, the spacecraft spent the next 2 days taking a movie covering five rotations of the planet. Meanwhile, telescopic monitoring from Earth was providing a context to assist in targeting the Voyager 2 observations. NASA had built a 3-meter-diameter infrared telescope on the summit of Mauna Kea in Hawaii specifically to make observations to support the Voyager project. Although this was unavailable for Voyager 1, it entered service just in time for Voyager 2's arrival.[90,91] Whereas the Pioneers had found few changes on Jupiter in the year between their flybys, the atmospheric structures changed significantly in the 4 months between the Voyager encounters. In particular, the Great Red Spot had drifted in longitude and its details had changed: the broad white band situated almost tangential to its southern boundary was now just a thin ribbon; the turbulent region to the west had become stretched and was orange rather than white; and a series of white hooks were projecting from there north toward the equator. In fact, the Great Red Spot now appeared to be detached from its surroundings, in hue at least, being more uniformly reddish-orange, rather reminiscent of its appearance to the Pioneers. The displacements of 100-kilometer-scale details during several axial rotations yielded measurements of the speeds of the high-altitude winds at a wide range of latitudes, with the greatest speed of 150 m/s occurring just north and south of the equator. A comparison with the record of Jovian meteorology as documented by drawings and telescopic pictures indicated that such streams were stable over decades.[92,93]

In March the solar wind reduced to a more 'normal' state, and after a 1-month far-encounter phase mostly devoted to Jovian meteorology Voyager 2 crossed the bow shock for the first time on 2 July at almost exactly the predicted distance of 99 Rj. However, conditions were so dynamic that no less than 11 crossings were reported over the next 3 days, the final one occurring at just 67 Rj. Moreover, the shock itself was thicker: whereas it had washed over Voyager 1 in less than 1 minute, one time it took 10 minutes to wash over Voyager 2. The Jovian magnetosphere, too, was less compressed, and the last of a number of crossings of the magnetopause occurred at a distance of 62 Rj.

Meanwhile, Voyager 2 had begun to make long-range observations of Io in search of volcano plumes. However, the large concentrations of sulfur and oxygen ions that the low-energy charged-particle instrument had noted in March had declined, raising the prospect that the volcanism had subsided. Then, on 4 July, an image taken at a

Jupiter imaged by Voyager 2 on 9 May 1979 from a distance of 46 million km. Note how much the planet's atmosphere has changed, in particular around the Great Red Spot, since Voyager 1.

range of 4.7 million km (comparable to the 8 March image by Voyager 1, but from a totally different perspective) showed the presence of at least one plume at an altitude of about 200 km. As Io loomed, it was observed that the heart-shaped feature that surrounded Pele had become more elliptical, and in the process some 10,000 km^2 of Io had been resurfaced – although to what depth was impossible to say. Moreover, the Pele eruption had evidently ceased, since its plume was no longer present. It had probably been responsible for the increased concentration of ions in the system, now diminishing. On the other hand, at least three of the volcanoes that were active in March were still erupting: Prometheus, Loki and Marduk. The 'lava lake' in Loki's caldera seemed to have changed shape along its northern rim, and Loki now had a twin-columned plume. A fan-shaped deposit had appeared next to Loki, although the vent that had produced it was evidently no longer active.[94,95]

Voyager 2's first close encounter with a Galilean satellite was with Callisto, and the early views of the hemisphere that had not been visible to Voyager 1 showed the moon to be essentially saturated by impact craters. In fact, one estimate said that it had the highest crater density in the solar system. However, this hemisphere had

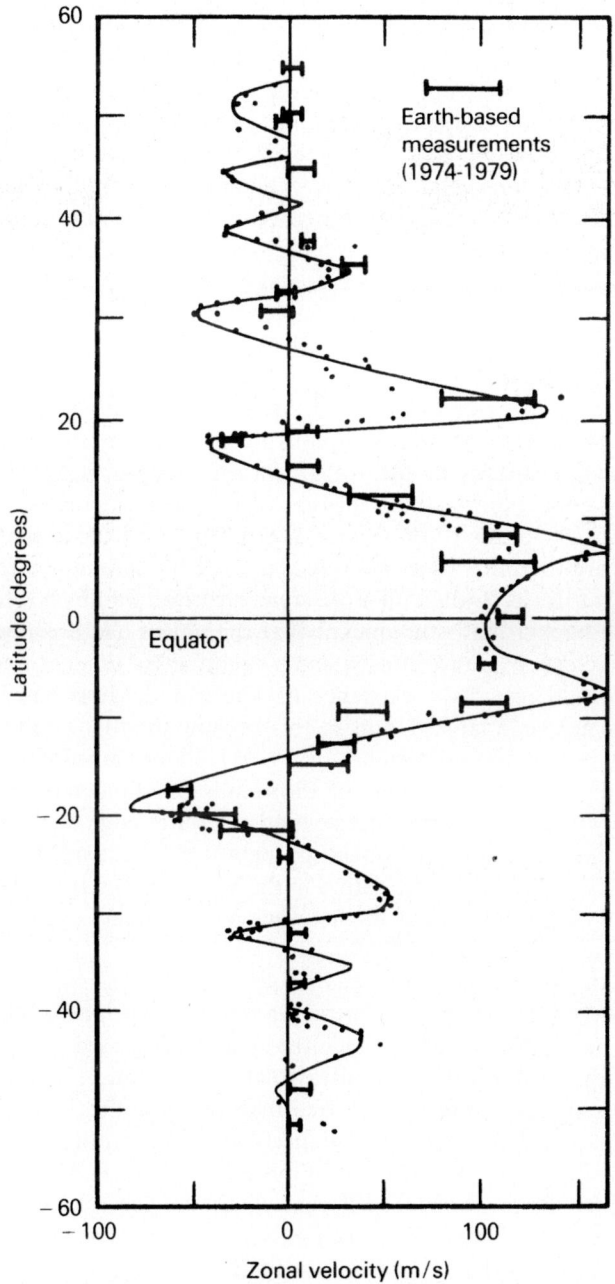

Comparing Voyager measurements and ground-based observations, scientists were able to compile this diagram of how the wind speed varies with latitude on Jupiter. The alternating jet streams manifest themselves as dark belts and bright zones.

been spared the large impacts that left multiple-ring structures. Images of Ganymede showed a 4,000-km-diameter dark feature bounded by bright grooved tracts that was later named Galileo Regio. This proved to match the location of a dark splotch that had been glimpsed by telescopic observers at times of exceptionally good 'seeing', and was also evident in the low-resolution Pioneer 10 pictures. On 8 July, Voyager 2 took some oblique-angle pictures of Jupiter's ring, revealing it to be a faint ribbon of a width of only a few thousand kilometers. Some 3 hours later, the spacecraft flew by Callisto at a range of 214,930 km, imaging it at a best resolution of 4 km per pixel. Overall, the imagery from the two missions showed there to be a global asymmetry, in that the very large craters were concentrated on the hemisphere seen by Voyager 1. The imagery gave an improved measurement of the diameter of the moon, and hence of its bulk density, the very low value of which indicated that a large fraction of the material was water ice.[96,97]

Twenty-three hours prior to its closest approach to Jupiter, Voyager 2 crossed the equatorial plane of the system at a planetocentric distance of 1 million km, and took long-exposure images of the ring from an edge-on perspective. When the pictures were studied months later, a point of light whose position did not match that of any star was present in one wide-angle frame exactly in-plane. The object had also been captured in a narrow-angle frame taken 5 minutes earlier, but in this case the fact that it was 'trailed' in a different direction and by a different number of pixels to the background stars meant that the object had a proper motion and was relatively close by, and consequently a hitherto unidentified moon. The pictures that would reveal the existence of Thebe and Metis had been taken by Voyager 1, but not yet analyzed, and so this became the first satellite to be found, as indicated by its provisional designation 1979J1. Later named Adrastea, it was 40 km in size and orbited several tens of thousands of kilometers farther out than Metis, which traveled in an orbit just beyond the outer edge of the ring; in fact, it seemed likely that there was an association between Adrastea, Metis and the ring. While no images of Adrastea were found in the Voyager 1 imagery, both Thebe and Metis were able to be identified in the Voyager 2 imagery, which enabled their orbital parameters to be refined.[98,99]

The busiest day of the encounter was 9 July, as within the space of several hours Voyager 2 made its closest approaches to Ganymede, Europa and the planet itself. Over a period of 4.5 hours during which the spacecraft passed within 62,130 km of Ganymede, it obtained infrared and ultraviolet spectra and 217 images. There were large dark ancient cratered terrains dotted by bright splashes of impact ejecta and circular palimpsests. The bright and intensely grooved terrains were suggestive of fracturing and slipping of the icy crust in more recent times. A series of concentric ridges on Galileo Regio looked as if they might be the relic of a multiple-ring structure whose central basin had been erased by the formation of the adjacent bright terrain. Between them, the two Voyagers had mapped about 80 per cent of the surfaces of Ganymede and Callisto at resolutions at least as good as 5 km per pixel. Overall, Ganymede seemed to have four main types of terrain: old dark cratered plains, grooved terrain of a variety of ages, a young impact basin with an ejecta blanket (discovered by Voyager 2 near the south pole) and young patches of

An image taken during Voyager 2's first Jovian ring-plane crossing. The ring is the faint diagonal band. The brighter star on the right is trailed more and at a different angle to the one on the left because the one on the right is actually the tiny satellite Adrastea.

smooth terrain. The Voyagers also discovered enigmatic chains of craters classified as catenae on both Ganymede and Callisto. Their origin remained a mystery until 1993, when the comet Shoemaker–Levy 9 was discovered orbiting Jupiter like a 'string of pearls' after having been shattered by tides during a close encounter with the planet. This must have happened frequently in Jupiter's history, and at times its moons must have been bombarded by trains of fragments, creating the catenae. The bulk density of Ganymede was slightly greater than that of Callisto. Finally, as the spacecraft flew by the moon the magnetometer noted disturbances in the ambient magnetic field.[100,101]

The hemisphere of Callisto viewed by Voyager 2 did not reveal any new multiple-ring structures.

A few hours after Ganymede, Voyager 2 aimed its scan platform to make the first close inspection of Europa. At a best resolution of 4 km, the images were almost a 10-fold improvement on those by its predecessor, whose trajectory had not favored this moon. Europa rivaled Io for the title of being Jupiter's most bizarre satellite. It appeared to be completely covered by an icy crust that in places had cracked like an egg, with the icy blocks remaining in place. The mostly white crust was marred by splotches of dark mottled terrain. Almost the entire surface was transected by brownish cracks that ran for hundreds, and in some cases even thousands of kilometers tracing small and great circles, often cutting across each other. There were also bright ridges that formed repeating arcs or cusps, each several hundred kilometers in length. Although there were no well-formed impact craters on view, there was a trio of small circular darks spots, one of which might be a multiple-ring structure, and a number of dark circular spots that might be craters too small to be resolved as such. The paucity of craters indicated that the Europan surface was relatively young in geological terms. The terminator indicated that the moon was almost devoid of relief, and extremely smooth, "like a billiards

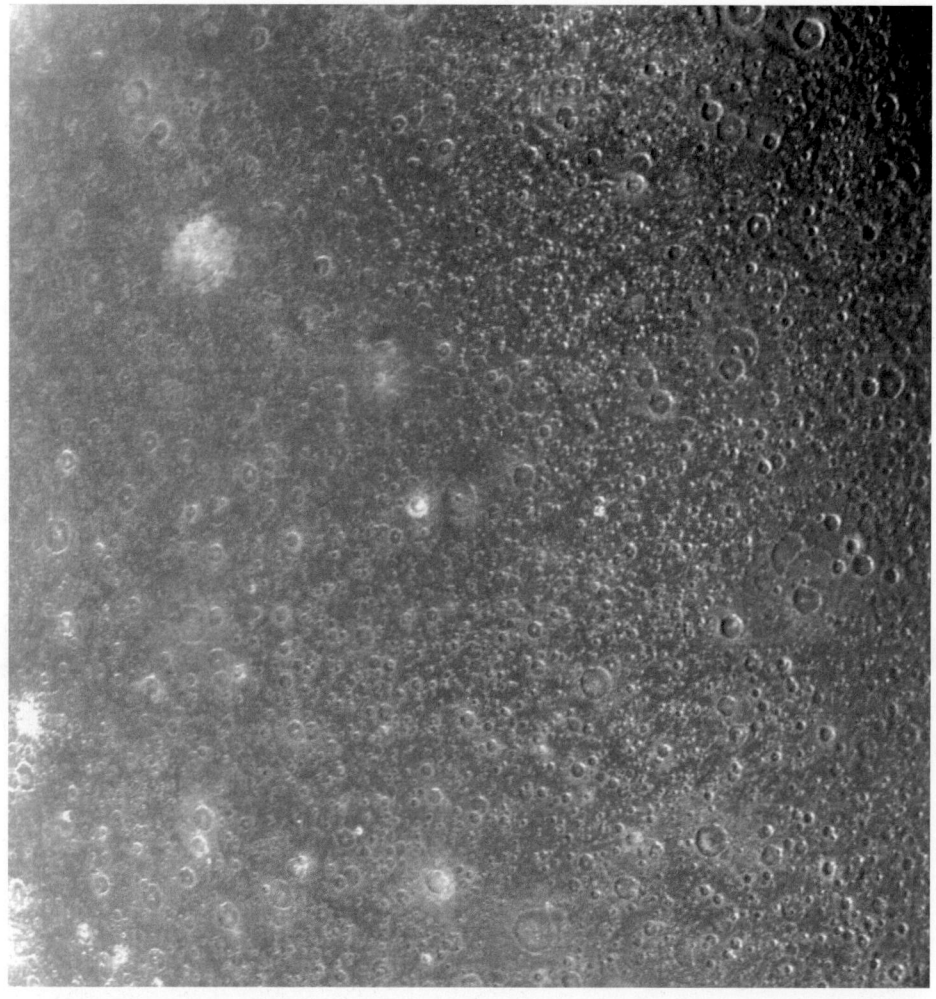

The surface of Callisto was deemed the most cratered yet found in the solar system.

ball''. About the only relief seemed to be some thin bright ridges that rose no more than a few hundred meters. It appeared that the crust was not rigid, but a soft shell of ice whose flexibility was maintained by heat from the gravitational tides of the planet and the other satellites. In comparison to Callisto and Ganymede, Europa showed a high bulk density, indicating that it contained a substantial amount of rock in its core. This implied that the surrounding mantle of ice might be 100 km thick, and if only the crust was solid then beneath this there would be a deep layer of slushy ice or possibly even liquid water. A variety of theories were advanced to explain the different types of cracks and ridges, involving various aspects of the manner in which the moon's interior may have evolved.[102,103] The closest point of

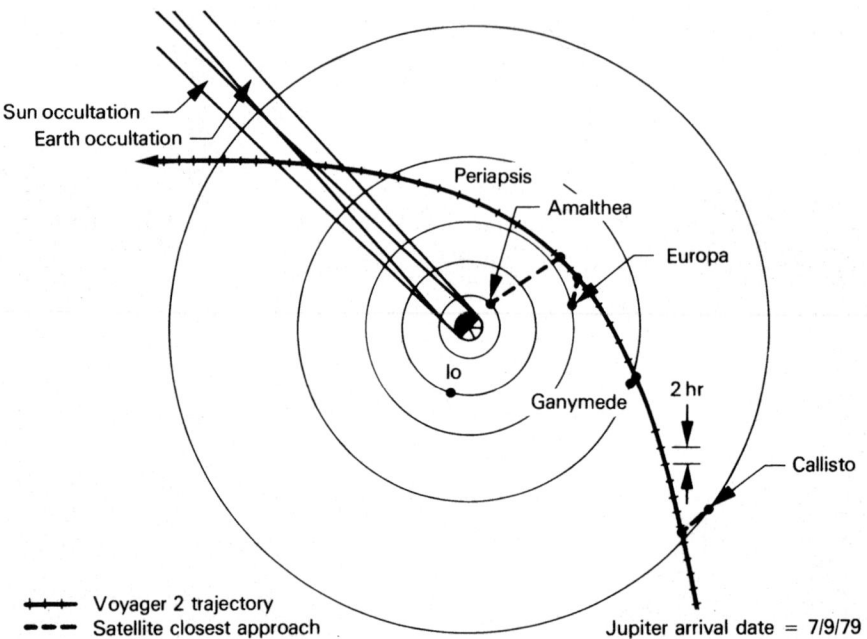

Voyager 2's passage through the Jovian system differed from that of Voyager 1, in that Callisto and Ganymede were encountered on the inbound leg, Europa was to be viewed at close range, but Io would not, and the spacecraft would not approach the planet as closely.

approach to the moon was 205,720 km. Within a matter of hours, Europa had been transformed from being a fuzzy ball with indistinct markings, to being the most likely place in the solar system (other than Earth) to find liquid water. If the gravitational tides also gave rise to volcanism on the ocean floor, then there might well be all the ingredients needed for the development of life.[104] As events were to transpire, this was to be humanity's last close look at this fascinating world – and indeed the Galileans in general – for almost two decades.

Even although Voyager 2 did not penetrate as deeply into the Jovian system as its predecessor, the fact that the radiation was more intense than expected prompted the central frequency of its damaged receiver to shift unpredictably, and when this broke the uplink the DSN had to retransmit commands at different frequencies in the hope that one would fall within the receiver's narrow bandwidth. In view of the harsh environment, it was also decided to temporarily power down the sensitive ultraviolet spectrometer. As the spacecraft continued to close on Jupiter it was able to view tiny Amalthea from a range of 500,000 km against the backdrop of the planet's clouds, which enabled its size and shape to be accurately determined, revealing it to be some $270 \times 170 \times 155$ km with a distinctly faceted or 'diamond-like' shape, and with its long axis maintained radial to the planet.[105]

The spacecraft's Jupiter flyby occurred at 22:29 UTC on 9 July at a planetocentric

The hemisphere of Ganymede presented to Voyager 2 was dominated by the dark terrain of Galileo Regio, which is one of the few surface markings on this satellite that can be seen from Earth.

distance of 721,670 km, passing some 650,000 km from the cloud tops at a relative speed in excess of 20 km/s. The gravity-assist increased the eccentricity of its solar orbit to 1.34, sufficient to head off in the general direction of Saturn. The 76-minute-long maneuver that was performed some 2 hours after periapsis not only precisely targeted the Saturn encounter but also saved about 10 kg of hydrazine to preserve the option of continuing to Uranus, and possibly Neptune. (Recall that in order to recover from the greater than expected consumption of propellant owing to the underperformance of the thrusters, it had been decided to carry out this maneuver ahead of schedule.) By chance, the spacecraft made the burn at a time when its uplink was not receiving, owing to radiation. With the maneuver accomplished, it turned its attention to Io for a 10-hour 'volcano watch'. This was the ideal time to do so as the moon was on the far side of its orbit, about 1 million km away, and the evolving line of sight during the watch period would cause the position of the limb of the thin crescent to traverse a wide range of longitudes, with plumes from vents located there being even more prominent than on the historic 8 March image. Plumes from Amirani and Maui were seen on the limb illuminated

Craters and criss-crossing grooved terrain on Ganymede as imaged by Voyager 2.

by the setting Sun, while the 250-km-tall plume from Loki was seen in the early morning Sun. Meanwhile, the infrared spectrometer made thermal scans in search of 'hot spots' on the dark hemisphere. A combination of the imagery from Voyager 1 with the intermittent low-resolution imaging by Voyager 2, and the results of its 'volcano watch', mapped nearly all the prospective limb, and hence should have mapped all the active plumes rising to at least 100 km. However, no plume lower than 70 km was seen during either encounter, suggesting that perhaps large eruptions are more common than small ones.[106]

In the 4 months since Voyager 1's flyby, the relative geometry of Earth, Sun and Jupiter had evolved to such a degree that there would be very little overlap between the radio occultation during which Voyager 2's radio signal would 'sound' Jupiter's

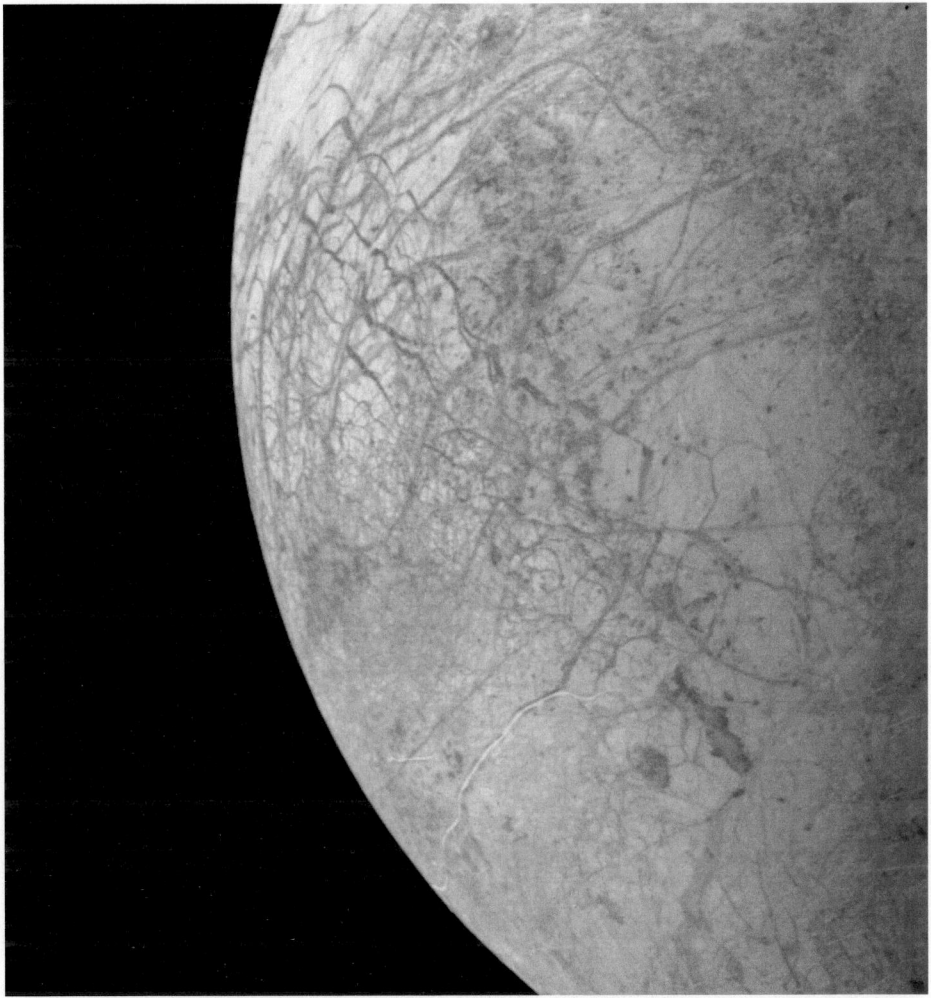

During its close flyby of Europa, Voyager 2 revealed an amazing network of dark cracks, spots and mottling but, remarkably, no obvious impact craters, suggesting that the moon had been resurfaced relatively recently, perhaps in the last several hundred million years.

atmosphere, and passing through the planet's shadow. On this occasion, therefore, the spacecraft would be able to report the data collected in the shadow cone in real time.

At 21:21 UTC on 10 July Voyager 2 passed behind Jupiter's limb as viewed from Earth, with the point of ingress over the evening side at a latitude of 66.7°S and the point of egress on the morning side at about 50.1°S, during which time the strong refraction maintained the signal for most of the time. However, it was not possible to

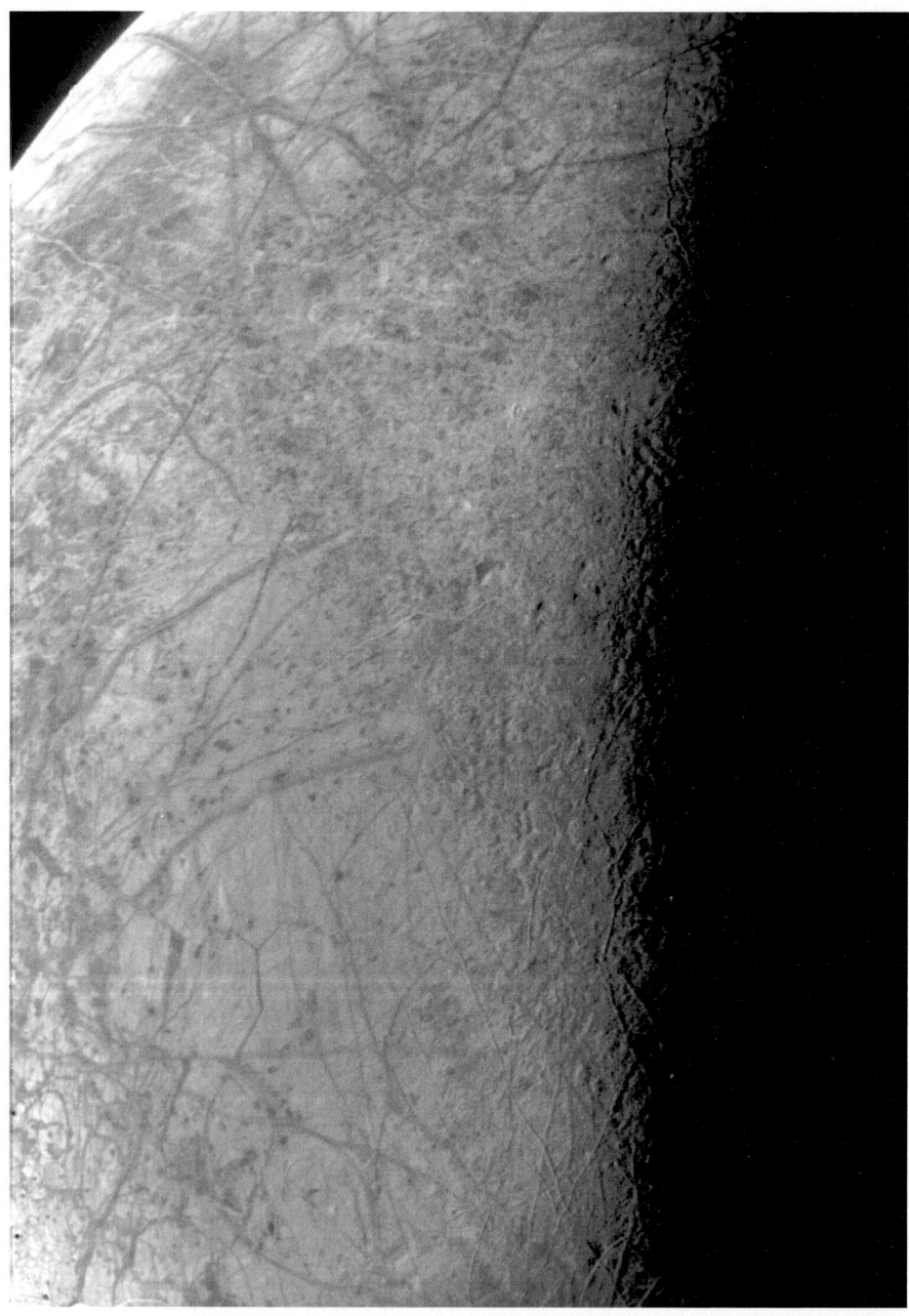

The terminator in this Voyager 2 view of Europa implied that the moon had little if any surface relief.

Amalthea passes in front of the cloudy disk of Jupiter. Such pictures helped to determine the dimensions and hence the volume and density of the planet's fifth moon.

get good atmospheric profiles owing to the uncertainty of the craft's position – due mostly to the targeting maneuver for Saturn, whose effects on the trajectory had not yet been accurately measured.[107] As it was about to exit the Earth occultation, the spacecraft slipped into Jupiter's shadow. Over a 24-minute interval the spacecraft took a series of six long-range pictures from its vantage point a few degrees below the equatorial plane to search for lightning and auroras at both poles, plus a spectacular view of the ring shining by forward-scattered sunlight and broken by the shadow of the planet, which was itself seen in silhouette by the sunlight refracted by the highest levels of its atmosphere. Narrow-angle pictures were also taken of the ring, but infrared scans failed to detect it. The optically-thin ring appeared to have a bright well-defined and fairly sharp outer section that spanned about 800 km wide, and a dimmer section some 5,200 km wide with an ill-defined inner edge. The interior of the ring appeared to contain material of ever-decreasing density that could possibly extend all the way down to the planet's atmosphere. Unfortunately, because all the long-exposure images were slightly blurred, it was not possible to measure the thickness of the ring more accurately than had been achieved using the single picture obtained by Voyager 1. The ring appeared to be surrounded by a 'halo' some 22,000 km in thickness, which was 1,000 times more than the ring itself. The brightness of the ring at differing angles of solar illumination established that it was composed primarily of micron-sized particles, but to account for the depletion of energetic charged particles noted by the Pioneers there would also require to be centimeter-sized pebbles. Orbiting so close to Jupiter, which strongly radiated thermal energy, this material, which was probably a silicate, had a reddish spectrum. On the basis that the ring was in a

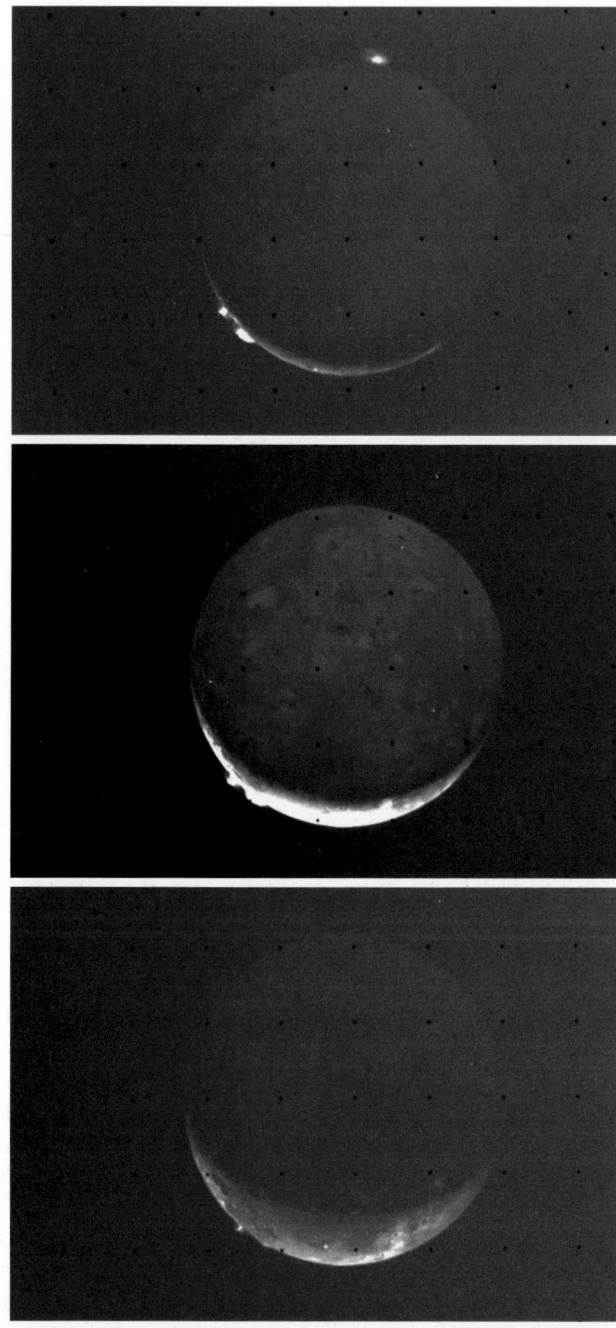

Three pictures of Io taken by Voyager 2 during its 'volcano watch'. The plumes of Amirani and Maui are on the left limb, while the plume of Loki is on the right limb on the last image. A bright spot on the crescent in the first picture marks the site of the volcano Masubi.

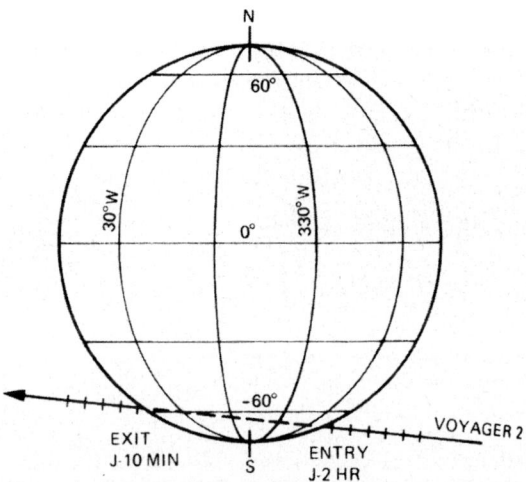

The trajectory of Voyager 2 as viewed from Earth took it behind the far southern hemisphere of Jupiter. In fact, the Jovian atmosphere refracted the signal sufficiently for contact to be maintained for most of the occultation.

'steady state' in which material is lost and replenished at the same rate, theories for its origin involved cometary and meteoritic debris, ejecta from the inner satellites, and volcanic particles from Io that were transported by magnetospheric forces.[108,109]

In addition to some 17,000 pictures, Voyager 2 provided a wealth of data on the Jovian system. In the ultraviolet, the Io torus was found to be twice as bright as in March, but its temperature had reduced by 30 per cent. It was found that the Jovian auroras were much brighter at the longitudes corresponding to where Io's flux tubes intercepted the atmosphere. During the flyby, the spacecraft was able to observe the planet occult the bright star Regulus in Leo, enabling the ultraviolet spectrometer to yield further data on the chemistry, temperature and structure of the atmosphere.[110] The infrared spectrometer provided scans of the ammonia abundance and the high-altitude temperatures for an entire rotation period in which the overall atmospheric structure of belts and zones was evident, the chilly temperature of the Great Red Spot confirmed that it stood high above the ammonia clouds, and the warmth of the elongated brown ovals confirmed that they were gaps in the ammonia cloud deck through which the interior was visible.[111]

After spending 2 weeks traveling 16 million km down Jupiter's magnetospheric tail, Voyager 2 made a long series of magnetopause crossings at distances between 169 and 279 Rj, then several bow-shock crossings at 283 Rj prior to emerging back into interplanetary space on 3 August. The encounter was officially concluded on 5 August. Between them, the two Voyagers had established a 7-month-long base for

While in the shadow of Jupiter, Voyager 2 took the pictures used to assemble this mosaic in order to determine the extent of the two ansae of the ring.

A closer look at part of Jupiter's ring. Note that its outer edge is clearly defined but its interior is diffuse. The image includes star trails.

the investigation of Jupiter's atmosphere, and an even longer base for studies of the Io torus.[112]

SATURN AND MYSTERIOUS TITAN

Although the interval between the arrival times of the two spacecraft at Jupiter had been 4 months, Voyager 1 received a greater gravity-assist by virtue of flying closer to Jupiter and would therefore arrive at Saturn 9 months ahead of its partner. A variety of activities were scheduled for the cruise across the gulf separating the two gas giants. First, all the scientific instruments were thoroughly recalibrated. It was at this point that the troublesome photopolarimeter was found to have become almost insensitive to light, and consequently no longer usable. A number of trajectory corrections were also performed. A maneuver on 9 April 1979 refined the approach to make a close flyby of Titan, and another maneuver on 10 October ensured that the spacecraft would not hit Titan! For the final targeting maneuver on 13 December 1979, Voyager 1 was commanded to adopt the requisite attitude for the maneuver, thus breaking contact with Earth, and to reacquire contact after the burn, but at the scheduled time for reacquisition there was no signal. Some hours later, the DSN detected a very weak signal. The next day the spacecraft was commanded to reorient itself to point its high-gain antenna at Earth, which it did, and it was discovered that the star sensor had erroneously locked onto alpha Centauri instead of Canopus.[113] At various times during the cruise, both spacecraft complemented their monitoring of interplanetary space with ultraviolet observations of a number of 'hot' stars and white dwarfs, the embers left by stars like the Sun at the end of their lives.[114,115,116] Meanwhile at JPL the detailed sequences required to make observations during the Saturn encounters were written, tested and rehearsed. With many satellites, rings and other phenomena, the Saturnian system offered twice as many objects for study as the Jovian system, with their times of closest approach all condensed into a period of a little over 24 hours rather than being distributed over almost 3 days. Moreover, owing to Saturn being much further from Earth, telemetry and data would be able to be returned only at the reduced rate of 44.8 kbps instead of 115.2 kbps.

The inbound plan called for, in rapid succession, a close flyby of Titan, a radio occultation by this moon, and a southbound crossing of the ring plane. The gravity-assist from the encounter with Saturn would deflect the trajectory north to provide a radio occultation while traversing the planet's shadow. This would be followed in succession by a radio occultation by the ring system and recrossing the ring plane between the orbits of Dione and Rhea, where it was hoped that these moons would have swept a zone clear of the particles that occupy the inner system.[117]

In early 1980 Saturn had passed through an equinox, and the Sun crossed the ring plane to illuminate the north face of the ring system. Voyager 1 commenced imaging on 23 August. As the vehicle's perspective of the geometry changed in the ensuing months, the rings 'opened' and the resolution increased to match, and then surpass, that of the best terrestrial telescopes. Meanwhile, in early January 1980 the radio-astronomy instrument had started to receive very-low-frequency radio bursts from the planet, modulated by its still-uncertain rotation period. The radio emissions appeared to be further modulated with a period matching that of the moon Dione, suggesting that this satellite was interacting with the magnetosphere, but why it should do this posed a mystery because this would require Dione, like Io, to be a source of plasma, albeit a much weaker one. In contrast to the approach to Jupiter, the particles and fields instruments did not detect any streams of energetic charged particles. Apparently, either the process in the Jovian magnetosphere that accelerated particles to speeds sufficient to escape the giant planet's gravity was not operating in the Saturnian system, or, if it was operating then it was much less efficient.

As Voyager 1 closed in, the classical rings were revealed to comprise many bands of narrow ringlets, there was a faint smear of material within the supposedly vacant Cassini Division that separated the outermost A ring from the B ring, and there were narrow gaps between broad bands within the C ring. An examination of the imagery of the B ring revealed dark streaks that radiated like the spokes of a wheel for distances of tens of thousands of kilometers. Since the rings were made up of individual particles, each pursuing its own orbit – with those further out following slower orbits than those closer in – any large-scale radial structure would be sheared apart within hours. But the 'spokes', as they became known, appeared to persist. Furthermore, they were much easier to see on one ansa of the ring system than on the other. The spacecraft was instructed to undertake additional imaging, to further investigate this phenomenon. Other ring features under study included an inconspicuous gap at the outer fringe of the A ring. The Lick Observatory issued

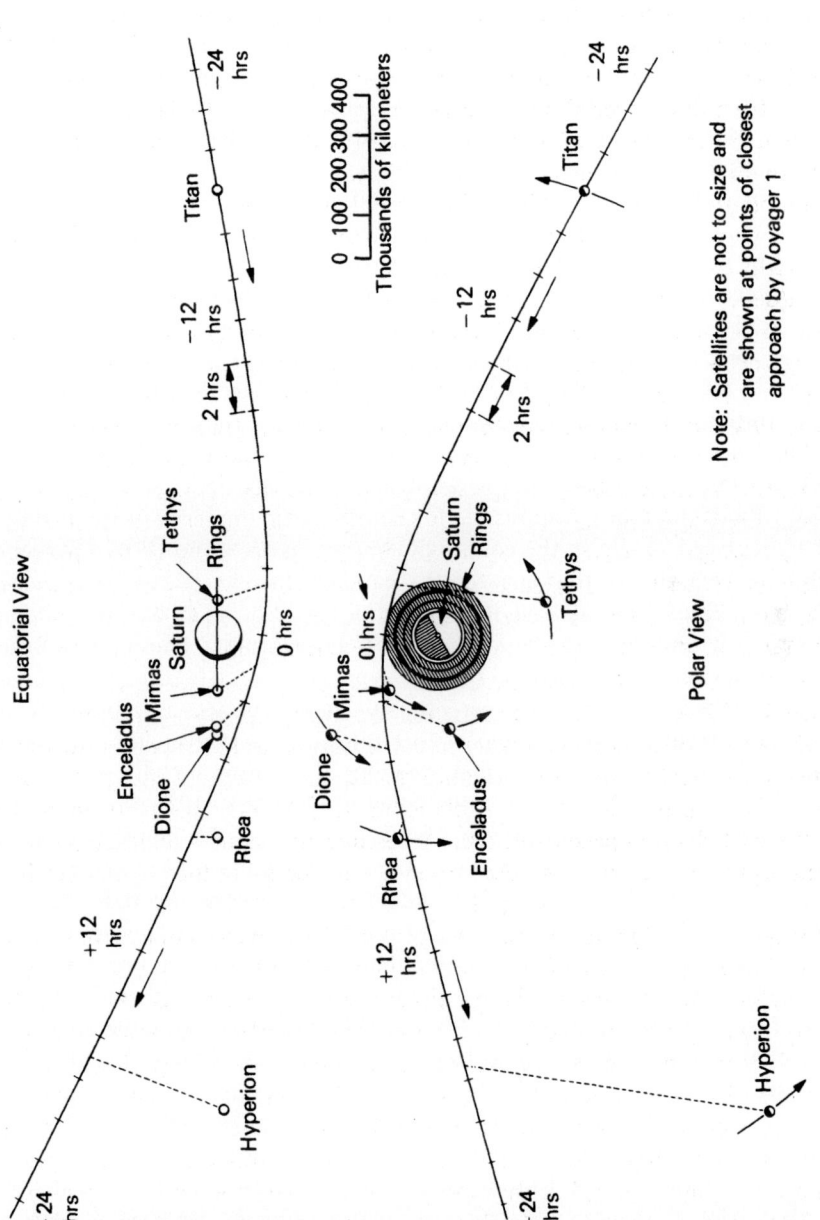

Voyager 1's trajectory through the Saturnian system.

a claim that this gap was discovered by J.E. Keeler in 1888 while testing its 36-inch refractor, but in fact it had been observed by several sharp-eyed astronomers in the preceding decades.[118,119] It had long been presumed that gaps in the ring system were caused by resonances with the satellites that orbit further out, but some of the gaps revealed by the high-resolution imagery did not correspond to resonances with the known satellites. This led to the suggestion that kilometer-sized objects might be orbiting in those gaps and 'sweeping' them clean. Observations were also made of several small satellites that had been glimpsed from Earth during the 1966 and 1980 ring-plane crossings.[120]

Imagery of Saturn's disk revealed light and dark spots similar to those on Jupiter, but the pervasive high altitude haze tended to mask them from telescopic observers. The infrared spectrometer measured atmospheric temperatures, and the ultraviolet spectrometer scanned the planet, its rings and the surrounding space for emissions from atomic hydrogen. On 25 October, an examination of the new ring 'movies' for spokes or other new phenomena identified the presence of two small moons, one in an orbit just short of that of the F ring and the other located just beyond it. Initially designated as 1980S26 and 1980S27, these moons were later named Pandora and Prometheus respectively; Prometheus being the one nearest Saturn. However, they became better known as the 'shepherd' moons, for their gravity was almost certainly responsible for confining the material of the F ring to such a narrow band. A dip in the counts of charged particles noted by Pioneer 11 may well have been due to that spacecraft's passage close by one of these moons. The known satellites were also under regular observation in order to determine their shape, size and density. Titan remained frustratingly bland, with heavy computer enhancement required to discern even hints of structure in its ruddy orange haze. On 6 November, Voyager 1 was to make a course correction of 1.52 m/s to nudge its trajectory 650 km closer to Titan, so as to convert the occultation that was to follow the flyby from a mere grazing of the moon's limb into a radio occultation that would 'probe' its atmosphere. However, JPL director Bruce Murray and his team were uncertain about performing this maneuver; if the spacecraft were to lose attitude in the same manner as it had on 13 December 1979, this would jeopardize the activities planned for its passage through the inner system. It was nevertheless decided to proceed, and the maneuver was accomplished without incident.[121,122] Later that day, the first pictures to resolve detail on distant Iapetus confirmed that the trailing hemisphere was indeed bright and the leading hemisphere was dark, although there was nothing in the imagery to indicate why this was so. Referring to this dichotomy, some writers have described it as the 'ying yang moon'. By this time, the resolution had improved to the degree that the material in the Cassini Division was seen to take the form of several narrow rings spaced hundreds of kilometers apart. A satellite was found orbiting between the A and the F rings. Initially designated 1980S28 and later named Atlas, this was about 100 km in size and seemed to be responsible for defining the A ring's sharp outer edge.[123] Searches continued for satellites traveling within the ring system. As the resolution improved, bright and dark spots, elongated clouds and a number of belts and zones became evident on Saturn. By 8 November, detailed

A distant Voyager 1 image of Saturn. Some of the moons are present as star-like points and dark spokes are visible on the left ansa of the B ring.

images of the F ring had revealed it to be no wider than 100 km and of non-uniform brightness, there being bright knots and concentrations, in addition to faint sections. The first medium-resolution images of Rhea, Tethys and Dione were taken on 10 November. There was a strange 200-km-diameter circular patch on the trailing hemisphere of Tethys which the illumination implied was not a crater, but it was not obvious what else it could be. Unfortunately, the spacecraft's trajectory meant that it would not be able to view this feature at high resolution. (Voyager 2 would find that the circular patch was indeed a crater.) Although the resolution was only 100 km per pixel, detail was also becoming evident on Rhea and Dione, and intriguing bright streaks raised the prospect that these satellites might have been geologically active at some time in the past. The first images of Hyperion were taken later that day, but little information could be gleaned from them because, at a range of some 5 million km, the moon spanned only a few pixels.

It had been hoped that as Voyager 1 closed in on Titan the increasing resolution of the imagery would reveal gaps in the haze through which it would be possible to catch a glimpse of the surface, but the atmosphere was at least as opaque as that of Mars during the 1971 dust storm. The frustration of the geologists was matched only by the delight of the atmospheric specialists when, in studying a view of the moon's limb, they saw a 'detached' layer about 100 km above the main haze. The southern hemisphere was fairly uniform, but the northern hemisphere was darker, redder and capped by an even darker polar hood that looked as if it might be a seasonal effect. As the Titan flyby loomed, spectrometry of the disk and limb became increasingly effective. It was also hoped that Saturn's magnetosphere would extend out beyond the orbit of Titan, in order that the spacecraft would be able to make a sensitive search for any intrinsic magnetic field that the moon might possess. The spacecraft was expected to encounter the bow shock on 10 November, but this did not occur until early on 12 November, at a planetocentric distance of 26.2 Rs – slightly further out than in the case of Pioneer 11. It then made five crossings of the magnetopause, the last at 22.9 Rs, just 3 hours before the Titan flyby. Although the preprogrammed Titan activities included imaging, the most interesting data was provided by the infrared and ultraviolet spectrometers. The foiled geologists proposed a mission to map the surface of Titan by radar, in the same manner as was planned for Venus. Many years after Voyager 1's passage through the Saturnian system, it was realized that the surface of Titan should be visible through a number of narrow windows in the near-infrared. When this was confirmed in the mid-1990s, the imagery taken by Voyager 1 through an orange-filter was reprocessed and it was found that there were some hints of surface detail; this had passed unnoticed at the time owing to the very low contrast. Correlating the positions of the surface markings over the interval of two decades proved that Titan's rotation period was synchronized with its orbital period.[124] At 05:41 UTC on 12 November, Voyager 1 flew by Titan at a speed of 17 km/s, passing about 4,000 km above the top of the haze. It entered the moon's shadow some 6 minutes later, and after a further minute the radio occultation began. While it was out of contact, Voyager 1 crossed south of the ring plane. Until it made the northbound crossing after the Saturn flyby, it would view the unilluminated face of the ring system. It re-emerged into daylight less than 15 minutes later. The radio occultation ended 43 seconds early, which meant that the craft was 200 km from where it was intended to be. The scan-platform sequence for the passage through the inner system had to be hastily revised to accommodate this discrepancy. Although the trajectory displaced the second ring-plane crossing to a point 1,500 km from the orbit of Dione, this was not a serious issue because that entire zone was expected to be clear of ring particles.[125]

In preparation for the Voyager 1 flyby, two models had been developed of Titan's atmosphere: one primarily of nitrogen with a high surface pressure, and the other of methane with a lower surface pressure. The occultation provided good temperature and pressure profiles. On the simplifying assumption that the atmosphere was made entirely of nitrogen, this gave a surface pressure as high as 1,600 hPa (more than 50

Haze layers over the north polar region of Titan.

per cent greater than on Earth) and a temperature of $-179°C$. Given the weak gravity and the depth of the atmosphere, the 'air' was almost five times as dense as at sea level on Earth. The occultation also measured the diameter of Titan at 5,140 km, placing it second only to Ganymede (5,280 km) as the solar system's largest satellite. It was, however, a matter of definition, as the atmosphere made the visible disk of Titan appear some 400 km larger. The ultraviolet spectrometer confirmed that the atmosphere is indeed dominated by nitrogen molecules, atoms and ions. Titan is the only body apart from Earth to possess an atmosphere that is rich in nitrogen. As it monitored the limb of the moon occulting the light of a star, the spectrometer also detected traces of methane and some more complex hydrocarbons. Owing to the relatively weak gravity, the highest layers of haze were at altitudes in excess of 400 km – several hundred kilometers above the visible atmosphere. There appeared to be two distinct layers of haze: the upper layer absorbed ultraviolet and the lower one absorbed visible light, this being the one observed in limb images. The surface was obscured by a thick blanket of aerosol. The model suggested that there might be methane clouds in the lower atmosphere. The infrared spectrometer confirmed the presence of nitrogen, and also detected various hydrocarbons and hydrogen cyanide, the simplest nitrile. These could be made by reacting nitrogen and methane in the presence of solar ultraviolet and the charged particles circulating in the Saturnian magnetosphere. In particular, when exposed to ultraviolet in the upper atmosphere, hydrogen cyanide would polymerize to form a variety of reddish-brown compounds and thereby give the atmosphere its characteristic hue. If this model was correct, then such compounds would make up most of the aerosol blanket, and might even drift down to the surface as a 'snow' of hydrocarbons. It was computed that a deposit of hydrocarbon ice at least 1 km in thickness could have built up over the ages. Given the pressures and temperatures present in the lower atmosphere, there might be a counterpart of the hydrological cycle on Earth, but with methane raining onto the surface and draining into lakes, seas or even an ocean, from which it would evaporate to complete the cycle. Titan was therefore the first solar system body apart from Earth to offer the prospect of sustaining a liquid on its surface. However, it had to be accepted that this was extrapolation from remote-sensing data, and to test the hypothesis it would be necessary to send a radar mapper or, ultimately, a lander.

The presence of nitrogen, methane and more complex hydrocarbons made Titan a laboratory for prebiotic chemistry. Hydrogen cyanide was particularly significant, because it is an intermediate molecule in the synthesis of amino acids. The Earth's early atmosphere may have been similar to that of Titan now, except that it would have been warm enough for liquid water to be stable on the surface. Unfortunately, the cryogenic conditions on Titan almost certainly mean that life has not developed there.

Finally, the particles and fields instruments found that Titan made a cavity in the magnetosphere in the manner that suggested that the satellite did not possess a global magnetic field.[126,127,128,129,130]

After Titan, Voyager 1 faced a busy 24 hours during which it was to make close inspections of most of the known moons, as well as the recently rediscovered Janus and Epimetheus. Although its trajectory would pass close to some of the other small satellites, their orbits were too poorly defined to permit targeted observations. From its vantage point south of the ring plane, the spacecraft saw the unilluminated face of the ring system. The parts that contained the densest concentrations of material and appeared brightest in sunlight were now the darkest, because the light could not pass through.

Some 16 hours after Titan, Voyager 1 flew within 415,670 km of Tethys, a moon with a diameter of 1,060 km that was not only heavily cratered but also possessed a 100-km-wide canyon on the Saturn-facing hemisphere that suggested that the moon had undergone intense geological activity at some time in the remote past. At 23:46 UTC on 12 November, Voyager 1 made its closest approach to Saturn, passing some 126,000 km above its cloud tops. This was soon followed by an 88,440-km flyby of Mimas, the innermost of the classical satellites. However, at the moment of closest approach the view was of the dark hemisphere, and by the time that the lighting had improved enough for imaging, the range had opened by a further 20,000 km. When the pictures of Mimas were received, they caused a sensation. Although the moon was only 390 km in diameter, its leading hemisphere contained a 130-km-diameter crater that had a raised rim and a central peak, the sight of which prominently displayed on the terminator prompted a comparison with the 'Death Star' of the George Lucas movie *Star Wars*. The crater was later named in honor of William Herschel, who discovered the moon. Its floor was about 10 km below the rim, and the central peak rose some 6 km above the floor. If the impact had been any more violent it would probably have shattered the moon. The remainder of the surface was heavily cratered, and although there were grooves on the Saturn-facing hemisphere (the area best imaged) extending for up to 90 km, there was little to suggest that the moon had undergone geological activity.[131,132]

Two views of Mimas. The picture on the left shows the enormous crater Herschel that prompted the nickname of the 'Death Star' moon. The picture on the right is a relatively high-resolution image of the Saturn-facing hemisphere showing a wealth of craters and hints of grooves (in particular near the terminator at bottom).

At 01:44 UTC on 13 November Voyager 1 flew behind Saturn's limb at 75°S, and the radio occultation provided good temperature and pressure profiles spanning an altitude range exceeding 200 km through the ionosphere and atmosphere. As the spacecraft entered the planet's shadow a few minutes later, limb observations by the ultraviolet spectrometer yielded spectra of hydrogen in this region. While it was out of contact, the spacecraft passed Enceladus at range of 202,040 km. The highest resolution imagery, however, was taken from a distance of over 600,000 km, with a resolution no better than 15 km per pixel. Nevertheless, at 500 km in diameter, Enceladus was clearly not the twin of Mimas that had been expected. No craters were evident down to the limiting resolution. In fact, the hints of lines prompted a comparison with Europa in the Jovian system. The fact that the orbital period of Enceladus was exactly half that of Dione suggested that gravitational tides had caused the crust of Enceladus to flex, but the scientists would have to await Voyager 2's arrival to get a closer look at this enigmatic moon.[133,134,135] It proved difficult to search for lightning and auroras on the night-side of the planet because, being brightly illuminated by the sunlit rings, it was not as dark as Jupiter's night-side. Nevertheless, auroras were clearly detected for the first time in the ultraviolet at latitudes over 80 degrees, suggesting that they were due to magnetospheric processes similar to those that occur on Earth, and quite unlike the Jovian auroras that are dominated by Io-induced effects. The ultraviolet spectrometer was also able to observe the hydrogen 'atmosphere' of the ring system, and a torus of atomic hydrogen between the orbits of Rhea and Titan that was very probably composed of atoms that leaked from the atmosphere of the latter.[136]

At 03:11 Voyager 1 emerged from behind Saturn at approximately 2°S, at a point within the eastern ansa of the ring system, and then passed in sequence behind the C, B and A rings, all the while operating its transmitter in the radio occultation mode to investigate their structure. Meanwhile, because from the spacecraft's point of view the rings passed in front of the Sun, the infrared instrument measured how sunlight passed through the rings. The radio occultation data suggested that the objects in the C ring were on average 2 meters in size, in the Cassini Division they were 8 meters, and in the A ring they were 10 meters, but these figures should not be taken at face value since the method would have been best suited to larger objects, and for every object of 1 meter or more in size there may have been tens of thousands of smaller ones with sizes ranging down to a few microns. No figures were obtained for the B ring because, in addition to being opaque to sunlight, it proved to be so dense that the radio signal could not pass through it.[137]

While in the shadow of the rings, Voyager 1 passed within 161,520 km of Dione. Marginally larger than Tethys, this 1,120-km-diameter satellite was revealed to be heavily cratered on its leading hemisphere, but to possess a variety of geological structures on its trailing hemisphere, including bright broad 'wisps', irregular valleys and faults. The wisps gave the appearance of being deposits of clean water ice that had extruded from the interior, perhaps as a result of the resonance with Enceladus. Lending support to this was the fact that the leading hemisphere was not uniformly cratered, suggesting that some areas may have been resurfaced at some time.[138,139]

After Voyager 1 had crossed back north of the ring plane and was able once again to view the illuminated face of the rings, this time in close up, the spokes were brighter than the rings, which suggested that they were composed of micron-sized particles that efficiently forward-scattered sunlight, and that when they had been viewed from up-Sun their presence had been evident only from the shadows they cast on the rings.

Around the time of its outbound ring-plane crossing, Voyager 1 also succeeded in imaging the extremely tenuous but very broad E ring, the most concentrated part of which coincided with the orbit of Enceladus, prompting speculation that this moon might be the source of the particles which, given the way that they forward-scattered sunlight, were evidently very small. The first good images of the co-orbital satellites Janus and Epimetheus were obtained. Both were first seen by French astronomers in 1966, at a time when the rings were edge-on from Earth, and were recovered at the

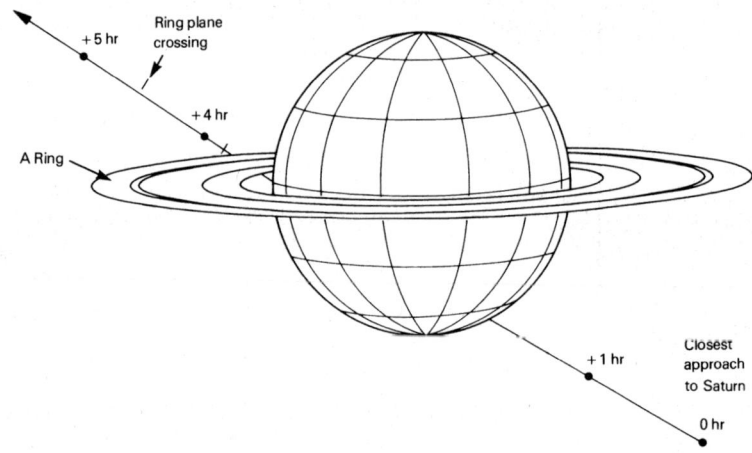

Viewed from Earth, in between making its closest approach to Saturn and crossing back north of the ring plane, Voyager 1 was occulted first by the planet and then by an oblique section of the ring system.

next such alignment in 1980. One of them was also the object with which Pioneer 11 almost collided. Both moons proved to be small irregular bodies with cratered surfaces. Janus was 160 × 200 km in size, Epimetheus was 100 × 140 km, and their major axes were aligned with the planet. Their orbital periods differed by just a few seconds, the slower one orbiting Saturn just 50 km farther out than the faster one. However, in a remarkable gravitational dance, every 4 years of so the inner one catches up with the outer one and, instead of colliding, the inner one is accelerated and climbs into the higher orbit while the outer one is retarded and drops into the lower orbit – in effect they swap places. The next 'swap over' was predicted to occur in 1982.[140,141,142]

Images of the F ring taken at about this time with a resolution of 15 km per pixel showed it to be a complex structure comprising a broad diffuse band and at least two narrow strands that gave the appearance of having been 'braided' together, although it was not possible to say that this was actually so, since the strands could have been in different elliptical orbits that were inclined at slightly different angles. In addition, the brighter components displayed knots and kinks, as well as clumps. The scientists found it difficult to explain how such structures might have formed, let alone predict their stability.

Voyager 1's final close encounter was with 1,500-km-diameter Rhea, the second largest of Saturn's retinue. The angle of the departure trajectory provided a view of the north polar region from a range of 73,980 km, showing this moon to be the most heavily cratered and least interesting of all the satellites that had been inspected. At a resolution of about 1.3 km per pixel, the leading hemisphere and polar regions were completely saturated with craters, giving it the appearance of an icy analog of

Although the leading hemisphere of Dione was heavily cratered, it showed grooves and other signs of a past internal activity.

Mercury or the Moon. However, whereas cratering in the inner solar system was due mostly to asteroids traveling in planet-crossing orbits, in the outer solar system the principal population of impactors would have been short-period comets. The earlier long-range imagery had shown the trailing hemisphere to possess a pattern of bright markings on a darker surface, similar to Dione, but unfortunately this region was in darkness at the time of closest approach. The infrared spectrometer observed the temperature of the sunlit part of the moon as it passed into the planet's shadow,

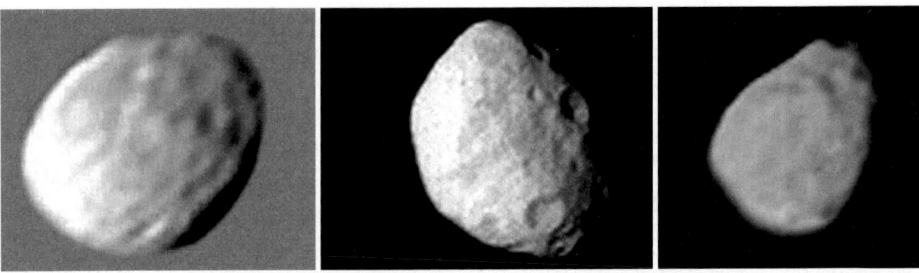

Voyager pictures of Janus.

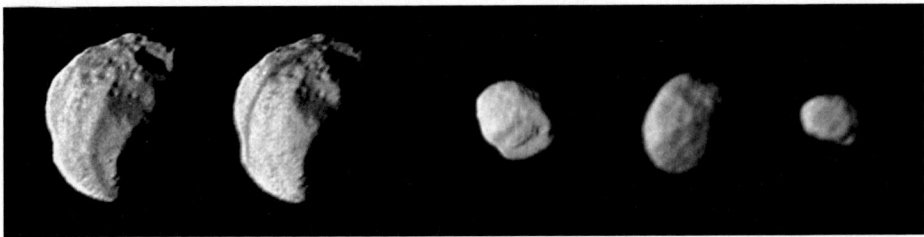

Voyager pictures of Epimetheus. Note that the two views on the left caught the curved shadow of the F ring moving across the face of the moonlet.

and the inferred thermal inertia revealed that although much of the surface was solid ice there were also patches of light frost.[143,144,145]

As it left the Saturnian system on 14 November, Voyager 1 turned its attention once again to Iapetus, which it had already inspected on the inbound leg. The range was never less than 3 million km, but in addition to the dichotomy between the hemispheres, the new images revealed the presence of a 200-km-diameter circular structure on the bright terrain adjacent to the dark hemisphere. This looked as if it might be an impact crater. There were also a number of unresolved dark spots. On its passage through the Saturnian system, Voyager 1 had imaged all the already known satellites except Phoebe, and had discovered three small moons. Voyager 2's trajectory would enable it to make closer inspections of Enceladus, Tethys, Iapetus, Hyperion and Phoebe.[146,147] As it departed, Voyager 1 looked down from its vantage point far above the plane of the system and took one of the most famous sequences of the mission, showing the dark side of Saturn and its shadow projecting on the wide open ring system. From this angle, the spokes appeared bright, indicating that they were particles that were efficient at forward-scattering sunlight. The fluctuating magnetopause washed across the spacecraft five times on 14 November at distances between 42.7 and 46.9 Rs, before it finally re-entered the interplanetary medium on 16 November at 77.4 Rs.[148]

In comparison to Pioneer 11, the scientific results of Voyager 1's passage through the Saturnian system were overwhelming. Saturn had long been known to be a much weaker radio source than Jupiter – radio-astronomy satellites orbiting Earth and the

One of the most amazing of Voyager 1's images showed the individual strands of the F ring, with braids, kinks and knots.

Moon had detected it only rarely, if at all – but Voyager 1 provided the first detailed observations of it in this part of the spectrum. Almost 9 months of data indicated a periodicity of 10 hours 39 minutes and 24 ($\pm$ 7) seconds that was presumed to match the rotation of the planet's core.[149] The radio-astronomy instrument also discovered a new type of radio burst, with a wavelength that was too long to pass through the ionosphere and therefore must have originated from a source above this altitude. It was suggested that these bursts were caused by invisible lightning in the rings, where billions of small particles might become electrically charged by interactions with plasma trapped in the magnetosphere.[150] The infrared spectrometer and radiometer determined the helium-to-hydrogen molecular number ratio for Saturn's cloud tops at 6 per cent; as against 13 $\pm$ 4 per cent for Jupiter, which matched the solar abundance of 13 $\pm$ 2 per cent. In terms of the mass fraction, the helium was 11 $\pm$ 3 per cent; as against 19 $\pm$ 5 per cent for Jupiter, which was close to the expected 20 per cent. Since both planets would have formed with near-solar abundances, this implied a depletion of helium in Saturn's outer atmosphere. This is thought to be a consequence of the fact that Saturn was sufficiently cool for its atmosphere to fractionate, which was not possible for Jupiter, owing to it having a much hotter

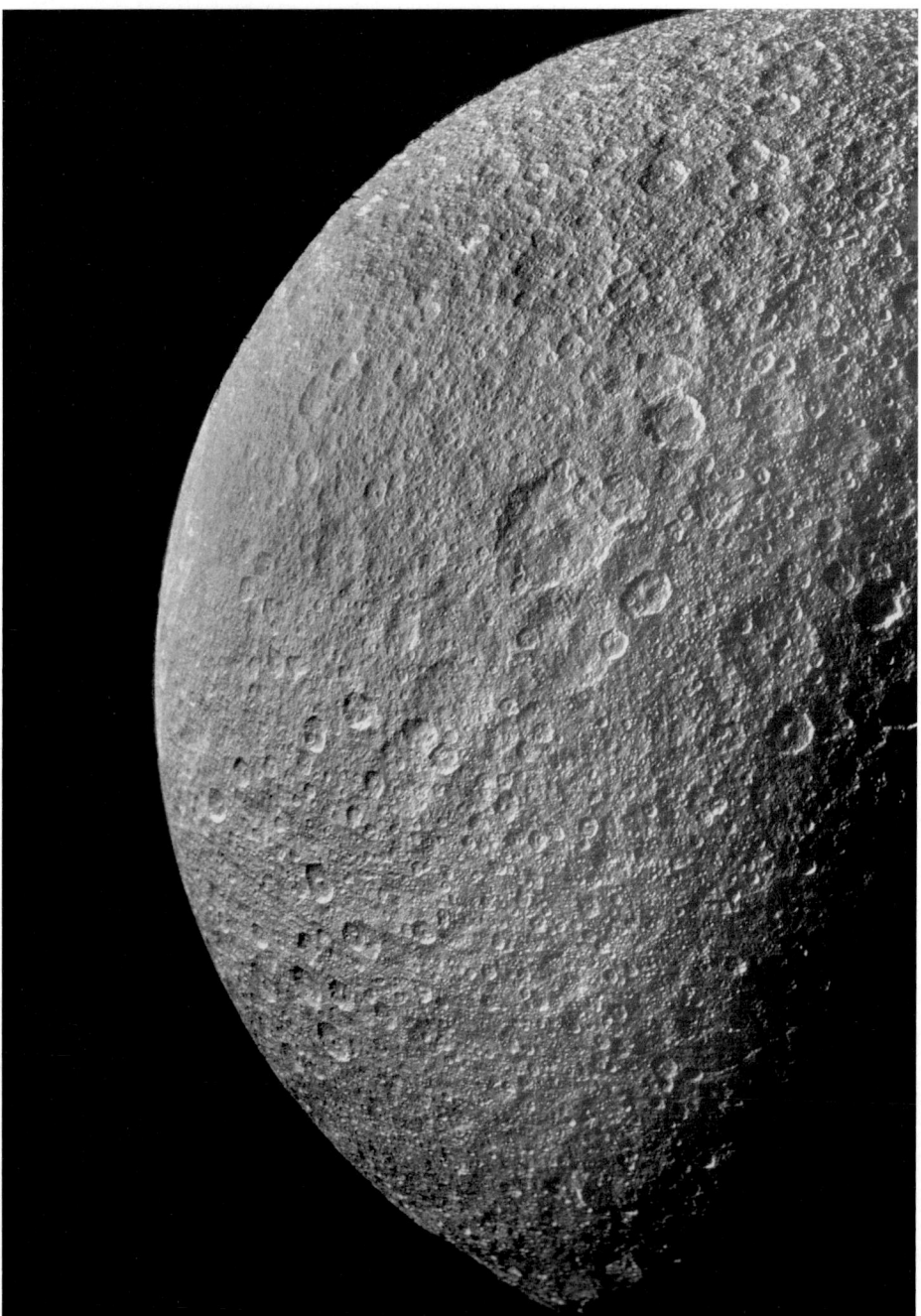

The heavily cratered north polar region of Rhea.

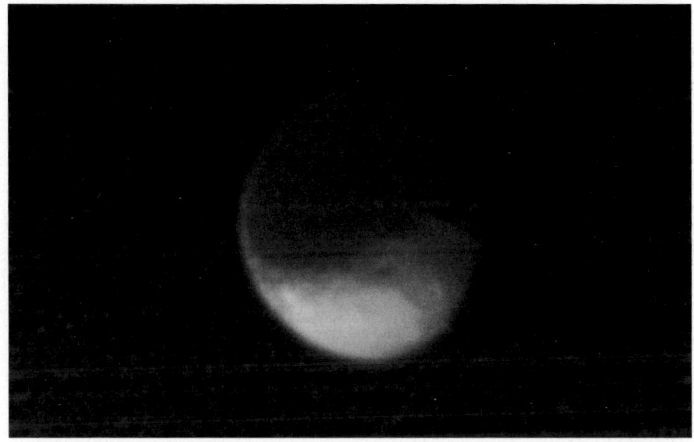

A distant Voyager 1 image of the 'ying yang moon' Iapetus, showing the dichotomy between the dark leading hemisphere and the bright trailing one. Some large craters are evident on the bright area near the terminator.

core. This also explained another measurement by the instrument. As in the case of Jupiter, Saturn was warmer than it should be from solar heating alone. In fact, Saturn radiated to space 1.8 times as much energy as it received; as against 1.7 for Jupiter. But whereas Jupiter was heated by its ongoing gravitational collapse, the mechanism for Saturn appeared to be due to the fact that as the helium condensed in the upper atmosphere and the droplets 'rained' down into the interior, their gravitational potential was converted into heat. The spectrometer also measured the composition of the visible atmosphere where, in addition to the hydrogen and helium, it detected ammonia and methane and also traces of complex compounds involving carbon, hydrogen and phosphorus. The radiometer's temperature profiles were in excellent agreement with those derived from the radio occultations.[151] Although both Jupiter and Saturn were 'gas giants', the structure of their atmospheres proved to be different. On Jupiter the alternating zonal winds produced a series of narrow belts and zones, but much of Saturn's atmosphere within a 70-degree-wide belt centered on the equator was a single jet stream that flowed eastward at speeds as high as 480 m/s, or about two-thirds of the speed of sound in that environment. The imagery obtained during the long approach to Jupiter had enabled atmospheric features to be tracked, but Saturn's atmosphere was relatively bland. Nevertheless, the results gained from tracking the few features that were evident were consistent with the scant data from telescopic observations. A structure more like that of Jupiter was present at the higher latitudes of both hemispheres, and in fact most of the atmospheric features seen were in the Northern Temperate Belt. Also present at high latitudes were anticyclonic spots similar to, but smaller than, the ovals on Jupiter.

Contrary to terrestrial and Pioneer 11 observations, Voyager 1 found that the ring system was highly dynamic. Perhaps the most surprising discovery was the spokes.

Significantly, their motion appeared to be related to the planetary magnetic field. In fact, the distance from the center of the planet at which they became the narrowest was the distance at which the orbital period matched that of the magnetic field, but the available data could not show whether the motion of the spokes was dominated by gravitational or magnetic forces. One hypothesis was that spokes formed when micron-sized ring particles passed through the zone where the planet's shadow fell on the rings, and somehow became electrically charged and levitated away from the ring plane. This explained why the spokes were most well-defined on that section of the ring system that had just emerged from the shadow. In sunlight, the charge slowly dissipated, and the particles had settled by the time they approached the shadow again. It was also suggested that it was the discharging of such particles that caused the radio bursts. However, theoreticians had difficulty building a single framework that could explain all the observed characteristics, leaving the precise cause of the spokes a mystery.

A great many images were taken of the classical ring system. The C ring, which appeared to be the most regularly ordered, contained at least two gaps, each several hundred kilometers wide: one gap may have been due to an orbital resonance with Titan; the other contained within it an enigmatic ringlet several kilometers in width that was in an eccentric orbit. Although small moons were evidently responsible for the sharp edges of the rings, there was no obvious reason for the inner edge of the C ring to be as sharp as it was. In terms of structure and brightness, the B ring was dramatically different to the C ring. While there was an abrupt boundary between them, this was not marked by a gap, or even by an orbital resonance. The B ring was divided into four equally wide zones of differing opacity. Its outer boundary was marked by the Cassini Division, which had long been known to match a strong resonance with the innermost and smallest of the classical moons, Mimas. At least five broad bands and a second eccentric ringlet were found in the Cassini Division, exhibiting a wealth of structure. The A ring appeared to be thinner and more homogeneous. The Encke Division contained at least two narrow ringlets, made mostly of small particles. Immersed in the outer ring were five alternating dark and bright lanes, at least four of which were evidently associated with the superposition of two or more resonances with classical moons. The first feature seemed to be a ring that was slightly inclined relative to the mean ring plane, possibly the result of interactions with Mimas. The sharp outer edge of the A ring was due to the presence of the moonlet Atlas, orbiting 3,600 km further out.

Voyager 1's observations of the F ring from various angles revealed it to be made of small particles, possibly similar to those in Jupiter's tenuous ring. In addition to discovering the two moonlets responsible for shepherding the ring, it resolved it into separate strands that comprised bright clumps and knots. The complex structure of the F ring was probably a gravitational effect of the shepherds, but the resolution of the imagery was insufficient to study these interactions in detail. It was decided to revise Voyager 2's encounter sequence to conduct follow-up observations.

Long-exposure pictures of the region between Saturn and the C ring revealed this to be occupied by a multitude of narrow rings of widths ranging from 100 km to the limiting resolution, that might have extended down to the atmosphere of the planet.

Two unprecedented views of Saturn as a crescent phase taken as Voyager 1 drew away and was once again able to view the illuminated face of the ring system: a wide-angle view (top) and a narrow-angle close up of the shadow of the ring system on the planet.

Interestingly, a telescopic study in 1969 had reported a tenuous ring in this zone, but the structures seen by Voyager 1 were too faint to be seen from Earth. Nevertheless, it was decided to use the originally assigned name of the D ring. Another faint ring was found close to the orbit of Janus and Epimetheus and named the G ring. Images taken in the days immediately prior to the encounter confirmed telescopic reports of an asymmetry in brightness between the two ansae, and revealed that the particles in the various rings were subtly different in color.[152,153]

Good measurements of the diameters and densities of the Saturnian satellites were obtained, indicating that their bulk compositions were mainly water ice with only a small fraction of rock or metal. Judging by their albedos, even the small satellites seemed to be mostly ice. However, DSN tracking of the spacecraft's flyby of Titan showed that this moon was composed of approximately 50 per cent rock, very likely as a core. The spacecraft's second-closest flyby was with Rhea, enabling its mass to be determined more accurately than had been possible from the Pioneer 11 data. It was remarkable that in the Jovian system the densities of the moons decreased with planetocentric distance, but the situation appeared to be reversed in the case of the classical satellites of Saturn.[154]

The gravity-assist from the Saturn flyby had further increased the eccentricity of Voyager 1's orbit to 3.73, which was equivalent to a speed with respect to the Sun of about 3.5 AU/year, thereby making it the fastest of the four spacecraft that were in the process of departing from the Sun's realm at that time. The constraints imposed by the Titan flyby caused Voyager 1 to leave on a path inclined at about 35 degrees above the ecliptic. For the coming years, possibly decades, its mission would consist of monitoring the interplanetary medium. It would also provide engineering support for its twin, which, given that the first spacecraft had achieved the goals set for the project, was free to attempt the challenging Grand Tour of the outer solar system. But there were no guarantees. C.E. Kohlhase, in charge of mission planning, gave Voyager 2 a 65 per cent chance of reaching Uranus in 1986 in a functional state. No one was offering a figure for Neptune in 1989. Given this rather pessimistic outlook, some scientists were of the view that it would be better not to try for Uranus and Neptune, and instead to make follow-up observations of Titan, but in January 1981 NASA was granted permission to venture beyond Saturn and the sense of adventure was irresistible.[155]

THE FINAL ONE–TWO PUNCH

The decision to attempt the Grand Tour required Voyager 2 to approach Saturn on a trajectory that would pass well north of Titan's orbit, and then make its one and only crossing of the ring plane at about the time of closest approach to Saturn, doing so at a point just outside the classical ring system – a route that Pioneer 11 had shown to be survivable. Luckily, this trajectory through the Saturnian system did not preclude Voyager 2 making observations designed to complement those of its sister. During its approach, it would make movies to study the spokes on the ring system, and the F ring would also receive attention. On the way in, it would be able to inspect Iapetus from a range of 1 million km. While at 665,000 km the Titan encounter would not be particularly close, the spacecraft would make observations to study the layers of haze in the upper atmosphere. Once deep inside the system, it would undertake targeted studies of some of the recently discovered minor satellites, and of the classical ones that were not previously well presented, the most eagerly awaited being the 87,000-km flyby of Enceladus. Furthermore, exploiting their experience with Voyager 1, the scientists were able to choose more appropriate exposure times and filter settings to make better observations of Saturn's weather system. And Voyager 2 offered a new opportunity. The photopolarimeter on Voyager 1 had been crippled by the radiation in the Jovian system. Although this instrument on Voyager 2 had suffered problems, tests during the post-Jovian cruise had established that it was still capable of certain functions, and it had therefore been decided to attempt unique observations of the planet's atmosphere and the ring system. In fact, from the spacecraft's viewpoint the rings would pass in front of the star delta Scorpii, and the photopolarimeter was to monitor how much of the star's light passed through the rings in order to map the distribution and sizes of the particles at a resolution of 100 meters, which was much better than imagery could attain. This observation was considered to be so important that the photopolarimeter had been allocated exclusive use of the scan platform for 2.5 hours as the spacecraft made its Saturn flyby.

Voyager 2 commenced its Saturn encounter on 5 June 1981 at a distance of some 77 million km, with the 10-week 'observatory' phase during which the atmospheric features of the planet were monitored. The ring system was now more 'open', and better illuminated, and therefore appeared much brighter. Although the resolution of the narrow-angle camera was a modest 2,000 km per pixel, the spokes were almost immediately evident, and a number of movies were taken to study their evolution as they emerged from the shadow of the planet. The scientists were surprised to be able to see structures in the atmosphere that had not been apparent to Voyager 1 until that spacecraft had been just a few days from its encounter. Meanwhile, the DSN's tracking indicated that Voyager 2 would have to make a course correction to move the closest point of approach to Saturn some 900 km closer to the planet, and on 19 August this was achieved without incident. Two days later, Voyager 2 turned its attention to the double-faced Iapetus, and by the time of closest approach on 23 August it had taken imagery with a resolution as good as 18 km that showed the bright side of the moon to be so heavily cratered as to resemble Rhea. Unfortunately, no detail could be seen on the dark hemisphere. While there was nothing to indicate the source of the dark material, the fact that it was also present on the floors of some of the craters adjacent to the dark hemisphere raised the possibility that it was of endogenic origin; perhaps a dark hydrocarbon-rich fluid that leaked out from the interior in a similar manner to which silicate lava welled up through fractures to cover the floors of many craters on the Moon. The mass and dimensions of Iapetus implied it to be even less dense than the inner icy satellites, suggesting that it was largely made of methane, ammonia and hydrocarbons. This supported the contention that the dark material was of endogenic origin, but if this was the case why was the leading hemisphere so comprehensively flooded? Alternatively, the material was of exogenic origin. Phoebe had a retrograde orbit and a dark surface,

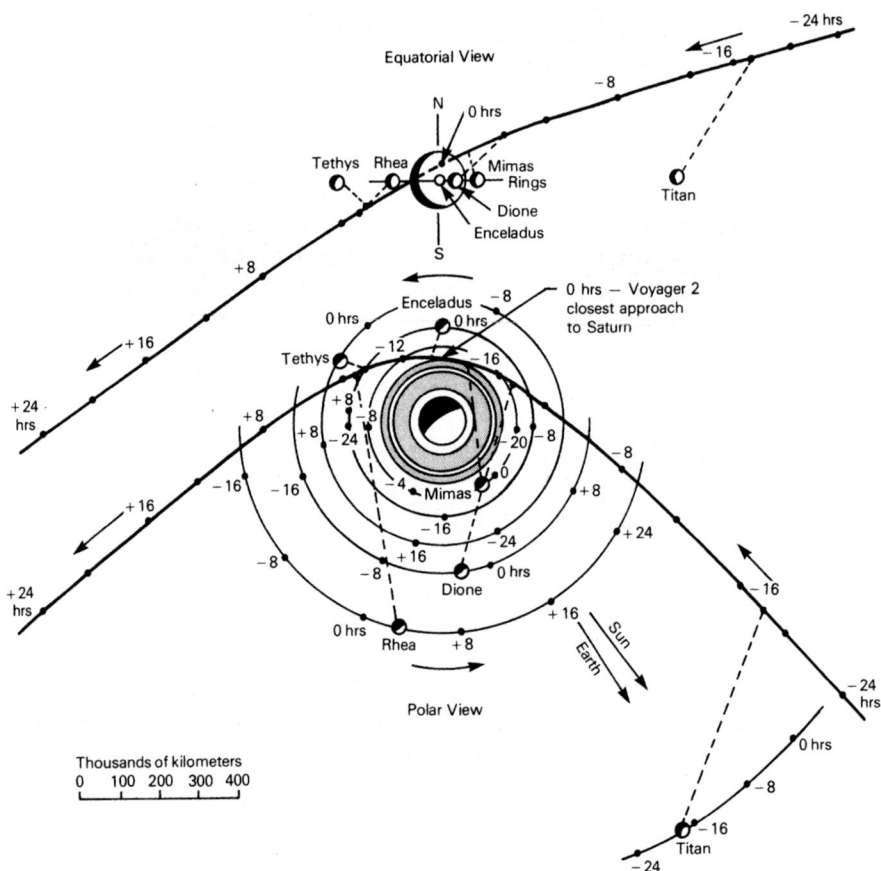

Voyager 2's trajectory through the Saturnian system. Phoebe, Iapetus and Hyperion orbit too far out to be shown.

and dust ejected by micrometeoroid impacts would tend to spiral into the system and be swept up by the leading hemisphere of Iapetus. The infrared radiometer found the dark hemisphere of Iapetus to be warmer than a highly reflective icy surface, but it was logical that a material as dark as coal would soak up solar energy.[156,157,158]

Later on 23 August Voyager 2 took its first picture of Hyperion, orbiting between Iapetus and Titan, from a range of 1.2 million km, revealing it to have a surprisingly elongated shape. In fact, the initial view happened to capture it edge-on, and images taken the following day during the 480,000-km flyby revealed it to be shaped like a hamburger with dimensions of roughly 400 × 250 × 200 km, and to have an icy surface pocked by craters. The largest crater observed was about 120 km across. However, an arcuate scarp hinted at the relic of a crater 200 km in diameter, and prompted the suggestion that Hyperion might be a fragment of a larger satellite that was destroyed by a catastrophic impact. What was immediately clear was that

One of the best images of Iapetus taken by Voyager 2. Note that dark material is visible on the floors of some of the craters on the bright terrain.

Hyperion's longest axis was not pointed toward Saturn for gravitational stability. In fact, an analysis of all the imagery was unable to determine either the orientation of the spin axis or the rotation rate. At the time, this difficulty was attributed to the short interval available for observation, but a theoretical model published some years later provided a more fascinating explanation: the dominating presence of Titan had forced Hyperion into an eccentric orbit that completed three revolutions in the time it took Titan to make four orbits, and these perturbations, in combination with Saturn's tidal forces, had given rise to a precarious balance in which the orientation of Hyperion's spin axis and its rotational period could dramatically vary over an interval as brief as several orbits – making this the first case of 'chaotic' rotation identified in the solar system. In fact, the model suggested that in just two orbits of Saturn – about 6 weeks – the moon could transform from a state in which it was essentially not rotating, to one in which it had a rotational period of 10 days, with the orientation of its axis varying on a continuous basis. This meant that even if Voyager 1's trajectory had enabled it to take pictures of Hyperion with sufficient resolution to enable its rotational state to be measured, it would have been impossible to link this to the state of rotation observed by Voyager 2. Furthermore, the model indicated that Hyperion would remain in this chaotic state. Although the observations provided by Voyager 2 covered too short a time base to prove or disprove this mathematical theory, a long series of telescopic observations in 1987 specifically to test the idea led to the conclusion that this moon "is not in any regular/periodic rotation state". Since then several other examples of chaotic behavior have been discovered in the solar system.[159,160,161]

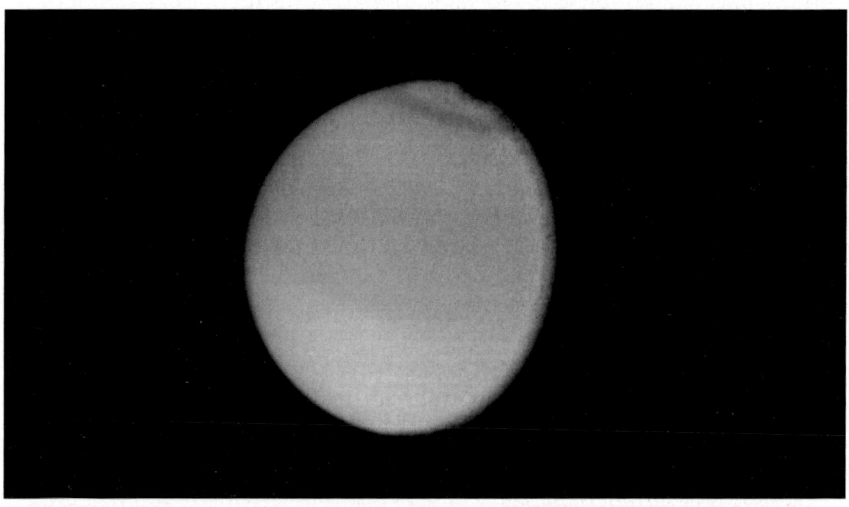

Four views of the tumbling irregular 'hamburger moon' Hyperion.

At the time they were received, the pictures of Hyperion were scrutinized for any asymmetry in its albedo that would corroborate the idea that Iapetus had swept up dark material. Although the absence of such variation was then considered not to lend support to this theory, the subsequent discovery that Hyperion's rotation was chaotic confirmed that there could be no such asymmetry.[162,163]

The trajectory of Voyager 2 was particularly well suited to viewing Titan's north polar hood.

On 24 August, Voyager 2 crossed the bow shock on three occasions between 31.7 and 28.1 Rs. Unfortunately, this time there was not another spacecraft sunward to warn of gusts in the solar wind, and hence the location of the bow shock could not be accurately predicted. Indeed, several hours later, a gust in the solar wind caused the bow shock to retreat faster than the spacecraft was approaching the planet, leaving the craft back in the interplanetary medium. The fifth and final passage was made early on 25 August, at about 22 Rs. It crossed the magnetopause at 18.6 Rs. The pressure of the solar wind was so high that the magnetopause was inside the orbit of

Titan. It was about this time that the spacecraft made its closest approach to Titan, at a range of 500,000 km. The vantage point provided a view of the moon's northern hemisphere, ideal for observing the north polar hood, and then the night-side with the high-altitude haze layers back-lit on the limb. There was no measurable change to the dichotomy of the northern hemisphere being darker than the southern one since Voyager 1, but Voyager 2 found the altitude of the visible haze layer to be some 50 km higher in the south. The spacecraft also used the available functionality of its polarimeter to measure the size of the aerosols in the haze layers.

Being unable to see its surface, the geologists on the science team had temporarily lost interest in Titan; they longed for a closer look at Enceladus to see if it really did bear any similarity to Europa, and as Voyager 2 penetrated further into the system it returned increasingly fascinating pictures of this icy moon.

Frustratingly, the high-resolution pictures of the F ring had so far failed to show any of the kinks, clumps and braids that had been observed by Voyager 1. However, Helene, Telesto and Calypso were imaged. These small moons had been discovered telescopically when the rings were edge-on to Earth in 1980. Given the preliminary designation of 1980S6, Helene shared its orbit with Dione, leading it by 60 degrees. Previously only the Trojan asteroids that lead and trail Jupiter in its orbit of the Sun were known to occupy such gravitationally stable positions, whose existence had been inferred in 1772 by the French mathematician J.L. de Lagrange. Telesto and Calypso respectively lead and trail Tethys. Unfortunately, the ranges were such that these satellites were barely resolvable. However, it was evident that they were all in the 30 to 40 km size range.[164]

Pictures of Tethys were also obtained showing features as small as 5 km in size. The dominant feature was an impact crater (called Odysseus) almost 400 km across, making it even larger in relative terms than Herschel on Mimas. But in contrast to the sharply defined Herschel, the rim of Odysseus had slumped and the floor of the crater had risen almost to re-establish the moon's spherical profile, making it appear much less conspicuous. The Tethys flyby revealed that the vast canyon that was seen on the Saturn-facing hemisphere by Voyager 1 actually girdled at least 75 per cent of

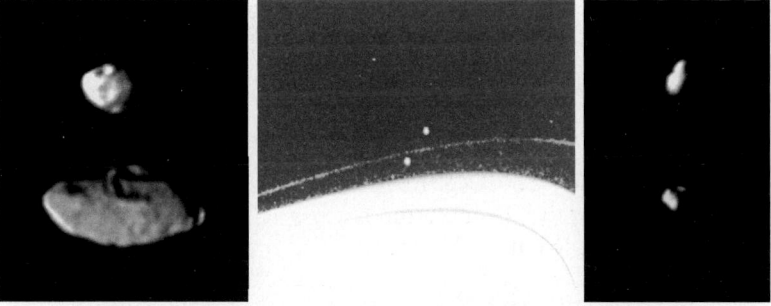

Voyager 2 was able to image four of Saturn's minor satellites: Pandora (upper left) and Prometheus (lower left), which are the outer and inner shepherds of the F ring respectively (central image); and Calypso (top right) and Helene (lower right).

the circumference. It was named Ithaca Chasma. One suggestion was that it was a fracture that formed as the interior ice froze and expanded; but it was not clear why this should produce a single vast canyon rather than a network of fractures all across the entire surface. Another suggestion was that since the canyon seemed to occur in an equatorial relationship to an axis that had one pole at the center of Odysseus, the canyon might have been produced as a side-effect of the massive impact. Voyager 2 took three temperature measurements of the surface. Like other Saturnian satellites, Tethys gave hints of having undergone internal activity at some point in its history, with some older-looking regions being darker and more heavily cratered.[165,166,167]

About 3.5 hours prior to Voyager 2's closest approach to Saturn, its platform was oriented to point the photopolarimeter to where the star delta Scorpii would emerge from behind the planet's disk, and the instrument began to take data. No sooner had the star appeared than it passed behind the ring system, and for the next 2 hours the instrument measured its brightness 100 times per second in order to provide a radial sample across, in turn, the D, C, B, A and finally F rings at a resolution equivalent to

A full-disk image of Tethys by Voyager 2, and (top) the only high-resolution picture of this moon, showing a section of its cratered limb.

about 100 meters. The 40-km-wide main strand of the F ring comprised no fewer than 10 ringlets. Furthermore, in some places there were clearly defined regions that were no larger than 3 km wide and had abrupt inner and outer edges. But the agent that confined the material so tightly was not evident. Because the photopolarimeter had sampled only a very narrow section running across the ring, nothing could be inferred about how the material varied along its circumference. The data showed there to be multiple ringlets within the Encke Division, but Voyager 2 imagery had shown just one. There were wave-like patterns on the A ring inward of the Encke Division and spiral waves on the B ring caused by satellite resonances. The absence of very narrow gaps implied that there were few if any moonlets present in the rings that were beyond the resolution of the cameras. The results also placed much more accurate constraints on the thickness of the rings than could be measured telescopically at times when the system was edge-on to Earth: the material could not be thicker than 200 meters where the rings yielded to gaps.[168] The ultraviolet spectrometer also monitored the occultation of delta Scorpii through Saturn's atmosphere, and the infrared spectrometer made thermal scans of the A, B and C rings.[169,170]

Although Voyager 2 flew within 310,000 km of Mimas, it was busy observing the stellar occultation, and after that sequence was concluded it swung its scan platform onto Enceladus. After Voyager 1, scientists had come to believe that this moon, like Europa, had a sparsely cratered icy surface, but Voyager 2's long-range imagery had shown craters. Now the closer views showed large areas that were totally devoid of craters. At least six types of terrain could be discerned in the imagery from a range of 119,000 km at a resolution of 2 km. Judging from the sizes and numbers of their craters, two terrains were ancient. Three others terrains were plains on which the cratering implied a range of intermediate ages. Finally, there was the youthful craterless plain that bore a complex network of ridges and grooves. Apparently, several times in the moon's history material had extruded from the interior and resurfaced vast areas; the most recent episode evidently having occurred during the last hundred million years. Although there was no evidence of volcanic vents, it was speculated that there might still be residual activity and that it was this that was seeding the E ring with particles of ice. The infrared spectrometer made five thermal scans of Enceladus, showing its temperature to be similar to that of Tethys.[171,172,173]

At 01:21 UTC on 26 August, Voyager 2 flew within 101,000 km of Saturn's cloud tops. About 20 minutes later it passed Enceladus at a range of 87,140 km, and was to image it at high resolution over the following 2 hours. Only 36 minutes after its closest approach to Saturn, the spacecraft passed behind the planet's late-afternoon limb at 36.5°N, with the radio occultation 'sounding' the Northern Temperate Belt, and then entered the planet's shadow. During the 90 minutes in which the spacecraft was out of contact, it was to pass through the ring plane at 2.86 Rs in order to depart on a trajectory that would lead to an encounter with Uranus. While Pioneer 11 had shown the path to be survivable, there was always the risk that Voyager 2 might be disabled by a chance impact and the mood at JPL was tense. In fact, as the spacecraft passed through the ring plane at a speed of 13 km/s it detected a remarkable

As Voyager 2 flew over the illuminated face of the ring system it provided detailed images the dark spokes on the B ring.

phenomenon. For several minutes before and after the crossing, the spacecraft was hit by thousands of micron-sized dust grains which, although too small to cause damage, made 'puffs' of plasma as they were vaporized and these were detected by both the plasma wave and radio-astronomy instruments. When the taped telemetry was transmitted to Earth, it became evident that the attitude control jets had been fired many times during the ring-plane crossing to overcome the perturbations resulting from the impacts of the dust grains. From its duration, the 'Ring Plane Event' indicated that the G ring, which the spacecraft traversed, was some 1,500 km thick, which was considerably thicker than had been inferred from imagery.[174] The spacecraft re-emerged from the almost diametrical occultation just before dawn at 31°S, and the radio signal provided information on conditions in a 100-m/s eastward jet. Apart from a variation of less than 10°C at high altitude, the temperature profiles were remarkably similar, and in agreement with the corresponding observations by Voyager 1.[175]

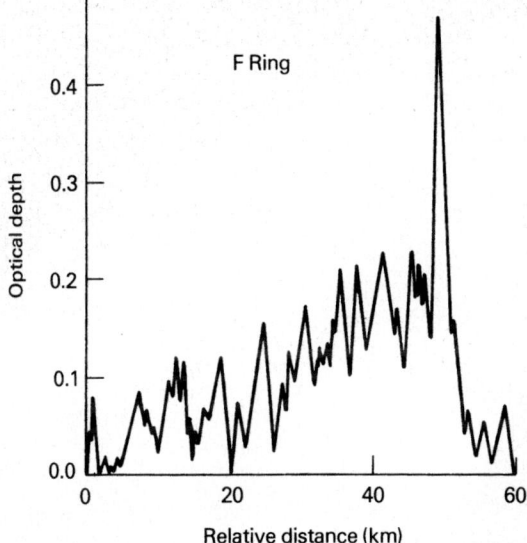

This profile by the photopolarimeter on Voyager 2 of how a star was occulted by the F ring revealed that the primary strand was accompanied by a number of faint ringlets.

The reacquisition of the radio signal from Voyager 2 as it emerged from behind Saturn spurred a deafening round of applause in the control room at JPL, as it meant that the spacecraft had survived the ring-plane crossing. However, the elation turned to despair when the telemetry indicated that the scan platform was not pointing in the right direction. Its mechanism to slew in elevation was functional, but that for slewing in azimuth had jammed. On diagnosing the fault, the spacecraft computer had halted all sequences that required the use of the platform. To make the situation worse, the position in which the system had come to rest was allowing sunlight to enter the optics of the sensitive instruments. The first task was therefore to use the functioning mechanism to turn the platform to protect the instruments. While this was underway, Voyager 2 flew by Tethys at a range of 93,000 km, the scheduled high-resolution observations abandoned. Meanwhile, the images that had been taped in Saturn's shadow were replayed. Most had been spoiled by incorrect targeting as a result of a calibration error in the attitude control system, and showed blank sky, but one picture of Tethys taken from a range of 120,000 km showed the moon's cratered limb at resolution of 2 km. However, wide-angle images of the rings had been successfully acquired, including views that resolved the F ring into one bright and four faint strands with longitudinal variations, and a single picture of the ring system dramatically foreshortened because it was taken just before the vehicle crossed the ring plane. Other images taken for this sequence probed the region between the E ring and the orbit of Titan, and confirmed that there were no hitherto unidentified rings occupying this region – at least none as prominent as the E ring.[176]

Some 55 minutes after the ring-plane crossing the azimuth mechanism had seized.

Voyager 2 made a close flyby of Enceladus. This is one of the best images showing icy cratered terrains on the right and young striped terrain on the left.

In fact, two other pictures were taken before the computer intervened, and by chance one happened to be aimed at a section of the Encke Gap in the A ring, and showed a narrow ring running near one of the gap's edges. The lost data included: close views of the limbs of Enceladus and Tethys; infrared spectrometer observations scheduled for while the spacecraft was in Saturn's shadow; infrared measurements to monitor ring material as it entered Saturn's shadow in order to determine its thermal inertia; an ultraviolet spectrometer observation of the rings occulting the Sun; stereoscopic imaging of the F ring; an outbound 1-hour photopolarimeter observation of the rings occulting the star beta Tauri; and a particles and fields roll maneuver.[177,178]

After the taped data had been downloaded, the engineers set out to determine just what had afflicted the scan platform and, if possible, to try to develop a procedure to reinstate the lost functionality. The worry, of course, was that they would find that the drive had been damaged as a result of ingesting particles during the ring-plane crossing, in which case it would probably be irrecoverable. Although the azimuthal mechanism responded to a test by smoothly slewing through 10 degrees, its motion was erratic during a second test. To further complicate the situation, the passage through Saturn's shadow had upset the subtle model used to predict the temperature of the faulty receiver, which meant that commands had to be repeatedly transmitted at slightly different frequencies in order to 'hit' the narrow band, and, of course, the

This breathtaking oblique view of the illuminated face of Saturn's ring system was taken by Voyager 2 shortly before it crossed the ring plane. It shows an entire ansa, but in a foreshortened perspective that magnifies the F ring (near the bottom of the frame) and progressively compresses features ranging across to the other side. The bright streaks on the B ring are the spokes forward-scattering sunlight.

pace of events was slowed by the lengthy round-trip light-travel time resulting from the distance. If the scan platform were to prove to be unusable, it would seriously diminish the science expected from Uranus – if indeed the vehicle were to survive the trip – because the scan platform would have to be locked and the entire spacecraft slewed in order to aim the instruments. On 28 August, as the analysis of the problem continued, scientists and controllers decided to make a series of small azimuthal steps, in the hope of returning Saturn to the field of view of the camera in order to undertake a planned outbound sequence.

Finally, early on 29 August, after almost 3 days without any pictures having been taken, an image was received showing Saturn's shadow projecting on the rings, seen from a vantage point at a high southern latitude. Additional pictures of the receding planet were taken as engineers struggled to regain full control of the scan platform in the hope of being able to attempt the final objective of the Saturnian encounter on 4 September: a distant flyby of Phoebe, the outermost known moon and probably an interloper in the system. In the middle of all this, the spacecraft had to perform a maneuver to refine its trajectory for Uranus. Voyager 2 flew by Phoebe at a range of 1.5 million km, and took pictures that showed it to be a spheroidal object 200 km in diameter. Its surface was generally dark, with an albedo of less than 5 per cent, but there were also intriguing bright features. Although the color of this material did not exactly match that on the leading hemisphere of Iapetus, this did not in itself rule out Phoebe as the source of that material. Having an orbital eccentricity of 0.16, Phoebe could not rotate synchronously. A study of the unresolved bright features indicated a

Despite the problem with its scan platform, as Voyager 2 withdrew from Saturn it was able to document the unilluminated face of the rings at an oblique angle. From this perspective the B ring appears dark and the Cassini Division is bright. A bright clump is evident in the F ring.

rotational period of about 9 hours.[179] After the Phoebe observations, use of the scan platform was discontinued pending a decision on its future status, and the Saturnian encounter officially ended on 25 September.

As with every planetary encounter, that of Voyager 2 at Saturn had resolved some issues and raised new ones. Owing to the unusual trajectory in which it approached from the north, made a single ring-plane crossing, and then departed heading south, Voyager 2 was better able than its sister to observe Saturn at high latitudes, and its perspective of the north polar atmospheric structure gave wind-speed profiles. The scan platform problem prevented imaging the southern hemisphere at a comparable resolution, with the result that the best resolution of 50 km per pixel in the north was several times better than that in the southern hemisphere. The appearance of the atmosphere had changed between the two Voyager encounters, with bands changing

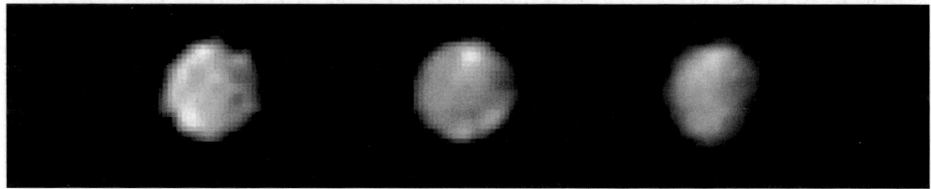

Long-range views of Phoebe by Voyager 2.

color, or splitting. The wealth of detail in the Voyager 2 imagery gave Saturn's atmosphere a striking similarity to that of Jupiter. A number of anticyclonic oval spots were seen that might be as long-lived as those on Jupiter, in addition to white, brown and red ovals that received informal names such as 'Big Bertha', 'Anne's Spot', etc. One of these, 'Brown Spot 1', some 5,000 km in length, was observed to interact with a pair of white spots over an interval of several days. Some of these features were still conspicuous when on the limb, which suggested that they were at altitudes above the haze layers that obscured most atmospheric details. Irregular and short-lived clouds were present only at intermediate latitudes, as was a feature that appeared to be peculiar to Saturn – a dark wavy ribbon that had anticyclonic and cyclonic vortices at its southern and northern edges, respectively. Wind speed measurements suggested that although these features were at high altitudes, their 'roots' were very deep in the atmosphere.

The color coverage of the ring system by Voyager 2 was at higher resolution than that by its predecessor, revealing, for example, the C ring to be distinctly bluer than the main rings, possibly owing to differences in the compositions of their particles, or perhaps to a greater number of particles of a size suitable for scattering light of that wavelength. The morphologic features such as gaps and ringlets also displayed color differences. High-resolution imaging sequences tracked some of the spokes over an entire orbit, passing through the planet's shadow to re-emerge into sunlight. Furthermore, whereas the narrower spokes – which, being so well-defined were probably the most recently created – rotated synchronously with the magnetic field, the broader wedges that had made several orbits tended to follow gravity-dominated trajectories. During the early approach, before the spacecraft had to pursue its hectic schedule of satellite observations, it took pictures covering an entire revolution appropriate to the Cassini Division to undertake a search for moonlets orbiting there, but the results showed that there was nothing larger than 10 km. Two apparently discontinuous rings that varied in brightness along their circumference were located in the Encke Division. (It was not possible to tell from the Voyager 2 imagery whether these were separate rings or bright arcs on a single eccentric ring, but Voyager 1 clearly showed them as isolated bands.) Imaged at high resolution immediately after the ring-plane crossing, one of these rings had a clumpy appearance, possibly due to perturbations from an unseen satellite orbiting within the Encke Division. In fact, a satellite residing in this gap would explain the wavelike patterns on the adjacent A ring. Various parts of the F ring were regularly observed for kinks and clumps – in particular near the shepherd satellites, and in

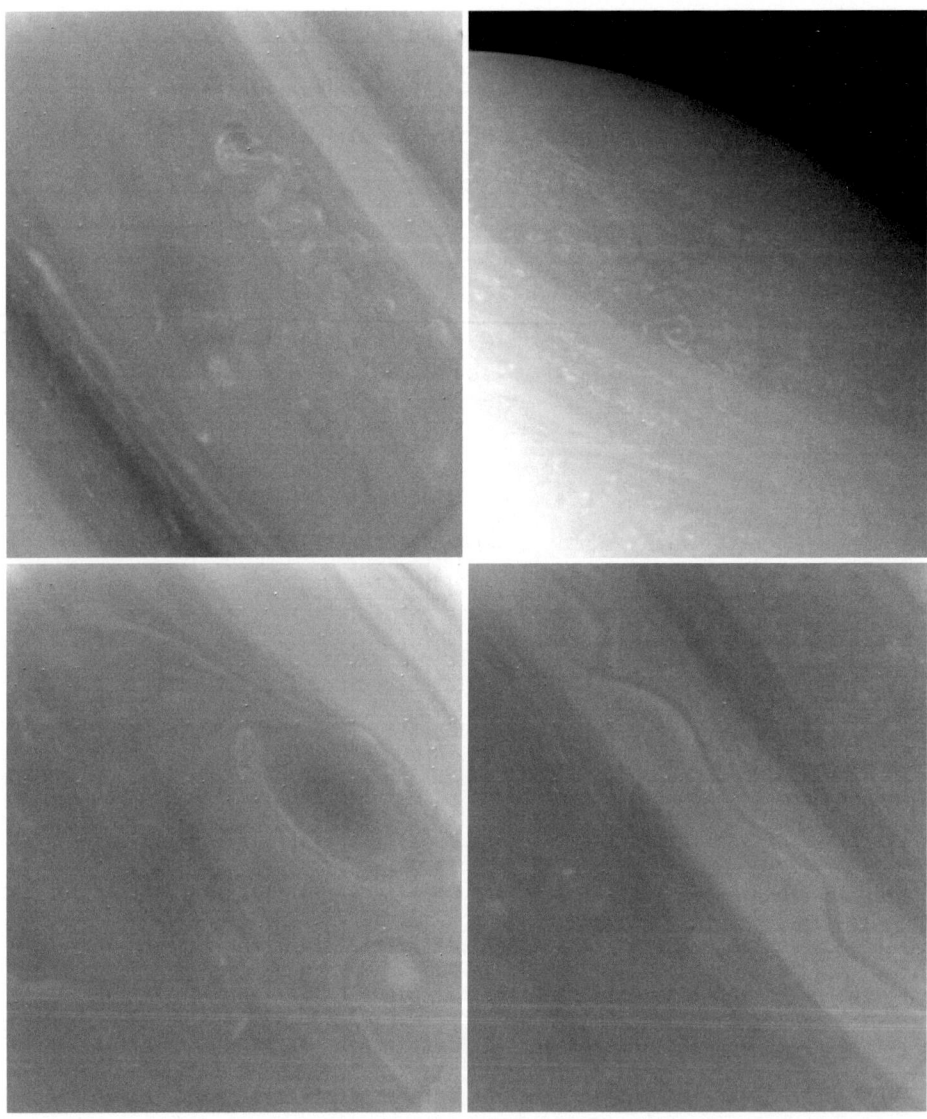

Structures in the atmosphere of Saturn: a plume of rising material (top left); the mottled polar regions (top right); a dark spot (lower left); and a wavy dark current (lower right).

the vicinity of where these moons were most recently at their least separation. Although they were initially not evident, Voyager 2 eventually observed kinks in the F ring, but other than in one small section there was no evidence of the braiding seen by its predecessor.[180] Most clumps persisted with little or no changes during the interval of the encounter, but both spacecraft saw knots suddenly appear, rapidly brighten and then disperse over several days – possibly clouds of dust blasted off unresolvable bodies within the F ring that had been hit by interplanetary particles several centimeters in size. A related theory was that the spokes on the B ring were clouds of electrically charged particles released by collisions between the objects that orbit in this dense ring.[181]

Subsequent analysis of the imagery from both Voyager spacecraft led to a number of reports between 1981 and 1995 of possible small moons. These included further moonlets at the Langrangian points of Tethys and Dione, and possibly also Mimas. Moreover, a dip in the counts of charged particles might have been caused by a small satellite in an orbit similar to that of Mimas. Although the paucity of observations meant that most of these 'discoveries' could not be verified, this did not mean that they were spurious. In 2004 the next spacecraft to reach the Saturnian system – the Cassini orbiter – found an object that proved to be 1981S14 (one of the candidates arising from the Voyager 2 image analysis); it was named Pallene.[182,183,184,185,186] Furthermore, in 1990 Mark R. Showalter of the Ames Research Center analyzed the perturbations seen on the edges of the Encke Division and, reasoning that they were caused by a small satellite in the gap, calculated its position and tracked it down in Voyager 2 images taken several days apart. To induce the observed gravitational perturbations, its mass indicated that it had to be about 20 km in size. It had not previously been recognized because even in the best images it spanned only a handful of pixels. It was initially designated 1981S13, then named Pan.[187] None of these small satellites could readily be observed from Earth, owing to the glare of the rings. But when the rings were edge-on to Earth in 1995 a new generation of sensitive instruments on large telescopes, including the Hubble Space Telescope, allowed the recovery of all the confirmed satellites except Pan. Atlas was about 25 degrees ahead of the predicted position, but that was not unreasonable given that it was an uncertainty of only 0.4 second in its orbital period projected over the 14-year interval. Prometheus was instead found mysteriously lagging by 20 degrees. There was a small object trailing behind Prometheus, but a detailed check of the Voyager 2 imagery did not reveal anything, and it (as with several other objects reported in 1995) was almost certainly a short-lived clump of F ring material. The 1995 ring-plane crossing also provided the first opportunity for terrestrial telescopes to observe the faint G ring.[188,189,190]

Of course, the other instruments on Voyager 2 had also collected very useful data. Some polarimetry of Iapetus, Rhea, Enceladus and Phoebe was obtained, but the problem with the scan platform meant that the coverage of these satellites was rather less comprehensive than planned. The infrared spectrometer gave good temperature profiles of Saturn's atmosphere, but it was unable to 'sound' the stratosphere owing to a hardware issue. The ultraviolet spectrometer confirmed the presence of a north polar aurora spanning all longitudes. Concerning the satellites and their interactions with the Saturnian environment, Voyager 2 identified another torus of hydrogen and oxygen ions, products of the dissociation of water molecules, in between the orbits of Dione and Tethys. Radio tracking of the spacecraft gave estimates for the masses of Tethys, Iapetus and Mimas – the latter because it is gravitationally related to Tethys. These results *reversed* the trend inferred from Voyager 1 in which the densities of the satellites increased with increasing planetocentric range. However, Titan did not fit the new trend. Because the frequencies of the plasma waves near Saturn were in the same range as sound waves in Earth's atmosphere, the scientists fed their data into a synthesizer as a demonstration for the public, and the eerie music that resulted was likened by a journalist to an electronic version of Bach.[191,192,193,194,195]

The gravity-assist from the Saturn flyby had further increased the eccentricity of Voyager 2's solar orbit to 3.45, and put it on course for an encounter with Uranus in early 1986. The engineers were confident that they would be able to restore the scan platform to at least partial functionality long before that time.

DULL PLANET, INCREDIBLE MOON

As soon as tracking by the DSN had accurately determined Voyager 2's post-Saturn trajectory, the spacecraft made a maneuver to adjust its aim for Uranus. The nominal flight plan provided opportunities for four 'tweaks' during the interplanetary cruise to eliminate residual trajectory errors and to refine the schedule for the forthcoming encounter to satisfy scientific requirements such as solar and radio occultations. The first tweak on 13 November 1984 was a 1.54-m/s burn to move the point of closest approach to Uranus 40,000 km nearer the planet. Two maneuvers about 1 month and about 5 days prior to the encounter would make last-minute trajectory refinements to arrange the most favorable illumination for the encounter. In flying beyond Saturn, the navigators at JPL were faced with the issue that had plagued the Soviet missions to Mars in previous decades – the fact that the ephemerides for the outer planets had uncertainties of several thousand kilometers, as against several hundred for Saturn. It was decided to make the scientific sequences sufficiently flexible to enable them to fit the actual situation when the encounter loomed. Another problem would be to determine the position and trajectory of the spacecraft. Although two of the three DSN stations were in the northern hemisphere, at that time both Uranus and Neptune were south of the celestial equator. It would therefore be necessary to place a greater reliance on optical navigation, in which the vehicle would take a series of images of its next destination against the background of stars in order to refine its ephemeris. In addition, the position of the spacecraft in the celestial sphere would be accurately determined by the novel technique of using arrays of radio telescopes to measure its offset with respect to astronomical radio sources.[196] Despite the spacecraft being at a distance from Earth in excess of 3 billion km, JPL was able to determine its position in space to within about 20 km.[197,198]

Although it was in deep space, Voyager 2 was upgraded to enable it to undertake the mission beyond Saturn. First, the attitude control system was modified. Since the radius of Uranus's orbit of the Sun is twice that of Saturn, the illumination at Uranus would be only one-quarter of that at Saturn, which in turn meant that exposure times for taking pictures would be four times as long. In order to reduce smearing of the images, the logic for the attitude control system was revised to fire the thrusters in shorter bursts to reduce the chance of the vehicle drifting in attitude, and all possible sources of perturbation were investigated to ensure that such activities were not scheduled at times when the spacecraft was ultra-sensitive. The new techniques were tested on Voyager 1, which, although reporting useful data on the solar wind, was, in a sense, expendable. The nature of the Uranian system posed a challenge. Whereas for both Jupiter and Saturn the planes in which their satellites orbited were only slightly inclined to the plane of the ecliptic, with the result that the spacecraft met them one by one over an interval of several days, the plane of the Uranian system was almost perpendicular and the spacecraft was approaching it like a dart thrown at a bull's-eye. Further, as a result of two gravity-assists, the spacecraft was traveling much faster than it would be if it had followed a slow elliptical interplanetary transfer, and hence would have very little time to make observations. Even with the new thruster logic, therefore, the spacecraft's sheer speed might cause 'motion blurring'. It was decided to have the spacecraft rotate at a rate that would maintain the camera firmly fixed on its target. This 'panning' method had been introduced by Voyager 1 in order to obtain high-resolution images of Rhea. And by lucky coincidence (in view of the spacecraft substitution made prior to launch) the vidicons on Voyager 2 were about 50 per cent more sensitive than those of its twin, and hence better suited to imaging in the dim illumination far from the Sun. The fact that there would be so little time available for making observations in the Uranian system was exacerbated by the problem with the scan platform. A lot of time and money went into understanding precisely what had happened, and to devise constraints on the future use of the scan platform. JPL built 86 full-scale models of the Viking-heritage actuator in order to evaluate the influence of parameters such as the temperature of the gears, their speed, and the applied torque. It was realized that frantic slewing to address a multitude of targets at the time of the closest approach to Saturn had caused an almost total loss of lubrication in the azimuth drive. (Lubrication in the harsh environment of space is an art akin to black magic!) Without lubrication, and subjected to intensive operation, the high-speed gear train had heated, deformed within its very tight tolerances, and seized. The problem had not occurred using the original Viking actuator because on those missions the mechanisms were used less extensively and the missions were not as long. Fortunately, by allowing the scan platform some time to 'rest', the lubricant seeped back into the gear. It was therefore reasoned, and tests confirmed, that it could safely be operated at its slower slewing speeds. A maximum speed of one-third of the fastest possible rate was imposed. Nevertheless, just in case the actuator should seize again, various self-protection and contingency procedures were uploaded to prevent a complete loss of data.[199,200]

Voyager 2's power management was also studied. As usual, the trajectory was to pass both behind Uranus as seen from Earth and through the planet's shadow. At the same time as it was transmitting a high-powered carrier signal for the radio occultation to 'sound' the planet's atmosphere, the spacecraft was to take images of the night-side and of the recently discovered ring system in silhouette, storing the imagery on tape for later replay to Earth. The entire occultation sequence was thoroughly analyzed to ensure that the simultaneous activities would not exceed the available power, and a rehearsal was conducted during the cruise phase to confirm that the self-protection system would not intervene and interrupt the observations by turning off some systems.

Finally, the DSN was upgraded. If only the largest and most favorably situated antennas were used, the theoretical data rate would be 14.4 kbps. The existing 64-meter-diameter antennas were near the effective limit, as the performance of larger dishes would be degraded by gravity loads and wind gusts. However, by pooling Canberra's primary antenna with its two 34-meter dishes, and also with the 64-meter radio telescope of the nearby Parkes Observatory, made available especially for the purpose, the data rate of the received signal was able to be increased to 29.9 kbps. This was not the first time that Parkes had been used in support of space missions: it had assisted in the Apollo 11 lunar mission in 1969, and was to be the main antenna to track ESA's Giotto spacecraft to comet Halley several weeks after Voyager 2's Uranian flyby. A test during the Voyager 1 Saturnian encounter had demonstrated the feasibility of 'arraying' several antennas. To further improve the effectiveness of a given data rate, the hardware that was built into the spacecraft's data management system to provide data redundancy facilitated a very efficient Reed–Solomon data codification and error compensation system. Another increase was obtained by using an image compression algorithm by which only the difference in brightness between adjacent pixels was returned, rather than the absolute brightness of each single pixel. In this way, each transmitted pixel would require on average only three instead of eight bits. This was the first time the technique had been attempted on a planetary mission. In contrast to the Reed–Solomon coding algorithm, which was provided by dedicated hardware, image compression had to be implemented by reprogramming the data management system. Despite having a tone-deaf receiver and a sticky scan platform, Voyager 2 was nevertheless a lucky ship, because the image compression algorithm required the use of some capabilities of the data management system that a hardware failure had denied to its sister. Had data compression not been used, the spacecraft would have been able to return only about 60 images per day from the Uranian system instead of 200. The prime flight data computer was programmed to deal with all non-imaging data, while its backup processed imagery for transmission. However, in case of trouble, a single program was prepared that would enable either computer to handle all the data, albeit at a much slower pace. And, as at Jupiter and Saturn, to guard against the total loss of the command receiver a program was provided to enable Voyager 2 to autonomously perform a basic set of observations and transmit the results to Earth.[201,202]

Uranus was discovered by William Herschel in 1781, but even after 200 years of observation very little was known about the planet, and even less about its satellites apart from their orbits. It was not until the invention in the late 1970s of infrared telescopes and detectors that were sufficiently sensitive to take spectra of such faint objects that it became possible to measure the surface temperatures of the satellites and, with assumptions as to their albedos, their sizes. The most prominent feature of their spectra was water ice, but owing to the temperature at that distance from the Sun, other ices, particularly ammonia, were also expected. In 1981 their diameters were measured to within a few tens of kilometers, and the manner in which they influenced one another's orbits enabled the masses and densities of all five of the satellites known at that time to be calculated. The results showed a trend in which the densities of the moons increased with planetocentric distance. Thus, Oberon and Titania appeared to be mostly rock coated with a thin layer of surface ice, whereas the densities of Umbriel and Ariel suggested that they were similar to Saturn's icy satellites. The diameter of Miranda, the innermost moon, was not accurately known but, against the trend, it appeared have the highest density – a fact that made the geologists particularly eager to study it.[203] The observed perturbations of Miranda's orbit suggested the presence of at least one other satellite. It would probably be too close to the planet to be seen from Earth, but if it existed then Voyager 2 should have no difficulty in resolving it.

On 10 March 1977 astronomers had set out to observe Uranus occult a star in order to determine profiles for the temperature and composition of the planet's atmosphere, and were surprised to observe the star 'wink' five times in rapid succession half an hour before its passage behind the planet, and again afterwards. The possibility of an instrumental problem was ruled out by the fact that this pattern was observed by two telescopes. When it was noted that the timing of the two anomalous sequences was almost symmetrical with respect to the actual occultation, it was realized that the planet had a system of very narrow rings. Remarkably, a paper positing that both Jupiter and Uranus ought to have originally had a ring system was rejected by the planetary journal *Icarus* in 1972 because a referee questioned its theoretical basis.[204] Further observations in the following years showed that Uranus possessed at least nine rings. In order of increasing planetocentric distance, these were designated 6, 5, 4, alpha, beta, eta, gamma, delta and epsilon. They proved to be much thicker in the 'vertical' direction than the rings of Saturn, ranging between 1 and 100 km. Their widths varied along their circumference, with epsilon, the largest, varying in width between 20 and almost 100 km. Their orbits were slightly eccentric, and in some cases were slightly inclined to the mean plane of the system. Their narrowness argued for the existence of shepherding moons like those that bounded the F ring of Saturn. Again, if such satellites existed, then Voyager 2 should have no difficulty in resolving them. Infrared observations revealed that the rings had low reflectivity and hence must be quite dark. Nevertheless, the sensitive electronic detectors introduced in the 1980s enabled them to be photographed. In view of their dark composition, and the fact that the constituent particles were probably of very small size, the rings were expected to be almost invisible during Voyager 2's approach phase, and therefore the observations intended to try to study them in detail were scheduled for when they would shine by forward-scattering sunlight.[205]

As noted earlier, the spin axis of Uranus was inclined by 95 degrees to its orbital plane. This meant that the planet underwent an unusual seasonal cycle during the 84 years that it took to orbit the Sun. At some times it would 'roll' along its orbit, maintaining one or other of its poles facing sunward. It would reach southern summer solstice in 1985, with the result that as Voyager 2 made its approach it would see only the southern hemisphere. From the spacecraft's perspective, therefore, the view would change little on the inbound leg, and most of the action was scheduled for just before it crossed the equatorial plane. In fact, the closest encounters with the various satellites would all occur in the space of 6 hours. There would be little to see on the outbound leg, because the northern part of the system was in darkness. If Uranus were to prove to possess a magnetic field, then scientists were eager to find out if the magnetic axis was tilted at a similar angle to the rotational axis, in which case at this time in its orbit the planet would be facing a magnetic pole to the solar wind, which would enable the plasma to penetrate deeply into the outer atmosphere to give rise to unusual auroras and other related phenomena.[206] Although the location of the point of closest approach at a planetocentric distance of just over 100,000 km was defined by the gravity-assist required to fly on to Neptune, the timing of this event could be selected to optimize observations of the Uranian system. In particular, it was decided to make a close pass by Miranda inbound, and have the spacecraft pass behind the rings and the planet at a time that would enable the antennas in Australia to monitor the radio occultations.[207]

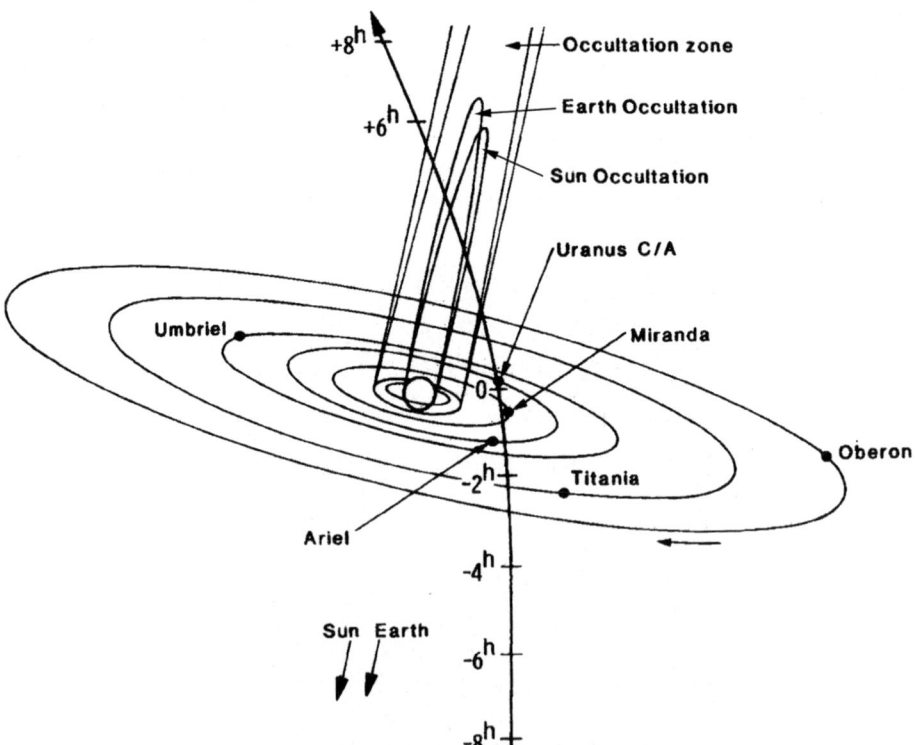

The geometry of Voyager 2's encounter with Uranus in the hours nearest to closest approach. Owing to the extreme tilt of the system relative to the ecliptic, the spacecraft appeared to be speeding toward a giant bull's-eye target.

The first navigational pictures of Uranus were taken in June 1985. They showed the planet and its four outer moons against the background of stars, and enabled the navigators to check the trajectory and the scientists to refine the ephemerides for the satellites. In October the near-encounter sequence was successfully rehearsed, which cleared the way to use the scan platform. The distant observatory phase started on 4 November at a range of just over 100 million km. At that time the spacecraft was 2.88 billion km from Earth, and speeding toward its target at 15 km/s. Its radio signal took 2 hours 25 minutes to reach Earth. During the observatory phase the spacecraft was to take five movies, each of some 300 images taken through a variety of filters over a period of about 38 hours – equivalent to about two Uranian days – in order to monitor the dynamics of the atmosphere and, if possible, determine the true rotational period; telescopic studies had produced conflicting estimates ranging from 10 hours to 24 hours, with the 'best' estimate being 15–17 hours. In addition, three ultraviolet scans were to be made each day.[208] The first movie sequence began on 6 November. The narrow-angle camera resolved the planet as a pale green disk, but, frustratingly, it was devoid of detail; evidently Uranus was not going to yield its secrets easily. To an atmospheric scientist, however, the fact that the atmosphere was featureless was itself a valuable observation. A few days later, long exposures managed to capture a trace of the outermost and brightest member of the ring system – the epsilon ring. Until then, scientists had been skeptical that the rings would be seen so early in the mission. As better images were taken later in the month, epsilon proved to be more prominent in some sections than in others, correlating with the predicted locations of the broadest and narrowest portions respectively.

By late November 1985 Voyager 2's unfiltered long-exposure images were showing the epsilon ring. Note that the brightness of the ring does not appear uniform around its whole circumference.

The other instruments were also returning data. The fact that the radio-astronomy instrument had not yet detected radio emissions from the planet tended to imply that if the Uranian system possessed a magnetosphere, then its radiation environment could not pose a threat to the spacecraft. Communications entered a hiatus in early December, as Uranus passed through solar conjunction. JPL had historically avoided scheduling planetary encounters near solar conjunctions, but the window for the gravity-assist to Neptune required it, and a 10-day outage was the consequence. On the other hand, it provided an opportunity to use the spacecraft's double-frequency signals to study the solar corona. A development shortly after contact was restored served as a reminder of just how little was known about Uranus: discrepancies between the predicted and the actual trajectories obliged the navigators to conclude that, notwithstanding 200 years of telescopic observations, the mass of the planet had been underestimated by 0.3 per cent, which, all things considered, was a fairly large error. This correction brought the calculated trajectory into line with the actual one; but it was not the one desired, and a 2.1-m/s maneuver on 23 December displaced the point of closest approach several hundred kilometers further from Uranus to re-establish the gravity-assist for Neptune. When the DSN determined the spacecraft's trajectory after the correction, it proved to be so close to the one required that for the first time in the mission the final targeting maneuver (the one scheduled for 1 week before the flyby) was canceled. In addition to avoiding an always-risky loss of contact while the craft turned to the best orientation to make a burn, this decision would permit continuous tracking.[209,210]

Uranus continued to appear featureless. If there was any detail to be seen, then it must be of very low contrast and was being masked by the 'zero phase' glare of the Sun, as this was almost directly behind the spacecraft. Only in early December was a mathematical model developed which enabled the glare on the planet's disk to be subtracted from the images, but while this method revealed some details, these were extremely small and only a few per cent brighter than their surroundings. Even when the first clouds were spotted, the camera team were uncertain whether they were real or image-processing artifacts.[211] On 30 December long-exposure images of the rings revealed a new satellite orbiting between Miranda and the epsilon ring. Initially designated 1985U1, it was later named Puck. Because the encounter was still several weeks away, mission planners were confident that they would be able to schedule an opportunity to image this object at close range. Two other satellites were spotted on 3 January 1986, another on 9 January, and a further three on 13 January. Initially designated 1986U1 through 1986U6, they were subsequently named Portia, Juliet, Cressida, Rosalind, Belinda and Desdemona respectively.

Early in January 1986 the two instruments with the most troubled records – the infrared spectrometer and the photopolarimeter – began to act erratically, but their engineers were confident that they would work satisfactorily at closest approach. The far-encounter phase began on 10 January. It called for another two movies of Uranus's atmosphere and many pictures to search for additional satellites. A final test of the scan platform was also made.[212] When the disk of the planet filled the frame of the narrow-angle camera about a week before closest approach, the wide-angle camera took over full-disk imaging and the narrow-angle camera observed the

A distant picture of Uranus showing a bright cloud (near 11 o'clock) and traces of banding in the atmosphere. The south pole is facing the spacecraft, with the result that the equatorial zone is around the rim of the disk.

few individual atmospheric features at an ever-increasing resolution. An issue that developed 6 days out caused black-and-white stripes to spoil the compressed images, but when this fault was identified as a defective memory bit in the data management system's computer, it was readily rectified.[213] Scientists had begun to doubt that Uranus had a magnetic field of its own, but 5 days out radio emissions and charged particles were detected, indicating that it had not only a magnetosphere but also radiation belts. Furthermore, the radio emission was modulated, suggesting that the magnetic field was inclined to the rotational axis. Two further moons were spotted on 20 January, apparently shepherding the epsilon ring. Designated 1986U7 and 1987U8, they were later named Cordelia and Ophelia respectively. Furthermore, by this time six of the nine known rings were showing in long-exposures.

Voyager 2 began the 4-day-long near-encounter phase on 22 January, but until the hectic hours just before it crossed the plane of the system, Uranus remained the only viable target for observations. The first images of the principal satellites were taken in December, but it was not until a few days before the encounter that they started to reveal surface markings and, remarkably, each proved to very distinctive in its own way. Oberon initially showed only patches of darker and lighter terrain; craters were not evident until 22 January, the day before the best images of this satellite would be obtained. Titania was more somber and only slightly mottled,

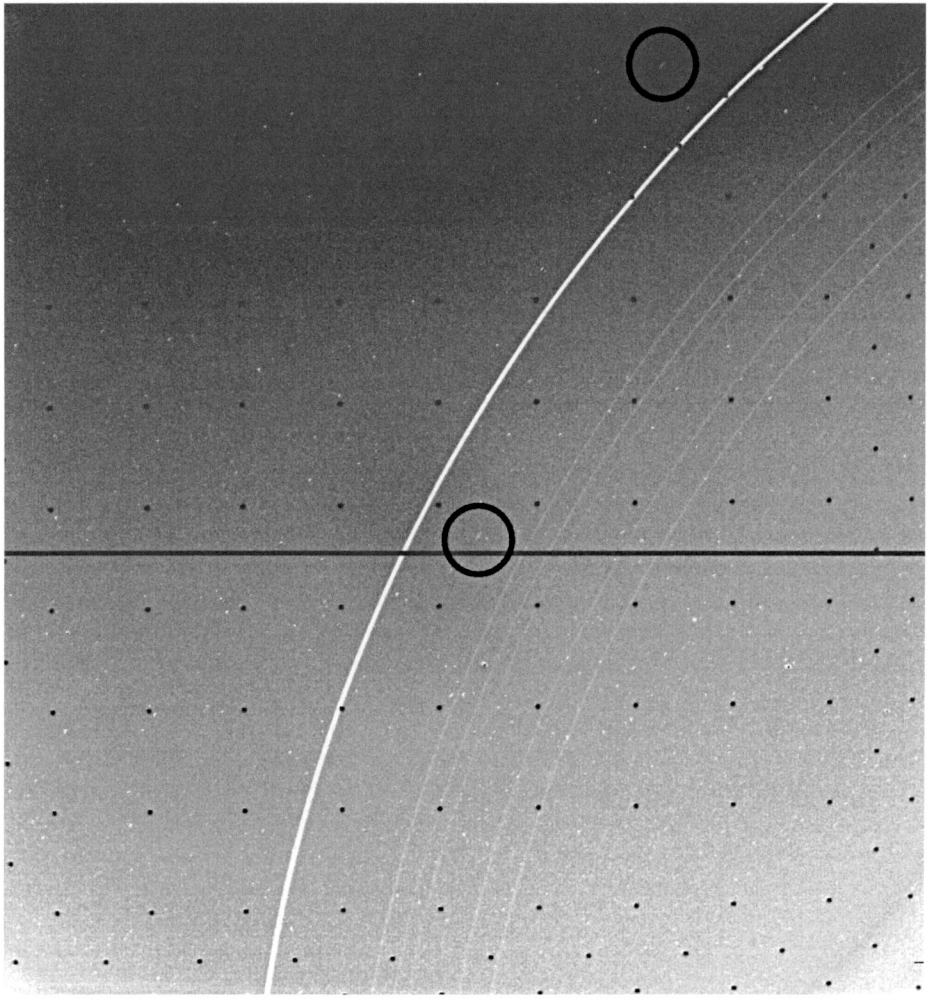

Most of Uranus's rings are visible in this picture. The two circled trails are the tiny moonlets Ophelia (at top, beyond the epsilon ring) and Cordelia (inside the epsilon ring).

until features became evident a few hours prior to closest approach. Umbriel was by far the darkest of the large moons, and almost the only feature that could be seen on its bland surface was a slightly brighter ring on the limb that might have been a crater. Ariel was the most variegated. From the moment its disk was resolved, Ariel displayed dark and bright areas similar to those of Oberon, but there were also bright lanes that may have been ridges or faults. Discovered by G.P. Kuiper in 1948, Miranda was the smallest of the moons known prior to the mission. As soon as its disk was resolved, it displayed an unusual patchwork of albedo features. On 23

January Voyager 2 was given the final pointing angles and timings for the schedule of observations, based on the trajectory derived from the inbound tracking. By this point, the relative positions of the spacecraft and Miranda were known to within 50 km, which was half the tolerance needed to point the scan platform at it accurately during the flyby.[214] Even at this late stage, only the epsilon ring and the broadest parts of the alpha ring were showing any detail. During the last 3 days of the approach, several sequences, each of hundreds of images, had been taken to search for additional shepherd satellites, but nothing larger than 10 km was discovered. However, on the eve of the encounter 1986U9, later named Bianca, was discovered. Apart from Puck, none of the new moons appeared to exceed a few tens of kilometers in size. In some images Cordelia was resolved into a score of pixels, but the others were seen only as dim streaks or, at best, as 'moving stars'. All were in circular orbits close to the equatorial plane, between Miranda and the epsilon ring – with the exceptions of Ophelia, the outer shepherd of the epsilon ring, which was in a slightly eccentric orbit, and Cordelia, the inner shepherd. The initial impression was that, apart from shepherds, the Uranian retinue had none of the strange characteristics of the Saturnian satellites.[215] However, observations over the ensuing decades would prove differently. When the Voyager 2 imagery was reanalyzed by Erich Karkoschka in 1999, a new small moonlet was discovered orbiting near Belinda. It was designated 1986U10, then named Perdita; an apt moniker, since this means lost in Latin! When the Hubble Space Telescope recovered it in 2003 it became evident that Perdita suffers complex perturbations from both Belinda and Rosalind, and that its orbit must have evolved considerably during historical times. The Hubble Space Telescope found another satellite orbiting between Miranda and Puck. Named Mab, it was traced back to a barely captured moving speck in four Voyager 2 images. Furthermore, the Hubble Space Telescope also revealed a second system of two extremely faint rings between the known ring system and the classical moons; these, too, were barely present in a sequence of images taken by Voyager 2 on its outbound leg, shining by forward-scattered sunlight.[216,217]

The busiest day of the encounter was 24 January 1986. Approaching more or less along the Sun–Uranus line, Voyager 2 crossed the planet's bow shock only once, at a planetocentric distance of 23.7 Ru, 10 hours prior to closest approach. From 18 Ru the spacecraft was fully immersed in the magnetosphere. An image 8 hours out that was originally assigned to Miranda had been retargeted to obtain a single picture of Puck, which was 500,000 km away and conveniently located on the same side of the planet as Miranda, which was itself close to where the spacecraft would intersect the plane of the system. Despite the relatively low resolution of 9 km, the image showed Puck to be about 190 km in size and roughly spheroidal. At least one crater 45 km in diameter was visible near the terminator. Intriguingly, Puck was much darker than Umbriel, the darkest of the large satellites; in fact, it was comparable to the particles that formed the rings. The fact that the other newly discovered moons appeared to be just as dark suggested that the preliminary estimates of their diameters, which had been based on an assumed albedo, would prove to be too small.[218] Pictures taken in the final hours of the approach also revealed a ring that had not been detected from Earth. In fact, Cordelia was found to orbit just inside this ring, which was initially

This is the only high-resolution picture of a new Uranian satellite to be obtained by Voyager 2. Its shows Puck, the largest of the newly discovered moons and at 190 km across one of the smallest quasi-spherical objects of the solar system. A large crater is evident. The first replay of this and other imagery was lost to rain interference, but fortunately the pictures were able to be safely downloaded before the tape had to be rewritten.

designated 1986U1R and then named lambda. Intriguingly, it showed hints of radial structure. At 13 hours to the planetary flyby, the photopolarimeter monitored the flickering of the star sigma Sagittarii (Nunki) over a period of 30 minutes during an almost tangential occultation behind the epsilon and delta rings. Whereas terrestrial measurements using this technique could resolve structures a few kilometers in size, the spacecraft, being closer, could achieve a resolution of just 10 meters.

Voyager 2 now turned its attention to the large satellites. Owing to the brief time that they would be in sight and the limit imposed on the rate that the scan platform could slew, it would be possible to take only a relatively small number of pictures, which would be stored on tape for replay the next day. The first target was Titania, which was passed at a range of 365,200 km some 3 hours before the planetary flyby. Oberon was 1 hour later at a range of 470,600 km. Barely 9 minutes later was Ariel at a range of 127,000 km. When it had time during this phase, the spacecraft was to snap pictures of Umbriel, although the 325,000-km closest approach would not be until 3 hours after the planetary flyby. If this phase of the encounter could have been shown by real-time television, in the style of the Mariner 10 flybys of Mercury, the slewing field of view would have presented a dizzying spectacle. But the star of the show was Miranda, which the spacecraft passed just 8 minutes before the ring-plane crossing and 55 minutes ahead of the planetary flyby. In fact, at only 28,260 km this was the nearest that the spacecraft had come to any moon of any planet to date. A series of eight narrow-angle images were taken, mostly of the terminator, which, owing to the geometry of the system, traced the equatorial region rather than pole to pole. To compensate for the relative speed of about 20 km/s, the spacecraft broke its link with Earth and rolled to hold the camera pointing at the target, then immediately resumed contact to enable the mass of Miranda to be inferred from the manner in which it deflected the spacecraft's trajectory. The ring-plane crossing was made at a planetocentric distance of 116,000 km. Even although the trajectory passed at least 60,000 km beyond the outermost known ring, the plasma instrument detected 'puffs' from 30 to 50 impacts per second by microscopic particles. The fact that the peak rate did not occur exactly at the time of ring-plane crossing, but was displaced by about 280 km, suggested that there was a 4,000-km-thick sheet of dust orbiting Uranus in a plane slightly inclined to the equator.[219,220,221]

At 17:59 UTC Voyager 2 made its flyby at a planetocentric distance of 107,000 km. Shortly after this, the scan platform slewed to enable the photopolarimeter to monitor the 18-minute occultation of beta Persei (Algol) by the rings at a resolution of 100 meters – intermediate between that of a terrestrial observation and that taken by the spacecraft during the inbound occultation of Nunki. Owing to the fact that Algol was a variable star, the instrument had been carefully calibrated during the interplanetary cruise. So favorable was the geometry that the outbound occultation provided two slices of each ring, in addition to the slices of the epsilon and delta rings obtained on the inbound occultation, thereby providing information on the eccentricities of the orbits and the distribution of matter along the rings. The outer edge of the epsilon ring was found to be sharply defined, and to be at most 150 meters thick. Features resembling the waves raised in Saturn's rings by satellites were also apparent. The observation proved that Uranus's rings had extremely variable thicknesses and opacities around their circumferences. For example, the eta ring did not occult Algol on ingress, which was evidently at a point where it was depleted of material, while narrow incomplete arcs were present all across the sampled region, including just outside the epsilon ring. In fact, ring 6, the innermost member of the main system, was found to have a companion of similar width and opacity.[222] About 90 minutes after the planetary flyby, the spacecraft passed behind the epsilon ring as seen from Earth, although in reality it was 150,000 km from the ring. This marked the start of the radio-occultation sequence. Since the radio would transmit a strong carrier signal for the next 3.5 hours, the spacecraft taped its observations during this period. Less than 1 hour into the occultation sequence, the spacecraft entered the planet's shadow, and moments later passed behind its limb at 2°S (owing to the tilt of the system, this latitude also approximated the line of the terminator). The craft was able to compensate for the refraction by turning slightly, and thus maintain almost continuous contact during this period. After spending 80 minutes in the planet's shadow and 86 minutes behind its disk, Voyager 2 reappeared at 7°S, giving a second atmospheric profile. While it was on the night-side the spacecraft had taken several very-long-exposure images, but these failed to show any lightning, or indeed an aurora in the north polar region, which had been in darkness for several decades. The radio occultations appear to have probed a cloud deck of methane ice crystals, and deep into the region below, but without precise knowledge of the composition the occultation data could only be used to put constraints on the temperature profile. Nevertheless, fluctuations in the signal strength indicated considerable atmospheric turbulence. The radio-occultation sequence concluded with a second sampling of the ring system, this time in reverse order. A number of structures were apparent across the span of the rings, including a particularly dense region near the outer edge of the epsilon ring. Although the occultation data suggested that the rings consisted mostly of particles in the centimeter-to-meter range, this was because the method was most sensitive to particles of this size; later optical observations would establish that there was also a great deal of dust.[223]

Uranus proved to have an unusually complex magnetic field, being inclined at an angle of about 60 degrees to the planet's rotational axis, and it had both quadrupole and octupole components as large as the dipole. On the assumption that the field was similar to that of a bar magnet, this meant that the magnet was not at the planet's center but was displaced from it by about one-third of the planet's radius. A number of theories were offered, including the prospect that the region in which convection generated the dynamo that created the field was located nearer the surface than in the case of Jupiter. If this is true, it might reflect some aspect of Uranus's internal constitution. The offset field had a number of consequences. For example, the field at the surface nearest the dipole was over 10 times stronger than that at the antipodal position. Moreover, the magnetosphere would also have a complex shape, with a rotating bulge corresponding to the extension of the dipole's axis, and near the equatorial plane the magnetosphere would be sufficiently large as to encompass the rings and satellites for most of the time.[224,225] The charged-particle detectors found radiation belts circling the planet of an intensity sufficient to be the cause of the dark appearance of rings and of satellites orbiting within them. In fact, high-energy particles impinging on methane ice surfaces would erode hydrogen and leave behind a residue of carbon, but there was no evidence to prove that there was methane ice either in the rings or on the surfaces of the satellites. Alternatively, the dark material might be a carbon-rich mix of ices and rock similar to a carbonaceous chondrite meteorite. However, in contrast to the Jovian and Saturnian systems, the Uranian environment was devoid of the heavier charged particles such as ionized water, helium and oxygen.[226,227]

Having identified the orientation of the magnetic axis, it was possible to interpret the modulation in the radio emissions. The source regions were near the magnetic poles, but on the way in only one pole was visible and the received-strength of the signal varied as the planet rotated on its axis over a period of 17.24 hours. This value surprised the scientists who had inferred a shorter period by tracking the few clouds that were visible in the 25–70°S latitude range, since this implied that the atmosphere was 'super-rotating'.[228] Images enhanced to highlight atmospheric structure showed that despite the unusual geometry that resulted in the south pole facing the Sun for several decades, the bands of clouds were centered on the pole rather than on the subsolar point, at which solar heating would be greatest. Depending on their latitude, the clouds circled the planet in periods ranging between 14 and 17 hours, indicating that the planet's rotation played a greater role than solar heating in driving the meteorology. In this respect, despite its unusual tilt, Uranus's weather was similar to Earth's. On Jupiter there were thousands of individual details to track in order to measure wind speeds, and on Saturn there were hundreds, but on Uranus there were only a few. The longest-lived were tracked for up to 15 rotations, but the smaller features at intermediate latitudes persisted at most several days. The rotation rate at the highest latitude was determined by tracking a single feature that was visible for several days in images taken using a violet filter. Owing to the degree of image processing, features at low latitude (i.e. around the limb of the disk during the spacecraft's approach) were considered to be unreliable indicators. Nevertheless, an extrapolation of the observed wind speeds against latitude raised the prospect that the regions nearer the equator took longer to circle the planet than the modulation of the magnetic field. There was a brownish violet-absorbing haze present over the south pole.[229,230] The helium concentration in the visible atmosphere was measured by the infrared spectrometer, and found to be greater than for both Jupiter and Saturn. As the spacecraft made its flyby, the infrared radiometer was able to make two scans of the temperatures on a pole-to-pole basis. Intriguingly, in the lower atmosphere the temperatures of the poles were similar, even though one was in full sunlight and the other was in continuous darkness. Furthermore, the temperatures at high latitudes in the southern hemisphere were similar to those near the equator. However, there was a band within 10 to 15 degrees of 40°S that was several degrees cooler, and this was undoubtedly related to the fact that it was in this latitude range that the clouds were most apparent. Although there was a similar 'cold band' in the northern hemisphere, the darkness precluded visual observations. However, the models were at odds to explain why the equator had the same temperature as the poles. The available data could only put constraints on a possible internal heat source as being at most several per cent of the amount of heat received from the distant Sun. Uranus was evidently very different to Jupiter and Saturn.[231,232]

The ultraviolet spectrometer made observations of sunlight filtering through the atmosphere while crossing the planet's shadow, and also of the two stellar occultations, to determine the extent and temperature of the outer atmosphere, which proved to be mostly atomic and molecular hydrogen. This revealed the presence of an unexpected gaseous envelope that extended to the rings, and was sufficiently dense to influence the motions of the particles residing in them. Although auroras were expected at the sunlit pole, their presence – if they were indeed present – was overwhelmed by an 'electroglow', a phenomenon seen at both Jupiter and Saturn in which atomic and molecular hydrogen is excited by solar radiation and emits in the ultraviolet. The detection of Uranus in terrestrial ultraviolet observations had first suggested the presence of auroras, but now it seemed that what had been observed in this way was electroglow. An aurora was detected on the night-side (at ultraviolet but not visual wavelengths) and monitored for almost two local days as it rotated with the planet in a location several tens of degrees from the offset magnetic pole.[233]

A study of how Uranus rotated and how its gravitational field influenced the path of Voyager 2 suggested a two-layer model with a small rocky core and a very dense atmosphere of liquid water and various gases. At the low temperatures prevailing in the upper atmosphere, almost every detail of the deeper structure was masked by an outer haze of hydrogen and helium.[234,235,236]

Most of the imagery from the near-encounter phase was downloaded from tape

One of the best images of Oberon, the outermost of Uranus's satellites. Note the indication of a mountain on the limb near 8 o'clock.

on the day after the flyby. To general relief, the precautions to steady the spacecraft and reduce smearing had worked successfully, even in the difficult case of Miranda.

The best images of Oberon were taken at a distance of 660,000 km, and therefore had a resolution no better than 10 km per pixel. The surface appeared to be covered with ancient craters of diameters ranging from just over 100 km down to the limit of resolution. The fact that the craters were quite well preserved suggested that internal activity had played at best a marginal role. Although there were several faults, there was nothing to suggest widespread tectonism. The most likely traces of endogenic activity were dark spots on the floors of some of the craters, possibly resulting from eruptions through fractures opened by the impact. There was a large mountain some 20 km tall projecting over the limb, but it may have been the central peak of a crater that was itself not visible. Unlike Oberon, Titania, the largest of the retinue, showed signs of tectonism. It was heavily cratered, but the craters appeared smaller than on Oberon, and probably younger. Drop-faulted tracts known as grabens transected the cratered landscape, possibly produced as the interior cooled and caused the water ice of which the moon was made to expand and open cracks on the surface. Meanwhile, cryovolcanism had erased most of the oldest craters, leaving only the relics of a few large basins. Small patches of terrain that seemed to be flat and relatively crater-free suggested that cryovolcanism had continued for a long period of time. The grabens

The paucity of large craters and the presence of substantial faults on the largest of Uranus's satellites, Titania, suggests that its surface was active in ancient times.

themselves gave the appearance of being relatively young, having only a few craters superimposed. The photopolarimeter revealed the surface to be covered with a fluffy layer of regolith, indicating that it was old enough to have been ground down to fine dust.[237]

Although Umbriel had the darkest albedo of the classical satellites, it was brighter than Puck and the other newly identified objects. The total absence of bright rays of ejecta made it seem virtually featureless at long range prompting speculation that it was Europa-like, but on close inspection it was shown to be saturated with large craters. Perhaps bright ice was excavated and then darkened by the rain of high-energy particles, or perhaps for some reason it was dark to start with. However, there were at least two bright features: one in the interior of the 80-km-diameter crater that had been sighted on the limb during the approach phase, and the other was the central peak of another crater. Perhaps comet-like venting had sprayed the entire surface with carbon-rich material and the bright spots marked the active regions. It was difficult to formulate a rigorous theory on the basis of so little data.

Having small craters and large tectonic faults, Ariel resembled Titania, but it had only one-third of the crater density. The endogenic activity that had eliminated

Umbriel is the darkest and most heavily cratered of Uranus's moons but, remarkably, there are no bright rays. In fact, only two bright features are visible: the collar inside the crater at the top and the central peak of the crater near the terminator.

the most ancient craters had lasted much longer on Ariel than on Titania. In fact, Ariel had the youngest surface of all the large satellites. Its faults were deeper and more complex than those of Titania, and were also sometimes accompanied by furrows. Parallel linear patterns on the floors of some of the largest faults might mark the axis of extrusion as the faults progressively opened. In some places, the erupted material had buried the pre-existing landscape. The melting point of water ice was hundreds of degrees higher than the ambient temperatures in the Uranian system, but a small fraction of ammonia in the ice would have acted as an efficient antifreeze. There were the relics of several large and ancient craters but, as on the icy moons of Jupiter and Saturn, they had slumped as the material relaxed over the ages. Overall, Ariel displayed some of the characteristics of Europa, Ganymede and Enceladus.

Miranda was not only the most striking of the Uranian satellites, it was by far the strangest object yet seen in the solar system. It appeared to be a bizarre assemblage of the valleys and layered deposits found on Mars, the grooved terrain of Ganymede and the compression faults of Mercury. Its surface was mostly made of rolling hills that were peppered by small craters. But juxtaposed with these mundane features were three oval regions 200–300 km wide (called 'ovoids') that were lightly cratered and therefore even younger than the hilly terrain. The most striking characteristic of the ovoids was that they contained a series of concentric grooves and scarps that in some places made sharp turns – in one case so sharply as to resemble a gigantic chevron. Transecting both the hills and ovoids were faults with terraces, and in some places cliffs standing up to 20 km tall. In the moon's weak gravity, an object thrown off the crest of such a cliff would take fully 10 minutes to reach the base. The craters

Criss-crossed by faults and furrows, Ariel has the youngest surface of the satellites in the Uranian system. The craters are relatively few and small, with the larger ones being subdued and remarkably similar in appearance to those of Jupiter's Ganymede.

in the ovoids were either bright or filled by dark material similar to that on Oberon. Although there were small flat pools of material on at least one ovoid, it looked as if endogenic activity had not played a significant role in shaping Miranda – the process that had shaped this small moon had been much more interesting! As Laurence A. Soderblom, the deputy leader of the imaging team, put it, "If you can imagine taking all the geologic forms in the solar system and putting them on one object, you've got it in front of you." Two hypotheses were proposed to explain this unprecedented landscape. One argument was that at some point in its history Miranda must have been destroyed, and the fragments re-accreted to form a jumble in which some of the denser material that was originally near the core became integrated into the surface, adjacent to fragments of less dense ice. In fact, a statistical study of the population of comets gravitationally linked to Uranus found that during their histories the inner satellites should have had at least one impact sufficiently energetic to shatter them into pieces, and so it is likely that Miranda and Ariel, and possibly Umbriel, were re-accreted at least once. In the second hypothesis, the appearance of ovoids was due to vast agglomerates of ice that worked their way to the surface when Miranda was still warm, producing huge blisters that slowly settled to create the pattern of concentric wrinkles. In fact, this could be regarded as a case of the arrested development of the familiar process of differentiation in which dense material sinks to the core and light material rises to the surface, and because the process was not completed it left the moon frozen in an intermediate state. The key was the age of the ovoids, and of course this was controversial, with some scientists offering the relatively youthful age of 1 billion years, while others said they had to be much older.

Voyager 2 looked for structures similar to Io's flux tube near Miranda, but found none. None of the instrument scans designed to detect gaseous envelopes around the satellites detected anything significant; and imagery that was taken 2 days after the planetary flyby – when most of the moons were thin crescents – detected no forward scattering sunlight to suggest gaseous emissions from their surfaces. The character of many of the Uranian satellites posed the question of what heat source could have prompted endogenic activity in such small bodies? One possibility was resonances between the orbits of the satellites. This mechanism was undoubtedly

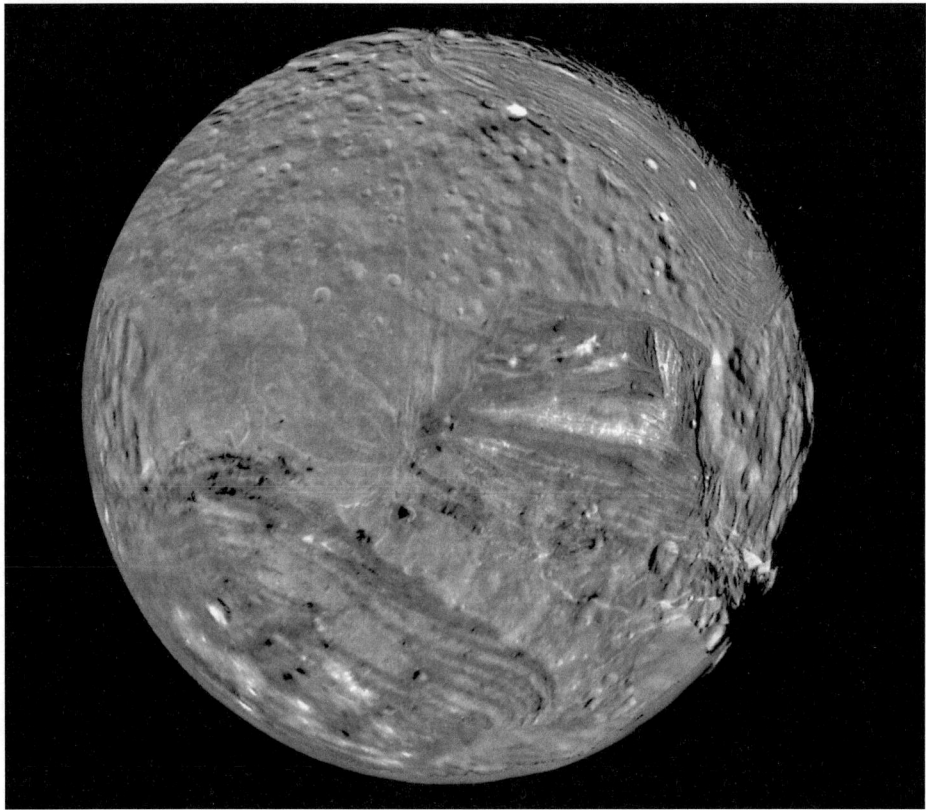

The processes that shaped the remarkable moon Miranda, with its juxtaposition of rolling cratered plains, parallel faults, chevrons and ovoids, remain a mystery. Note the vast cliff on the terminator.

responsible for heating Io, and might possibly explain the activity on Enceladus. Although there were currently no such resonances, they may have existed in the past and might again in the future. The DSN's tracking of the spacecraft provided good measurements of the masses of Uranus (because it was the dominant member of the system) and Miranda (because the spacecraft passed so close to it). The trajectory had not taken the spacecraft close enough to the other satellites to enable their masses to be measured in this way, but they could be estimated from accurate knowledge of their orbits, which was derived from the optical navigation imagery and from computational models of their mutual perturbations. Remarkably, the densities thereby obtained indicated that even though the planet had formed farther from the Sun than did Saturn, its satellites consisted of at least 30 per cent rock.[238,239,240,241,242,243,244]

Some astonishing images were taken around the time of the planetary flyby. One, taken from a position less than 10 degrees south of the ring system 27 minutes prior to crossing the equator, showed the most conspicuous rings as narrow dark ribbons silhouetted against Uranus. Sixteen minutes later, by which time the viewing angle had reduced to 5 degrees, a four-frame mosaic showed the entire ring system with a rapidly changing perspective. This sequence also revealed the presence of a hitherto unsuspected broad band of material more than 2,000 km wide located inside ring 6. It was designated 1986U2R.

Scientists eagerly awaited the downlinking of a single 96-second-long wide-angle frame that had been taken when Voyager 2 was 225,000 km beyond the ring plane

The rings of Uranus are silhouetted against the planet's atmosphere in this image by Voyager 2 taken near the time of its ring-plane crossing.

The epsilon ring is viewed against space in a picture taken just after the ring-plane crossing. Two other rings are just visible nested inside epsilon.

and in the planet's shadow, to record the rings when backlit by the Sun. Showing hundreds of diffuse bands of dust that were invisible under other illumination, it was a masterpiece. At first glance, the wealth of detail was reminiscent of an image of Saturn's rings. The epsilon ring appeared to be almost isolated, perhaps owing to the moonlets that shepherded it, and the other known rings also looked remarkably free of dust. The brightest feature in this light was ring lambda, meaning that it was made mostly of microscopic dust. Such a large amount of dust in the rings prompted the question of the source, for this must be continuously supplying material to make up for the thinning of the system by various processes, including the vast rarefied outer atmosphere of the planet which, in itself, would erode it in a matter of centuries and leave only the largest particles behind.[245,246]

On the outbound leg, Voyager 2 made four crossings of the neutral current sheet

This long-exposure image of the night-side of the Uranian rings shows that there are bands of dust between the known rings that were not evident when viewing the day-side. Epsilon, the outermost ring, appears almost isolated from the rest of the series, while the next one inward, lambda, seems to be the brightest, meaning that it is rich in small dust particles.

that marked the boundary between the two polarities of the magnetosphere (positive and negative, or north and south). A well-formed magnetotail extended for at least 6 million km 'downstream'. The magnetopause was encountered at a planetocentric distance of less than 80 Ru, and the bow shock was crossed no less than seven times at distances over 162 Ru. The final bow-shock crossing occurred on 29 January at a distance of 227.7 Ru, and the spacecraft re-entered the realm of the solar wind.[247]

While the spacecraft had been making its approach to the Uranian system, people at JPL noticed that NASA intended to launch a Space Shuttle during the most hectic days of the encounter, and asked the agency to delay that mission by several days to ensure that no tracking facilities assigned to support Voyager would be withdrawn in the event of an emergency involving the Shuttle, but because this particular mission had a high public profile owing to the presence on the crew of a high-school teacher, NASA refused. There was even a plan that some commands could be issued by the Shuttle to JPL for relaying to Voyager. To the relief of JPL, the launch of the Shuttle was delayed by technical issues and the weather. On 28 January, as JPL prepared to give the final press science briefing of the Uranian encounter, Challenger lifted off for mission STS-51L. It had been a frigid night at the Kennedy Space Center on the Atlantic coast of Florida, and the launch had been postponed several hours to allow the Sun to melt the multitude of icicles that had formed on the service structure. The severe chill had also reduced the resiliency of the rubber O-rings in place to seal the joints between the segments of the solid rocket boosters, and one such ring did not form a tight seal. As the pressure built up the hot gas began to leak through the joint and erode the ring. Just over 1 minute later, the leak had developed into a flame that acted like a blowtorch and cut through the lower strut that held that booster tight against the central tank of propellants for Challenger's engines. The booster pivoted on its upper strut, struck and crushed the tank of liquid oxygen, which caused an explosion that engulfed Challenger in a fireball even as the asymmetric aerodynamic loads caused it to disintegrate. This loss grounded the remaining Shuttle fleet for almost 3 years and prompted a complete rethinking of America's access to space.[248]

Never before had the difference between human and robotic space missions been so striking. As a cheap robotic emissary was in the 9th year of its exploration of the outer solar system, seven human lives had been lost attempting what was regarded as a routine flight to deploy a communication satellite. It was suggested that some of the newly discovered Uranian satellites be named in honor of the lost astronauts, but the International Astronomical Unit, which had sole authority to name astronomical objects, decided to continue to use the nomenclature based on the literary works of Shakespeare and Pope, and the astronauts received asteroids with catalog numbers 3350 to 3356; with Christa McAuliffe, the teacher, having entry 3352, which was the only near-Earth object of the seven.[249]

The gravity-assist from the flyby of Uranus had further increased the eccentricity of Voyager 2's solar orbit to 5.81 and put it on course for Neptune. To this day, no one knows when Uranus will again be visited by a spacecraft from Earth.

A view of the night-side of Uranus taken on 25 January 1986 as Voyager 2 left the planet behind.

TO A BLUE PLANET

Although the Uranus flyby sent Voyager 2 in the direction of Neptune, a targeting maneuver was made on 14 February 1986 to refine the aim. In fact, lasting just over 2.5 hours and using an estimated 12 kg of hydrazine, this was the largest course correction that the spacecraft had ever made. The trajectory left open several options for the encounter, whose actual geometry and timing were still a matter of debate. A number of engineering tests and scientific observations were performed as the spacecraft cruised to Neptune. It took about 14 hours of particles and fields data per day, and, within the strict constraints imposed on slewing its scan platform, made 100 observations of ultraviolet sky sources per year, often in cooperation with terrestrial telescopes or with the long-lived International Ultraviolet Explorer satellite.

Meanwhile, Voyager 1 was reporting on particles and fields and conducting some engineering experiments. Soon after Voyager 2's Uranus encounter was completed, the imaging team suggested that Voyager 1 should attempt to take pictures of comet Halley's perihelion passage, but a study revealed that the comet would be so close to the Sun in the sky as to dazzle the optics of the narrow-angle camera. Undeterred, the engineers suggested that the spacecraft should first position itself to cause the shadow of the rim of its high-gain antenna to fall on the camera in order to reduce the glare. Although the comet had come and gone before this technique could be tested, it was decided to proceed because if it was successful then the spacecraft would be able to image certain dust structures that had been discovered several years earlier by IRAS (InfraRed Astronomy Satellite). A test in early 1987 showed that the parts of the spacecraft's structure that were illuminated reflected onto the rear of the high-gain antenna, causing a portion of the pictures to be overexposed. Nevertheless, the unaffected parts of the pictures provided some scientific data, although this appears not to have been published.[250,251]

The data-handling capacity of Voyager 2 could not be improved beyond what was done for Uranus, and the only possible software update was to enable the cameras to take the very long exposures required in the dim illumination at Neptune, where the Sun resembled (in size) our view of Venus. Thus any modification to increase the data rate would have to be made on Earth. The primary DSN antennas were enlarged from 64 meters to 70 meters, and low-temperature and low-noise amplifiers were installed to boost their sensitivities. To follow up on the successful arraying of radio telescopes for the Uranus encounter, NASA sought additional facilities to assist with the Neptune encounter. One such facility was the Very Large Array (VLA) in New Mexico, operated by the US National Radio Astronomy Observatory. It consisted of 27 independently steerable 25-meter-diameter antennas that could move on the three arms of a Y-shaped track to provide an effective collecting area equivalent to a pair of 70-meter antennas. The antennas required only minor modifications to receive the spacecraft's X-Band signal. The VLA and the Parkes radio telescope would almost totally offset the loss of gain due to the increased distance from Uranus to Neptune, and permit a 19.2-kbps real-time transmission data rate, or a 21.6-kbps playback of taped data. The situation would be able to be further improved by arraying the VLA with the DSN station at Goldstone, and the Parkes antenna with the DSN station at Canberra. With the addition of a 34-meter high-efficiency X-Band antenna, the DSN antenna in Spain would be able to yield a respectable 8.4 kbps data rate. In addition, the 64-meter Usuda radio telescope, built by the Japanese in order to track their two probes on flyby missions to comet Halley in 1986, was to help in collecting data during the Neptune radio occultation. In total, 38 antennas on four continents would be involved, with the spacecraft almost overhead in Australia at the time of closest approach.[252,253] Rolling the spacecraft to maintain the camera pointed at a target had worked well at Uranus, but two additional techniques were introduced to limit the degradation of resolution during the very long exposures required at Neptune. One method, referred to as nodding compensation, involved rocking the spacecraft to and fro between the attitude extremes at which it would lose contact with Earth. The other method involved orienting the spacecraft to enable the elevation drive of the scan platform (the one that had not suffered any problem) to slew to track a target as the spacecraft's attitude remained fixed. Unlike compensating for smearing by rolling the spacecraft, the fact that these two techniques could maintain high-gain contact with Earth meant that it would be possible to provide real-time transmission of pictures. The sequence for the Neptune encounter included a total of nine medium-rate slews of the scan platform.[254]

In the years leading to Voyager 2's encounter with Neptune, terrestrial telescopes utilizing state-of-the-art instruments such as CCD (charge-coupled device) cameras were able to obtain good images of the planet's atmosphere that showed bright polar regions, aerosol layers and several discrete clouds. The fact that all these features were conspicuous at methane absorption wavelengths was good news, because the spacecraft's cameras had methane filters.[255] When infrared measurements found that the temperatures of Uranus and Neptune were similar, despite Neptune being much further from the Sun, it was realized that the planet must have an internal source of heat; as for Jupiter and Saturn, but *unlike* Uranus – or to be specific, expressing their energy budgets as a ratio of how much heat they released relative to the energy that they received from the Sun, the figures were 1.7 for Jupiter, 1.8 for Saturn, but only 1.1 for Uranus. Because Neptune was slightly more massive and more compact than Uranus, its bulk density was significantly greater. Neptune was therefore expected to be intrinsically warmer than Uranus and, since the sunlight was that much weaker, it was expected to have a considerable energy budget. Neptune's weather system was therefore expected to be more active than that of Uranus. Attempts to measure the rotation rate by measuring the Doppler on opposing limbs were frustrated by the fact that the disk subtended such a tiny angle, but cyclic photometric variations provided a rotational period of 17 hours 43 minutes. Then in the early 1980s high-altitude clouds were detected crossing the disk at different rates, thereby confirming that the atmosphere rotated differentially.

The large moon, Triton, had been found in 1846, just a few weeks after Neptune itself was discovered. Nereid, a much smaller body, was identified in 1949 pursuing a distant elliptical orbit. The fact that Triton's orbit was retrograde had prompted the hypothesis that it had originated elsewhere and, in the process of being captured, had ejected from the system most of the moons present at that time, with Nereid either having been retained in a distant and eccentric orbit or having been captured at a later date. A spectrum of Triton taken in 1978 showed no evidence of water ice but instead showed that it had methane ice. This surprising result had a remarkable implication, because at Triton's distance from the Sun the temperature was too cold for water ice to sublime, but not for methane ice to do so, which suggested that the moon should have a tenuous atmosphere of methane. Furthermore, the spectrum also showed nitrogen, which might be present either in solid or liquid form. The moon's reddish hue hinted at the presence of complex hydrocarbons, but these were not seen by spectroscopy. At one point, it was thought that Triton might possess a global haze of similar composition to that of Titan, although owing to the low temperature this would be sufficiently transparent to enable the spacecraft to image the surface. Some scientists even expected to find an ocean of nitrogen. If Triton had indeed been captured, the morphology of the surface might show evidence of its tortured history. It would not be possible for Voyager 2 to make a close inspection of Nereid, about which little was known apart from the fact that its diameter was between 150 and 525 km, but one astronomer thought that it would be either unusually elongated (almost pencil-like), or double-sided like Iapetus.[256]

In the decade prior to Voyager 2's encounter, Neptune was observed during more than 20 stellar occultations. In addition to providing data on the atmosphere of the planet, the occultation records were scrutinized for evidence of the presence of rings. In 25 per cent of the events the starlight was seen to flicker either prior to ingress or after egress but, remarkably, never on both legs. At the first event, in May 1981, the starlight dimmed to such a degree that it was suggested that it had been occulted by a satellite with a diameter of at least 80 km orbiting at a planetocentric distance of 60,000 km that had not been discovered directly owing to the glare from the planet. Five occultations in 1984–1986 were compatible with occultations by ring material, but when the results were integrated they seemed to indicate that Neptune possessed only short arcs rather than complete rings. While theorizing how ring arcs might be maintained – perhaps by complex gravitational interactions with unseen satellites – the Voyager 2 team considered how the presence of rings would affect their plans for the Neptune encounter.[257] Free of the requirement to plot a trajectory through the Neptunian system that would set up a future planetary encounter, the planners were able to design a trajectory to maximize the scientific output. The original plan was to skim a mere 1,300 km over the cloud tops to enable Neptune's gravity to deflect the trajectory for an 8,200-km flyby of Triton on the outbound leg. However, the point at which the vehicle would cross the equatorial plane was moved to a planetocentric distance of 71,000 km in order to avoid any ring material, and this had the result of increasing the flyby distances to 4,800 and 37,750 km respectively. If further studies were to identify hazards on this route before the encounter began, then the trajectory would be further revised as it was preferable to be sure of obtaining long-range data of Triton than to get none from a spacecraft that had been disabled shortly prior to a closer flyby. Nevertheless, dynamicists warned that the gravities of Neptune and Triton could jointly provide stable rings in orbits inclined to the equator – possibly even in polar orbit, although there was no evidence to show that this was so. On the

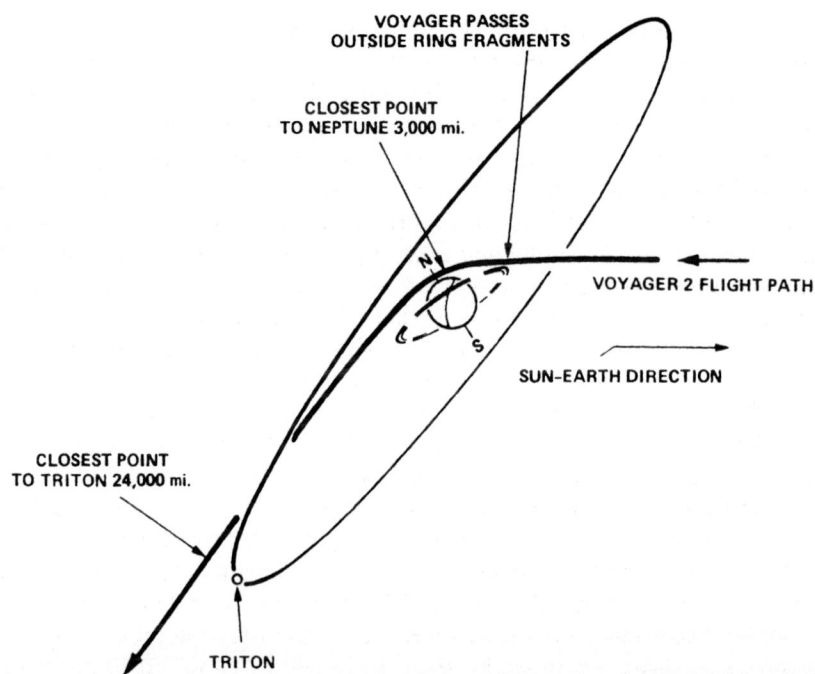

VOYAGER PASSES OUTSIDE RING FRAGMENTS

CLOSEST POINT TO NEPTUNE 3,000 mi.

VOYAGER 2 FLIGHT PATH

SUN-EARTH DIRECTION

CLOSEST POINT TO TRITON 24,000 mi.

TRITON

Voyager 2's path through the Neptunian system. Note that the spacecraft would just clear the ring system and be dramatically deflected to the south toward Triton. The orbit of Nereid is so remote that it cannot be included on this scale.

other hand, increasing the range of the Triton flyby would increase the coverage, reduce the scope for smearing and allow time to take more images through filters in order to produce color pictures. It was considered that such images would be useful because Triton had many characteristics in common with Pluto, which the Voyager Grand Tour was not going to visit. The planners had even managed to squeeze in occultations of the Earth and the Sun by both Neptune and Triton in order to study Neptune's atmosphere (and Triton's, if it proved to possess one). The new trajectory had the additional benefit of an improved viewing geometry for the antennas in Australia and Japan during the radio occultations. The radio signal would also be studied for signs of the rings. Further, the spacecraft would use its photopolarimeter to monitor the passage of sigma Sagittarii across the line of sight of the putative rings, in case there should be an occultation. And the particles and fields instruments would look for a Triton torus akin to that of Io. This planning also took account of several parameters that had recently been refined, in particular the discovery that the mass of Neptune had been overestimated by 1.6 per cent, and a refined ephemeris of Triton's orbit. On 13 March 1987 Voyager 2 executed a maneuver to adopt the new trajectory.[258,259,260] At this time, Neptune was the furthest known planet from the Sun, because the perihelion of Pluto's eccentric orbit was inside Neptune's orbit and Pluto was near that point.

As so little was known of Neptune, the schedule of scientific observations for the encounter was made flexible to be able to accommodate the last-minute discovery of satellites or rings that, in addition to being interesting in their own right, could also be used for optical navigation purposes. With an axial tilt of 29 degrees, the planet experienced seasons similar to those on Earth, but lasting 160 times as long. On the inbound leg, the spacecraft would view the southern hemisphere since the north pole would be in darkness. As usual, it was loaded with a backup encounter sequence in case its command receiver should fail completely. This called for basic imaging of the planet, particles and fields data, a radio occultation by Neptune and its rings, and imagery of Triton.[261,262,263,264]

The first navigation images of Neptune were taken in May 1988, when Voyager 2 was still more than 4 AU from the planet, but it was not until the end of the year that the first meaningful images were taken. Even although the planet was more than 300 million km away and the narrow-angle resolution was only 6,000 km per pixel, the view was better than that from Earth. Neptune appeared as a small bluish disk, and Triton as a pinkish dot. There were large bright clouds near the equator, a hint of dark belts near the pole, and crossing the center of the disk there was a dark blue spot that was strikingly reminiscent of Jupiter's Great Red Spot. The rate at which the clouds were moving gave a rotational period of 17–18 hours, confirming the recent telescopic observations. Imagery taken from a distance of 175 million km in early April 1989 enabled the dark spot's size to be measured as 13,000 km east-to-west and 6,500 km north-to-south, and it was promptly dubbed the Great Dark Spot. A check of the archives raised the prospect that it might date back to at least 1899, when astronomers made the first of a number of reports of having seen a dark feature, but we now know that both the Great Dark Spot and a smaller spot to the south that was referred to both as the Lesser Dark Spot and as D2 were short-lived, because they were absent when the Hubble Space Telescope took high-resolution images in 1994.[265,266,267] After making one of the final targeting maneuvers on 20 April 1989, Voyager 2 began the distant observatory phase on 5 June at a distance of just over 117 million km. A few weeks later the first new moon was discovered. It was designated 1989N1, and later named Proteus. Whereas the orbits of Triton and Nereid were both inclined to the equatorial plane of the planet, and retrograde and eccentric respectively, the orbit of this moon was prograde, circular and in-plane, raising the prospect of there being a family of such objects. The diameter of Proteus was estimated at anything from 200 to 640 km, depending on the value assumed for its albedo. Fortunately, the schedule was able to accommodate observations that would facilitate a closer scrutiny.[268] By early August, 1989N2 to 1989N4 had been found. These were later named Larissa, Despina and Galatea. They were no larger than several hundred kilometers in size, and, like Proteus, were in prograde, circular near-equatorial orbits. Intriguingly, all three were at planetocentric distances that raised the prospect that they were embedded within the rings (if indeed these existed) and possibly acting as shepherds. The discovery of a 'regular' system of satellites around Neptune seemed at odds with the scenario in which Triton was an interloper that had been captured, unless these moons and the rings were byproducts of that event. Extrapolating the orbits back in time established that Larissa had been responsible for the anomalous occultation in 1981.[269,270]

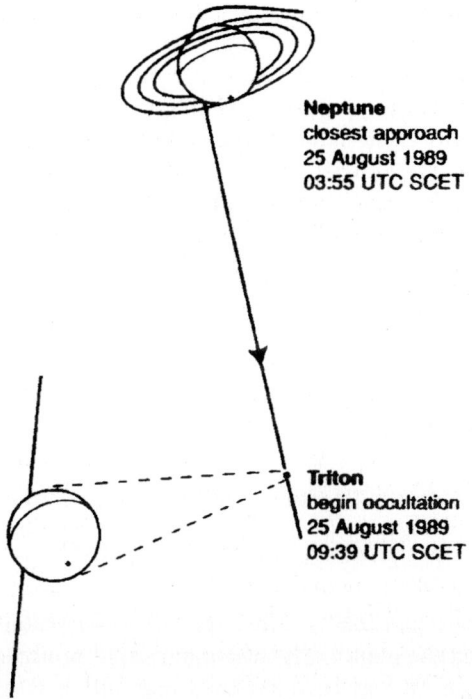

Neptune
closest approach
25 August 1989
03:55 UTC SCET

Triton
begin occultation
25 August 1989
09:39 UTC SCET

Voyager 2's path through the Neptunian system as viewed from Earth. Note that the spacecraft would be occulted by the outer ring on both the inbound and outbound legs, but by the intermediate ring only on the outbound leg. The innermost ring was obscured by the planet's atmosphere on both sides. Some hours later, the spacecraft would be occulted by Triton. SCET stands for Spacecraft Event Time, which is the actual time of the event; as distinct from Earth Received Time. (Reprinted with permission from Tyler, G.L., et al., "Voyager Radio Science Observations of Neptune and Triton", Science, 246, 1989, 1466–1473. Copyright 1989 AAAS)

Since Triton was, in a sense, to co-star with Neptune, Voyager 2 devoted much of the observatory phase to monitoring this moon. Although astronomers had thought it might be the largest satellite in the solar system, the long-range images established it to be only half as large as believed; with a diameter of 2,700 km it was smaller than the Moon and only just larger than Pluto. Its size had been overestimated because its albedo, at 70 per cent, proved to be much greater than expected. This, in turn, meant that the surface had to be significantly colder; so cold, in fact, as to rule out an ocean of liquid nitrogen. Although few surface details were evident at this range, there was a pinkish hue that some scientists thought might indicate radiation discoloration of methane ice. At first, spots were presumed to be clouds, but were soon realized to be surface detail, which meant that the atmosphere was clear. A prominent blue–white streak suggested that there might be areas of freshly deposited frost. [271,272]

Ongoing imaging of Neptune provided new insights into the weather system. The Great Dark Spot was at 22°S. Despite the strong anticyclonic winds at its periphery, bright cirrus-like clouds appeared to hover close beside it, possibly where methane droplets that were swept to high altitude by rising currents made ice crystals, thereby making them equivalent to the orographic clouds that form above mountain ranges on Earth and Mars. The Great Dark Spot was seen to stretch and contract as it rolled. At one point, a 'tail' was seen to extend to the west and then swing north and dissolve into a string of small 'spots and beads'. The Lesser Dark Spot was at about 55°S, and early on developed a bright core that persisted for the remainder of the encounter, possibly indicating where gas was billowing up from the depths. A small bright feature at a latitude between the Great Dark Spot and the Lesser Dark Spot was nicknamed the Scooter. It comprised a group of bright streak-like clouds. The Great Dark Spot was traveling around the planet in 18.3 hours, the Lesser Dark Spot took about 16 hours, and the Scooter was even faster; hence its name.[273,274]

An image taken on 11 August from 21 million km confirmed the presence of ring arcs. An arc spanning about 50,000 km was located just beyond the orbit of Galatea, and a 10,000-km arc was trailing Despina, several hundred kilometers from its orbit. The long arc was designated 1989N1R and the shorter one 1989N2R. Owing to the flexible schedule, it was possible to order specific observations of these arcs as the spacecraft approached the planet. The mechanism that produced ring arcs remained a total mystery, because it had been assumed that within at most a few years such concentrations would become evenly distributed to form a complete ring.[275,276] Soon thereafter 1989N5 and 1989N6 were spotted; estimated to be about 90 and 50 km across, they were named Thalassa and Naiad respectively. Unlike the other five new satellites, Naiad's orbit appeared to be slightly inclined to the equatorial plane. On 17 August, at a planetocentric distance of 470 Rn, the spacecraft detected for the first time emissions from Neptune's magnetosphere, thereby confirming that each of the solar system's giant planets possesses a magnetic field.[277] As the resolution of the imagery increased, it became evident that the rings were not arcs but circular. A full analysis would later establish that the shorter arc was actually the brightest part of an otherwise very faint ring that was named Le Verrier in honor of the French mathematician who had performed the calculation that led directly to the planet's discovery. The outer ring was named Adams after the Englishman who had made a similar calculation, although this had not played a part in prompting the search that found the planet. They were confined by the presence of the moons Despina and Galatea just inside their orbits, but their outer edges were unconstrained. Of the two, the Adams ring was the more interesting, because it contained three bright arcs. In homage to the connection of France with Neptune's discovery and also to the 200th anniversary of the French revolution, the clumps were named Liberté, Egalité, and Fraternité and spanned respectively 4, 4 and 10 degrees in longitude. When calculations found that these were responsible for all but one of the stellar occultations seen from Earth, this meant that the distribution of material had persisted for at least 5 years, in contrast to the transient clumps in the F ring of Saturn. An even fainter ring was found at a planetocentric distance of 42,000 km. It

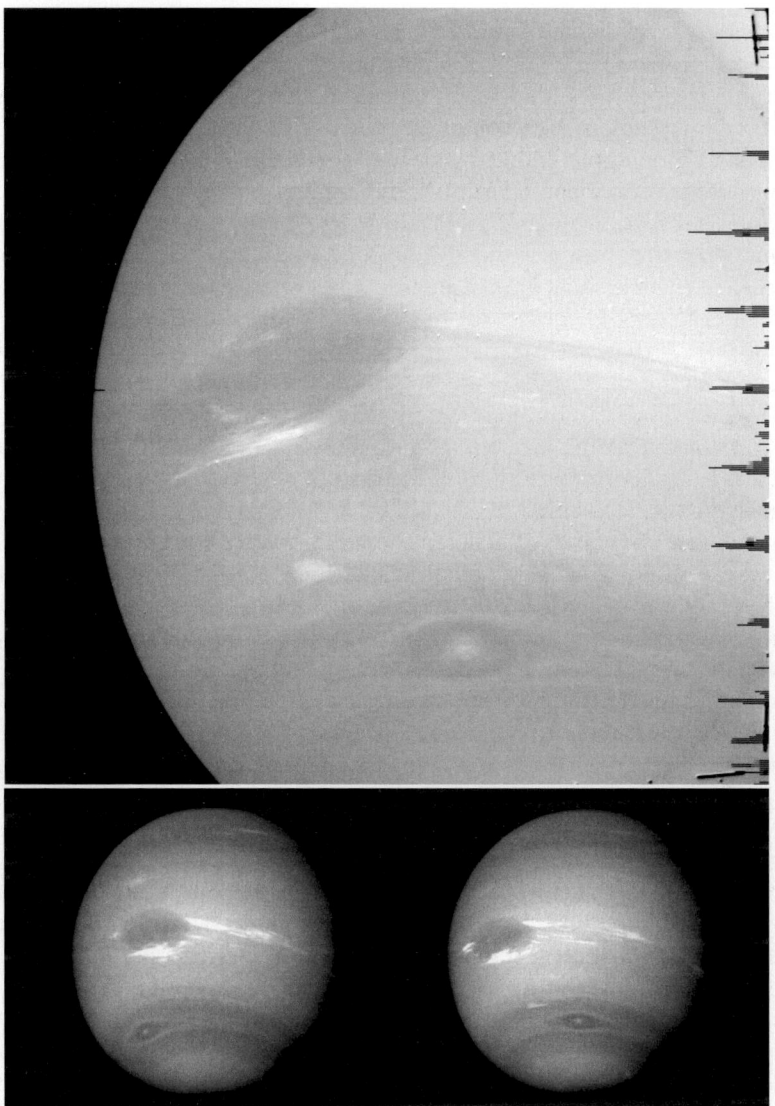

A 'raw' view (top) of Neptune taken by Voyager 2 as it approached the planet with the Great Dark Spot at 22°S, the bright Scooter, and the Lesser Dark Spot at 55°S. The streaks of bright cloud at the southern margin of the Great Dark Spot and the bright core of the Lesser Dark Spot were persistent. In the 17.6 hours that elapsed between the lower images the Great Dark Spot had made just under one rotation of the planet, and the Lesser Dark Spot made more than one rotation.

was named Galle, after the German astronomer who checked the position indicated by Le Verrier's calculation and confirmed the existence of the planet.[278,279] Even on the best inbound pictures, these rings had such a low signal-to-noise ratio that they were difficult to study. If they contained a significant amount of dust, as at Uranus, then the best observations would be when they were forward-scattering sunlight, and such images were scheduled to be taken after the planetary flyby.

The maneuver by Voyager 2 planned for 15 August was canceled, and the final targeting correction was performed 3 days prior to the flyby. It was made by the roll thrusters, which were known not to upset the thermal regime of the faulty command receiver, in order to preserve the option of sending last-minute commands to correct memory errors such as had ruined some of the Uranus images and timing errors such as Voyager 1 had suffered while flying through the Jovian radiation belts – although there was little prospect of the Neptunian system posing a serious radiation threat to the spacecraft. The 90-minute maneuver would be the final course correction, since the thrusters were subsequently to be used only to provide attitude control.[280]

At a planetocentric distance of 35 Rn, some 13 hours prior to the flyby, Voyager 2 encountered a well-defined bow shock. It had been expected that the magnetopause would be crossed within a few Neptune radii, but instead of such a distinct boundary the spacecraft found that the particle fluxes slowly diminished as the magnetic field increased. By a curious coincidence, the spacecraft had entered the magnetosphere from a direction that was almost pole-on and parallel to the 'lines of force' of the planetary field.[281,282] Almost exactly 5 hours before reaching Neptune, the spacecraft slewed its scan platform to enable the photopolarimeter to monitor an occultation of Nunki by the rings, measuring the light from the star 100 times per second. By sheer luck, the track grazed the leading end of the middle arc (Egalité) of the Adams ring, and this data provided a full profile at a resolution of several tens of meters, revealing it to be about 50 km in width, but with a dense core 10 km wide near its inner edge. The other rings were much more difficult to detect by this method, but a very thin feature was observed more or less matching the location of the Le Verrier ring.[283] One hour later, Voyager 2 came as close to Nereid as its trajectory allowed, but this was a very distant 4.65 million km, and at a resolution of 40 km per pixel it spanned only a score of pixels. Although the pictures suggested that Nereid was a spheroid about 340 km in diameter, the absence of striking albedo features meant that it was not possible to determine its rotation rate. Later observations from Earth raised the prospect that, like Saturn's Hyperion, it was rotating chaotically, but a more recent study showed conclusively that it rotates in a little less than 12 hours.[284,285,286]

Several narrow-angle pictures of Proteus were taken, the best of which was from a range of 146,000 km with a resolution a little better than 3 km per pixel. It showed Proteus to be an irregularly shaped ellipsoid of 436 × 401 km, with craters of various sizes but no sign of geological activity. Although larger than Nereid, it had not been previously spotted only because it was lost in Neptune's glare and, indeed, it was not isolated telescopically until the early 1990s. The imaging coverage was insufficient to enable its rotational rate to be measured, but it is very likely synchronized with its orbital period. Fair resolution images showed Larissa to be an irregular object about

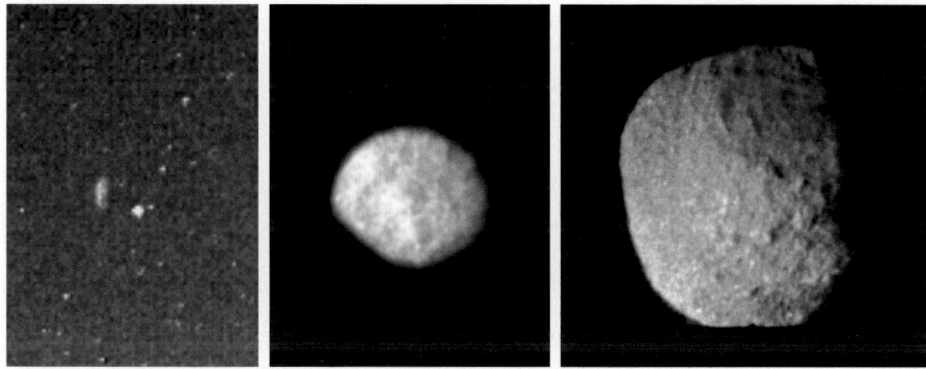

Left to right: a long-range view of Nereid (partially illuminated); the best image of Larissa (1989N2); and the best image of Proteus (1989N1), which, as the second largest of Neptune's moons, seemed at the time to be one the largest non-spherical bodies in the solar system. (Not to scale)

200 km across, and although Despina spanned only a few pixels it appeared to have an irregular shape. If Triton was a captured interloper, then the moons in the inner system would appear to be a second generation of satellites that re-accreted from the debris of the satellites that were shattered when Triton appeared.

The region of the rings was examined for additional moonlets, but nothing larger than 10 km was found – not even in recent re-analyses of the data. Nevertheless, as has proved to be the case for the other giant planets, Neptune has irregular satellites in distant orbits, but none of these was large enough, bright enough or close enough to have shown up in the Voyager 2 images.[287,288]

Some 2 hours before the planetary flyby, when the imaging resolution was tens of kilometers per pixel, shadows were observed on Neptune being cast by high-altitude cirrus onto the main cloud deck some 50–75 km beneath. This was the first time that shadows had been detected on any of the giant planets.

With 63 minutes to go, Voyager 2 crossed the ring plane for the first time heading north at a planetocentric distance of 85,500 km. Then at 03:56 UTC on 25 August (a few days after the 12th anniversary of its launch) it achieved the point of closest approach at a planetocentric distance of 29,240 km. Despite being almost 4.5 billion km (or just over 4 light-hours) from Earth, the final aiming point was only 30 km off target. After its trajectory had taken it to 79°N, the spacecraft passed behind the planet's limb at 61°N as viewed from Earth some 6 minutes after the flyby, and 49 minutes later it re-emerged at about 44°S, with the radio occultations providing two good temperature profiles. Furthermore, at both ingress and egress the spacecraft passed behind the Adams ring and, in egress only, behind the Le Verrier ring too, and the fact that the signal was not affected indicated that the particles were smaller than in the rings of either Saturn or Uranus. Unfortunately, the line of sight did not provide a radio occultation of any of the three arcs of the Adams ring.[289,290,291] Some 79 minutes after the flyby,

This image taken near the time of closest approach shows high-level clouds casting shadows on Neptune's main cloud decks below.

Voyager 2 re-crossed the ring plane at a planetocentric distance of 103,700 km, going south. The spacecraft had been expected to be struck by dust as it passed through the ring plane, as it had at both Saturn and Uranus, but in this case the bombardment started 2 hours prior to each crossing and continued for 2 hours afterward, with the impact rate on the inbound crossing peaking at 280 hits per second! In fact, the craft suffered a significant number of hits while it was tens of thousands of kilometers from the ring plane, and also while near the pole, but, as previously, the particles were too small to cause any damage.[292]

Beginning 2 hours after the planetary flyby, and for the next 5 hours, Voyager 2's attention was on Triton, the final moon on its itinerary. It mapped two-thirds of the moon's surface, but just one-half at a resolution as high as 800 meters per pixel. The coverage included the south polar region, much of the southern hemisphere and across the equator as far as 30°N. Southern spring had begun in 1960, and would last for another decade or so. The north pole was in darkness, and would remain so until about 2040. Triton was a moon unlike any yet seen. It had a very bright surface with subtle hues ranging from yellow to peachy-pink, and a variety of geological features. Most of the southern hemisphere was covered by a cap of nitrogen ice that had been deposited as frost the previous winter, but there were also various dark splotches, streaks, and fragmented terrains. A very bright belt of frost that extended from the edge of the cap north to the equator was superimposed on a complex terrain that was darker and more strongly colored. Most of the northern terrain extending to the terminator was darker and even more reddish. Because the south polar cap reflected almost 90 per cent of the sunlight reaching it, its temperature was 38K, making it the coldest place known in the entire solar system – in fact, it was so frigid that the infrared radiometer could barely detect it.[293] Intriguingly, there were a large number of dark streaks in the south polar region, and the fact that they were superimposed on all other features implied that they were the most youthful. A form of surface activity was immediately suspected, but it took almost a month for the scientists to realize that they were due to geyser activity. Even more astonishingly, there were a number of active sites where narrow plumes rose to altitudes of many kilometers before being deflected by the high-altitude winds and then blown horizontally for hundreds of kilometers, leaving dense dark trails. Despite the low temperatures, radiation-darkened carbonaceous material on the surface would absorb the heat from the Sun, perhaps sufficiently to increase the pressure of subsurface nitrogen ice and prompt explosive venting and geysers. The fact that all the geysers were located in that portion of the polar region that was in continuous sunlight was consistent with this hypothesis. The venting gas would carry with it particles of the dark material, which would then fall out to create the distinctive trails. After the volcanoes of Io, the geysers of Triton were only the second known case of current surface activity on a major solar system object other than Earth.[294]

The south polar region of Triton. This is the image in which dark geyser-like streaks (arrowed) were discovered.

Toward the equator there were a variety of terrain types, the most remarkable of which was a sparsely cratered plain that was peppered by closely spaced and almost circular dimples 30–40 km in diameter that so resembled the skin of a cantaloupe that it was soon labeled the cantaloupe terrain. It was crossed by long narrow lines that might mark where icy fluid had oozed from the ground. Frozen terraced pools were seen on the adjacent plains. Remarkably, water was the only liquid that had the right properties to create features with such characteristics, but at the present temperatures on Triton this must be in various phases of ice, and thus as hard as rock. The volcanism was probably induced by tidal heating after Triton had been captured and was slowly settling into a circular orbit. In fact, the moon must have suffered tremendous tidal effects as its axial rotation was synchronized with its retrograde orbit. The resurfacing during this time was probably so extensive that no trace of its original surface remains. The cantaloupe terrain appears to be the oldest surface unit. It is unique because only Triton suffered such a major thermal pulse after the period of bombardment that made large basins on every other solid surface in the solar system.[295]

Craters were generally rare, the largest being only 27 km in diameter. They were predominantly located on the leading hemisphere where, as a result of the retrograde orbital motion, impacts would occur at the greatest speed. Spots with dark cores and bright halos located near the eastern limb defied explanation. Limb imagery showed faint layers of cloud at an altitude of several kilometers, but it was not possible to tell whether this was a side-on view of geysers or an atmospheric temperature inversion that was causing gases to condense.[296,297]

At 09:10 UTC Voyager 2 reached its closest point of approach, 39,800 km from the moon's center. A 3-minute radio occultation shortly thereafter indicated that the atmosphere had a surface pressure of just 1.6 Pa – corresponding to 1.6×10^{-5} of that at sea level on Earth. The ultraviolet spectrometer observed both the solar occultation as the spacecraft passed through the moon's shadow and also an occultation of beta Canis Majoris, and found that the predominant atmospheric species were molecular and ionized nitrogen, with methane in trace amounts at lower altitudes. The manner in which the moon deflected the spacecraft's trajectory, as measured by tracking the radio signal, provided a bulk density of 2 g/cm^3, which was intriguing because it was the same as for Pluto.[298,299] A number of hypotheses have been advanced to explain how Triton might have been captured. One idea was that early in its history Neptune had an extended primordial atmosphere and that this enabled the interloper to decelerate and circularize its orbit, but a recent study has ruled out such a large envelope. There were also suggestions that the deceleration was due to gravitational interactions with an earlier generation of satellites – in one case it was suggested that Pluto had been a member of that system and was ejected by Triton – but these scenarios posed dynamical challenges. Nevertheless, it has recently been suggested that Triton was a member of a binary system like Pluto and its satellite Charon, and that when this 'double planet' approached Neptune at a slow speed the forces acting on the two components tended to draw the binary apart, with one member escaping and the other being captured.[300,301]

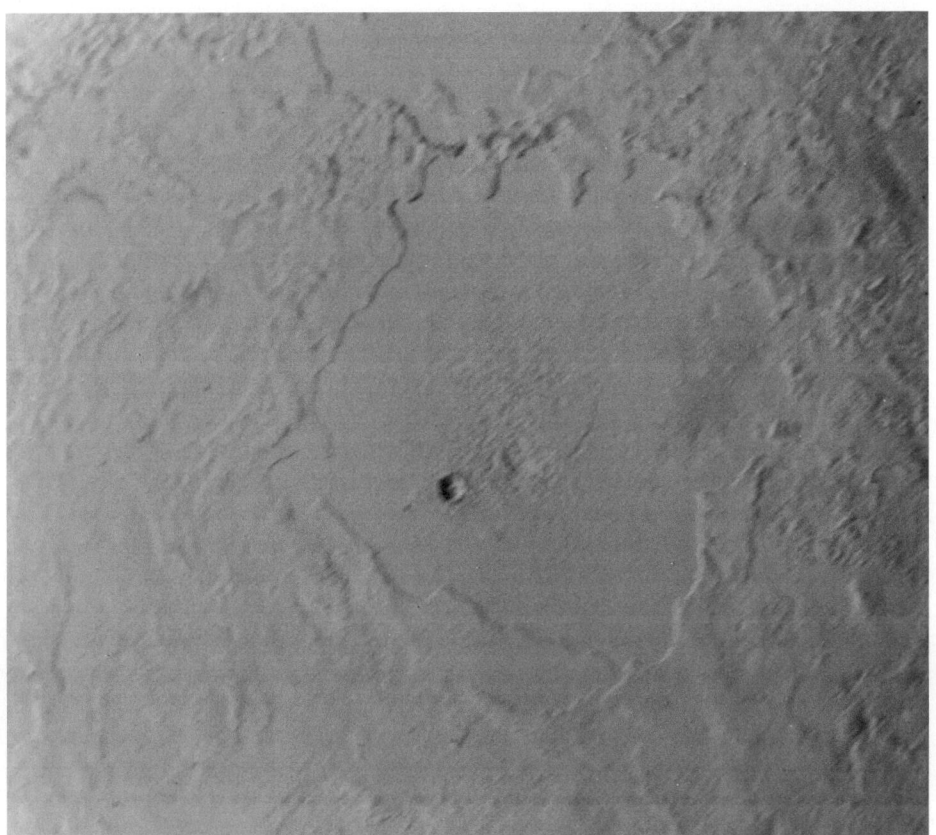

A high-resolution picture of Ruach Planitia on Triton, one of the best examples of a 'terraced pool'.

As the spacecraft left Triton behind, it was well south of the line from the Sun to Neptune and the lighting was not as favorable for imaging the ring system as it had been at Uranus, but a series of 111-second exposures caught Le Verrier, Galle and the three arcs of the Adams ring. On the morning of 27 August a pair of 591-second wide-angle exposures were taken about 90 minutes apart, one just to the left of the planet's disk and the other to the right in order to span the entire ring system shining by forward-scattered sunlight. These pictures revealed a great deal of detail that was not seen on the approach imagery owing to the lighting at that time. The Galle ring was resolved as a lane of dust more than 1,000 km wide. There was a broad sheet of material extending from Adams to Le Verrier, and this 'plateau' ring (also known as Lassell) had a denser inner rim that was named the Arago ring. Photometry showed that dust extended from the plateau ring down past Galle and some way toward the planet. That is, both Galle and Le Verrier were embedded in a broad sheet of dust, with a gap inward of Le Verrier where Despina orbited. Photometry also showed

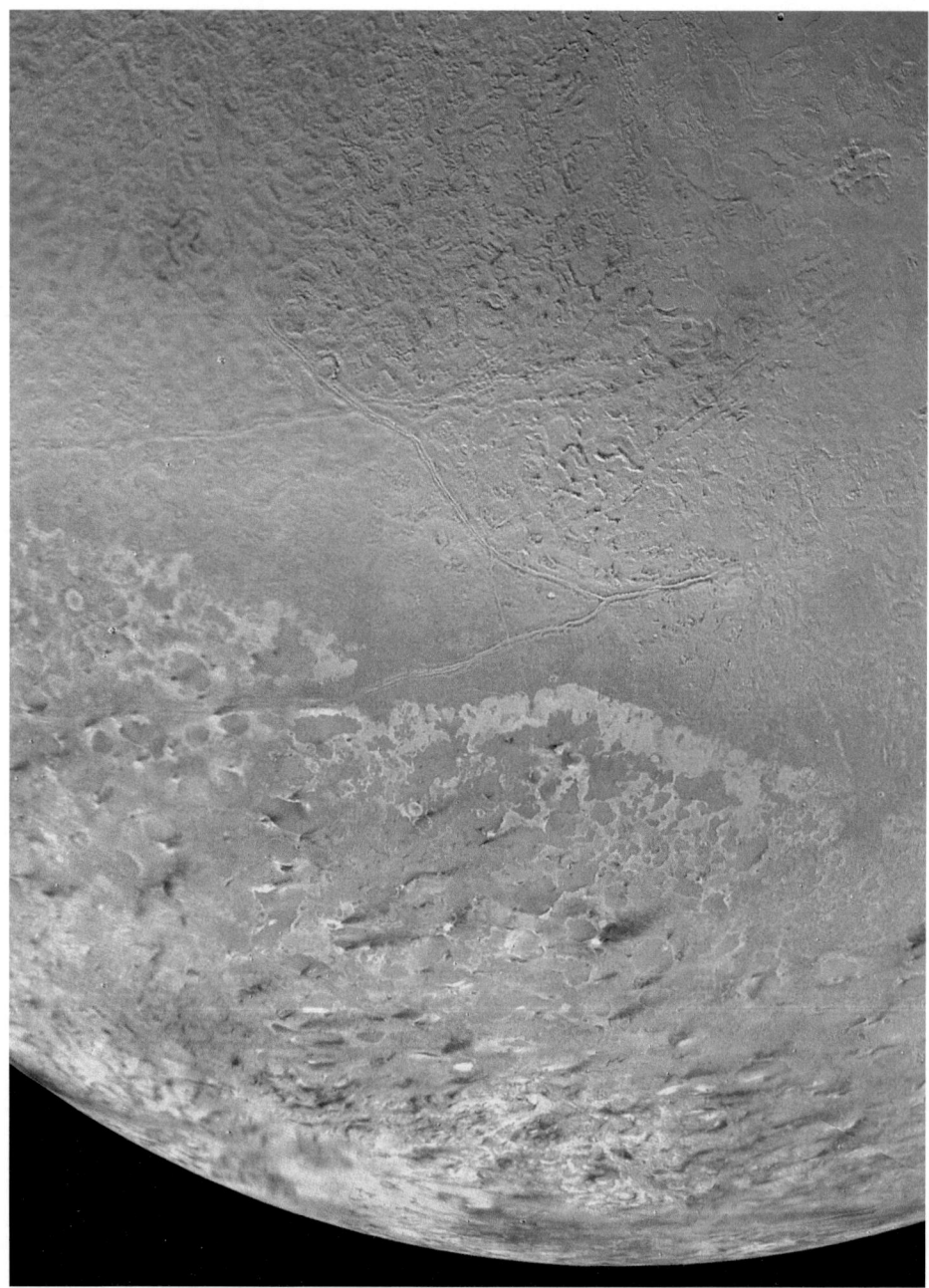

Triton's southern hemisphere showing the remarkable pitted 'cantaloupe' terrain in the upper half of the image. (This mosaic includes the area in which the geyser-like streaks were first identified.)

that only the outermost ring, Adams, was an isolated structure. Unfortunately the three arcs on Adams were on the opposite side of the planet at the time that each picture was taken and so are not present in the mosaic. Also evident in some of the images of the backlit rings was an narrow clumpy ring just inward of Adams, at about the same distance from the planet as the moon Galatea. An examination of the imagery for evidence of polar rings found nothing. Last-minute updates had been made to the encounter sequence to obtain high-resolution pictures of the arcs of the Adams ring within 13 hours of the planetary flyby, and a narrow-angle image of a portion of the trailing arc (Fraternité) gave a resolution of about 14 km per pixel, which was close to the actual width of the ring; this image also appeared to show clumps, but they were badly trailed by the rapid motion of the spacecraft and by the orbital motion of the material within the ring.

The spacecraft then swung its cameras back to Neptune, and the fact that features such as bright clouds and the Lesser Dark Spot were visible on the crescent proved that these were above most of the atmospheric haze (the corresponding views of the other giant planets had been bland). On 31 August, Voyager 2 returned a series of images showing the crescent of Triton progressing across the much larger crescent of Neptune, causing one scientist to exclaim in awe, "What a way to leave the solar system."[302]

Radio tracking of Voyager 2 throughout the encounter provided an accurate mass for the planet and for the total mass of the system. Several years later, these figures, together with those derived for the Uranian system, were used to reconsider the issue of how the two planets perturbed each other. It was a study of how Uranus initially lagged behind and then drew ahead of its predicted position that led to the discovery of Neptune. A further study that took Neptune into account prompted the search for Pluto. And because the estimated masses of Neptune and Pluto could not completely account for the perturbations of Uranus, some astronomers insisted that there was a 'Planet-X' still to be discovered. Using the new and much more reliable masses, the differences between the computed and observed positions of the planets in the outer solar system were reduced almost to zero. If there is another massive planet, it must be so much further out as not to exert significant influence on Uranus or Neptune.[303] In addition, the tracking enabled the masses of Nereid and Proteus to be inferred with some accuracy.[304]

Although Neptune's atmosphere rotated differentially, the periodicity of the radio emissions had indicated that its interior rotated in just over 16 hours, which implied that whereas the Lesser Dark Spot circled the planet in one local day, the Great Dark Spot, which took 18 hours, was being carried westward by a very strong wind. Tracking small clouds and features over an interval of several days proved difficult because the atmosphere was so dynamic that it was not always possible to recognize a given cloud in sequences of images, even if they were taken just a few hours apart.[305,306,307] The infrared spectrometer continuously scanned the night-side of the planet for 20 hours, yielding three temperature profiles and maps of the temperature field across a wide range of latitudes. At high altitudes, the temperatures were similar at the equator and the poles, but were colder at intermediate latitudes in a pattern reminiscent of Uranus. Hydrogen was the main constituent of the atmosphere, with

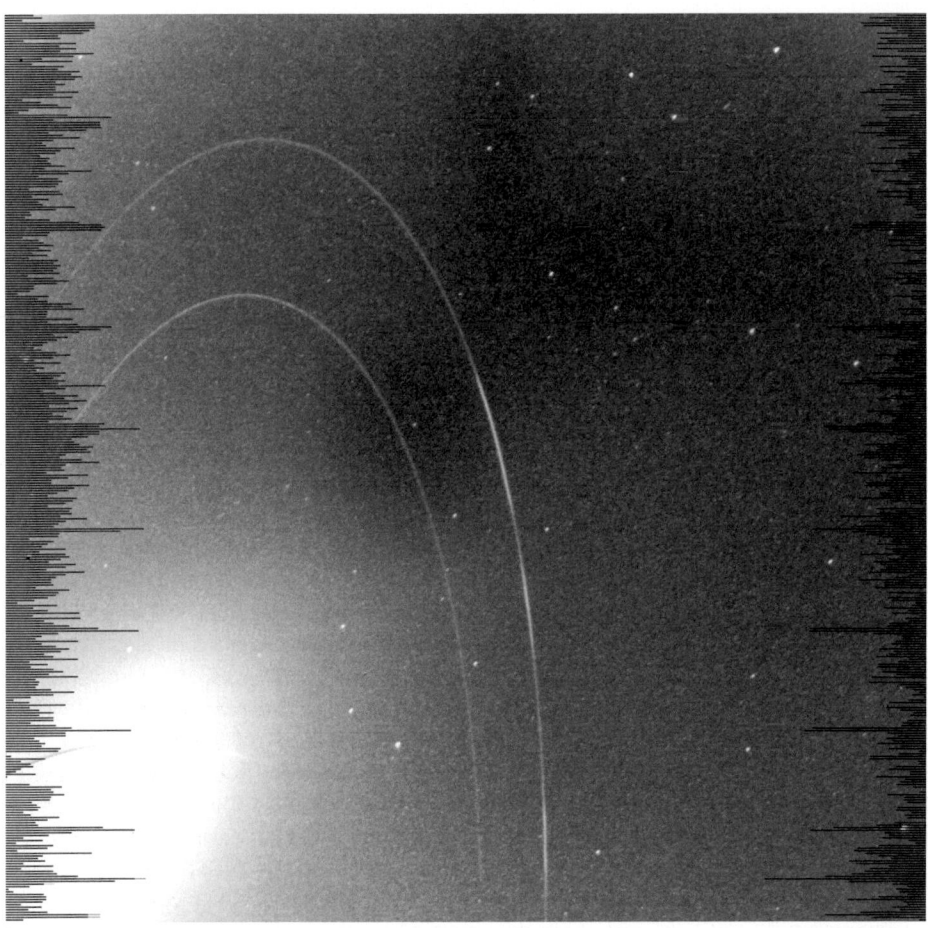

A 111-second exposure of the night-side of the Neptunian rings. This is one of the best pictures of the Liberté, Egalité and Fraternité arcs of the Adams ring. Orbital motion is clockwise. On some lines of the image there are missing pixels near the edges of the frame.

a helium mass fraction of about 20 per cent being consistent with an envelope that condensed from the solar nebula – as indeed had been the case for Uranus.[308,309] Methane in the upper atmosphere was responsible for Neptune's blue hue, but acetylene was also detected at deeper levels, and small amounts of ammonia might also have been present to cause a strong absorption during the radio occultation. The planet's energy budget was confirmed, with it emitting 2.3 times as much heat as it receives from the Sun. Given its small mass and size in comparison to Jupiter, this was a remarkable figure. This heat was one of the reasons that Neptune showed a more active weather system than Uranus.[310,311,312] It proved difficult to model the magnetosphere, and it was not until some days after the encounter that a preliminary

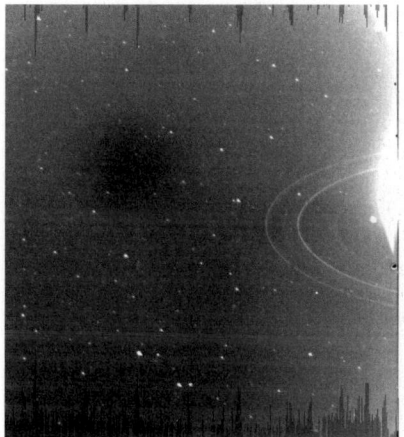 

This pair of 591-second images of the night-side of the Neptunian rings are the best available. Although they show all of the rings, including the faintest ones, the three arcs of the Adams ring are absent because they were on the right side when the left image was taken and on the left side when the right image was taken!

reconstruction became available. The difficulties derived from the fact that the spacecraft had come in over the south pole, with the result that its trajectory was parallel to the magnetic field lines. One totally unexpected finding was that, like Uranus, Neptune had a magnetic field whose axis was both inclined with respect to the rotational axis, in this case by 47 degrees, and was well represented as a dipole offset 10,000 km (more than half a planetary radius) from the center of the planet. One inference from this discovery was that the unusual characteristics of Uranus's magnetic field had nothing to do with the fact that the planet's spin axis was tilted so dramatically with respect to the plane of the orbit. The idea that Uranus had been in the act of reversing the polarity of its magnetic field at the time of the Voyager 2 flyby was also no longer tenable.[313] The spacecraft saw faint auroras on the night-side, but owing to the unusual configuration of Neptune's magnetic field these were not small ovals centered on the poles; they were distributed across broad regions, and might have been powered by gas and plasma that escaped from Triton.[314] For many years it was difficult to model the magnetic fields of Uranus and Neptune, but a recent interpretation as a dynamo involving a thin convective shell surrounding a stratified fluid sphere was able to replicate most of the observed characteristics – in particular, the tilt and the deviation from a strict dipole field. This theory, of course, also helps to place constraints on the internal structure of both planets.[315]

Voyager 2 remained within the confines of Neptune's magnetosphere for 28 hours following the planetary flyby. There was a well-developed magnetotail, but owing to the tilt of the magnetic axis this dramatically changed its characteristics over short intervals. A recognizable magnetopause crossing was made at 72 Rn, the event lasting more than 15 minutes. There were five bow-shock crossings over

On 31 August 1989 Voyager 2 took a sequence of the crescent of Triton crossing the crescent of Neptune, of which this image is part. Note that atmospheric structure can still be glimpsed on the planet.

the next day or so, and then the spacecraft finally departed the system on 28 August at 185 Rn.[316,317,318] The Neptune encounter was formally concluded on 2 October 1989.

Against the odds, Voyager 2 had survived 12 years in space and accomplished the Grand Tour of the outer solar system. Up to the Neptune encounter, the project had cost just under $1 billion, including the spacecraft, launch and flight operations; a sum that would fund only a few Space Shuttle flights, and by now there had been 30 such missions. Even although the Voyagers were performing as planned, the project was under annual threat, not only by the Reagan Administration's desire to reduce NASA's share of the federal budget but also by the agency's focus on operating the Space Shuttle and designing Space Station Freedom. Many scientific space projects fell by the wayside as a result of this pressure. At one point, the White House gave serious consideration to terminating funds for the DSN, whose only reason for existence, in the absence of any new planetary missions, was to support Voyager. Had it not been for a campaign by scientists and industries involved in the space business, Voyager 2 might have reached both Uranus and Neptune with us unable to hear its reports! In many ways, the 1980s marked the nadir of planetary exploration initiatives in the United States. Although a number of missions were scheduled for launch in the second half of the 1980s, they were grounded by the Challenger disaster in 1986, with the result that NASA launched no planetary missions between 1978 and 1989, and the first, the Magellan radar mapper for Venus, was dispatched only 4 months before Voyager 2's Neptune flyby. Fortunately, the Voyager project had provided sufficient data to keep the planetary scientists gainfully employed during the lean years.[319]

The gravity-assist from the Neptune flyby had further increased the eccentricity of Voyager 2's solar orbit to 6.3, and deflected it 48 degrees south of the ecliptic. From now on, the spacecraft would depart at an almost constant speed of 3 AU/year. This made it the second-fastest spacecraft heading out of the solar system, exceeded only by its sister.

By the end of the 1980s every planet in the solar system, with the exception of Pluto, had been the target of at least one reconnaissance by a robotic emissary from Earth. As this book was being written, the New Horizons mission was using a Jovian gravity-assist to arrange an encounter with the Pluto–Charon system in 2015. In August 2006, however, the International Astronomical Union made the controversial decision to demote Pluto from the status of a 'classical planet' to a 'dwarf planet', in recognition of the recent discovery that it was merely one of a population of Kuiper Belt Objects, some of which were much larger. Thus, it can be said, *ex-post*, that Voyager 2's flyby of Neptune did indeed complete the first phase of the exploration of the planets that began 27 years earlier with Mariner 2. It was not yet time for the Voyagers to retire, however.

THE LARGER PERSPECTIVE

While the world was awed by the unique scientific data provided by Voyager 2 from Uranus and Neptune, its twin was climbing out of the ecliptic plane and routinely reporting particles and fields data. However, two failures had deprived Voyager 1 of key hardware backups: in 1981 one of the two redundant memories of the scientific data computer failed, and in 1987 it lost the component of the X-Band transmitter that had been so difficult to manufacture a decade earlier. On the scientific side, the plasma instrument had been operating only in a degraded status since Saturn.[320] Apart from these issues, the spacecraft was healthy: in particular, the scan platform and its cameras were fully functional and the scientists and engineers were eager to put them to use.

Since the cameras would not make very capable telescopes, they were of little use for astronomical work; but it would be possible to exploit the unprecedented vantage point to record the Sun and all its planets in a single mosaic. In response to concern from NASA headquarters that the vidicon sensors would be damaged if they were to

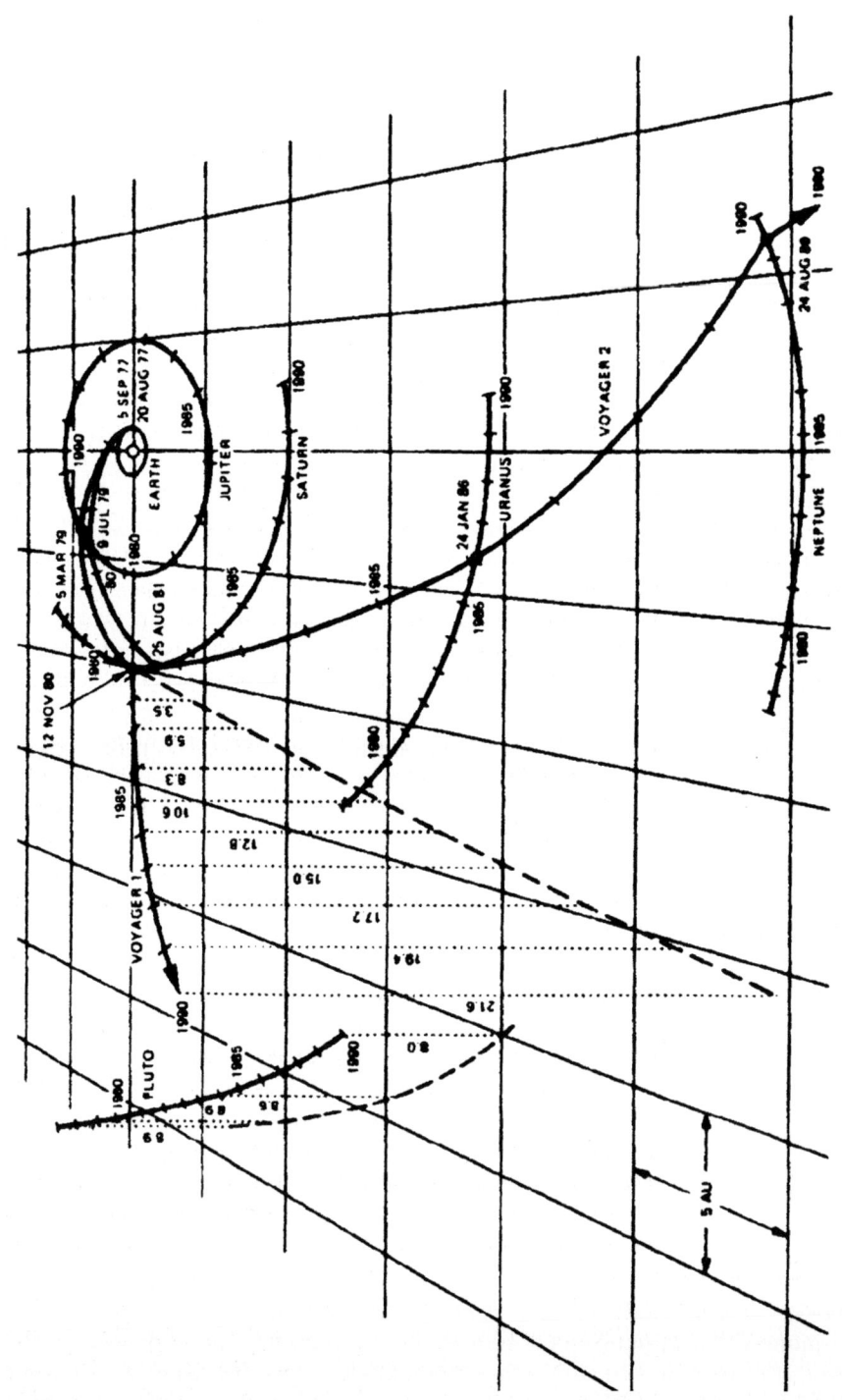

The trajectories of the two Voyagers as of 1990. Note in particular the striking angle at which Voyager 1 climbed above the ecliptic plane.

be aimed at the Sun, JPL argued that, in the absence of other targets, there was little point in preserving the imaging capability. The plan was to slew the camera to take a color picture centered on each planet – with the exception of Pluto, which was too faint – to show them in the mosaic at their relative positions around the Sun, with the stars as a backdrop. The images of the planets would be no more than bright dots, but their colors would be accurately represented.[321,322] On 14 February 1990, Voyager 1, now at a distance of some 40 AU from the Sun, took 64 images along a sinuous track that took in all of the targets. Unfortunately, Mercury was lost in the solar glare, and at a late stage in the process of drawing up the imaging sequence it was realized that the phase of Mars was such that it would not be visible through 3-color filters and the clear filter would have to be used instead, but by then it was too late to alter the sequence and consequently Mars did not appear in the mosaic.[323] Earth appeared as a small blue dot that just happened to be crossed by a ray of sunlight in the image. As astronomer and popularizer Carl Sagan wrote, "Look again at that dot. That's here. That's home. That's us." This picture has all the evocative power, if not the sheer beauty, of the more widely known pictures of Earth that were taken by the Apollo astronauts.[324] These were the last of some 67,000 pictures that the Voyagers had returned over an interval of 13 years, or as Project Scientist Ed Stone put it, they were the 'last light'. There was also a proposal to continue to monitor the vicinity of the Sun for a year in order to make a movie of Earth orbiting the Sun, but there were doubts about whether it would be feasible to image Earth for a complete orbit, and in any case the money was not available.[325,326]

After its final planetary encounter Voyager 2 joined its sister in reporting particles and fields data. This was formalized on 1 January 1990 by the start of the somewhat dramatically named Voyager Interstellar Mission (VIM), the budget for which could support only 20 people. The stated objectives were to characterize the interplanetary medium beyond the orbit of Neptune and to search for and investigate the transition from the heliosphere (i.e. the region of space around the Sun that is dominated by its magnetic field and the supersonic solar wind) to the interstellar medium. At some yet-to-be-determined distance the pressure of the increasingly rarefied solar wind must equal that of the interstellar medium, which comprises mostly hydrogen and helium, causing the solar wind to slow to subsonic speeds through a shock wave known as the termination shock. The supersonic interstellar wind was also believed to be slowed upon meeting the heliosphere and become subsonic through a bow shock. In the space between these two shock waves, the solar wind and the interstellar wind must meet at a boundary called the heliopause. Just as in the case of the magnetospheres of the giant planets, the heliosphere was expected to possess a complex boundary region, with, in turn, an inner termination shock, a heliopause, a turbulent heliosheath and an outer shock wave. It was also thought that owing to the motion of the Sun and is retinue around the galaxy, the heliosphere must have a long tail 'downwind' with respect to the local interstellar medium. There were widely differing estimates of the distance to the heliopause. Eugene Parker, the scientist who postulated the existence of the solar wind several years before it was actually observed by the Soviet Luna probes and measured by Mariner 2, placed it at between

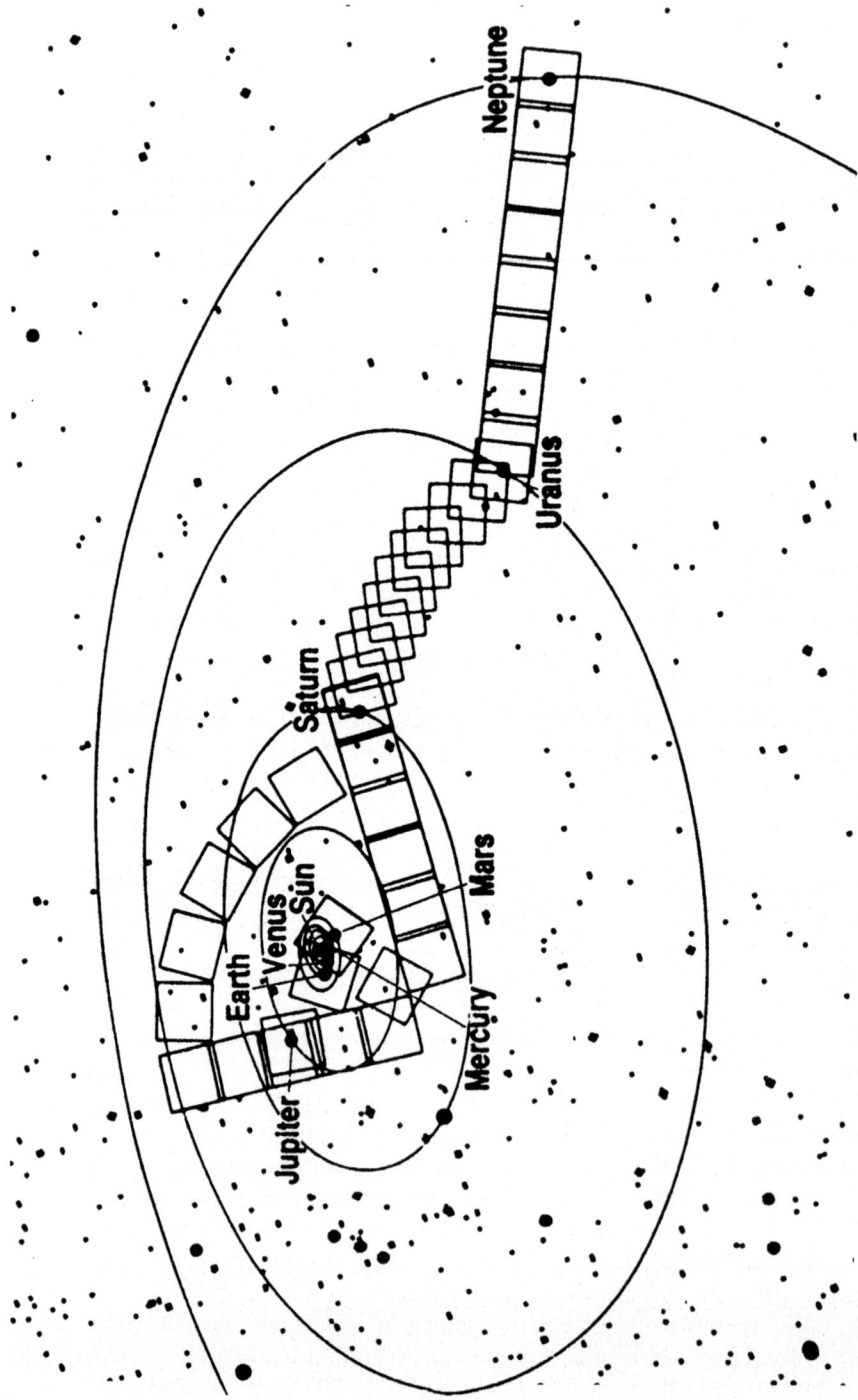

The very last images taken by the Voyager cameras were in a sequence to provide a 'portrait' of the solar system with the exception of Pluto, which would be too faint.

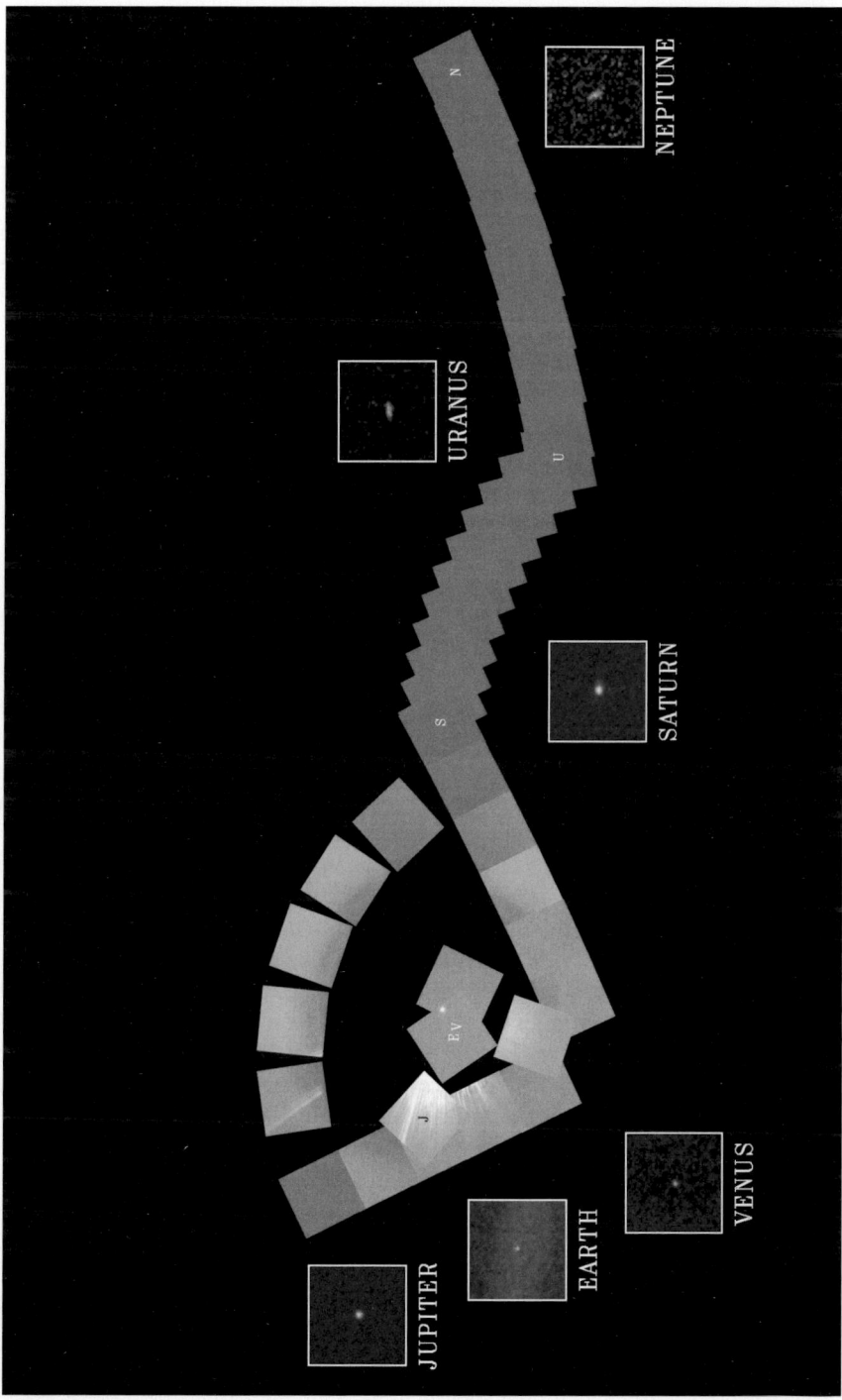

In the Voyager 'portrait' of the solar system all of the targeted planets were visible except Mercury, which was too close to the Sun, and Mars, which was showing its night-side to the camera.

40 and 50 AU from the Sun. Other early estimates put it inward of the orbit of Neptune. By the early 1990s, however, the termination shock was expected to be somewhere between 60 and 105 AU, with the heliopause somewhere between 116 and 177 AU. With good fortune, both would be able to be reached in order to provide an apt finale to one of the greatest voyages of exploration of all time. In fact, at the pace it was going, Voyager 1 would reach 100 AU from the Sun in August 2006 and 150 AU in 2020 – by which time its slower sister would be at 125 AU. Meanwhile, the actual existence of the heliopause was confirmed by satellites and spacecraft that observed the ultraviolet glow of interstellar gas colliding with the solar wind. As was to be expected, this glow was centered around the solar apex, the apparent direction of travel of the Sun in the galaxy.

Of the four spacecraft departing the solar system, Pioneer 10 was flying down the heliospheric tail and was not likely to reach the interstellar medium any time soon, but by sheer luck the other three were heading in the approximate direction of the solar apex. On approaching the boundary of the heliosphere, the modulation of galactic cosmic rays by the solar magnetic field and solar activity should decrease and then disappear altogether. The crossing of the termination shock would be noted by a magnetometer as a sudden increase in the strength of the magnetic field. What might lie beyond the heliopause was a matter of speculation – but that was precisely why the prospect of sampling it was so fascinating. To astronomers, the interstellar medium appeared to be dominated by gas clouds and by the low-density shells made by supernova explosions, which interact to create enormous complex and evanescent structures. Radio astronomy and X-ray observations of the local region had shown that the Sun was contained within one such low-density and high-temperature shell known as the 'Local Bubble', which was possibly formed 350,000 years ago by the supernova that transformed its progenitor into the 'Geminga' neutron star. Such an explosion in the vicinity of the Sun could have pushed the heliopause inward of the orbit of Uranus, and may even have left traces in the form of unusual isotopes on the surfaces of the satellites of the outer planets. The Milky Way is a spiral galaxy, and the Sun takes 250 million years to complete its roughly circular orbit, during which time it must pass through many dense clouds of gas and dust. It is possible that as the Sun penetrates such an environment the heliosphere is so compressed that it effectively 'switches off'. It is difficult to determine the effect that this might have had on terrestrial life. Despite being enclosed within the Local Bubble, the Sun and the neighboring alpha Centauri stellar system are both believed to be immersed in a smaller cloud of cold hydrogen dubbed the 'Local Fluff'.[327,328]

Barring a major hardware failure, the main operational limitation was the output of the RTGs, which was slowly declining. Contrary to popular belief, it was not just the radioactive decay of the plutonium dioxide that was responsible for the reduction of power, it was also (and mainly) the degradation of the thermocouples that turned the liberated heat into electricity. At launch the RTGs had yielded 450 W, but by the mid-1990s Voyager 1 was at 336 W and Voyager 2 was at 338 W. Nevertheless, it was thought that the decay rate of 5 W/year would permit the particles and fields instruments to be run until 2020 or thereabouts. Fuel was another factor: in the mid-1990s each spacecraft still had about 35 kg of hydrazine, which was being consumed at a rate of 6–8 grams per week. In addition to attitude control and maintaining the high-gain antenna pointing at Earth, fuel was used to make periodic 360-degree roll maneuvers to calibrate the magnetometer. By permitting less accurate pointing, it was estimated that the fuel supply would be sufficient to maintain attitude control for another 50 years. As the cameras, infrared spectrometer and photopolarimeter were no longer in use, the bandwidth of the downlink was so reduced that a data rate of 160 bps was sufficient for most science and engineering data for the interstellar mission. For once, receiving the downlink was not an issue, because the largest DSN antennas would be able to 'read' the signal until at least 2030. However, these antennas were routinely dedicated to higher priority missions, and the Voyagers were being serviced by the smaller 34-meter dishes. In addition to the six particles and fields experiments, the ultraviolet spectrometer on the scan platform was to perform up to four observations per week. In fact, this instrument had proved particularly suitable for taking spectra of variable and subluminous stars, supernova remnants and extragalactic sources. Furthermore, it was to observe the Sun in order to characterize changes in the solar ultraviolet flux, which can also affect the Earth's atmosphere. Also, being far from the Sun, the two spacecraft were well positioned to monitor the ultraviolet glow of interstellar gas penetrating the solar system. At one time, each spacecraft pointed its spectrometer in the direction of its sister to measure the density of interstellar hydrogen atoms along the line of sight. By 1992, the two spacecraft were using different stars for attitude determination – one in the northern sky and the other in the southern sky – with the result that the portion of the sky that was blocked by the structures on one spacecraft was visible to the instrument on the other, which further reduced the need to maneuver, and thereby saved fuel. The spacecraft stored ultraviolet data on tape and replayed it at 600 bps, and twice each year they replayed high-time-resolution plasma wave data at 1.4 kbps.[329] If by the turn of the century there should be insufficient power to run both the scan platform and the ultraviolet spectrometer, the plan was to turn off both. In addition to these instruments, serendipitous discoveries could still be made simply by tracking the spacecraft, as this would reveal the gravitational perturbations of any large masses lurking in the outer realms, including Nemesis, the putative companion star to the Sun, which (according to a theory that was well received by the popular press in the 1980s, although not accorded commensurate acceptance by the scientific community) was the cause of periodic mass extinctions of terrestrial life.

In order to prevent losing Voyager 2 if its command receiver were to fail before the spacecraft was switched off, an automatic sequence was stored in its memory to enable it to continue to operate. Voyager 1 was given similar instructions as a precaution. The software would assume that communications had failed if no instructions were received over an interval of 6 weeks. Once active, this autonomous sequence was to periodically calibrate the instruments, and turn them off one by one as necessary to save power, and place the spacecraft into a roll in order to maintain the high-gain antenna pointing at Earth until 2020 in the case of Voyager 1 and 2017 for Voyager 2.[330,331,332,333]

In July 1992 the Voyagers began to detect intense low-frequency radio emissions that lasted for several months. Each burst was probably a cloud of plasma that had

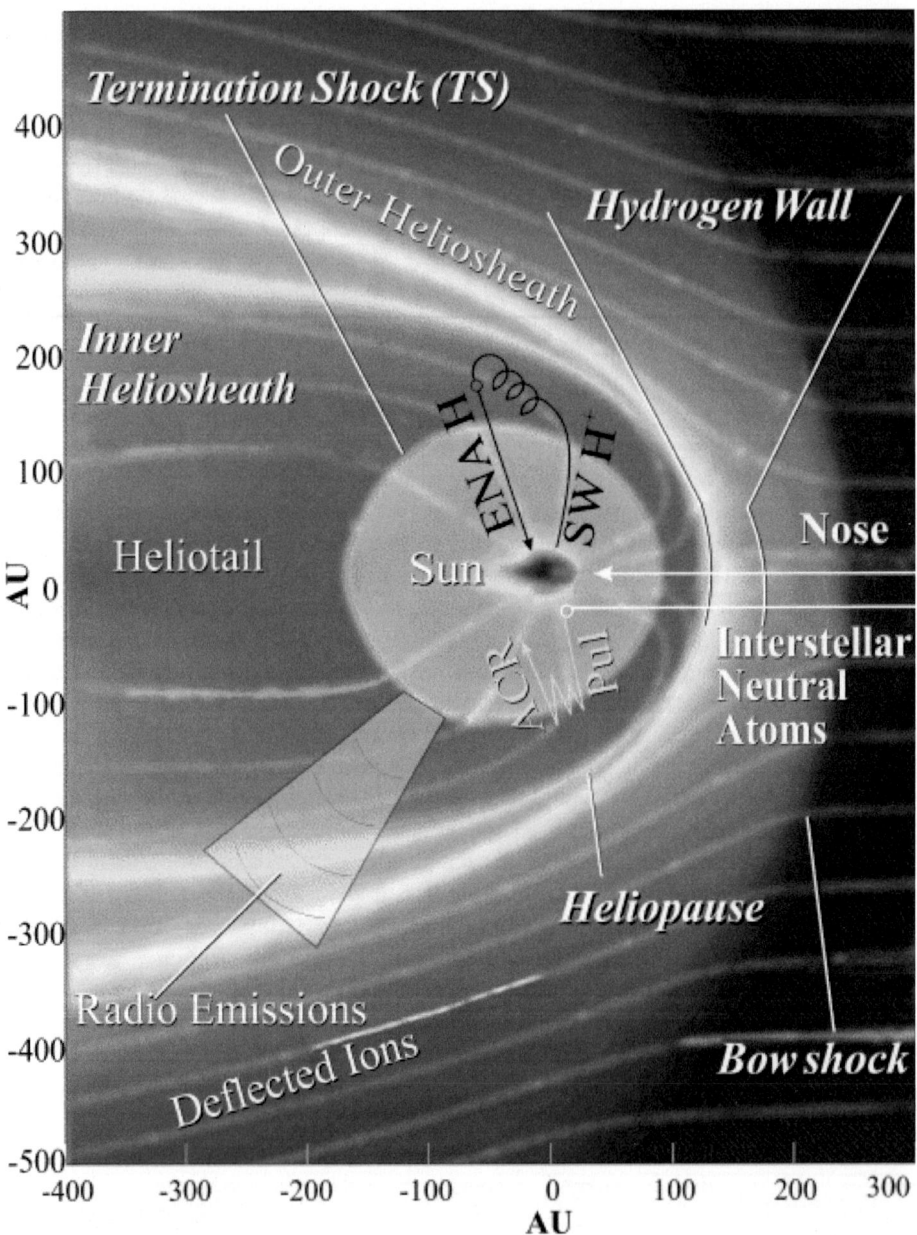

The structure of the heliopause as it is understood today. (Courtesy of David J. McComas and Gary P. Zank)

been ejected by the Sun about 400 days earlier and was interacting with interstellar gas at the heliopause. These observations helped to position the termination shock at between 87 and 133 AU.[334] The spacecraft also made 'on demand' observations. For example, Voyager 2 was one of several spacecraft that were used to study the outburst of Nova Cygni 1992, one of the best-observed novas in astronomical history and one of the brightest of the century. Its ultraviolet spectrometer was able to see the star 'turn on' in the far-ultraviolet part of the spectrum and then monitor its brightness over an interval of many days.[335] NASA considered reviving Voyager 2's camera in order to study comet Shoemaker–Levy 9, which had been discovered on 24 March 1993. As calculations showed, an encounter with Jupiter in July 1992 had caused the comet to break up. The fragments were strung out in a line, and on course to smash into Jupiter in July 1994! Frustratingly, the spectacle would not be visible from Earth as the impact site would be just beyond the limb; although it would come into view several minutes later. For some time it seemed that Voyager 2 would be the only spacecraft to have a direct view. Even though Jupiter's disk would span only a few pixels in the narrow-angle camera, long trailed exposures would be worthwhile because they would show each impact as a brief burst of light. Unfortunately, the imaging software had been erased after the Neptune flyby because the cameras had not been expected to be used again, and the team had dispersed. As revised computations of the comet's path became available in early 1994 it became apparent that the impacts would be visible to the Galileo spacecraft, which was heading for Jupiter and was much closer to it. Although Voyager 2 made ultraviolet and radio observations, it failed to detect anything that was related to the impacts. A subsequent analysis of the temperatures of the fireballs from the impacts measured by other means showed that they were just below the level of detectability of the Voyager 2 instruments.[336,337,338]

On 17 February 1998, after a 20-year chase, Voyager 1 overtook Pioneer 10 as the spacecraft most distant from the Sun – at a distance of 69.4 AU. As the race to the boundary of the solar system continued, however, the activities of the two Voyagers

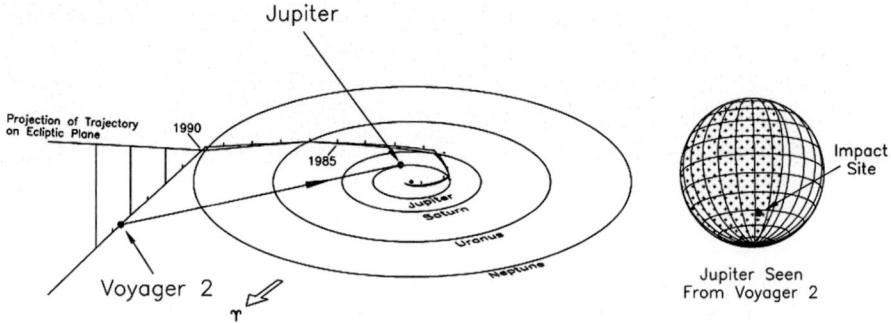

A projection of Voyager 2's line of sight to the location on Jupiter at which the fragments of comet Shoemaker–Levy 9 would impact. (Courtesy of Ronald J. Vervack, Jr)

steadily declined. That same year, Voyager 2's scan platform was turned off to save power. First, the platform heater was turned off, but as this had also been used to keep the sensor of the ultraviolet spectrometer warm, the only remaining functioning instrument on the platform had also to be switched off within weeks. Unfortunately, the strings of commands sent to the probe contained the erroneous command to turn off the exciter of the S-Band transmitter, with the result that contact was lost for 66 hours until, upon realized the error, JPL was able to reactivate the transmitter on 14 November.* Thus ended the saga of the scan platform that had been the cause of so much angst dating back to the first hours of the mission. The platform on Voyager 1 was to have been deactivated in 2000, but although it was decided to cease slewing it, the heater was left on in order to enable the ultraviolet spectrometer to continue to make observations.[339]

On 1 August 2002, at a distance of 85 AU from the Sun, Voyager 1 noted a large increase in the flux of energetic charged particles. Some of the scientists took this to indicate that the spacecraft had crossed the termination shock where the solar wind was slowed to subsonic speeds; however, to other scientists this data and the counts of cosmic rays indicated that the spacecraft was very close to the shock, but had yet to cross it. A few months later, following a surge of solar activity, the spacecraft was again immersed in the supersonic solar wind as the heliosphere inflated.[340] However, by this time Voyager's role as humanity's pathfinder into interstellar space was at risk. In January 2004, a year after the loss of Space Shuttle Columbia and her crew, President George W. Bush announced his 'Vision for Space Exploration', in which he called for an increased human presence in space, and in particular for the return of astronauts to the Moon as a precursor to a mission to Mars. Despite assurances that NASA would accommodate these activities within the existing budget, and that funding for scientific projects would be safe, the agency – as it had repeatedly done in the past to accommodate overruns on the human space program – was left with no option than to cancel some scientific projects, to delay others and to consider ending support for missions that were underway. On 16 December 2004, just as rumors of the demise of the Voyager Interstellar Mission were rife, the scientists revealed that at a distance of 94 AU from the Sun the magnetometer on Voyager 1 was reporting that the intensity of the magnetic field had tripled to values that had not been seen in 20 years. This was an unambiguous indication that the spacecraft had reached the termination shock, as the field strength would increase when the solar wind was compressed upon slowing to subsonic speeds. At the same time, plasma-wave oscillations were detected that mimicked those recorded shortly prior to penetrating the shock waves in front of planetary magnetospheres. This time no one doubted that Voyager 1 had indeed entered the heliosheath, although unfortunately the actual crossing of the termination shock had passed unobserved because the spacecraft was not being tracked at that time. Having lost its plasma experiment, Voyager 1 had not been able to monitor the speed of the solar wind, but matching

---

* In a similar accident in November 2006, another spurious command switched on the long-dormant infrared spectrometer, wasting power.

observations by Voyager 2 indicated that the pressure of the solar wind increased until mid-2004, then started to decline, and this allowed the termination shock to contract and wash across Voyager 1 several months later. As had so often happened, when a spacecraft observed a new phenomenon – even one that had been so eagerly anticipated – there were surprising discoveries. During the 1970s an anomalous component of the cosmic-ray flux had been discovered that scientists believed had been produced as interstellar gas reached the termination shock and was ionized and accelerated by the solar wind, but the in-situ data gave no sign of this phenomenon, implying that its source was further out.[341,342,343,344,345]

The next major event is expected around 2010, when Voyager 2 should reach the termination shock. The final milestone for Voyager 1 will be to exit the heliosphere and enter interstellar space. Its observations of the termination shock suggested that the heliopause should be located at about 125 AU, which the spacecraft will attain in the mid-2010s. To further reduce the power consumption, and thus extend the life of the spacecraft, the tape recorder will soon be turned off and gyroscopic operations will cease, including the roll maneuvers to calibrate the magnetometer. Without the gyroscopes it will no longer be possible to point the high-gain antenna accurately at Earth, but there will be no taped data to replay at a high data rate. In 2015, in a final power-reduction regime, the instruments will be turned off one by one as necessary in a sequence that will depend on their status and scientific productivity at that time. Around 2020 the output of the RTGs will no longer be sufficient to simultaneously run any instruments in addition to the core systems, and at 145–150 AU from the Sun both vehicles will suffer power starvation and their transmitters will fall silent – that is, unless the shortsightedness of NASA policy-makers has already ordered them to be switched off.[346,347] Even if the Voyagers reach the heliopause, they will provide data for only the locations they sample; they will not be able to investigate the shape of the heliosheath and the local processes. To put the Voyager data into context, NASA has decided to launch the Interstellar Boundary Explorer in 2008, a low-cost satellite with an imaging system to make an all-sky survey of the manner in which the energetic neutral atoms arriving from interstellar space interact with the heliosphere.[348]

Voyager 1 is traveling in the direction of the constellation Ophiucus, and in about 18,000 years it will be 63,000 AU from the Sun, or, to express it more conveniently, 1 light-year. And 20,000 years later it will travel within 1.64 light-years of the star AC + 79 388, some 2,000 years before Pioneer 11 passes it at a similar distance. Over the next 500,000 years Voyager 1 will perform a further three stellar encounters at distances in excess of 2 light-years. If Voyager 2's Neptune flyby had been at the initially intended position, its subsequent trajectory would have taken it within 0.8 light-year of the nearby star Sirius in Canis Major (currently the brightest star of the terrestrial sky), but after the target was moved to avoid the zone occupied by the rings, it will now only come as close as 4.30 light-years to the star in about 296,000 years. However, before that, in about 40,000 years, it will make a 1.7-light-year pass of Ross 248.[349,350,351] Unless humanity finds itself trapped within the solar system owing to the impracticability of interstellar travel, these emissaries may well outlive the species that dispatched them.

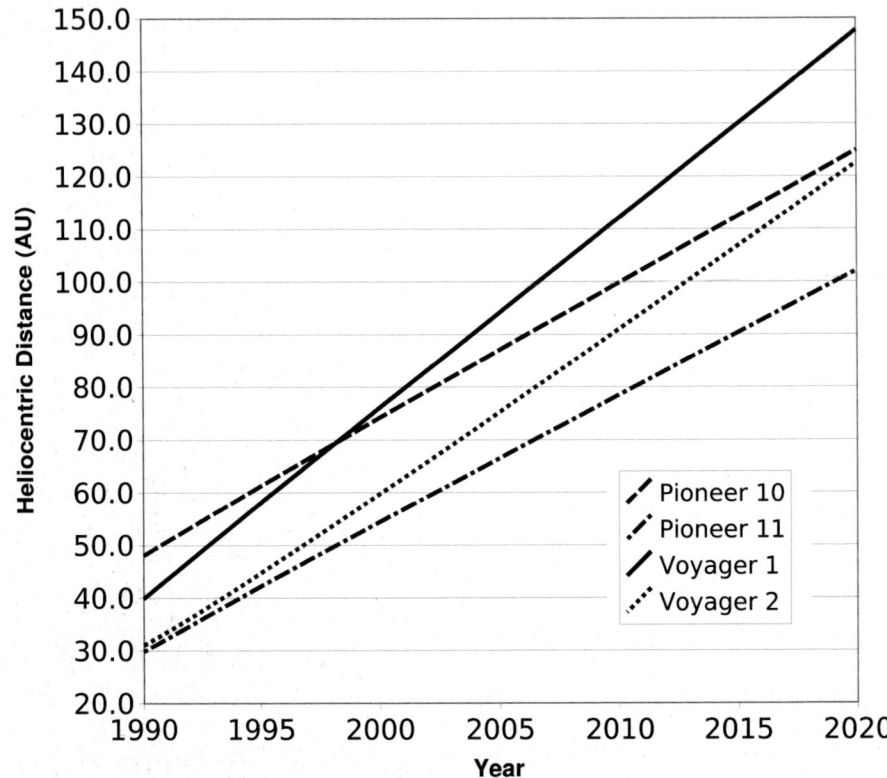

A plot of the heliocentric distances of the four spacecraft that are currently leaving the solar system. Note that the fastest, Voyager 1, overtook Pioneer 10 in 1998 and reached 100 AU in 2006. They will be joined by the New Horizons spacecraft after its Pluto flyby in 2015.

Star encounters by the Voyager spacecraft

Spacecraft	Date (Year)	Star Name	Distance (light-years)
Voyager 2	20279	Proxima Centauri	3.21
Voyager 2	20584	Alpha Centauri	3.47
Voyager 2	40170	Ross 248	1.65
Voyager 1	40272	AC+79 3888	1.64
Voyager 2	46348	AC+79 3888	2.76
Voyager 2	129671	DM+15 3364	3.46
Voyager 1	146193	DM+25 3719	2.35
Voyager 1	172867	Krueger 60	2.85
Voyager 2	298477	Sirius	4.30
Voyager 2	318323	DM−5 4426	3.97
Voyager 1	497482	DM+21 652	2.08

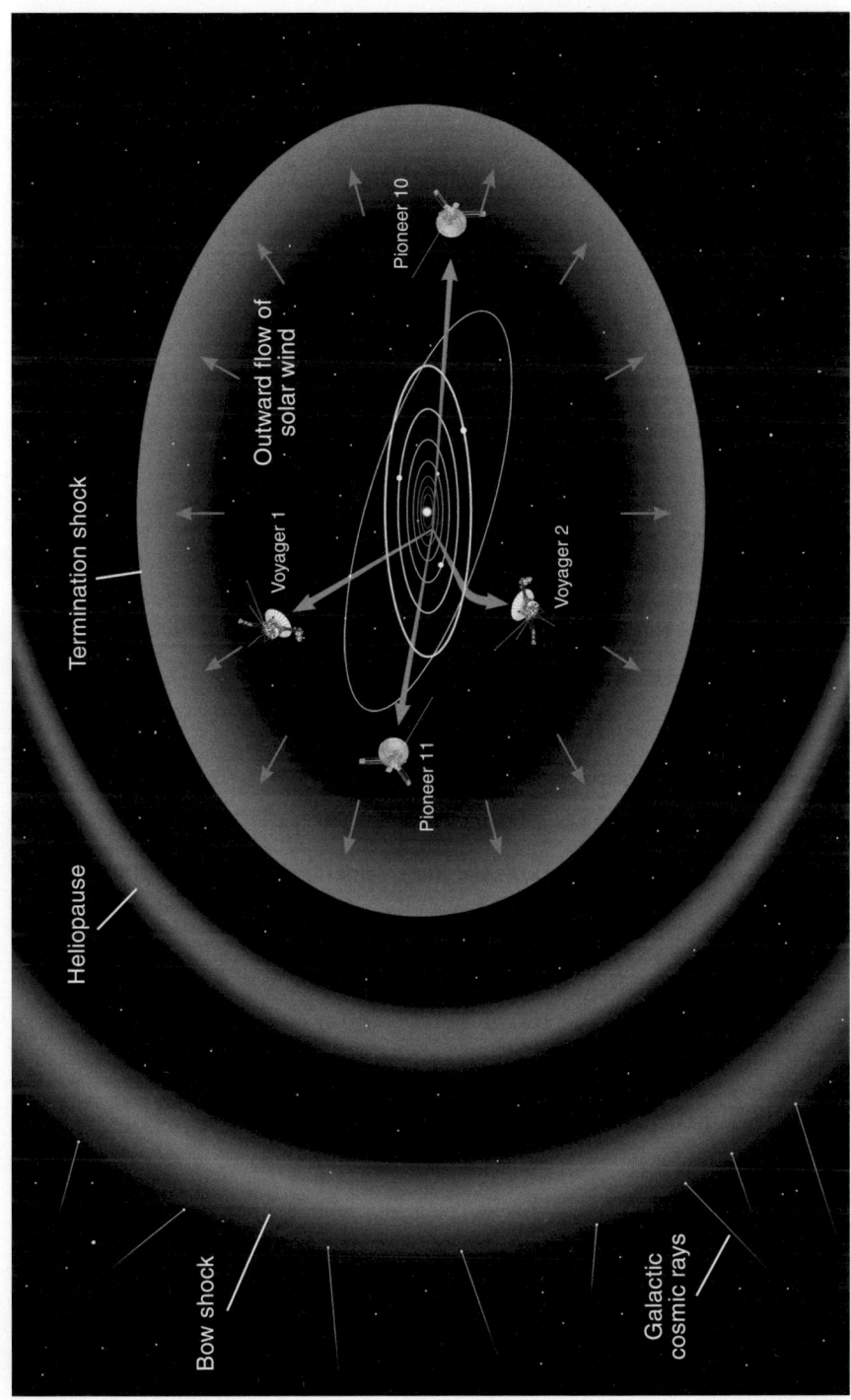

The heliosphere. (Courtesy of the Lunar and Planetary Institute)

SOURCES

Citations refer to the chapter references section at the end of the book.

1 Brown-1965	45 Murray-1989e	89 Hord-1979
2 NASA-1965f	46 Poynter-1984d	90 Kraemer-2000i
3 NASA-1966	47 Evans-2004e	91 Terrile-1979
4 Flandro-1966	48 Morrison-1980f	92 Ingersoll-1981
5 Hale-1970	49 Swift-1997i	93 Smith-1979b
6 Crocco-1956	50 Murray-1989f	94 Smith-1979b
7 Swift-1997a	51 Laeser-1986	95 Johnson-1983
8 CIA-1971	52 Morrison-1980g	96 Smith-1979a
9 CIA-1973b	53 Ingersoll-1981	97 Soderblom-1980
10 Evans-2004a	54 Smith-1979a	98 Jewitt-1979
11 Faget-1970	55 Edelson-1979	99 IAUC-3507
12 Parker-1971	56 Morrison-1980h	100 Smith-1979b
13 Hale-1970	57 Brush-1996a	101 Soderblom-1980
14 Swift-1997b	58 Smith-1979a	102 Smith-1979b
15 Swift-1997c	59 Hanel-1979a	103 Soderblom-1980
16 Gatland-1972j	60 Synnott-1980	104 Stone-1979b
17 Evans-2004b	61 Synnott-1981	105 Smith-1979b
18 Kraemer-2000i	62 Ness-1979b	106 Smith-1979b
19 Rubashkin-1997	63 Smith-1979a	107 Eshelman-1979b
20 Kraemer-2000i	64 Soderblom-1980	108 Smith-1979b
21 Kraemer-2000i	65 Johnson-1983	109 Pollack-1981
22 Morrison-1980a	66 Davidson-1999f	110 Sandel-1979
23 Poynter-1984a	67 Sagan-1976	111 Hanel-1979b
24 Swift-1997d	68 Miller-1953	112 Morrison-1980k
25 Swift-1997e	69 Eshelman-1979a	113 Swift-1997j
26 Swift-1997f	70 Eshelman-1979b	114 Holberg-1982
27 Moore-1974	71 Smith-1979a	115 Holberg-1980a
28 Hyde-1974	72 Soderblom-1980	116 Holberg-1980b
29 Swift-1997c	73 Broadfoot-1979	117 Stone-1981a
30 Edelson-1979	74 Smith-1979a	118 Sheehan-1998
31 Poynter-1984b	75 Soderblom-1980	119 Medkeff-1998
32 Wilson-1987u	76 Broadfoot-1979	120 Morrison-1982a
33 Morrison-1980b	77 Morrison-1980j	121 Swift-1997j
34 Swift-1997g	78 Morabito-1979	122 Murray-1989g
35 Morrison-1980c	79 Hanel-1979a	123 Morrison-1982a
36 Davidson-1999e	80 Hartline-1980	124 Richardson-2004
37 Evans-2004c	81 Soderblom-1980	125 Morrison-1982a
38 Smith-1977	82 Johnson-1983	126 Smith-1981
39 Poynter-1984c	83 Peale-1979	127 Owen-1982
40 Morrison-1980d	84 Ness-1979b	128 Broadfoot-1981
41 Evans-2004d	85 Broadfoot-1979	129 Hanel-1981
42 Dawson-2004h	86 Hanel-1979a	130 Tyler-1981b
43 Swift-1997h	87 Stone-1979a	131 Smith-1981
44 Morrison-1980e	88 Morrison-1980i	132 Johnson-1982

133 Smith-1981
134 Johnson-1982
135 Morrison-1982a
136 Broadfoot-1981
137 Tyler-1981b
138 Smith-1981
139 Johnson-1982
140 Smith-1981
141 Dollfus-1998a
142 Dollfus-1998c
143 Smith-1981
144 Johnson-1982
145 Hanel-1981
146 Smith-1981
147 Morrison-1982a
148 Bridge-1981
149 Kaiser-1980
150 Warwick-1981
151 Hanel-1981
152 Smith-1981
153 Pollack-1981
154 Tyler-1981b
155 Swift-1997k
156 Smith-1982
157 Hanel-1982
158 Tyler-1982
159 Peterson-1993
160 Klavetter-1989a
161 Klavetter-1989b
162 Smith-1982
163 Johnson-1982
164 Lamy-1980
165 Smith-1982
166 Johnson-1982
167 Hanel-1982
168 Lane-1982
169 Hanel-1982
170 Sandel-1982
171 Smith-1982
172 Johnson-1982
173 Hanel-1982
174 Warwick-1981
175 Tyler-1982
176 Smith-1982
177 Butrica-1998
178 Hanel-1982
179 Smith-1982
180 Smith-1982
181 Showalter-1998

182 IAUC-3651
183 IAUC-3656
184 IAUC-3660
185 IAUC-6162
186 Vogt-1982
187 IAUC-5052
188 Nicholson-1996
189 Bosh-1996
190 Kelly Beatty-1996
191 Lane-1982
192 Hanel-1982
193 Sandel-1982
194 Tyler-1982
195 Morrison-1982b
196 Stuart-1984
197 Laeser-1986
198 Gray-1982
199 Swift-1997l
200 Swift-1997m
201 Laeser-1986
202 Bartok-1986
203 Brown-1985
204 Brush-1996b
205 Cuzzi-1987
206 NASA-1986
207 Stone-1986
208 McLaughlin-1985
209 McLaughlin-1986a
210 AWST-1985
211 Ingersoll-1987
212 McLaughlin-1985
213 AWST-1986a
214 McLaughlin-1986a
215 Owen-1987
216 Showalter-2006
217 IAUC-7171
218 Smith-1986a
219 Gurnett-1986
220 Miner-1988
221 McLaughlin-1985
222 Lane-1986
223 Tyler-1986
224 Ness-1986
225 Ingersoll-1987
226 Ingersoll-1987
227 Johnson-1987
228 Ingersoll-1987
229 Smith-1986a
230 Ingersoll-1987

231 Hanel-1986
232 Ingersoll-1987
233 Broadfoot-1986
234 Ingersoll-1987
235 Smith-1986b
236 Smith-1986c
237 Lane-1986
238 Smith-1986a
239 Johnson-1987
240 Tyler-1986
241 Bridge-1986
242 Smith-1986b
243 Smith-1986c
244 Miner-1988
245 Smith-1986a
246 Cuzzi-1987
247 Ness-1986
248 Shayler-2000
249 Gore-1986
250 McLaughlin-1986b
251 Smith-2006
252 Bartok-1986
253 Mudgway-226-241
254 Stone-1989
255 Hammel-1989a
256 Brown-1985
257 Pandey-1987
258 Swift-1997n
259 Kohlhase-1986
260 Dobrovolskis-1988
261 NASA-1989
262 Stone-1989
263 McLaughlin-1986b
264 Smith-1989b
265 Smith-1989a
266 Beish-1993
267 Kelly Beatty-1995
268 IAUC-4806
269 IAUC-3608
270 IAUC-4824
271 Smith-1989c
272 Stone-1989
273 Hammel-1989b
274 Smith-1989a
275 IAUC-4830
276 AWST-1989a
277 Warwick-1989
278 Smith-1989a
279 Smith-1989c

280 Swift-1997o
281 Belcher-1989
282 Ness-1989
283 Lane-1989
284 Smith-1989a
285 Schaefer-2000
286 Grav-2003
287 Smith-1989a
288 Jewitt-2006
289 Tyler-1989
290 Smith-1989a
291 Stone-1989
292 Gurnett-1989
293 Conrath-1989
294 Soderblom-1990
295 Boyce-1993
296 Smith-1989a
297 Kinoshita-1989
298 Tyler-1989
299 Broadfoot-1989
300 Agnor-2006
301 Jewitt-2006
302 Smith-1989a
303 Standish-1993

304 Tyler-1989
305 Hammel-1989b
306 Smith-1989a
307 Warwick-1989
308 Lindal-1990
309 Conrath-1991
310 Conrath-1989
311 Tyler-1989
312 Broadfoot-1989
313 Ness-1989
314 Broadfoot-1989
315 Stanley-2004
316 Belcher-1989
317 Ness-1989
318 Stone-1989
319 Burrows-1998
320 Rudd-1997
321 Swift-1997p
322 AWST-1989b
323 Swift-1997q
324 Sagan-1997
325 Swift-1997r
326 Swift-1997s
327 Teske-1993

328 Verschuur-1993
329 Linick-1991
330 Rudd-1996
331 Rudd-1997
332 Cesarone-1983
333 Evans-2004f
334 Gurnett-1993
335 Starrfield-1994
336 Kelly Beatty-1994
337 Vervack-1994
338 Vervack-2006
339 Massey-2006
340 Krimigis-2003
341 Kerr-2005
342 Lawler-2005
343 Fisk-2005
344 Stone-2005
345 Burlaga-2005
346 Rudd-1996
347 Rudd-1997
348 McComas-2004
349 Cesarone-1983
350 Rudd-1997
351 Swift-1997t

Glossary

ABL: Automated Biological Laboratory

ACP: Advanced Cooperation Project

Aphelion: The point of maximum distance from the Sun of a solar orbit. Its contrary is the perihelion.

Apoapsis: The point of maximum distance from the central body of any elliptical orbit. This word has been used to avoid complicating the nomenclature, but a term tailored to the central body is often used. The only exceptions used herein owing to their importance were for Earth (apogee) and the Sun (aphelion). The contrary of apoapsis is periapsis.

Apogee: The point of maximum distance from the Earth of a satellite orbit. Its contrary is the perigee.

Astronomical Unit: To a first approximation the average distance between the Earth and the Sun is 149,597,870,691 ($\pm$ 30) meters.

AU: Astronomical Unit

Booster: Auxiliary rockets used to boost the lift-off thrust of a launch vehicle.

Bus: A structural part common to several spacecraft.

CIA: Central Intelligence Agency

CNES: Centre National d'Etudes Spatiales (the French National Space Studies Centre)

CNR: Centro Nazionale Ricerche (the Italian National Research Centre)

Conjunction: The time when a solar system object appears close to the Sun as seen by an observer. A conjunction where the Sun is between the observer and the object is called 'superior conjunction'. A conjunction where the object is between the observer and the Sun is called 'inferior conjunction'. See also opposition.

CONSCAN: Conical Scan

Cosmic velocities: Three characteristic velocities of spaceflight:

First cosmic velocity: Minimum velocity to put a satellite in a low Earth orbit. This amounts to some 8 km/s.

Second cosmic velocity: The velocity required to exit the terrestrial sphere of attraction for good. Starting from the ground, this amounts to some 11 km/s. It is also called 'escape' speed.

Third cosmic velocity: The velocity required to exit the Solar System for good.

Cryogenic propellants: These can be stored in their liquid state under atmospheric pressure at very low temperature; e.g. oxygen is a liquid below −183°C.

Deep Space Network: A global network built by NASA to provide round-the-clock communications with robotic missions in deep space.

Direct ascent: A trajectory on which a deep-space probe is launched directly from the Earth's surface to another celestial body without entering parking orbit.

DSN: Deep Space Network

DZhVS: Dolgozhivushaya Veneryanskaya Stanziya (long duration Venusian probe)

Ecliptic: The plane of the Earth's orbit around the Sun.

Ejecta: Material from a volcanic eruption or a cratering impact that is deposited all around the source.

ELDO: European Launcher Development Organization (became part of ESA)

EOS: Eole–Venus

ESA: European Space Agency

Escape speed: See Cosmic velocities

ESRO: European Space Research Organization (became part of ESA)

ESTEC: European Space Technology Center

Flyby: A high relative speed and short duration close encounter between a spacecraft and a celestial body.

FTU: FotoTelevisionnoye Ustroistvo (photo-television system)

GCMS: Gas Chromatograph Mass Spectrometer

GE: Gas Exchange

GRB: Gamma-Ray Bursts

GSFC: Goddard Space Flight Center

GSOC: German Space Operation Center

HST: Hubble Space Telescope

Hypergolic propellants: Two liquid propellants that ignite spontaneously on coming into contact, without requiring an ignition system. Typical hypergolics are hydrazine and nitrogen tetroxide.

IBEX: Interstellar Boundary Explorer

ICBM: InterContinental Ballistic Missile. A military strategic and usually nuclear-tipped missile with a range of at least 6,400 km. Many early space launchers were adapted from ICBMs.

IKI: Institut Kosmicheskikh Isledovanii (the Russian Institute for Cosmic Research)

IMP: Interplanetary Monitoring Platform

IRAS: InfraRed Astronomical Satellite

ISEE: International Sun–Earth Explorer

JPL: Jet Propulsion Laboratory (a Caltech laboratory under contract to NASA)

J–S–P: Jupiter–Saturn–Pluto trajectory

J–S–U–N: Jupiter–Saturn–Uranus–Neptune trajectory

J–U–N: Jupiter-Uranus-Neptune trajectory

KDU: Korrektiruyushaya Dvigatelnaya Ustanovka (course correction engine)

KSC: Kennedy Space Center

KTDU: Korrektiruyushaya Tormoznaya Dvigatelnaya Ustanovka (course correction and braking engine)

Lander: A spacecraft designed to land on another celestial body.

LaRC: Langley Research Center

Launch window: A time interval during which it is possible to launch a spacecraft to ensure that it attains the desired trajectory.

LR: Labeled Release

Lyman-alpha: The emission line corresponding to the first energy level transition of an electron in a hydrogen atom.

MBB: Messerschmitt Bölkov Blohm

MESO: Mercury Sonde

MIT: Massachusetts Institute of Technology

MJS: Mariner Jupiter–Saturn (later named Voyager)

MJU: Mariner Jupiter–Uranus

MSFC: Marshall Space Flight Center

MV: Mars–Venera (Soviet Mars and Venus probes)

MVM: Mariner Venus–Mercury (also named Mariner 10)

N-1: Nossitel 1 (Launcher 1, the Soviet moon rocket)

NAS: National Academy of Sciences

NASA: National Aeronautics and Space Administration

OAO: Orbiting Astronomical Observatory

Occultation: When one object passes in front of and occults another, at least from the point of view of the observer.

OOE: Out Of the Ecliptic

Orbit: The trajectory on which a celestial body or spacecraft is traveling with respect to its central body. There are three possible cases:

Elliptical orbit: A closed orbit where the body passes from minimum distance to maximum distance from its central body every semiperiod. This is the orbit of natural and artificial satellites around planets and of planets around the Sun.

Parabolic orbit: An open orbit where the body passes through minimum distance from its central body and reaches infinity at zero velocity in infinite time. This is a pure abstraction, but the orbits of many comets around the Sun can be described adequately this way.

Hyperbolic orbit: An open orbit where the body passes through minimum distance from its central body and reaches infinity at non-zero speed. This describes adequately the trajectory of spacecraft with respect to planets during flyby manoeuvres.

Opposition: The time when a solar system object appears opposite to the Sun as seen by an observer.

Orbiter: A spacecraft designed to orbit a celestial body.

P-L: Palomar–Leiden asteroid survey

PAET: Planetary Atmosphere Experiment Test

Parking orbit: A low Earth orbit used by deep-space probes before heading to their targets. This relaxes the constraints on launch windows and eliminates launch vehicle trajectory errors. Its contrary is direct ascent.

PAS: Plavalyuschaya Aerostatnaya Stantsiya (buoyant aerostatic station)

PEPP: Planetary Entry Parachute Program

Periapsis: The minimum distance point from the central body of any orbit. See also apoapsis.

Perigee: The minimum distance point from the Earth of a satellite. Its contrary is apogee.

Perihelion: The minimum distance point from the Sun of a solar orbit. Its contrary is the aphelion.

PR: Pyrolytic Release

PrOP: Pribori Otchenki Prokhodimosti (instrument for cross-country characteristics evaluation)

PVM: Pioneer Venus Multiprobe

PVO: Pioneer Venus Orbiter

'Push-broom' camera: A digital camera consisting of a single row of pixels, with the second dimension created by the motion of the camera itself.

RAE: Radio Astronomy Explorer

Rendezvous: A low relative speed encounter between two spacecraft or celestial bodies.

Resonance: A resonance in the solar system occurs when the rotational and orbital periods of a body are commensurate, or when they are so with another body. For example, most moons have resonant rotation and orbital periods, meaning that they complete one rotation in the same exact time it takes for them to complete one orbit. In another example, some of the gaps in Saturn's rings correspond to particles whose orbits would be resonant with the largest satellites.

Retrorocket: A rocket whose thrust is directed opposite to the motion of a spacecraft in order to brake it.

Rj: Jupiter radii (approximately 71,200 km)

Rn: Neptune radii (approximately 24,750 km)

Rover: A mobile spacecraft to explore the surface of another celestial body.

Rs: Saturn radii (approximately 60,330 km)

RTG: Radioisotope Thermal Generator

RTH: Radioisotope Thermal Heater

Ru: Uranus radii (approximately 25,600 km)

SERT: Space Electric Rocket Test

SETI: Search for Extraterrestrial Intelligence

SNAP: System for Nuclear Auxiliary Power

SNC: Shergottites–Nakhlites–Chassignites meteorites

Solar flare: A solar chromospheric explosion creating a powerful source of high energy particles.

SOREL: Solar Orbiting Relativity Experiment

Space probe: A spacecraft designed to investigate other celestial bodies from a short range.

Spectrometer: An instrument to measure the energy of radiation as a function of wavelenghts in a portion of the electromagnetic spectrum. Depending on the wavelength the instrument is called, e.g. ultraviolet, infrared, gamma-ray spectrometer etc.

Spin stabilization: A spacecraft stabilization system where the attitude is maintained by spinning the spacecraft around one of its main inertia axes.

SS: Surface to Surface missile

SSB: Space Science Board of the National Academy of Sciences

STAR: Self-Testing And Repairing

STS: Space Transportation System (the Space Shuttle)

Synodic period: The period of time between two consecutive superior or inferior conjunctions or oppositions of a solar system body.

Telemetry: Transmission by a spacecraft via a radio system of engineering and scientific data.

3-axis stabilization: A spacecraft stabilization system where the axes of the spacecraft are kept in a fixed attitude with respect to the stars and other references (the Sun, the Earth, a target planet etc.)

TOPS: Thermoelectric Outer Planet Spacecraft

TRW: Thompson Ramo Wooldridge Inc.

UDMH: Unsymmetrical DiMethyl Hydrazine

Ullage rockets: Small rockets, usually solid fueled, used to provide sufficient acceleration in weightlessness to force liquid propellants towards the pump intakes prior to starting a larger rocket engine.

UMVL: Universalnyi Mars, Venera, Luna (Universal for Mars, Venus and the Moon)

UTC: Universal Time Coordinated (essentially Greenwich Mean Time)

V2: Vergeltungswaffe 2 (vengeance weapon 2)

Vernier engines: Small attitude control engines mounted in clusters and firing trough the spacecraft's center of mass. Attitude control is obtained by simple differential throttling of the engines.

Vidicon: A television system based on resistance changes of some substances when exposed to light. It has been replaced by the CCD.

VIM: Voyager Interstellar Mission

VLA: Very Large Array

VOIR: Venus Orbiting Imaging Radar

VPM: Visual Polarimeter–Mars

Appendix 1

FINDING YOU WAY THROUGH THE SOLAR SYSTEM: A CELESTIAL MECHANICS PRIMER[*]

Hohmann Transfer Orbits

The easiest way to travel from one planet to another within the solar system is by way of a Hohmann transfer orbit. Although attributed to Walter Hohmann in Germany in 1923, the Russian mathematician Vladimir Vetchinkin had invented it several years previously. Assuming the orbits of the two planets to be circular and co-planar, Hohmann showed that the optimum way to travel between them (i.e. the least expensive in terms of the energy requirements) was by an ellipse that was tangent to the orbit of the departure planet at departure and tangent to the orbit of the arrival planet at arrival. Thus, after traveling half of its orbit, the spacecraft would reach its target. Of course, the approximation of these 'bi-tangent' orbits is purely theoretical, and a spacecraft for an exterior planet usually departs slightly after the perihelion of the transfer ellipse and arrives slightly before or after aphelion. The same applies for a flight to an interior planet if the word perihelion is substituted for aphelion, and vice versa. In practice, the target can be reached when the spacecraft has subtended an angle of either less than 180 degrees around the Sun, or more than 180 degrees. If less than 180 degrees, the orbit is called a type-1 transfer; if more than 180 degrees it is a type-2 transfer. Each has its merits: a type-1 orbit reduces the duration, while a type-2 reduces the relative speed with respect to the target at arrival – which might be desirable if an orbital insertion burn is intended, or if a landing is to be attempted. In any case, it must be ensured that the target planet and the spacecraft arrive at the arrival point simultaneously, which is ensured by the use of a proper 'launch window'.

[*] While no mathematics is used in this appendix, a basic knowledge of vector algebra is needed.

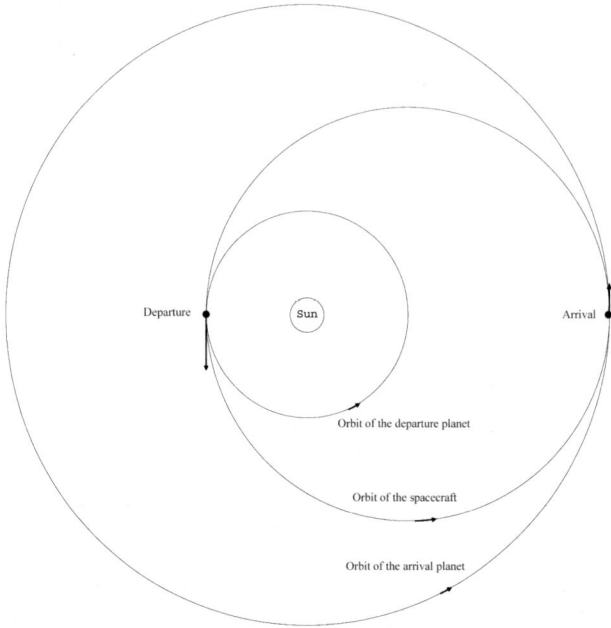

A tangential Hohmann transfer orbit.

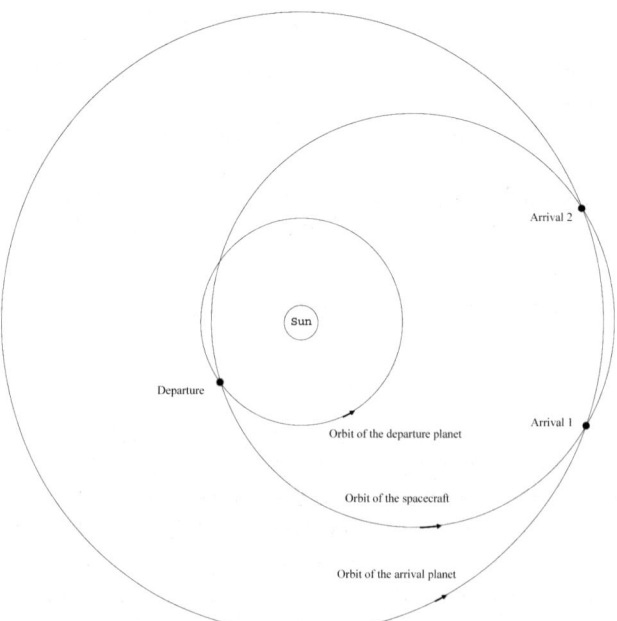

Type-1 and Type-2 transfer orbits.

Different targets have different requirements in terms of the amount by which the spacecraft must increase or decrease its speed in order to reach them, the duration of the flight and so on; the following table summarizes some of these requirements.

Hohmann transfer ellipses from Earth

Target	Ellipse semimajor axis (AU)	Velocity change needed at departure* (km/s)	Time of flight (Years)	Relative Speed at arrival (km/s)
Mercury**	0.694	−7.53	0.289	9.6
Venus	0.862	−2.5	0.400	2.7
Mars**	1.262	2.95	0.709	2.6
Jupiter	3.102	8.79	2.731	5.6
Saturn	5.270	10.29	6.048	5.4
Uranus	10.091	11.28	16.03	4.7
Neptune	15.529	11.65	30.60	4.1
Pluto**	20.220	11.81	45.46	3.7
Escape Sun		12.34	–	–

* Negative values are in the opposite sense to the Earth's orbital motion. Positive values are in the same sense.
** On average, because the orbits of these planets are markedly elliptical.

Note a few things about the data in the table:

1 Planets beyond Saturn cannot be reached with a Hohmann orbit within a reasonable time.
2 Contrary to intuition, distant Jupiter and nearby Mercury have similar departure speed requirements.
3 Venus and Mars have similar requirements in terms of speed at departure, but a flight to Mars takes almost twice as long as a flight to Venus.

Of course, these values are valid only for a bi-tangent co-planar Hohmann transfer orbit; real orbits yield longer flight times and larger speeds at departure and arrival.

Gravity-Assists

Consider the case of a spacecraft that catches up with a planet from 'behind'. Due to the gravity of the planet, the path of the spacecraft will be deflected through an angle that depends on the mass of the planet, on the distance of closest approach, and of the square of the relative speed between the spacecraft and the planet (Vin or Vout). Note that the velocity of the spacecraft is unchanged in modulus, only in direction. The velocity of the spacecraft with respect to the Sun (i.e. its heliocentric velocity) is the vectorial sum of the velocity of the planet (Vp) and of the spacecraft with respect to the planet (Vin or Vout). Hence summing Vp with Vin and with Vout gives the heliocentric velocity of the spacecraft before and after the encounter. Notice that the

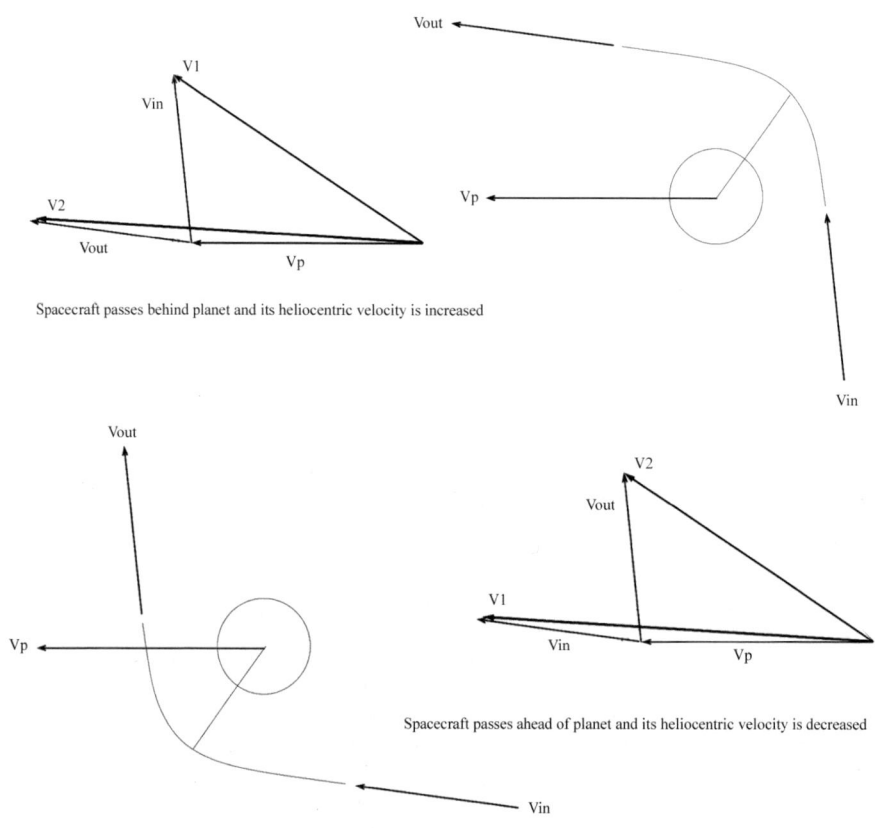

Spacecraft passes behind planet and its heliocentric velocity is increased

Spacecraft passes ahead of planet and its heliocentric velocity is decreased

How gravity-assist 'slingshots' work

velocity before the encounter (V1) is smaller than the velocity afterwards (V2). Now consider what happens when the spacecraft passes in front of the planet instead of behind it. In this case, the heliocentric velocity is reduced; this technique can be used to slow a spacecraft down. Such 'slingshots' have been used extensively in solar system exploration, first when Mariner 10 used Venus to reach Mercury, then when Pioneer 11 used Jupiter to reach Saturn, but most spectacularly to enable Voyager 2 to reach Uranus and Neptune – the latter mission would have been otherwise impracticable using Hohmann transfers.

Appendix 2

SOVIET PLANETARY PROBE DESIGNATIONS

Code	Role	Record
Korolyov's probes		
1M	Mars flyby/lander	2 failures
1V	Venus lander	Canceled
1VA	Venus flyby	2 failures including Venera 1
2MV-1	Venus lander	2 failures
2MV-2	Venus flyby	1 failure
2MV-3	Mars lander	1 failure
2MV-4	Mars flyby	2 failures including Mars 1
3MV-1A	Test	2 failures
3MV-1	Venus lander	2 failures including Zond 1
3MV-2	Venus flyby	Not flown
3MV-3	Mars/Venus lander	1 failure (Venera 3)
3MV-4	Mars/Venus flyby	Zond 3, plus 3 failures including Zond 2 and Venera 2
Lavochkin's probes		
1F*	Mars/Phobos orbiter	Fobos 1, 2
1M	Mars flyby/lander	Canceled
1V	Venus lander	Venera 4 plus 1 failure
2M	Mars orbiter	Atmospheric capsule deleted, 2 failures
2V	Venus lander	Venera 5 and 6
3MS	Mars orbiter	Mars 4 and 5, plus 1 launch failure
3MP	Mars lander	Mars 2, 3, 6 and 7
3V	Venus lander	Venera 7 and 8, plus 2 failures
4M	Mars lander/rover	Canceled
4NM	Mars lander/rover	Canceled

4V-1	Venus orbiter/lander	Venera 9, 10, 11 and 12
4V-1M	Venus orbiter/lander	Venera 13 and 14
4V-2*	Venus orbiter/radar	Venera 15 and 16
5VK*	Venus–Halley flyby	Vega 1 and 2
5VP*	Venus balloon carrier	Canceled
5VS*	Venus orbiter	Canceled
5M	Mars sample return	Canceled
5NM	Mars sample return	Canceled
DZhVS	Venus lander	Canceled
YuS*	Jupiter, Saturn, solar probe	Canceled

*   Will be covered in later volumes.

Appendix 3

REPORTED SPACECRAFT DISCOVERIES OF PLANETARY SATELLITES

Preliminary Designation	Official Name	Planet	Spacecraft
1979J1	Adrastea	Jupiter	Voyager 2
1979J2	Thebe	Jupiter	Voyager 1
1979J3	Metis	Jupiter	Voyager 1
1979S1	Epimetheus (?)	Saturn	Pioneer 11
1979S2	Janus	Saturn	Pioneer 11
1979S3	Unconfirmed	Saturn	Pioneer 11
1979S4	Unconfirmed	Saturn	Pioneer 11
1979S5	Unconfirmed	Saturn	Pioneer 11
1979S6	Unconfirmed	Saturn	Pioneer 11
1980S26	Pandora	Saturn	Voyager 1
1980S27	Prometheus	Saturn	Voyager 1
1980S28	Atlas	Saturn	Voyager 1
1980S33	Telesto	Saturn	Voyager 1
1980S34	Unconfirmed	Saturn	Voyager 1
1981S6	Unconfirmed	Saturn	Voyager 2
1981S7	Unconfirmed	Saturn	Voyager 2
1981S8	Unconfirmed	Saturn	Voyager 2
1981S9	Unconfirmed	Saturn	Voyager 2
1981S10	Unconfirmed	Saturn	Voyager 2
1981S11	Unconfirmed	Saturn	Voyager 2
1981S12	Unconfirmed	Saturn	Voyager 2
1981S13	Pan	Saturn	Voyager 2
1981S14	Pallene	Saturn	Voyager 2
1981S15	Unconfirmed	Saturn	Voyager 2
1981S16	Unconfirmed	Saturn	Voyager 2
1981S17	Unconfirmed	Saturn	Voyager 2

1981S18	Unconfirmed	Saturn	Voyager 2
1981S19	Unconfirmed	Saturn	Voyager 2
1985U1	Puck	Uranus	Voyager 2
1986U1	Portia	Uranus	Voyager 2
1986U2	Juliet	Uranus	Voyager 2
1986U3	Cressida	Uranus	Voyager 2
1986U4	Rosalind	Uranus	Voyager 2
1986U5	Belinda	Uranus	Voyager 2
1986U6	Desdemona	Uranus	Voyager 2
1986U7	Cordelia	Uranus	Voyager 2
1986U8	Ophelia	Uranus	Voyager 2
1986U9	Bianca	Uranus	Voyager 2
1986U10	Perdita	Uranus	Voyager 2
1989N1	Proteus	Neptune	Voyager 2
1989N2	Larissa	Neptune	Voyager 2
1989N3	Despina	Neptune	Voyager 2
1989N4	Galatea	Neptune	Voyager 2
1989N5	Thalassa	Neptune	Voyager 2
1989N6	Naiad	Neptune	Voyager 2

An additional satellite of Uranus, 2003U1, Mab, is present in the Voyager images, but it was not recognized until a check was made following its discovery in 2003 by the Hubble Space Telescope.

Appendix 4

CHRONOLOGY OF SOLAR SYSTEM EXPLORATION 1952–1982

Date	Event
20 September 1952	Eric Burgess and Charles A. Cross present their "The Martian Probe" paper
4 October 1957	Sputnik 1, the first artificial satellite is launched
1 February 1958	Explorer 1, the first US satellite is launched
2 January 1959	Luna 1, the first "artificial planet" is launched
11 March 1960	Launch of Pioneer 5, the first interplanetary probe
10 October 1960	The first Mars probe is launched, but lost due to rocket failure
19 May 1961	Venera 1 passes Venus, but it had failed in February
14 December 1962	Mariner 2 passes Venus and returns data
19 June 1963	Mars 1 passes Mars, but it had failed in March
15 July 1965	Mariner 4 passes Mars and returns data
1 March 1966	Venera 3 impacts on Venus, two weeks after falling inert
22 August 1967	The Voyager Mars landing mission is canceled
18–19 October 1967	Venera 4 enters Venus's atmosphere and Mariner 5 passes the planet
31 July 1969	Mariner 6 passes Mars
5 August 1969	Mariner 7 passes Mars
15 December 1970	Venera 7 lands on Venus
14 November 1971	Mariner 9 enters orbit around Mars
2 December 1971	Mars 3 lands on Mars
4 December 1973	Pioneer 10 passes Jupiter
29 March 1974	Mariner 10 passes Mercury
15 March 1975	Helios 1 passes within 0.30 AU of the Sun
22 October 1975	Venera 9 lands on Venus and returns pictures
20 July 1976	Viking 1 lands on Mars
4 December 1978	Pioneer Venus Orbiter enters orbit around Venus
9 December 1978	The Pioneer Venus Multiprobes enter Venus's atmosphere

5 March 1979	Voyager 1 passes Jupiter
9 July 1979	Voyager 2 passes Jupiter
1 September 1979	Pioneer 11 passes Saturn
12 November 1980	Voyager 1 passes Saturn
26 August 1981	Voyager 2 passes Saturn
1 March 1982	Venera 13 lands on Venus and returns color pictures

Related milestones

13 June 1983	Pioneer 10 crosses the orbit of Neptune
24 January 1986	Voyager 2 passes Uranus
25 August 1989	Voyager 2 passes Neptune
16 December 2004	Voyager 1 reaches the heliospheric termination shock

Appendix 5

PLANETARY LAUNCHES 1960–1981

Launch Date	Name	Main Target	Launcher	Nation
11 March 1960	Pioneer 5	Solar orbit	Thor–Able IV	USA
10 October 1960	(1M No.1)	Mars	8K78 Molniya	USSR
14 October 1960	(1M No.2)	Mars	8K78 Molniya	USSR
4 February 1961	(1VA No.1)	Venus	8K78 Molniya	USSR
12 February 1961	(Venera 1)	Venus	8K78 Molniya	USSR
22 July 1962	(Mariner 1)	Venus	Atlas–Agena B	USA
25 August 1962	(2MV-1 No.1)	Venus	8K78 Molniya	USSR
27 August 1962	Mariner 2	Venus	Atlas–Agena B	USA
1 September 1962	(2MV-1 No.2)	Venus	8K78 Molniya	USSR
12 September 1962	(2MV-2 No.1)	Venus	8K78 Molniya	USSR
24 October 1962	(2MV-4 No.1)	Mars	8K78 Molniya	USSR
1 November 1962	(Mars 1)	Mars	8K78 Molniya	USSR
4 November 1962	(2MV-3 No.1)	Mars	8K78 Molniya	USSR
19 February 1964	(3MV-1A No.4A)	Venus	8K78 Molniya	USSR
27 March 1964	(3MV-1 No.5)	Venus	8K78 Molniya	USSR
2 April 1964	(Zond 1)	Venus	8K78 Molniya	USSR
5 November 1964	(Mariner 3)	Mars	Atlas–Agena D	USA
28 November 1964	Mariner 4	Mars	Atlas–Agena D	USA
30 November 1964	(Zond 2)	Mars	8K78 Molniya	USSR
18 July 1965	Zond 3	Lunar flyby	8K78 Molniya	USSR
12 November 1965	(Venera 2)	Venus	8K78M Molniya	USSR
16 November 1965	(Venera-3)	Venus	8K78M Molniya	USSR
23 November 1965	(3MV-4 No.6)	Venus	8K78M Molniya	USSR
16 December 1965	Pioneer 6	Solar orbit	Thor–Delta E	USA
17 August 1966	Pioneer 7	Solar orbit	Thor–Delta E1	USA
12 June 1967	Venera 4	Venus	8K78M Molniya	USSR
14 June 1967	Mariner 5	Venus	Atlas–Agena D	USA

17 June 1967	(1V No.311)	Venus	8K78M Molniya	USSR
13 December 1967	Pioneer 8	Solar orbit	Thor–Delta E1	USA
8 November 1968	Pioneer 9	Solar orbit	Thor–Delta E1	USA
5 January 1969	Venera 5	Venus	8K78M Molniya	USSR
10 January 1969	Venera 6	Venus	8K78M Molniya	USSR
25 February 1969	Mariner 6	Mars	Atlas–Centaur	USA
27 March 1969	(2M No.521)	Mars	8K82K Proton-K/D	USSR
27 March 1969	Mariner 7	Mars	Atlas–Centaur	USA
2 April 1969	(2M No.522)	Mars	8K82K Proton-K/D	USSR
27 August 1969	(Pioneer E)	Solar orbit	Thor–Delta L	USA
17 August 1970	Venera 7	Venus	8K78M Molniya	USSR
22 August 1970	(3V No.631)	Venus	8K78M Molniya	USSR
9 May 1971	(Mariner 8)	Mars	Atlas–Centaur	USA
10 May 1971	(3MS No.170)	Mars	8K82K Proton-K/D	USSR
19 May 1971	Mars 2	Mars	8K82K Proton-K/D	USSR
28 May 1971	Mars 3	Mars	8K82K Proton-K/D	USSR
30 May 1971	Mariner 9	Mars	Atlas–Centaur	USA
3 March 1972	Pioneer 10	Jupiter	Atlas–Centaur	USA
27 March 1972	Venera 8	Venus	8K78M Molniya	USSR
31 March 1972	(3V No.671)	Venus	8K78M Molniya	USSR
6 April 1973	Pioneer 11	Jupiter	Atlas–Centaur	USA
21 July 1973	(Mars 4)	Mars	8K82K Proton-K/D	USSR
25 July 1973	Mars 5	Mars	8K82K Proton-K/D	USSR
5 August 1973	(Mars 6)	Mars	8K82K Proton-K/D	USSR
9 August 1973	(Mars 7)	Mars	8K82K Proton-K/D	USSR
3 November 1973	Mariner 10	Mercury	Atlas–Centaur	USA
10 December 1974	Helios 1	Solar orbit	Titan IIIE–Centaur	USA/FRG
8 June 1975	Venera 9	Venus	8K82K Proton-K/D	USSR
14 June 1975	Venera 10	Venus	8K82K Proton-K/D	USSR
20 August 1975	Viking 1	Mars	Titan IIIE–Centaur	USA
9 September 1975	Viking 2	Mars	Titan IIIE–Centaur	USA
15 January 1976	Helios 2	Solar orbit	Titan IIIE–Centaur	USA/FRG
20 August 1977	Voyager 2	Jupiter	Titan IIIE–Centaur	USA
5 September 1977	Voyager 1	Jupiter	Titan IIIE–Centaur	USA
20 May 1978	Pioneer Venus Orbiter	Venus	Atlas–Centaur	USA
8 August 1978	Pioneer Venus Multiprobe	Venus	Atlas–Centaur	USA
12 August 1978	International Cometary Explorer*	P/Giacobini –Zinner	Delta 2914	USA
9 September 1978	Venera 11	Venus	8K82K Proton-K/D-1	USSR
14 September 1978	Venera 12	Venus	8K82K Proton-K/D-1	USSR
30 October 1981	Venera 13	Venus	8K82K Proton-K/D-1	USSR
4 November 1981	Venera 14	Venus	8K82K Proton-K/D-1	USSR

* Will be covered in a later volume.

Missions in parentheses are missions that failed, but the status of Mars 2, 3 and 5 is disputed.

Chapter references

[Acunha-1980] Acunha, M.H., Ness, N.F., "The Magnetic Field of Saturn: Pioneer 11 Observations", Science, 207, 1980, 444–446

[Adelman-1986] Adelman, B., "The Question of Life on Mars", Journal of the British Interplanetary Society, 39, 1986, 256–262

[Agnor-2006] Agnor, C.B., Hamilton, D.P., "Neptune's Capture of its Moon Triton in a Binary-Planet Gravitational Encounter", Nature, 441, 2006, 192–194

[Ajello-1979] Ajello, J.M., Witt, N., Blum, P.W., "Four UV Observations of the Interstellar Wind by Mariner 10: Analysis with Spherically Symmetric Solar Radiation Models", Astronomy and Astrophysics, 73, 1979, 260–271

[Anderson-1965] Anderson, H.R., "Mariner IV Measurements near Mars: Initial Results", Science, 149, 1965, 1226–1228

[Anderson-1968a] Anderson, J.D., et al., "Determination of the Mass of Venus and Other Astronomical Constants from the Radio Tracking of Mariner V", Astronomical Journal, 73, 1968, S2

[Anderson-1968b] Anderson, J.D., Efron, L., Pease, G.E., "Mass, Dynamical Oblateness, and Position of Venus as Determined by Mariner V Tracking Data", Astronomical Journal, 73, 1968, S162–S163

[Anderson-1968c] Anderson J.D., et al., "Radius of Venus as Determined by Planetary Radar and Mariner V Radio Tracking Data", The Astronomical Journal, 73, 1968, S162

[Anderson-1975] Anderson, J.D., et al., "Experimental Test of General Relativity using Time-Delay Data from Mariner 6 and Mariner 7", The Astrophysical Journal, 200, 1975, 221–233

[Anderson-1976] Anderson, D.L., et al., "The Viking Seismic Experiment", Science, 194, 1976, 1318–1321

[Anderson-1978] Anderson, J.L., et al., "Venus in Motion", The Astrophysical Journal Supplement Series, 36, 1978, 275–284

[Anderson-1980] Anderson, J.D., et al., "Pioneer Saturn Celestial Mechanics Experiment", Science, 207, 1980, 449–453

[Anderson-1985] Anderson, J.D., Mashhoon, B., "Pioneer 10 Search for Gravitational Waves – Limits on a Possible Isotropic Cosmic Background of Radiation in the MicroHertz Region", The Astrophysical Journal, 290, 1985, 445–448

[Anderson-1998] Anderson, J.D., et al., "Indication from Pioneer 10/11, Galileo and Ulysses Data for an Anomalous, Weak, Long-Range Acceleration", Arxiv gr–qc/9808081 preprint

[Anderson-2001] Anderson, J.D., et al., "Study of the Anomalous Acceleration of Pioneer 10 and 11", Arxiv gr–qc/0104064 preprint

[Andreeva-1976] Andreeva, L.P., et al., "Digital Processing of the Venus Surface Panoramas", Soviet Astronomy Letters, 2, 1976, 119–120

[Armstrong-1985] Armstrong, J.W., Estabrook, F.B., Wahlquist, H.D., "A Search for Sinusoidal and Burst VLF Gravitational Waves Using Pioneer 10 and 11 Spacecraft Tracking Data", Bulletin of the American Astronomical Society, 17, 1985, 872

[Armstrong-1987] Armstrong, J.W., Estabrook, F.B., Wahlquist, H.D., "A Search for Sinusoidal Gravitational Radiation in the Period Range 30-2000 Seconds", The Astrophysical Journal, 318, 1987, 536–541

[Arvidson-1974] Arvidson, R.E., Mutch, T.A., Jones, K.L., "Craters and Associated Aeolian Features on Mariner 9 Photographs: an Automated Data Gathering and Handling System and Some Preliminary Results", Earth, Moon and Planets, 9, 1974, 105–114

[Arvidson-1978] Arvidson, R.E., Binder, A.B., Jones, K.L., "The Surface of Mars", Scientific American, March 1978, 76–91

[Arvidson-1983] Arvidson, R.E., et al., "Three Mars Years: Viking Lander 1 Imaging Observations", Science, 222, 1983, 463–468

[A&A 1964] "Astro Notes – Deep Space", Astronautics & Aeronautics, December 1964, 75

[AWST-1957] "USAF Launches Artificial Meteors", Aviation Week, 2 December 1957, 34

[AWST-1960a] "Pioneer V Deep Space Reports Parallel Earlier Radiation Data", Aviation Week, 28 March 1960, 32

[AWST-1960b] "Pioneer Signals May Be Received from 75-Million mi. Distances", Aviation Week, 11 April 1960, 33

[AWST-1960c] "Pioneer Switched to 150-Watt Unit", Aviation Week, 16 May 1960, 34

[AWST-1961] "Mars Flyby Satellite Could Drop Instrumented Capsule", Aviation Week, 26 June 1961, 101

[AWST-1964] "Project Beagle Mars Mission Proposed", Aviation Week & Space Technology, 13 July 1964, 48–54

[AWST-1965a] "New Mariner 4 Photos Planned", Aviation Week & Space Technology, 30 August 1965, 31

[AWST-1965b] "Mariner 4 Test", Aviation Week & Space Technology, 13 September 1965, 28

[AWST-1967a] "Venus 4 Underscores U.S. Delay", Aviation Week & Space Technology, 23 October 1967, 26

[AWST-1967b] "Fouled Antenna Impairs Venus 4 Mission", Aviation Week & Space Technology, 30 October 1967, 24–25

[AWST-1969a] "Redesigned Equipment Aids Soviet Missions to Venus", Aviation Week & Space Technology, 26 May 1969, 22

[AWST-1969b] "Soviet Admit Venus 5, 6 Problems", Aviation Week & Space Technology, 23 June 1969, 70–71

[AWST-1970] "Enhanced Photo Shows Mars' Moon", Aviation Week & Space Technology, 25 May 1970, 61

[AWST-1971a] "Soviets Say Venus 7 Transmitted 23 Min. Data on Planet's Surface", Aviation Week & Space Technology, 1 February 1971, 22

[AWST-1971b] "Soviets Impact Capsule on Mars", Aviation Week & Space Technology, 6 December 1971, 20

[AWST-1971c] "Soviets Land TV on Mars, Blame Failure on Wind, Dust", Aviation Week & Space Technology, 13 December 1971, 20

[AWST-1971d] "Mars Lander May Have Sunk in Dust", Aviation Week & Space Technology, 20 December 1971, 23

[AWST-1972a] "Soviets Glean Broad Martian Data", Aviation Week & Space Technology, 3 January 1972, 13–15

[AWST-1972b] "Dual Antenna Used by Soviets to Transmit Venus–Earth Data", Aviation Week & Space Technology, 31 July 1972, 18

[AWST-1972c] "Soviets End Mars 2, 3 Missions; TV Problem Cause Unexplained", Aviation Week & Space Technology, 4 September 1972, 21

[AWST-1972d] "New Instrument Array Aided Soviet Venus 8 Mission Lander", Aviation Week & Space Technology, 18 September 1972, 23

[AWST-1973] "Mars Viking Rover Feasibility Studied", Aviation Week & Space Technology, 23 July 1973, 14

[AWST-1975a] "Second Viking Launched Prior to Thunderstorm", Aviation Week & Space Technology, 15 September 1975, 20

[AWST-1975b] "Delays Peril Helios-2 Mission", Aviation Week & Space Technology, 22 September 1975, 18

[AWST-1975c] "Helios-1 Second Perihelion Finds Higher Temperatures than First", Aviation Week & Space Technology, 29 September 1975, 19

[AWST-1979] "Pioneer Obtains Saturn Moon Data", Aviation Week & Space Technology, 10 September 1979, 24

[AWST-1985] "Voyager Spacecraft Detects Outermost Ring of Uranus as it Nears Encounter", Aviation Week & Space Technology, 25 November 1985, 25

[AWST-1986a] "Voyager 2 Managers Resolve Imagery Problem Before Uranus Encounter", Aviation Week & Space Technology, 27 January 1986, 24–25

[AWST-1989a] "Voyager 2 Images of Neptune Confirm Presence of Partial Rings Near Moon Orbit", Aviation Week & Space Technology, 21 August 1989, 21

[AWST-1989b] "Voyager Spacecraft Beginning New, Interstellar Part of Mission", Aviation Week & Space Technology, 9 October 1989, 117–118

[Axford-1968] Axford, W.I., "Observations of the Interplanetary Plasma", Space Science Review, 8, 1968, 331–365

[Baker-1975] Baker, A.L., et al., "The Imaging Photopolarimeter Experiment on Pioneer 11", Science, 188, 1975, 468–472

[Baker-1977a] Baker, D., "Behind the Viking Scene. 1. The Mars Orbit Insertion Anomaly", Spaceflight, February 1977, 75–77

[Baker-1977b] Baker, D., "Behind the Viking Scene. 2. Landers 1 and 2 Site Selection", Spaceflight, March 1977, 109–112

[Baker-1977c] Baker, D., "Behind the Viking Scene. 3. The Viking 2 Attitude Anomaly", Spaceflight, April 1977, 146–148

[Baker-1977d] Baker, D., "Behind the Viking Scene. 4. Lander 1 Operations", Spaceflight, May 1977, 166–168

[Baker-1977e] Baker, D., "Behind the Viking Scene. 5. Lander 2 Operations", Spaceflight, June 1977, 232–240

[Baldwin-1949] Baldwin, R., "The face of the Moon", University of Chicago Press, 1949

[Ball-1999] Ball, A.J., Lorenz, R.D., "Penetrometry of Extraterrestrial Surfaces: an Historical Overview". Paper presented at the International Workshop on Penetrometry in the Solar System, Graz, 1999

[Barath-1964] Barath, F.T., et al., "Microwave Radiometers", Science, 139, 1963, 908–909

[Barath-1964] Barath, F.T., et al., "Mariner 2 Microwave Radiometer Experiment and Results," The Astronomical Journal, 69, 1964, 49–58

[Barmin-1983] Barmin, I.V., Shevchenko, A.A., "Soil-Scooping Mechanism for the Venera 13 and Venera 14 Unmanned Interplanetary Spacecraft", Cosmic Research, 21, 1983, 118–122

[Barth-1970a] Barth, C.A., "Mariner 5 Measurements of Ultraviolet Emission from the Galaxy". In Hoziaux L. and Butler, H.E. (eds.), "Proceedings from IAU Symposium no. 36: Ultraviolet Stellar Spectra and Related Ground-Based Observations", 1970, 334–340

[Barth-1970b] Barth, C.A., "Mariner 6 Measurements of the Lyman-Alpha Sky Background," The Astrophysical Journal, 161, L181–L184. According to the NASA NSSDC Internet site Mariner 6 UV scans included the area of recently discovered comet Kohoutek (known as C/1969 O1, 1969b, 1970III). However, according to Principal Investigator Charles A. Barth no such observations were made (Barth, C.A., Personal communication with the author, 8 June 2004)

[Barth-1971] Barth, C.A., et al., "Mariner 6 Ultraviolet Spectrum of Mars Upper Atmosphere". In: Sagan, C., Owen, T.C., Smith, H.J., "Proceedings from IAU Symposium no. 40: Planetary Atmospheres", 1971, 253–256

[Barth-1974] Barth, C.A., "The Atmosphere of Mars", Annual Review of Earth and Planetary Sciences, 1974, 333–367

[Bartok-1986] Bartok, C.D., "Catching the Whispers from Uranus", Aerospace America, May 1986, 44–48

[Basilevsky-1992] Basilevsky, A.T., "Venera 9, 10 and 13 Landing Sites as Seen by Magellan", paper presented at the XXIII Lunar and Planetary Science Conference, Houston, 1992

[Becker-2004] Becker, S., "Rise of the Machines: Telerobotic Operations in the U.S. Space Program", Quest, 11 No.4, 2004, 14–39

[Beech-1999] Beech, M., et al., "Satellite Impact Probabilities: Annual Showers and the 1965 and 1966 Leonid Storms", Acta Astronautica, 44, 1999, 281–292

[Beish-1993] Beish, J., "Neptune's Spot", Sky & Telescope, April 1993, 6–7

[Belcher-1989] Belcher, J.W., et al., "Plasma Observations Near Neptune: Initial Results from Voyager 2", Science, 246, 1989, 1478–1483

[Bertaux-1976] Bertaux, J.L., et al., "Interstellar Medium in the Vicinity of the Sun: A Temperature Measurement Obtained with Mars-7 Interplanetary Probe", Astronomy & Astrophysics, 46, 1976, 19–29

[Besser-2004] Besser, B.P., "Austria's History in Space", Noordwijk, ESA, 2004, 31–33

[Beuf-1964] Beuf, F. G., "Martian Entry Capsule", Astronautics & Aeronautics, December 1964, 30–37

[Bille-2004] On early satellites see: Bille, M., Lishock, E., "The First Space Race", Texas A&M University Press, 2004

[Blamont-1987a] Blamont, J., "Venus Devoilée" (Venus Unveiled), Paris, Editions Odile Jacob, 1987, 113–114 (in French)

[Blamont-1987b] ibid. 140–142

[Blamont-1987c] ibid., 211–212. Blamont does not reveal whether the comet was Schwassmann–Wachmann 2 (perihelion on 17 March 1981) or Schwassmann–Wachmann 3 (perihelion on 3 September 1979). The latter would seem more likely. Furthermore, the observation is attributed to Venera 13 "that had flown by Venus in December 1978".

[Blamont-1987d] ibid., 105–107

[Blamont-1987e] ibid., 139–172

[Bogard-1983] Bogard, D.D., Johnson, P., "Martian Gases in an Antarctic Meteorite?", Science, 221, 1983, 651–654

[Bogovalov-1984] Bogovalov, S.V., et al., "Short-Period Pulsations in Solar Hard X-Ray Bursts Recorded by Venera 13, 14", Soviet Astronomy Letters, 10, 1984, 286–288

[Bogovalov-1985] Bogovalov, S.V., et al., "Directionality of Solar Flare Hard X-Rays: Venera 13 Observations", Soviet Astronomy Letters, 11, 1985, 322–324

[Bohlin-1973] Bohlin, R.C., "Mariner 9 Ultraviolet Spectrometer Experiment", Astronomy & Astrophysics, 28, 1973, 323–326

[Born-1975] Born, G.H, Duxbury, T.C., "The Motions of Phobos and Deimos from Mariner 9 TV Data", Celestial Mechanics, 12, 1975, 77–88

[Borrowman-1983] Borrowman, G., "Pioneer Close Encounter", Spaceflight, February 1983, 68

[Bosh-1996] Bosh, A.S., Rivkin, A.S., "Observations of Saturn's Inner Satellites During the May 1995 Ring-Plane Crossing", Science, 272, 1996, 518–521

[Botvinova-1973] Botvinova, V.V., et al., "Photometric Data from some Photographs of Mars Obtained with the Automatic Interplanetary Station 'Mars 3'". In: Proceedings of the Symposium on Exploration of the Planetary System, Torun, Poland, September 5–8, 1973, 287–292

[Boyce-1993] Boyce, Joseph M., "A Structural Origin for the Cantaloupe Terrain of Triton", Proceedings Lunar & Planetary Sciences Conference, March 1993, 165–166

[Boyer-1961] Boyer, C., Camichel, H., "Observations photographiques de la planète Venus", Annales d'Astrophysique, 24, 1961, 531–535

[Brandt-1974] Brandt, J.C., Maran, S.P., "Preliminary Results on Comet Kohoutek:

Interaction with the Solar Wind", In Solar wind three; Proceedings of the Third Conference, Pacific Grove, March 25–29, 1974, 415–420

[Bridge-1981] Bridge, H.S., et al., "Plasma Observations Near Saturn: Initial Results from Voyager 1", Science, 212, 1981, 217–224

[Bridge-1986] Bridge, H.S., et al., "Plasma Observations Near Uranus: Initial Results from Voyager 2", Science, 233, 1986, 89–93

[Broadfoot-1979] Broadfoot, A.L., et al., "Extreme Ultraviolet Observations from Voyager 1 Encounter with Jupiter", Science, 204, 1979, 979–982

[Broadfoot-1981] Broadfoot, A.L., et al., "Extreme Ultraviolet Observations from Voyager 1 Encounter with Saturn", Science, 212, 1981, 206–211

[Broadfoot-1986] Broadfoot, A.L., et al., "Ultraviolet Spectrometer Observations of Uranus", Science, 233, 1986, 74–79

[Broadfoot-1989] Broadfoot, A.L., et al., "Ultraviolet Spectrometer Observations of Neptune and Triton", Science, 246, 1989, 1459–1466

[Brooks-2005] Brooks, M., "13 Things that don't Make Sense", New Scientist, 19 March 2005, 30–37

[Brown-1965] Brown, H., Taylor, J., "Navigator Study of Electric Propulsion for Unmanned Scientific Missions – Volume 1 Mission Analysis", NASA CR–54324, 1965

[Brown-1985] Brown, R.H., Cruikshank, D.P., "The moons of Uranus, Neptune and Pluto", Scientific American, July 1985, 38–47

[Brush-1996a] Brush, S.G., "Fruitful Encounters: The Origin of the Solar System and of the Moon from Chamberlin to Apollo", Cambridge University Press, 1996, 168–169

[Brush-1996b] ibid., 165–168

[Bugos-2000] Bugos, G.E., "Atmosphere of Freedom: Sixty Years at the NASA Ames Research Center", Washington, NASA, 2000, 151–152

[Bukata-1969] Bukata, R.P., et al., "Neutron Monitor and Pioneer 6 and 7 Studies of the January 28, 1967 Solar Flare Event", Solar Physics, 10, 1969, 198–221

[Burgess-1953] Burgess, E., Cross, C.A., "The Martian Probe", Journal of the British Interplanetary Society, 12, 1953, 72–74

[Burgess-1966] Burgess, E., "Are There Canals on Mars", Spaceflight, February 1966, 46–47

[Burgess-1982a] Burgess, E., "By Jupiter", New York, Columbia University Press, 1982, 145–148

[Burke-1992] Burke, J.D., "Past US Studies and Developments for Planetary Rovers". In: "Missions, Technologies et Conception des Vehicules Mobiles Planetaires", Toulouse, Cépaduès, 1993.

[Burlaga-1968] Burlaga, L.F., "Micro-Scale Structures in the Interplanetary Medium", Solar Physics, 4, 1968, 67–92

[Burlaga-2005] Burlaga, L.F., et al., "Crossing the Termination Shock into the Heliosheath: Magnetic Fields", Science, 309, 2005, 2027–2029

[Burrows-1998] Burrows, W.E., "This New Ocean", New York, The Modern Library, 1998, 502–504

[Butrica-1996a] Butrica, A.J., "To See the Unseen – A History of Planetary Radar Astronomy", Washington, NASA, 1996, 47

[Butrica-1996b] ibid., 27–53

[Butrica-1996c] ibid., 212–215

[Butrica-1996d] ibid., 118–120

[Butrica-1996e] ibid., 162–170

[Butrica-1996f] ibid., 171–176

[Butrica-1998] Butrica, A.J., "Voyager: The Grand Tour of Big Science". In: "From Engineering Science to Big Science: The NACA and NASA Collier Trophy Research Project Winners", Washington, NASA, 1998, 269–270

[Cahill-1963] Cahill, L.J. Jr., "Magnetic Field Measurements in Space", Space Science Review, 1, 1963, 399–414

[Caraveo-1995] Caraveo, P., "Mille Lampi non Squarciano il Buio" (One Thousand Bursts do not Light the Dark), l'Astronomia, July 1995, 14–26 (in Italian)

[Carlier-1995] Carlier, C., Gilli, M., "The First Thirty Years at CNES: the French Space Agency 1962–1992", CNES, Paris, 1995, 141

[Caroubalos-1974] Caroubalos, C., Steinberg, J.L., "Evidence of Solar Bursts Directivity at 169 MHz from Simultaneous Ground Based and Deep Space Observations (STEREO-1 Preliminary Results)", Astronomy and Astrophysics, 32, 1974, 245–253

[Carr-1976] Carr, M.H., "The Volcanoes of Mars", Scientific American, January 1976, 32–43

[Carr-1980] Carr, M.H., et al., "Viking Orbiters View of Mars", Washington, NASA, 1980, 163–164

[Cesarone-1983] Cesarone, R.J., Sergeyevsky, A.B., Kerridge, S.J., "Prospects for the Voyager Extra-Planetary and Interstellar Mission", paper AAS 83–308

[Chertok-2007] Chertok, B., "Rockets and People: Creating a Rocket Industry", Washington, NASA, 2007, 563–588

[CIA-1969] "National Intelligence Estimate: The Soviet Space Program", CIA NIE 11-1-69, 19 June 1969

[CIA-1971] "National Intelligence Estimate – Soviet Space Programs", Central Intelligence Agency NIE 11-1-71, 1 July 1971, 20–21

[CIA-1973a] CIA Directorate of Science and Technology, "Soviet Space Events in 1972", FMSAC-STIR/73-8, May 1973, 29–30

[CIA-1973b] "National Intelligence Estimate – Soviet Space Programs", Central Intelligence Agency NIE 11-1-73, 20 December 1973, 17

[Chapman-1969] Chapman, C.R., Pollack, J.R., Sagan, C., "An Analysis of Mariner-4 Cratering Statistics", The Astronomical Journal, 74, 1969, 1039–1051

[Chase-1963] Chase, S.C., Kaplan, L.D., Neugebauer, G., "Infrared Radiometer", Science, 139, 1963, 907–908

[Chase-1974] Chase, S.C., et al., "Pioneer 10 Infrared Radiometer Experiment: Preliminary Results", Science, 183, 1974, 315–317

[Chassefière-1986] Chassefière, E., et al., "Atomic Hydrogen and Helium Densities of the Interstellar Medium Measured in the Vicinity of the Sun", Astronomy and Astrophysics, 160, 1986, 229–242

[Clark-1960a] Clark, E., "Pioneer V Transmits Deep Space Data", Aviation Week, 21 March 1960, 28–29

[Clark-1960b] Clark, E., "Vega Study Shows Early NASA Problems", Aviation Week, 27 June 1960, 62–68

[Clark-1976] Clark, B.C., et al., "Inorganic Analyses of Martian Surface Samples at the Viking Landing Sites", Science, 194, 1976, 1283–1288

[Clark-1986] Clark, P.S., "The Soviet Mars Programme", Journal of the British Interplanetary Society, 39, 1986, 3–18

[Cline-1979a] Cline, T.S., et al., "Helios 2-Vela-Ariel 5 Gamma-Ray Burst Source Position", The Astrophysical Journal, 229, 1979, L47–L51

[Cline-1979b] Cline. T.S., et al., "Gamma-Ray Burst Observations from Helios 2", The Astrophysical Journal, 232, 1979, L1–L5

[Coffeen-1974] Coffeen, D.L., "Polarization Measurements of Jupiter at 103° Phase Angle", Bulletin of the American Astronomical Society, 6, 1974, 387

[Coleman-1962] Coleman, P.J. Jr., et al., "Interplanetary Magnetic Fields", Science, 138, 1962, 1099–1100

[Colin-1977a] Colin, L., Hall, C.F., "The Pioneer Venus Program", Space Science Reviews, 20, 1977, 283–306

[Colin-1977b] Colin, L., Hunten, D.M., "Pioneer Venus Experiment Descriptions", Space Science Reviews, 20, 1977, 451–525

[Colin-1979] Colin, L., "Encounter with Venus", Science, 203, 1979, 743–745

[Collins-1971] Collins, S.A., "The Mariner 6 and 7 Pictures of Mars", Washington, NASA, 1971

[Conrath-1989] Conrath, B., et al., "Infrared Observations of the Neptunian System", Science, 246, 1989, 1454–1459

[Conrath-1991] Conrath, B., et al., "The Helium Abundance of Neptune from Voyager Measurements", Journal of Geophysical Research Supplement, 96, October 1991, 18907–18919

[Corliss-1965a] Corliss, W.R., "Space Probes and Planetary Exploration", Princeton, Van Nostrand, 1965, 246–247

[Corliss-1965b] ibid., 387–389

[Corliss-1965c] ibid., 389–392

[Corliss-1965d] ibid., 392–394

[Corliss-1965e] ibid., 250–251

[Corliss-1965f] ibid., 485–502

[Counselman-1979] Counselman, C.C. III, et al., "Venus Winds are Zonal and Retrograde Below the Clouds", Science, 205, 1979, 85–87

[Covault-1976a] Covault, C., "Rover Pushed for 1979 Mars Mission", Aviation Week & Space Technology, 11 February 1974, 56–59

[Covault-1976b] Covault, C., "USSR Planetary Missions Awaiting Evaluation", Aviation Week & Space Technology, 28 June 1976, 60–62

[Covault-1979] Covault, C., "Pioneer 11 to Picture Saturn This Week", Aviation Week & Space Technology, 20 August 1979, 19–20

[Craig-1985] Craig, R.A., "The Pioneer Venus Extended Mission", Spaceflight, December 1985, 445–450

[Crocco-1956] Crocco, G.A., "Giro Esplorativo di un anno Terra-Marte-Venere-Terra", paper presented at the VII International Astronautical Congress, Rome, 1954. (In Italian. English translation: "One-Year Exploration-Trip Earth-Mars-Venus-Earth")

[Cruishank-1976] Cruishank, D.P., Morrison, D., "The Galilean satellites of Jupiter", Scientific American, May 1976, 108–116

[Cunningham-1988a] The diameter of Nike is from IRAS infrared observations, reported in Cunningham, C.J., "Introduction to Asteroids", Richmond, Willmann-Bell, 1988, 149. The diameter mentioned in "Pioneer Odissey" was 24 km.

[Cutts-1976] Cutts, J.A., et al., "North Polar Region of Mars: Imaging Results from Viking 2", Science, 194, 1976, 1329–1337

[Cuzzi-1987] Cuzzi, J.N., Esposito, L.W., "The Rings of Uranus", Scientific American, July 1987, 42–48

[Danielson-1964] Danielson, R.E., et al., "Mars Observations from Stratoscope II", The Astronomical Journal, 69, 1964, 344–352

[Davidson-1999a] Davidson, K., "Carl Sagan: A Life", New York, John Wiley & Sons, 1999, 117–120

[Davidson-1999b] ibid., 179–181 and Davidson, K., Personal communication with the author, 7 December 2004

[Davidson-1999c] ibid., 241–243

[Davidson-1999d] ibid., 276–281

[Davidson-1999e] ibid., 303–310

[Davidson-1999f] ibid., 87

[Dawson-2004a] Dawson, V.P. Bowles, M.D., "Taming Liquid Hydrogen: The Centaur Upper Stage Rocket 1958–2002", Washington, NASA, 2004, 120

[Dawson-2004b] ibid., 123–125

[Dawson-2004c] ibid., 125–131

[Dawson-2004d] ibid., 131–133

[Dawson-2004e] ibid., 147–151

[Dawson-2004f] ibid., 151–154

[Dawson-2004g] ibid.,145–147

[Day-2001] Day, D.A., "Early American Ferret and Radar Satellites", Spaceflight, July 2001, 288–291

[Day-2007] Day, D.A., personal correspondence with the author, 20 January 2007

[DeVorkin-1992] DeVorkin, D.H., "Science with a Vengeance", New York, Springer, 1992, 276–278

[DiGregorio-2004] DiGregorio, B.E., "Life on Mars? 27 Years of Questions", Sky & Telescope, February 2004, 40–45

[Dixon-1974] Dixon, W., "The Pioneer Spacecraft as a Probe Carrier". In: Proceedings of the Outer Planet Probe Technology Workshop, May 21–23, 1974, NASA CR–137543

[Dobbins-2002] Dobbins, T., Sheehan, W., "A Rare Opportunity to Glimpse Saturn's 'Lost Ring'", Sky & Telescope, February 2002, 102–107

[Dobrovolskis-1988] Dobrovolskis, A.R., Steimancameron, T.Y., Borderies, N.J.,

"Stability of Polar Rings Around Neptune", Bulletin of the American Astronomical Society, 20, 1988, 861

[Dolginov-1976] Dolginov, Sh.Sh., "Preliminary Magnetic-Field Measurements Near Pericenter of the Venera 9 Orbit", Soviet Astronomy Letters, 2, 1976, 34–35

[Dolginov-1987] Dolginov, Sh.Sh., "What We Have Learned about the Martian Magnetic Field", Earth, Moon and Planets, 37, 1987, 17–52

[Dollfus-1983] Dollfus, A., Deschamps, M., Ksanfomaliti, L.V., "The Surface Texture of the Martian Soil from the Soviet Spacecraft Mars-5 Photopolarimeters", Astronomy & Astrophysics, 123, 1983, 225–237

[Dollfus-1998a] Dollfus, A., "History of Planetary Science. The Pic di Midi Planetary Observation Project: 1941–1971", Planetary and Space Science, 46, 1998, 1037–1073

[Dollfus-1998b] Dollfus, A., "50 Ans d'Astronomie" (Fifty Years of Astronomy), Les Ulis, EDP Sciences, 1998, 124–126 (in French)

[Dollfus-1998c] ibid., 100

[Dollfus-2004] Dollfus, A., personal correspondence with the author, 23 March 2004

[Doty-1975] Doty, L., "MBB Stresses Broader Satellite Usage", Aviation Week & Space Technology, 22 September 1975, 19

[Dunne-1974] Dunne, J.A., "Mariner 10 Venus Encounter", Science, 183, 1974, 1289–1291

[Duxbury-1975] Duxbury, T.C., "Pioneer Imaging of the Galilean Satellites", Bulletin of the American Astronomical Society, 7, 1975, 379

[Duxbury-1978a] Duxbury, T.C., "Phobos Transit of Mars as Viewed by the Viking Cameras", Science, 199, 1978, 1201–1202

[Duxbury-1978b] Duxbury, T.C., Veverka, J., "Deimos Encounter by Viking: Preliminary Imaging Results", Science, 201, 1978, 812–814

[Dyal-1990] Dyal, P., "Pioneers 10 and 11 Deep Space Missions", NASA document TM–102269, 1990

[Dyer-1980] Dyer, J.W., "Pioneer Saturn", Science, 207, 1980, 400–401

[Edelson-1979] Edelson, R.E., et al., "Voyager Telecommunications: The Broadcast from Jupiter", Science, 204, 1979, 913–921

[Edenhofer-1984] Edenhofer, P., "Plasma-Fernerkundung mit Laufzeitmessungen (Korona-Sondierung)". In: Porsche, H. (ed.) "10 Jahre Helios – 10 Years Helios", 115–117 (in German and English)

[Elson-1978a] Elson, B.M., "Pioneer's Generation of Data Flow from Venus Begins", Aviation Week & Space Technology, 11 December 1978, 22–23

[Elson-1978b] Elson, B.M., "Venus Data Surprise Scientists", Aviation Week & Space Technology, 18 December 1978, 8–9

[Elson-1979a] Elson, B.M., "Scientists Begin Processing, Analyzing Venus Data", Aviation Week & Space Technology, 1 January 1979, 38–40

[Elson-1979a] Elson, B.M., "Pioneer Takes Look at a Saturn Moon", Aviation Week & Space Technology, 3 September 1979, 18–19

[Elson-1979b] Elson, B.M., "Pioneer's Brush with Disaster Detailed", Aviation Week & Space Technology, 17 September 1979, 20–21

[Elson-1979c] Elson, B.M., "Pioneer Returns Extensive Saturn Data", Aviation Week & Space Technology, 10 September 1979, 22–24

[Epstein-1974] G. Epstein, "Expérience de Radioastronomie Stéréo 1 Embarquée sur la Sonde Soviétique Mars III" (Stereo 1 Radioastronomy Experiment mounted on the Soviet Probe Mars 3), L'Onde Electrique, 1974, 281–291 (in French)

[ESA-1979] "Thirty-First SOL meeting : Paris from 03/05 to 04/05/1979", Document ESA 4218, 25 June 1979

[Eshelman-1979a] Eshelman, V.R., et al., "Radio Science with Voyager 1 at Jupiter: Preliminary Profiles of the Atmosphere and Ionosphere", Science, 204, 1979, 976–978

[Eshelman-1979b] Eshelman, V.R., et al., "Radio Science with Voyager at Jupiter: Initial Voyager 2 Results and a Voyager 1 Measure of the Io Torus", Science, 206, 1979, 959–962

[Espace Information-1979] "Le Programme Eole" (The Eole Programme), Espace Information, No. 16, October 1979, 3 (in French)

[ESRO-1966] Cost and other details of the GSFC Jupiter probe are taken from "Development costs of a Jupiter probe and of spacecraft in ESRO programme based on costing methods developed by the IIT Research Institute and Goddard Space Flight Centre (GSFC)", document ESRO 272, 16 June 1966

[ESRO-1969] Mariner costs are mentioned in the ESRO document 5141, "Mercury fly-by study by Bolkow", 21 April 1969

[ESRO-1970] "Helios project", ESRO document 7123, containing correspondence, memoranda etc. dated between 8 October 1970 and 6 September 1973

[Estulin-1981] Estulin, I.V., et al., "Three Gamma-Ray Bursts Recorded by Venera 11, Venera 12, and Prognoz 7", Soviet Astronomy Letters, 7, 1981, 12–14

[Evans-1979] Evans, W.D., et al., "Gamma–Ray Burst Observations by Pioneer Venus Orbiter", Science, 205, 119-121

[Evans-2004a] Evans, B., with Harland, D.M., "NASA's Voyager Missions: Exploring the Outer Solar System and Beyond", Chichester, Springer-Praxis, 2004, 49

[Evans-2004b] ibid., 40–49

[Evans-2004c] ibid., 241–244

[Evans-2004d] ibid., 58–61

[Evans-2004e] ibid., 65

[Evans-2004f] ibid., 236–240

[Ezell-1984a] Ezell, E.C., Ezell, L.N., "On Mars: Exploration of the Red Planet 1958–1978", Washington, NASA, 1984 , 25–30

[Ezell-1984b] ibid., 35–39 and 43–49

[Ezell-1984c] ibid., 86

[Ezell-1984d] ibid., 31–50

[Ezell-1984e] ibid., 77–80

[Ezell-1984f] ibid., 83–104

[Ezell-1984g] ibid., 66–74

[Ezell-1984h] ibid., 104–110

[Ezell-1984i] ibid., 110–119

[Ezell-1984j] ibid., 156–157
[Ezell-1984k] ibid., 157–159
[Ezell-1984l] ibid., 159
[Ezell-1984m] ibid., 159–167
[Ezell-1984n] ibid., 289
[Ezell-1984o] ibid., 288–297
[Ezell-1984p] ibid., 121–153
[Ezell-1984q] ibid., 155–201
[Ezell-1984r] ibid., 396–397
[Ezell-1984s] ibid., 203–242
[Ezell-1984t] ibid., 243–276
[Ezell-1984u] ibid., 277–317
[Ezell-1984v] ibid., 374–380
[Ezell-1984w] ibid., 400–414
[Ezell-1984x] ibid., 410
[Ezell-1984y] ibid., 412
[Faget-1970] Faget, M.A., Davis, H.P., "Space-Shuttle Applications", paper presented at the Third Conference on Planetology and Space Mission Planning, New York, October 1970
[Fan-1960] Fan, C.Y., Meyer, P., Simpson, J.A., "Rapid Reduction of Cosmic-Radiation Intensity Measured in Interplanetary Space", Physical Review Letters, 5, 1960, 269–271
[Farmer-1976] Farmer, C.B., Davies, D.W., LaPorte, D.D., "Mars: Northern Summer Ice Cap – Water Vapor Observations from Viking 2", Science, 194, 1976, 1339–1341
[Farquhar-1972] Farquhar, R.W., Ness, N.P., "Two Early Missions to the Comets", Astronautics & Aeronautics, October 1972, 32–37
[Farquhar-1999] Farquhar, R.W., "The use of Earth-return trajectories for missions to comets", Acta Astronautica, 44, 1999, 607–623
[Fillius-1974] Fillius, R.W., McIlwain, C.E., "Radiation Belts of Jupiter", Science, 183, 1974, 314–315
[Fillius-1975] Fillius, R.W., "Radiation Belts of Jupiter: A Second Look", Science, 188, 1975, 465–467
[Fillius-1980] Fillius, W, Ip, W.H., McIlwain, C.E., "Trapped Radiation Belts of Saturn: First Look", Science, 207, 1980, 425–431
[Fimmel-1976a] Fimmel, R.O., Swindell, W., Burgess, E., "Pioneer Odyssey", Washington, NASA, 1976, 39–45
[Fimmel-1976b] ibid., 49–58
[Fimmel-1976c] ibid., 187–188
[Fimmel-1976d] ibid., 67
[Fimmel-1976e] ibid., 95–97
[Fimmel-1976f] ibid., 68–69
[Fimmel-1976g] ibid., 101–102
[Fimmel-1976h] ibid., 102–103
[Fimmel-1976i] ibid., 115

[Fimmel-1976j] ibid., 139–160

[Fimmel-1976k] ibid., 82–83

[Fimmel-1976l] ibid., 83

[Fimmel-1976m] ibid., 88

[Fimmel-1980a] Fimmel, R.O., Swindell, W., Burgess, E., "Pioneer – First to Jupiter, Saturn, and Beyond", Washington, NASA, 1980, 84–85

[Fimmel-1980b] ibid., 87–89

[Fink-1976a] Fink, D.E., "JPL Shapes Broad Planetary Program", Aviation Week & Space Technology, 9 August 1976, 37–43

[Fink-1976b] Fink, D.E., "Viking Successes Spur Rover Mission", Aviation Week & Space Technology, 27 September 1976, 40–42

[Fisk-2005] Fisk, L.A., "Journey into the Unknown Beyond", Science, 309, 2005, 2016–2017

[Fjeldbo-1971] Fjeldbo, G, Kliore, A.J., "The Neutral Atmosphere of Venus as Studied with the Mariner V Radio Occultation Experiments", The Astronomical Journal, 76, 1971, 123–140

[Fjeldbo-1975] Fjeldbo, G. et al., "The Pioneer 10 Radio Occultation Measurements of the Ionosphere of Jupiter", Astronomy & Astrophysics, 39, 1975, 91–96

[Flandro-1966] Flandro, G.A., "Fast Reconnaissance Missions to the Outer Solar System Utilizing Energy Derived from the Gravitational Field of Jupiter", Astronautica Acta, 16, 1966, 329–337

[Flight-1963] "Flight to Mars", Flight International, 3 January 1963, 26–29

[Flight-1964a] "Cosmos 27 Up", Flight International, 9 April 1964, 587

[Flight-1964b] "Soviet Probe Heading for Venus?", Flight International, 9 April 1964, 587

[Flight-1965a] "Jodrell Bank Tracks Zond 2", Flight International, 25 February 1965, 303

[Flight-1965b] Flight International, 20 May 1965, 808

[Flight-1972] "Pioneer 7 Revived", Flight International, 28 September 1972, 437

[Flight-1974a] "Mars Probe Misses Target", Flight International, 28 February 1974, 271

[Flight-1974b] "Mars 5 and 6 Flight Analysed", Flight International, 4 April 1974, 439–440

[Flight-1987] Flight International, 28 March 1987, 135

[Flight-1990] "Pioneer 11 Leaves Solar System", Flight International, 7 March 1990, 13

[Flight-1997] "NASA Plans to Retire Pioneer 10 Shortly", Flight International, 22 March 1997, 25

[Florenskii-1982] Florenskii, K.P., et al., "Analysis of the Panoramas of the Venera 13 and Venera 14 Landing Sites", Soviet Astronomy Letters, 8, 1982, 233–234

[Florensky-1977] Florensky, C.P., et al., "First Panoramas of the Venusian Surface". In: Proceedings of the VIII Lunar Science Conference, 1977, 2655–2664

[Florensky-1983] Florensky, C.P., et al., "Redox Indicator 'Contrast' on the Surface of Venus", paper presented at the XIV Lunar and Planetary Science Conference, Houston, 1983

[Forney-1963] Forney, R.G., "Mariner II – Attitude Control System", paper presented at the XIV International Astronautical Congress, Paris, 1963

[Forney-1997] Forney, F.B., Kirkland, L.E., "Calibration of Mariner Mars 6/7 Infrared Spectrometers", paper presented at the XXVII Lunar and Planetary Science Conference, Houston, 1997

[Foukal-1977] Foukal, P.V., Mack, P.E., Vernazza, J.E., "The Effects of Sunspots and Faculae on the Solar Constant", Astrophysical Journal, 234, 1977, 952–959

[Fountain-1974] Fountain, J.W., et al., "Jupiter's Clouds: Equatorial Plumes and Other Cloud Forms in the Pioneer 10 Images", Science, 184, 1974, 1279–1281

[Fountain-1978] Fountain, J.W., "Cloud Motion on Jupiter from Pioneer 10 Imagery", Bulletin of the American Astronomical Society, 10, 1978, 564

[Frank-1963] Frank, L.A., Van Allen, J.A., Hills, H.K., "Charged Particles", Science, 139, 1963, 905–907

[Gangopadhyay-1989] Gangopadhyay, P. Ogawa, H.S., Judge, D., "Evidence of a Nearby Solar Wind Shock as Obtained from Distant Pioneer 10 Ultraviolet Glow Data", The Astrophysical Journal, 336, 1989, 1012–1021

[Garvin-1981] Garvin, J.B., "Dust Cloud Observed in Venera 10 Panorama of Venusian Surface: Inferred Surface Processes", paper presented at the XII Lunar and Planetary Science Conference, Houston, 1981

[Garvin-1984a] Garvin, J.B., Head, J.W., Zuber, M.T., "Venus: the Nature of the Surface from Venera Panoramas", paper presented at the XV Lunar and Planetary Science Conference, Houston, 1984

[Gatland-1964] Gatland, K.W., "Spacecraft and Boosters", Los Angeles, Aero Publishers, 1964, 9–18

[Gatland-1972a] Gatland, K.W., "Robot Explorers", London, Blanford Press, 1972, 189–192

[Gatland-1972b] ibid., 169–170

[Gatland-1972c] ibid., 170–176

[Gatland-1972d] ibid., 178–181

[Gatland-1972e] ibid., 182–183

[Gatland-1972f] ibid., 183–184

[Gatland-1972g] ibid., 210–220

[Gatland-1972h] ibid., 202–209

[Gatland-1972i] ibid., 232–239

[Gatland-1972j] ibid., 228–229

[Gatley-1974] Gatley, I., et al., "Infrared Observations of Phobos from Mariner 9", The Astrophysical Journal, 190, 1974, 497–503

[Gazis-1995] Gazis, P.R., Barnes, A., Mihalov, J.D., "Pioneer and Voyager Observations of Large-Scale Spatial and Temporal Variations in the Solar Wind", Space Science Reviews, 72, 1995, 117–120

[Gehrels-1974] Gehrels, T., "The Imaging Photopolarimeter Experiment on Pioneer 10", Science, 183, 1974, 318–320

[Gehrels-1980] Gehrels, T., "Imaging Photopolarimeter on Pioneer Saturn", Science, 207, 1980, 434–439

[Gel'man-1979] Gel'man, B.G., et al., "Venera 12 Analysis of Venus Atmospheric Composition by Gas Chromatography", Soviet Astronomy Letters, 5, 1979, 116–118

[Giorgini-2005] Giorgini, J., personal correspondence with the author, 24 January 2005

[Giragosian-1966] Giragosian, P. A., Parker, M. S., "Systems Considerations for a Planetary Entry Probe", paper presented at the XVII International Astronautical Congress, Madrid, 1966

[Goddard-1920] Goddard, R.H., "Report Concerning Further Developments", March 1920

[Goldstein-1969] Goldstein, R.M., "Superior Conjunction of Pioneer 6", Science, 166, 1969, 598–601

[Gore-1986] Gore, R., "Uranus: Voyager Visits a Dark Planet", National Geographics, August 1986, 178–195

[Grahn-2000] Grahn, S., "Jodrell Bank's Role in Early Space Tracking Activities", S. Grahn internet website

[Grav-2003] Grav, T., Holman, M.J., Kavelaars, J.J., "The Short Rotation Period of Nereid", Arxiv astro–ph/0306001 preprint

[Gray-1982] Gray, D.L., Cesarone, R.J., Van Allen, R.E., "Voyager 2 Uranus and Neptune Targeting", paper AIAA–82–1476

[Greenstadt-1963] Greenstadt, E.W., "Effect of Solar Activity Regions on the Interplanetary Magnetic Field," Astrophysical Journal, 137, 1963, 999–1002

[Greenstadt-1966] Greenstadt, E.W., "Final Estimate of the Interplanetary Magnetic Field at 1 A.U. from Measurements made by Pioneer V in March and April, 1960", Astrophysical Journal, 145, 1966, 270–295

[Gregory-1973a] Gregory, W.H., "Planetary Mission Competition Stiffens", Aviation Week & Space Technology, 23 July 1973, 12–13

[Gregory-1973b] Gregory, W.H., "Comet Exploration Mission Face Complex Hurdles", Aviation Week & Space Technology, 20 August 1973, 76–79

[Gregory-1973c] Gregory, W.H., "Comet Mission Aims at Nuclei Studies", Aviation Week & Space Technology, 3 September 1973, 42–45

[Gringauz-1976] Gringauz, K.I., et al., "Preliminary Measurements of the Plasma near Venus with the Venera 9 Satellite", Soviet Astronomy Letters, 2, 1976, 32–34

[Gross-1966] Gross, F.R., "Buoyant Probes into the Venus Atmosphere", Journal of Spacecraft, 3, 1966, 582–587

[Grün-1984] Grün, E., Rechtig, H., Kissel, J., "Das Mikrometeoritenexperiment auf Helios". In: Porsche, H. (ed.) "10 Jahre Helios – 10 Years Helios", 58–63 (in German and English)

[Gurnett-1984] Gurnett, D.A., Anderson, R.R., "Plasma Waves in the Solar Wind: 10 Years of HELIOS Observations". In: Porsche, H. (ed.) "10 Jahre Helios – 10 Years Helios", 100–105 (in German and English)

[Gurnett-1986] Gurnett, D.A., et al., "First Plasma Wave Observations at Uranus", Science, 233, 1986, 106–109

[Gurnett-1989] Gurnett, D.A., et al., "First Plasma Wave Observations at Neptune", Science, 246, 1989, 1494–1498

[Gurnett-1993] Gurnett, D.A., et al., "Radio Emission from the Heliopause Triggered by an Interplanetary Shock", Science, 262, 1993, 199–203

[Hack-1993] Hack, F., et al., "Precise Localization of Gamma-Ray Bursts from the 2nd Interplanetary Network", paper presented at the 182nd meeting of the American Astronomical Society, 1993

[Hajos-2005] Hajos, G.A., et al., "An Overview of Wind-Driven Rovers for Planetary Exploration", Paper AIAA 2005–0244

[Hale-1970] Hale, D.P., "Grand Tour Missions to the Outer Solar System with Saturn (Intermediate 20)", paper presented at the Third Conference on Planetology and Space Mission Planning, New York, October 1970

[Hammel-1989a] Hammel, H.B., "Neptune Cloud Structure at Visible Wavelengths", Science, 244, 1989, 1165–1167

[Hammel-1989b] Hammel, H.B., et al., "Neptune's Wind Speeds Obtained by Tracking Clouds in Voyager Images", Science, 245, 1989, 1367–1369

[Hanel-1979a] Hanel, R., et al., "Infrared Observations of the Jovian System from Voyager 1", Science, 204, 1979, 972–976

[Hanel-1979b] Hanel, R., et al., "Infrared Observations of the Jovian System from Voyager 2", Science, 206, 1979, 952–956

[Hanel-1981] Hanel, R., et al., "Infrared Observations of the Saturnian System from Voyager 1", Science, 212, 1981, 192–200

[Hanel-1982] Hanel, R., et al., "Infrared Observations of the Saturnian System from Voyager 2", Science, 215, 1982, 544–548

[Hanel-1986] Hanel, R., et al., "Infrared Observations of the Uranian System", Science, 233, 1986, 70–74

[Hanner-1978] Hanner, M., Leinert, C., Pitz, W., "UBV Surface Brighness Photometry of the Milky Way in Scorpius from the Space Probe Helios 1", Astronomy and Astrophysics, 65, 1978, 245–249

[Hargraves-1976] Hargraves, R.H., et al., "Viking Magnetic Properties Investigation: Further Results", Science, 194, 1976, 1303–1309

[Harford-1997] Harford, J., "Korolev: How one Man Masterminded the Soviet Drive to Beat America to the Moon", New York, John Wiley & Sons, 1997, 151–152

[Harrington-1965] Harrington, J.V., "Study of a Small Solar Probe (Sunblazer), Part I: Radio Propagation Experiment", NASA PR–5255–5, 1 July 1965

[Hartle-1983] Hartle, R.E. and Taylor H.A., "Identification of Deuterium Ions in the Ionosphere of Venus", Geophysical Research Letters, 10, October 1983, 965–968

[Hartline-1980] Hartline, B.K., "Voyager Beguiled by Jovian Carrousel", Science, 208, 1980, 384–386

[Hartmann-1974] Hartmann, W.K., Raper, O., "The New Mars: the Discoveries of Mariner 9", Washington, NASA, 1974

[Harvey-2006a] Harvey, B., "The Mars 6 Landing, 12th March 1974", Accepted for publication in the JBIS

[Harvey-2006b] Harvey, B., "Mikhail Tikhonravov (1900–1974): His Contribution to the Soviet Lunar and Interplanetary Programme", Journal of the British Interplanetary Society, 59, 2006, 266–272

[Harvey-2007a] Harvey, B., "Russian Planetary Exploration: History, Development, Legacy and Prospects", Chichester, Springer–Praxis, 2007, 94

[Harvey-2007b] ibid., 130

[Helton-1975] Helton, M.R., "Encounter Strategies Available for the First Mission to Saturn", Paper AIAA 75–1139

[Herr-1970] Herr, K.C., et al., "Martian Topography from the Mariner 6 and 7 Infrared Spectra", The Astronomical Journal, 75, 1970, 883–894

[Herriman-1966] Herriman, A.G., "Mariner IV Television – Spacecraft Photography in Planetary Astronomy", paper presented at the XVII International Astronautical Congress, Madrid, 1966

[Hess-1976a] Hess, S.L., et al., "Mars Climatology from Viking 1 after 20 Sols", Science, 194, 1976, 78–81

[Hess-1976b] Hess, N.L., et al., "Early Meteorological Results from the Viking 2 Lander", Science, 194, 1976, 1352–1353

[Hick-1991] Hick, P., Jackson, B.V., Schwenn, R., "Synoptic Maps for the Heliospheric Thomson Scattering Brightness as Observed by the Helios Photometers", Astronomy and Astrophysics, 244, 1991, 242–250

[Hill-1962] Hill, G., "Venus Probe is Believed Succeeding", the New York Times, 28 August 1962, 1

[Hilton-1985] Hilton, D., et al., "Construction of the 'Mars Ball' Prototype Exploration Vehicle", Bulletin of the American Astronomical Society, 17, 1985, 697

[Hoffman-1979] Hoffman, J.H., et al., "Venus Lower Atmospheric Composition: Preliminary Results from Pioneer Venus", Science, 203, 23 February 1979, 800–802

[Holberg-1980a] Holberg, J.B., et al., "Extreme-UV and Far-UV Observations of the White Dwarf HZ 42 from Voyager 2", The Astrophysical Journal, 242, 1980, L119–L123

[Holberg-1980b] Holberg, J.B., Forrester, W.T., Broadfoot, A.L., "Voyager 2 Ultraviolet Spectrometer Observations of Extreme Ultraviolet Emission from the White Dwarf G191 B2B", Bulletin of the American Astronomical Society, 12, 1980, 872

[Holberg-1982] Holberg, J.B., et al., "Voyager Absolute Far-Ultraviolet Spectrophotometry of Hot Stars", The Astrophysical Journal, 257, 1982, 656–671

[Hollweg-1968] Hollweg, J.V., "A Statistical Ray Analysis of the Scattering of Radio Waves by the Solar Corona", The Astronomical Journal, 73, 1968, 972–982

[Hollweg-2004] Hollweg, J.V., Personal communication with the author, 5 August 2004

[Hord-1979] Hord, C.W., et al., "Photometric Observations of Jupiter at 2400 Angstroms", Science, 206, 1979, 956–959

[Horowitz-1976] Horowitz, N.H., Hobby, G.L., Hubbard, J.S., "The Viking Carbon Assimilation Experiment: Interim Report", Science, 194, 1976, 1321–1322

[Horowitz-1977] Horowitz, N.H., "The Search for Life on Mars", Scientific American, November 1977, 52–61

[Houtkooper-2006] Houtkooper, J.M., Schulze-Makuch, D., "A Possible Biogenic

Origin for Hydrogen Peroxide on Mars: The Viking Results Reinterpreted", Arxiv physics/0610093 preprint

[Howard-1974] Howard, H.T., et al., "Venus: Mass, Gravity Field, Atmosphere, and Ionosphere as Measured by the Mariner 10 Dual-Frequency Radio System", Science, 183, 1974, 1297–1301

[Hufbauer-1991a] Hufbauer, K., "Exploring the Sun: Solar Science since Galileo", The Johns Hopkins University Press, Baltimore, 1991, 222–225

[Hufbauer-1991b] ibid., 232–236

[Hufbauer-1991c] ibid., 236–239

[Hughes-1974] Hughes Aircraft Company, "Pioneer Mars Surface Penetrator Mission: Mission Analysis and Orbiter Design", NASA CR–137568, August 1974

[Hughes-1977] Hughes, D.W., "The Direct Investigation of Comets by Space Probes", Journal of the British Interplanetary Society, 30, 1977, 2–14

[Hughes-1978] Hughes Aircraft Company, "Pioneer Venus Case Study in Spacecraft Design", New York, AIAA, 1978

[Humes-1975] Humes, D.H., et al., "Pioneer 11 Meteoroid Detection Experiment: Preliminary Results", Science, 188, 1975, 473–474

[Humes-1980] Humes, D.H., et al., "Impact of Saturn Ring Particles on Pioneer 11", Science, 207, 1980, 443–444

[Hunter-1967] Hunter, G.S., "Venus Atmosphere Found Refractive", Aviation Week & Space Technology, 30 October 1967, 22–23

[Huntress-2002] Huntress, W.T. Jr., Moroz, V.I., Shevalev, I.L., "Lunar and Planetary Robotic Exploration Missions in the 20th Century", Space Science Reviews, 107, 2003, 541–649

[Hutchings-1983] Hutchings, E. Jr. "The Autonomous Vikings", Science, 219, 1983, 803–808

[Hyde-1974] Hyde, J., "The Marriner [sic] Spacecraft as a Probe Carrier". In: Proceedings of the Outer Planet Probe Technology Workshop, May 21–23, 1974, NASA CR–137543

[IAUC-3356] "International Astronomical Union Circular No. 3356", 11 May 1979. Other than by Venera 11 and 12 and Prognoz 7, the 5 March 1979 gamma-ray burst was detected by the Pioneer Venus Orbiter, ISEE-3, Helios 2 and by the US military satellites Vela 5A, 5B and 6A.

[IAUC-3483] "International Astronomical Union Circular No. 3483", 6 June 1980

[IAUC-3507] "International Astronomical Union Circular No. 3507", 26 August 1980

[IAUC-3608] "International Astronomical Union Circular No. 3608", 29 May 1981

[IAUC-3651] "International Astronomical Union Circular No. 3651", 17 December 1981

[IAUC-3656] "International Astronomical Union Circular No. 3656", 8 January 1982

[IAUC-3660] "International Astronomical Union Circular No. 3660", 27 January 1982

[IAUC-4806] "International Astronomical Union Circular No. 4806", 7 July 1989

[IAUC-4824] "International Astronomical Union Circular No. 4824", 2 August 1989

[IAUC-4830] "International Astronomical Union Circular No. 4830", 11 August 1989

[IAUC-5052] "International Astronomical Union Circular No. 5052", 16 July 1990

[IAUC-6162] "International Astronomical Union Circular No. 6162", 14 April 1995

[IAUC-7171] "International Astronomical Union Circular No. 7171", 18 May 1999

[Ingersoll-1971] Ingersoll, A.P., Leovy, C.B., "The Atmosphere of Mars and Venus", Annual Review of Astronomy and Astrophysics, 9, 1971, 147–182

[Ingersoll-1975] Ingersoll, A.P., et al., "Pioneer 11 Infrared Radiometer Experiment: The Global Heat Balance of Jupiter", Science, 188, 1975, 472–473

[Ingersoll-1976a] Ingersoll, A.P., "The Atmosphere of Jupiter", Solar System Reviews, 18, 1976, 603–639

[Ingersoll-1976b] Ingersoll, A.P., "The Meteorology of Jupiter", Scientific American, March 1976, 46–56

[Ingersoll-1980] Ingersoll, A.P., "Pioneer Saturn Infrared Radiometer: Preliminary Results", Science, 207, 1980, 439–443

[Ingersoll-1981] Ingersoll, A.P., "Jupiter and Saturn", Scientific American, December 1981, 90–111

[Ingersoll-1987] Ingersoll, A.P., "Uranus", Scientific American, January 1987, 38–45

[Ip-1980] Ip, W.-H., "New Progress in the Physical Studies of the Planetary Rings", Space Science Reviews, 26, 1980, 97–109

[Israel-1974] Israel, G., et al., "Testing Gravitation Theories by Means of Heliocentric Probe". In: "Proceedings of the International School of Physics, Enrico Fermi, Course LVI", New York, Academic Press, 1974

[Istomin-1979] Istomin, V.G., Grechev, K.V., Kochnev, V.A., "Venera 11 and 12 Mass Spectrometry of the Lower Venus Atmosphere", Soviet Astronomy Letters, 5, 1979, 113–115

[Istomin-1982] Istomin, V.G., Grechnev, K.V., Kochnev, V.A., "Mass Spectrometry on the Venera 13 and Venera 14 Landers: Preliminary Results", Soviet Astronomy Letters, 8, 1982, 211–215

[Ivanov-1992] Ivanov, M.A., "Venera 13 and 14 Landing Sites: Geology from Magellan Data", paper presented at the XXIII Lunar and Planetary Science Conference, Houston, 1992

[Jackson-1990] Jackson, B.V., Benensohn, R.M., "The Helios Spacecraft Zodiacal Light Photometers Used for Comet Observations and Views of the Comet West Bow Shock", Earth Moon and Planets, 48, 1990, 139–163

[Jastrow-1968] Jastrow, R., "The Planet Venus: Information Received from Mariner V and Venera 4 is Compared", Science, 160, 1968, 1403–1410

[Jewitt-1979] Jewitt, D.C., Danielson, G.E., Synnott, S.P., "Discovery of a New Jupiter Satellite", Science, 206, 1979, 951

[Jewitt-2006] Jewitt, D., Sheppard, S.S., Kleyna, J., "The Strangest Satellites in the Solar System", Scientific American, August 2006, 23–29

[Johnson-1982] Johnson, T.V., Soderblom, L.A., "The Moons of Saturn", Scientific American, January 1982, 100–117

[Johnson-1983] Johnson, T.V., Soderblom, L.A., "Io", Scientific American, December 1983, 56–67

[Johnson-1987] Johnson, T.V., Brown, R.H., Soderblom, L.A., "The Moons of Uranus", Scientific American, April 1987, 48–60

[Jokipii-1995] Jokipii, J.R., McDonald, F.B., "Quest for the Limits of the Heliosphere", Scientific American, April 1995, 59–63

[JPL-1975a] "Mariner Venus-Mercury 1973 Project Final Report: Volume II Extended Mission – Mercury II and III Encounters", Pasadena, JPL Technical Memorandum 33–734, 1 December 1975, 9

[JPL-1975b] ibid., 9–21

[JPL-1976a] "Mariner Venus-Mercury 1973 Project Final Report: Volume I Venus and Mercury I Encounters", Pasadena, JPL Technical Memorandum 33–734, 15 September 1976, 5

[JPL-1976b] ibid., 23

[Judge-1974] Judge, D.L., Carlson, R.W., "Pioneer 10 Observations of the Ultraviolet Glow in the Vicinity of Jupiter", Science, 183, 1974, 317–318

[Judge-1980] Judge, D.L., Wu, F.-M., "Ultraviolet Photometer Observations of the Saturnian System", Science, 207, 1980, 431–434

[Kaiser-1980] Kaiser, M.L., et al., "Voyager Detection of Nonthermal Radio Emission from Saturn", Science, 209, 1980, 1238–1240

[Kayser-1984a] Kayser, S.E., Barnes, A., Mihalov, J.D., "The Far Reaches of the Solar Wind: Pioneer 10 and Pioneer 11 Plasma Results", The Astrophysical Journal, 285, 1984, 339–346

[Kayser-1984b] Kayser, S., Stone, R., "The HELIOS Radio Astronomy Experiment". In: Porsche, H. (ed.) "10 Jahre Helios – 10 Years Helios", 111–114 (in German and English)

[Keating-1979] Keating, G.M., Tolson, R.H., Hinson, E.W., "Venus Thermosphere and Exosphere: First Satellite Drag Measurements of an Extraterrestrial Atmosphere", Science, 203, 1979, 772–774

[Keenan-1975] Keenan, D.W., "The Galactic Orbit of Pioneer 10", Bulletin of the American Astronomical Society, 7, 1975, 466

[Kehr-1984] Kehr, J., Hiendlmeier, G., "HELIOS-Bodenbetrieb". In: Porsche, H. (ed.) "10 Jahre Helios – 10 Years Helios", 183–188 (in German and English)

[Kellogg-1984] Kellogg, P.J., "Evidence Concerning the Generation Mechanism of Solar Type III Radio Bursts". In: Porsche, H. (ed.) "10 Jahre Helios – 10 Years Helios", 106–110

[Kelly Beatty-1994] Kelly Beatty, J., Levy, D.H., "Awaiting the Crash", Sky & Telescope, January 1994, 40–44

[Kelly Beatty-1995] Kelly Beatty, J., "Hubble's Worlds", Sky & Telescope, February 1995, 20–25

[Kelly Beatty-1996] Kelly Beatty, J., "Rings of Revelation", Sky & Telescope, August 1996, 30–33

[Kelly Beatty-2001] Kelly Beatty, J., "Fade to Black", Air & Space, June–July 2001, 48–53

[Kemurdjian-1990] Kemurdjian, A.L., "From the Moon Rover to the Mars Rover", Planetary Report, July/August 1990, 4–11

[Kemurdjian-1992] Kemurdjian, A.L., et al., "Soviet Developments of Planet Rovers in Period of 1964–1990". In: "Missions, Technologies et Conception des Vehicules Mobiles Planetaires", Toulouse, Cépaduès, 1993

[Kennel-1977] Kennel, C.F., Coroniti, F.V., "Jupiter's Magnetosphere", Annual Review of Astronomy and Astrophysics, 15, 1977, 389–436

[Kerr-2005] Kerr, R.A., "Voyager 1 Crosses a New Frontier and May Save Itself from Termination", Science, 308, 2005, 1237–1238

[Kerzhanovich-1980] Kerzhanovich, V.V., et al., "Venera 11 and Venera 12: Preliminary Evaluations of Wind Velocity and Turbulence in the Atmosphere of Venus", The Moon and the Planets, 23, 1980, 261–270

[Kerzhanovich-1982] Kerzhanovich, V.V., et al., "Wind Velocities Estimated from Venera 13 and Venera 14 Doppler Measurements: Initial Results", Soviet Astronomy Letters, 8, 1982, 225–227

[Kerzhanovich-2003] Kerzhanovich, V., Pichkhadze, K., "Soviet Venera and Mars: First Entry Probe Trajectory Reconstruction and Science", paper presented at the International Workshop on Planetary Probe Atmospheric Entry and Descent Trajectory Analysis and Science, Lisbon, 6–9 October 2003

[Kinard-1974] Kinard, W.H., et al., "Interplanetary and Near-Jupiter Meteoroid Environment: Preliminary Results from the Meteoroid Detection Experiment", Science, 183, No. 4122, 25 January 1974, 321–322

[Kinoshita-1989] Kinoshita, J., "Neptune", Scientific American, November 1989, 82–85

[Kirkland-1998] Kirkland, L.E., Forney, P.B., Herr, K.C., "Mariner Mars 6/7 Infrared Spectra: new Calibrations and a Search for Water Ice Clouds", paper presented at the XXIX Lunar and Planetary Science Conference, Houston, 1998

[Klavetter-1989a] Klavetter, J.J., "Rotation of Hyperion. I – Observations", Astronomical Journal, 97, 1989, 570–579

[Klavetter-1989b] Klavetter, J.J., "Rotation of Hyperion. II – Dynamics", Astronomical Journal, 98, 1989, 1855–1874

[Klein-1976] Klein, H.P., et al., "The Viking Biological Investigation: Preliminary Results", Science, 194, 1976, 99–105

[Kliore-1965] Kliore, A., et al., "Occultation Experiment: Results of the First Direct Measurement of Mars's [sic] Atmosphere and Ionosphere", Science, 149, 1965, 1243–1248

[Kliore-1968] Kliore, A.J., et al., "Atmosphere of Venus as Observed by the Mariner V S-Band Radio Occultation Experiment", Astronomical Journal, 73, 1968, S21

[Kliore-1973] Kliore, A., "Radio Occultation Exploration of Mars". In: Proceedings of the Symposium on Exploration of the Planetary System, Torun, Poland, September 5–8, 1973, 295–316

[Kliore-1974a] Kliore, A., et al., "Preliminary Results on the Atmospheres of Io and Jupiter from the Pioneer 10 S-Band Occultation Experiment", Science, 183, 1974, 323–324

[Kliore-1974b] Kliore, A.J., et al., "The Atmospheres of Io and Jupiter from the Pioneer 10 S-Band Radio Occultation Experiment", Bulletin of the American Astronomical Society, 6, 1974, 388

[Kliore-1975] Kliore, A., et al., "Atmosphere of Jupiter from the Pioneer 11 S-Band Occultation Experiment: Preliminary Results", Science, 188, 1975, 474–476

[Kliore-1980] Kliore, A.J., et al., "Vertical Structure of the Ionosphere and Upper Neutral Atmosphere of Saturn from the Pioneer Radio Occultation", Science, 207, 1980, 446–449

[Knap-1977] Knap, P., "Mars, Zond", Letectví + Kosmonautika, 53, 1977, 791–792 (in Czech)

[Knollenberg-1979] Knollenberg, R.G., Hunten, D.M., "Clouds of Venus: Particle Size Distribution Measurements", Science, 203, 23 February 1979, 792–795

[Kohlhase-1986] Kohlhase, C.E., Frampton, R.V., Gerschultz, J.W:, "Towards Neptune", Spaceflight, January 1986, 10–15

[Kolcum-1963] Kolcum, E.H., "Mariner Reveals 800F Venus Temperature", Aviation Week & Space Technology, 4 March 1963, 30–31

[Kolosov-1985] Kolosov, M.A., Savich, N.A., Yakovlev, O.I., "Spacecraft Radio-physical Ivestigations of the Sun and Planets". In Kotelnikov, V.A. (ed.), "Problems of Modern Radio Engineering and Electronics, Moscow", Nauka, 1985, 64–102

[Koppes-1982a] Koppes, C.R., "JPL and the American Space Program", Yale University Press, 1982, 106

[Koppes-1982b] ibid., 126

[Koppes-1982c] ibid., 126–128

[Koppes-1982d] ibid., 165–166

[Koppes-1982e] ibid., 171

[Koppes-1982f] ibid., 169–170

[Koppes-1982g] ibid., 193–196

[Koppes-1982h] ibid., 196–197

[Koppes-1982i] ibid., 197–200

[Koppes-1982j] ibid., 200–202

[Koppes-1982k] ibid., 218–221

[Koppes-1982l] ibid., 221–226

[Koppes-1982m] ibid., 226–232

[Kovaly-1970] Kovaly, J.J., "Radar Techniques for Planetary Mapping with Orbiting Vehicle", paper presented at the Third Conference on Planetology and Space Mission Planning, New York, October 1970

[Kovtunenko-1992] Kovtunenko, V., et al., "Prospects for Using Mobile Vehicles in Missions to Mars and Other Planets". In: "Missions, Technologies et Conception des Vehicules Mobiles Planetaires", Toulouse, Cépaduès, 1993

[Kraemer-2000a] Kraemer, R. S., "Beyond the Moon: A Golden Age of Planetary Exploration 1971–1978", Washington, Smithsonian Institution Press, 2000, 44–61

[Kraemer-2000b] ibid., 62–64

[Kraemer-2000c] ibid., 90–118

[Kraemer-2000d] ibid., 80–89

[Kraemer-2000e] ibid., 119–162

[Kraemer-2000f] ibid., 130

[Kraemer-2000g] ibid., 146

[Kraemer-2000h] ibid., 202–220

[Kraemer-2000i] ibid., 163–182

[Kraemer-2000j] ibid., 186–188

[Krimigis-2003] Krimigis, S.M., et al., "Voyager 1 Exited the Solar Wind at a Distance of ~85 AU from the Sun", Nature, 426, 2003, 45–48

[Kronk-1984a] Kronk, G.W., "Comets: A Descriptive Catalog", Hillside, Henslow, 1984, 227–228

[Kronk-1984b] ibid., 274–275

[Kronk-1999] Kronk, G.W., "Cometography – A Catalog of Comets. Volume 1: Ancient–1799", Cambridge University Press, 1999, 447–451

[Ksanfomaliti-1976] Ksanfomaliti, L.V., "Infrared Radiometry and Photometry with Venera 9 and 10", Soviet Astronomy Letters, 2, 1976, 29–31

[Ksanfomaliti-1979] Ksanfomaliti, L.V., et al., "Electrical Discharges in the Atmosphere of Venus", Soviet Astronomy Letters, 5, 1979, 122–126

[Ksanfomaliti-1982a] Ksanfomaliti, L.V., "The Low Frequency Electromagnetic Field in the Venus Atmosphere: Evidence from Venera 13 and Venera 14", Soviet Astronomy Letters, 8, 1982, 230–232

[Ksanfomaliti-1982b] Ksanfomaliti, L.V., et al., "Microseisms at the Venera 13 and Venera 14 Landing Sites", Soviet Astronomy Letters, 8, 1982, 241–242

[Ksanfomaliti-1982c] Ksanfomaliti, L.V., et al., "Acoustic Measurements of the Wind Velocity at the Venera 13 and Venera 14 Landing Sites", Soviet Astronomy Letters, 8, 1982, 227–229

[Kumar-1978] Kumar, S., Broadfoot, A.L., "Evidence from Mariner 10 of Solar Wind Flux Depletion at High Ecliptic Latitudes", Astronomy and Astrophysics, 69, 1978, L5–L8

[Kumar-1979] Kumar, S., et al., "The Lyman-Alpha Observations of Comet Kohoutek from Mariner 10", The Astrophysical Journal, 232, 1979, 616–623

[Kunow-1984] Kunow, H., Wibberenz, G., "Die Schnellen Individualisten im Sonnensystem". In: Porsche, H. (ed.) "10 Jahre Helios – 10 Years Helios", 124–148 (in German and English)

[Kurt-1971] Kurt, V.G., "Results of Astronomical Studies in the Far UV Region". In: Labuhn F. and Lüst R. (eds.), "Proceedings from IAU Symposium no. 41: New Techniques in Space Astronomy", 1971, 219–232

[Kutzer-1984] Kutzer, A., "Die Helios-Missionen". In: Porsche, H. (ed.) "10 Jahre Helios – 10 Years Helios", 38–47 (in German and English)

[Lamy-1980] Lamy, P.L., Mauron, N., "The New Satellite Dione B and Outer Ring of Saturn", Bulletin of the American Astronomical Society, 12, 1980, 728–729

[Landgraf-2002] Landgraf, M., et al., "Origins of Solar System Dust Beyond Jupiter", Arxiv astro–ph/0201291 preprint

[Lane-1982] Lane, A.L., et al., "Photopolarimetry from Voyager 2: Preliminary Results on Saturn, Titan and the Rings", Science, 215, 1982, 537–543

[Lane-1986] Lane, A.L., et al., "Photometry from Voyager 2: Initial Results from the Uranian Atmosphere, Satellites, and Rings", Science, 233, 1986, 65–70

[Lane-1989] Lane, A.L., et al., "Photometry from Voyager 2: Initial Results from the Neptunian Atmosphere, Satellites, and Rings", Science, 246, 1989, 1450–1454

[Langereux-1971] Langereux, P., "Le Satellite 'Eole' Sera Lancé le 18 Aout" (The 'Eole' Satellite Will Be Launched on 18 August), Air et Cosmos, No. 398, 24 July 1971, 12–13 (in French)

[Langereux-1972] Langereux, P., "L'Experience Eole Est un Très Grand Succès" (The Eole Experiment is a Big Success), Air et Cosmos, No. 425, 4 March 1972, 16–17 (in French)

[Langereux-1974] Langereux, P., "Hélios Vers le Soleil" (Helios Toward the Sun), Air et Cosmos, 30 November 1974, 42–43 (in French)

[Lantranov-1996] Lantranov, K., "Na Mars! – Chast' 2" (To Mars! – Part 2), Novosti Kosmonavtiki, No. 21, 1996, page unknown (in Russian)

[Lantranov-1999] Lantranov, K., Hendrickx, B., "Mars-69: the Forgotten Mission to the Red Planet", Quest, 7, No. 2, 1999, 26–31

[Lardier-1992a] Lardier, C., "L'Astronautique Soviétique" (Soviet Astronautics), Paris, Armand Colin, 1992, 116–117 (in French)

[Lardier-1992b] ibid., 117–118

[Lardier-1992c] ibid., 271

[Lardier-1992d] ibid., 272

[Lardier-1992e] ibid., 272–273

[Lardier-1992f] ibid., 273

[Lardier-1992g] ibid., 278

[Lardier-1992h] ibid., 274

[Lawler-2005] Lawler, A., "NASA Plans to Turn Off Several Satellites", Science, 307, 2005, 1541

[Lazarus-1970] Lazarus, A.J., Oglivie, K.W., Burlaga, L.F., "Interplanetary Shock Observations by Mariner 5 and Explorer 34", Solar Physics, 13, 1970, 232–239

[LeBorgne-1983] Le Borgne, J.F., "Interpretation of the Event in the Plasma Tail of Comet Bradfield 1979 X on 1980 February 6", Astronomy and Astrophysics, 123, 1983, 25–28

[Leighton-1965] Leighton, R.B., et al., "Mariner IV Photography of Mars: Initial Results", Science, 149, 1965, 627–630

[Leighton-1971] Leighton, R.B., et al., "Mariner 6 and 7 Television Pictures: Preliminary Analysis", In: Sagan, C., Owen, T.C., Smith, H.J., "Proceedings from IAU Symposium no. 40: Planetary Atmospheres", 1971, 259–294

[Leinert-1978] Leinert, C., et al., "Search for a Dust Free Zone around the Sun from the Helios 1 Solar Probe", Astronomy and Astrophysics, 64, 1978, 119–122

[Leinert-1980] Leinert, C., et al., "The Plane of Symmetry of Interplanetary Dust in the Inner Solar System", Astronomy and Astrophysics, 82, 1980, 328–336

[Leinert-1981] Leinert, C., et al., "The Zodiacal Light from 1.0 to 0.3 A.U. as Observed by the Helios Space Probes", Astronomy and Astrophysics, 103, 1981, 177–188

[Leinert-1984] Leinert, C., Pitz, E., Link, H., "Zodiakallicht – Ein Abbild der Interplanetaren Staubwolke". In: Porsche, H. (ed.) "10 Jahre Helios – 10 Years Helios", Oberpfaffenhofen, DFVLR, 1984, 50–57 (in German and English)

[Leinert-1989] Leinert, C., Pitz, E., "Zodiacal Light Observed by Helios throughout Solar Cycle No.21: Stable Dust and Varying Plasma", Astronomy and Astrophysics, 210, 1989, 399–402

[Leovy-1977] Leovy, C.B., "The Atmosphere of Mars", Scientific American, July 1977, 34–43

[LePage-1993] LePage, A. J., "The Mystery of Zond 2", Journal of the British Interplanetary Society, 46, 1993, 401–404

[Levin-1976] Levin, G.V., Straat, P.A., "Viking Labeled Release Biology Experiment: Interim Results", Science, 194, 1976, 1322–1329

[Levy-1969] Levy, G.S., et al., "Pioneer 6: Measurement of Transient Faraday Rotation Phenomena Observed during Solar Occultation", Science, 166, 1969, 596–598

[Lewis-1960] Lewis, C., "Pioneer V Provides New Scientific Data", Aviation Week, 9 May 1960, 32–33

[Lindal-1990] Lindal, G.F., et al., "The Atmosphere of Neptune – Results of Radio Occultation Measurements with the Voyager 2 Spacecraft", Geophysical Research Letters, 17, September 1990, 1733–1736

[Linick-1991] Linick, S.H., Holberg, J.B., "The Voyager Ultraviolet Spectrometers – Astrophysical Observations from the Outer Solar System", Journal of the British Interplanetary Society, 44, 1991, 513–520

[Lissov-2004] Lissov, I., posting to the FPSpace discussion group, 29 May 2004

[Lockheed-1967] "Starlet/Starlite System Technical Description", Lockheed Missiles & Space Company, December 1967

[Lorenz-2006] Lorenz, R.D., "Spin of Planetary Probes in Atmospheric Flight", Journal of the British Interplanetary Society, 59, 2006, 273–282

[Lozier-2005] Lozier, D., personal correspondence with the author, 11 January 2005

[Luhmann-1994] Luhmann, J.G., Pollack, J.B., Colin, L., "The Pioneer Mission to Venus", Scientific American, April 1994, 90–97

[Malin-2005] Malin, M.C., "Hidden in Plain Sight: Finding Martian Landers", Sky & Telescope, July 2005, 42–46

[Marcus-2006] Marcus, G., "The Pioneer Rocket", Quest, 13 No.4, 2006, 26–30

[Mari-1962] Mari, D., "Il Monitore Solare" (The Solar Monitor), Oltre il Cielo, 100, 16 March 1962, 550 (in Italian)

[Mariani-1984] Mariani, F., et al., "Rome/GSFC Magnetic Field Experiment: A Summary of Results". In: Porsche, H. (ed.) "10 Jahre Helios – 10 Years Helios", 90–99 (in German and English)

[Mariani-2005] Mariani, F., personal correspondence with the author, 29 April 2005

[Marov-1974] Marov, M.Ya., "Vénus", La Recherche, November 1974, 927–939 (in French)

[Marov-1976] Marov, M.Ya., Lebedev, V.N., Lystsev, V.E., "Preliminary Estimates of the Aerosol Component in the Atmosphere of Venus", Soviet Astronomy Letters, 2, 1976, 98–100

[Marov-1978] Marov, M.Ya., "Results of Venus Missions", Annual Review of Astronomy and Astrophysics, 16, 1978, 141–169

[Massey-2006] Massey, E.B., Personal communication with the author, 3 December 2006

[Matrossov-2004] Matrossov, S., Personal communication with the author, 31 October and 26 December 2004

[Matrossov-2006] Matrossov, S., Personal communication with the author, 19 November 2006

[Mazets-1979] Mazcts, E.P., et al., "Venera 11 and 12 Observations of Gamma-Ray Bursts – The Cone Experiment", Soviet Astronomy Letters, 5, 1979, 87–90

[Mazets-1981] Mazets, E.P., et al., "Catalog of Cosmic Gamma-Ray Bursts from the Konus Experiment Data.", Astrophysics and Space Science, 80, 1981, 3–83 (part I and II), 85–117 (part III), 119–143 (part IV)

[Mazur-1978] Mazur, P., et al, "Biological Implications of the Viking Mission to Mars", Space Science Reviews, 22, 1978, 3–34

[McComas-2004] McComas, D., et al., "The Interstellar Boundary Explorer (IBEX)". In: Florinski, V., Pogorelov, N.V., Zank, G.P. (eds.), "Physics of the Outer Heliosphere: Third International IGPP Conference, Riverside, CA", AIP, 2004

[McElroy-1975] McElroy, M.B., Yung, Y.L., "The Atmosphere and Ionosphere of Io", The Astrophysical Journal, 196, 1975, 227–250

[McElroy-1976] McElroy, M.B., "Composition and Structure of the Martian Upper Atmosphere: Analysis of Results from Viking", Science, 194, 1976, 1295–1298

[McKibben-1985] McKibben, R.B., Pyle, K.R., Simpson, J.A., "Changes in Radial Gradients of Low-Energy Cosmic Rays Between Solar Minimum and Maximum: Observations from 1 to 31 AU", The Astrophysical Journal, 289, 1985, L35–L39

[McLaughlin-1985] McLaughlin, W.I, Wolff, D.M., "Voyager at the Seventh Planet", Spaceflight, November 1985, 403–409

[McLaughlin-1986a] McLaughlin, W., "A Voyager Diary", Spaceflight, March 1986, 123–126

[McLaughlin-1986b] McLaughlin, W., "Cruising to Neptune", Spaceflight, November 1986, 405–406

[Medkeff-1998] Medkeff, J., "Ring-Division Discoverers", Sky & Telescope, July 1998, 12

[Melbourne-1976] Melbourne, W.G. "Navigation Between the Planets", Scientific American, June 1976, 58–74

[Melin-1960] Melin, M., "Pioneer V and the Scale of the Solar System", Sky & Telescope, December 1960, 337. (Reprinted in: Page, T, Page, L.W. (ed.), "Wanderers in the Sky", New York, Macmillan, 1965, 135–136)

[Merat-1974] Merat, P., Pecker, J.-C., Vigier, J.-P., "Possible Interpretation of an Anomalous Redshift Observed on the 2292 MHz Line Emitted by Pioneer-6 in the Close Vicinity of the Solar Limb", Astronomy & Astrophysics, 30, 1974, 167–174

[Mihalov-1975] Mihalov, J.D., "Pioneer 11 Encounter: Preliminary Results from the Ames Research Center Plasma Analyzer Experiment", Science, 188, 1975, 448–451

[Mihalov-1987] Mihalov, J.D., et al., "Observation by Pioneer 7 of He(+) in the distant coma of Halley's comet", Icarus, 71, 1987, 192–197

[Miller-1953] On the Miller-Urey experiment see: Miller, S.L., "A Production of Amino Acids Under Possible Primitive Earth Conditions", Science, 117, 1953, 528–529

[Minear-1978] Minear, J., Friedman, L., "Future Exploration of Mars", Astronautics and Aeronautics, April 1978, 18–27

[Miner-1988] Miner, E.D., Stone, E.C., "Voyager at Uranus", Journal of the British Interplanetary Society, 41, 1988, 49–62

[Mitchell-2004a] "Soviet Telemetry Systems", D.P. Mitchell internet website

[Mitchell-2004b] "Remote Scientific Sensors", D.P. Mitchell internet website

[Mitchell-2004c] "Inventing the Interplanetary Probe", D.P. Mitchell internet website

[Mitchell-2004d] "Soviet Space Cameras", D.P. Mitchell internet website

[Mitchell-2004e] "Plumbing the Atmosphere of Venus", D.P. Mitchell internet website

[Moog-1973] Moog, R.D., et al., "Qualification Flight Tests of the Viking Decelerator System", Paper AIAA 73–457

[Moore-1974] Moore, J., "Uranus Science Planning". In: Proceedings of the Outer Planet Probe Technology Workshop, May 21–23, 1974, NASA CR–137543

[Morabito-1979] Morabito, L.A., et al., "Discovery of Currently Active Extra-terrestrial Volcanism", Science, 204, 1979, 972

[Moroz-1979] Moroz, V.I., et al., "Venera 11 and 12 Descent-Probe Spectro-photometry: the Venus Dayside Sky Spectrum", Soviet Astronomy Letters, 5, 1979, 118–121

[Moroz-1982] Moroz, V.I., et al., "The Venera 14 and Venera 14 Spectrophotometry Experiments", Soviet Astronomy Letters, 8, 1982, 219–223

[Morrison-1980a] Morrison, D., Samz, J., "Voyage to Jupiter", Washington, NASA, 1980, 25

[Morrison-1980b] ibid., 26–31

[Morrison-1980c] ibid., 28–29

[Morrison-1980d] ibid., 33–45

[Morrison-1980e] ibid., 47–48

[Morrison-1980f] ibid., 50–51

[Morrison-1980g] ibid., 50–56

[Morrison-1980h] ibid., 56–61

[Morrison-1980i] ibid., 63–86

[Morrison-1980j] ibid., 86–91

[Morrison-1980k] ibid., 93–115

[Morrison-1982a] Morrison, D., "Voyages to Saturn", Washington, NASA, 1982, 50–93

[Morrison-1982b] ibid., 96–135

[Morton-2000] Morton, O., "Mars Air: How to Built the First Extraterrestrial Airplane", Air & Space, December 1999–January 2000, 34–42

[Mudgway-2001a] Mudgway, D.J., "Uplink-Downlink A History of the Deep Space Network 1957–1997", Washington, NASA, 2001, 114–116

[Mudgway-2001b] ibid., 212–215

[Mudgway-2001c] ibid., 330–331

[Mudgway-2001d] ibid., 47

[Mudgway-2001e] ibid., 87–91

[Mudgway-2001f] ibid., 206–207

[Mudgway-2001g] ibid., 104–105

[Mudgway-2001h] ibid., 207–211

[Mukhin-1982] Mukhin, L.M., et al., "Venera 13 and Venera 14 Gas-Chromatography Analysis of the Venus Atmosphere Composition", Soviet Astronomy Letters, 8, 1982, 216–218

[Murray-1966] Murray, B.C., Davies, M.E., "A Comparison of U.S. And Soviet Efforts to Explore Mars", Science, 151, 1966, 945–954

[Murray-1973] Murray, B.C., "Mars from Mariner 9", Scientific American, January 1973, 48–63

[Murray-1974] Murray, B.C., et al., "Venus: Atmospheric Motion and Structure from Mariner 10 Pictures", Science, 183, 1974, 1307–1315

[Murray-1975] Murray, B.C., "Mercury", Scientific American, September 1975, 58–68

[Murray-1989a] Murray, B., "Journey into Space", New York, W.W. Norton & C., 1989, 50–51

[Murray-1989b] ibid. 55

[Murray-1989c] ibid. 56

[Murray-1989d] ibid., 99–100

[Murray-1989e] ibid., 147–148

[Murray-1989f] ibid., 153–155

[Murray-1989g] ibid., 165–166

[Murrow-1968] Murrow, H.N., McFall, J.C. Jr., "Summary of Experimental Results Obtained from the NASA Planetary Entry Parachute Program", Paper AIAA 68–934

[Mutch-1976a] Mutch, T.A., et al., "The Surface of Mars: The View from the Viking 1 Lander", Science, 193, 1976, 791–801

[Mutch-1976b] Mutch, T.A., et al., "The Surface of Mars: The View from the Viking 2 Lander", Science, 194, 1976, 1277–1283

[Mutch-1978] Mutch, T.A., Jones, K.L., "The Martian Landscape", Washington, NASA, 1978

[MVM-14] "Mariner Venus/Mercury 1973 Status Bulletin No.14", 23 January 1974

[MVM-15] "Mariner Venus/Mercury 1973 Status Bulletin No.15", 1 February 1974

[MVM-17] "Mariner Venus/Mercury 1973 Status Bulletin No.17", 5 February 1974

[MVM-18] "Mariner Venus/Mercury 1973 Status Bulletin No.18", 6 February 1974

[MVM-19b] "Mariner Venus/Mercury 1973 Status Bulletin No. 19 Part 2", 7 February 1974

[MVM-20] "Mariner Venus/Mercury 1973 Status Bulletin No. 20", 19 February 1974

[MVM-21] "Mariner Venus/Mercury 1973 Status Bulletin No. 21", 15 March 1974

[MVM-22] "Mariner Venus/Mercury 1973 Status Bulletin No. 22", 18 March 1974

[MVM-23] "Mariner Venus/Mercury 1973 Status Bulletin No. 23", 25 March 1974

[MVM-24] "Mariner Venus/Mercury 1973 Status Bulletin No.24", 26 March 1974

[MVM-28] "Mariner Venus/Mercury 1973 Status Bulletin No.28", 4 April 1974

[MVM-29] "Mariner Venus/Mercury 1973 Status Bulletin No.29", 3 May 1974

[MVM-30] "Mariner Venus/Mercury 1973 Status Bulletin No.30", 8 May 1974

[MVM-31] "Mariner Venus/Mercury 1973 Status Bulletin No.31", 15 May 1974

[MVM-32] "Mariner Venus/Mercury 1973 Status Bulletin No.32", 7 June 1974
[MVM-33] "Mariner Venus/Mercury 1973 Status Bulletin No.33", 3 July 1974
[MVM-34] "Mariner Venus/Mercury 1973 Status Bulletin No.34", 28 August 1974
[MVM-36] "Mariner Venus/Mercury 1973 Status Bulletin No.36", 23 September 1974
[MVM-37] "Mariner Venus/Mercury 1973 Status Bulletin No.37", 12 March 1975
[MVM-38] "Mariner Venus/Mercury 1973 Status Bulletin No.38", 3 April 1975
[NASA-1963] "Mariner II Reports", NASA Facts, NF B–4–63, 1963
[NASA-1964] "Mariner 4 Press Kit", Washington, NASA, 29 October 1964
[NASA-1965a] "Mariner-Venus 1962 Final Project Report", Washington, NASA, 1965, 11–15
[NASA-1965b] ibid., 25–39
[NASA-1965c] ibid., 69
[NASA-1965d] ibid., 313–337
[NASA-1965e] ibid., 87–120
[NASA-1965f] "Asteroid Belt and Jupiter Flyby Mission Study – Final Report", NASA CR–64621, 28 February 1965
[NASA-1966] "A Study of Jupiter Flyby Missions – Final Technical Report", NASA CR–76461, 17 May 1966
[NASA-1967a] "Mariner-Mars 1964 Final Project Report", Washington, NASA, 1967
[NASA-1967b] "Summary of the Voyager Program", NASA Office of Space Science and Applications, January 1967
[NASA-1969a] "Mariner '69 Press Kit", Washington, NASA, 18 July 1969
[NASA-1969b] "Mariner '69 Results", Washington, NASA, 11 September 1969
[NASA-1969c] "Mariner Six and Seven Mission Report", Washington, NASA MR–6, 29 October 1969
[NASA-1971a] "Pioneers 6 to 9", NASA Ames Research Center, Educational Data Sheet 503, July 1971
[NASA-1971b] "Mariner Mars 1971 Press Kit", NASA, 30 April 1971
[NASA-1971c] "Pioneer H Jupiter Swingby Out of the Ecliptic Mission Study", NASA TM–108108, 1971
[NASA-1974a] "Pioneer Outer Planets Orbiter", NASA TM–108622, December 1974
[NASA-1974b] "Pioneer Mars Mission Study: Executive Summary", NASA TM–108688, August 1974
[NASA-1975] "Viking Press Kit", Washington, NASA, 1975
[NASA-1986] "Voyager 1986 Press Kit", Washington, NASA, 1986
[NASA-1989] "Voyager 2 Neptune Encounter Press Kit", Washington, NASA, 1989
[Navarro-Gonzáles-2006] Navarro-Gonzáles, R., et al., "The Limitations on Organic Detection in Mars-Like Soils by Thermal Volatilization-Gas Chromatography-MS and Their Implications for the Viking Results", Proceedings of the National Academy of Sciences of the United States of America, 103, 2006, 16089–16094
[Ness-1970] Ness, N.E., "Magnetometers for Space Research", Space Science Review, 11, 1970, 459–554

[Ness-1979a] Ness, N.F., "The Magnetic Fields of Mercury, Mars and the Moon", Annual Review of Earth and Planetary Sciences, 1979, 249–288

[Ness-1979b] Ness, N.F., et al., "Magnetic Field Studies at Jupiter by Voyager 1: Preliminary Results", Science, 204, 1979, 982–987

[Ness-1986] Ness, N.F., et al., "Magnetic Fields at Uranus", Science, 233, 1986, 85–89

[Ness-1989] Ness, N.F., et al., "Magnetic Fields at Neptune", Science, 246, 1989, 1473–1478

[Neubauer-1984] Neubauer, F.M., Musmann, G., Dehmel, G., "Ergebnisse der Magnetfeld-Experimente E2 und E4 an Bord von Helios 1 und Helios 2". In: Porsche, H. (ed.) "10 Jahre Helios – 10 Years Helios", 80–89 (in German and English)

[Neugebauer-1962] Neugebauer, M., Snyder, C.W., "Solar Plasma Experiment", Science, 138, 1962, 1095–1097

[Neugabauer-1971] Neugebauer, G., et al., "Mariner 1969 Infrared Radiometer Results: Temperatures and Thermal Properties of the Martian Surface", The Astronomical Journal, 76, 1971, 719–728

[Neugebauer-1997] Neugebauer, M., "Pioneers of Space Physics: A Career in the Solar Wind", Journal of Geophysical Research, 102, 1997, 26887–26894

[Neumann-1966] Neumann, T. W., "System Design of Automated Laboratory Payloads for Planetary Research", paper presented at the XVII International Astronautical Congress, Madrid, 1966

[NewScientist-1974] "Mercury's Moon that Wasn't", New Scientist, 5 September 1974, 602

[Nicholson-1996] Nicholson, P.D., et al., "Observation of Saturn's Ring-Plane Crossings in August and November 1995", Science, 272, 1996, 509–515

[Niel-1976] Niel, M., et al., "The French-Russian 3 Satellite Gamma-Burst Experiment", Astrophysics and Space Science, 42, 1976, 99–102

[Nieto-2001] Nieto, M.M., et al., "The Anomalous Trajectories of the Pioneer Spacecraft", Arxiv hep–ph/0110373 preprint

[Nieto-2005] Nieto. M.M., Turyshev, S.G., Anderson, J.D., "Directly Measured Limit on the Interplanetary Matter Density from Pioneer 10 and 11", Arxiv astro–ph/0501626 preprint

[Nicks-1985a] Nicks, O.W., "Far Travelers: The Exploring Machines", Washington, NASA, 1985, 3–5

[Nicks-1985b] ibid., 33–40

[Nicks-1985c] ibid., 40–46

[Nicks-1985d] ibid., 46–47

[Nicks-1985e] ibid., 47–49

[NSSDC-2004a] NASA NSSDC Internet site, Venera 2 and 3 proton flux data

[NSSDC-2004b] NASA NSSDC Internet site, Venera 13 and 14 proton flux data

[Null-1967] Null, G.W., "A Solution for the Sun-Mars Mass Ratio Using Mariner IV Doppler Tracking Data", The Astronomical Journal, 27, 1967, 1292–1298

[Null-1976] Null, G.W., "The Gravity Field of Jupiter and its Satellites from Pioneer 10 and Pioneer 11 Tracking Data", The Astronomical Journal, 81, 1976, 1153–1161

[Null-1981] Null, G.W., et al., "Saturn Gravity Results Obtained from Pioneer 11 Tracking Data and Earth-Based Saturn Satellite Data", The Astronomical Journal, 86, 1981, 454–468

[Oberg-1981] Oberg, J.E., "Red Star in Orbit", New York, Random House, 1981, 39–49

[Opik-1950] Opik, E.J., "Mars and the Asteroids", Irish Astronomical Journal, 1, 1950, 22–24

[Opik-1951] Opik, E.J., "Collision Probabilities with the Planets", Proceedings Irish Academy, 54A, 1951, 165–199

[Owen-1982] Owen, T., "Titan", Scientific American, February 1982, 98–109

[Owen-1987] Owen, W.M., Synnott, S.P., "Orbits of the Ten Small Satellites of Uranus", The Astronomical Journal, 93, 1987, 1268–1271

[Oyama-1979] Oyama, V.I., et al., "Venus Lower Atmospheric Composition: Analysis by Gas Chromatography", Science, 203, 1979, 802–805

[Pandey-1987] Pandey, A.K., Mahra, H.S., "Possible Ring System of Neptune", Earth, Moon, and Planets, 37, 1987, 147–153

[Park-1964] Park, R.A., "Intercepting a Comet", Astronautics & Aeronautics, August 1964, 54–58

[Parker-1971] Parker, P.J., "'Grand Tour' Spacecraft Computer", Spaceflight, March 1971, 88

[Peale-1979] Peale, S.J., Cassen, P., Reynolds, R.T., "Melting of Io by Tidal Dissipation", Science, 203, 1979, 892–894

[Pearl-1973] Pearl, J., et al., "Results from the Infrared Spectroscopy Experiment on Mariner 9". In: Proceedings of the Symposium on Exploration of the Planetary System, Torun, Poland, September 5–8, 1973, 293–294

[Peebles-1981] Peebles, C., "The Martian Rovers", Spaceflight, 1981, 202–204

[Perminov-1999a] Perminov, V.G., "The Difficult Road to Mars: A Brief History of Mars Exploration in the Soviet Union", Washington, NASA, 1999, 7–8.

[Perminov-1999b] ibid., 8–10

[Perminov-1999c] ibid., 11–18

[Perminov-1999d] ibid., 19–33

[Perminov-1999e] ibid., 34–60

[Perminov-1999f] ibid., 61–66

[Perminov-1999g] ibid., 67–74

[Perminov-2001] Perminov, V, Morosov, V., "Proyekt Dolgozhivushey Veneryans-koy Stantsiy" (Project of the Long-Duration Venusian Probe), Novosti Kosmonavtiki, No. 8, 2001, page unknown (in Russian)

[Perminov-2002] Perminov, V., "Tak Poznavalis' Taini Veneri" (Thus were the secrets of Venus revealed), Novosti Kosmonavtiki, No. 12, 2002, page unknown (in Russian)

[Perminov-2004] Perminov, V., "Perviye Otechestvyenniye Radiolokatsionniye Karti Veneri" (The first national radar maps of Venus), Novosti Kosmonavtiki, No.9, 2004, page unknown (in Russian)

[Perminov-2005] Perminov, V., "Aerostaty v Nyeve Veneri: K 20-Letniyu Poleta AMS Vega" (Aerostats in the atmosphere of Venus: on the 20th Anniversary of the Flight of the Vega Probe), Novosti Kosmonavtiki, August 2005, 60–63

[Peters-1987] Peters, G.J., et al., "Pioneer 10 Observations of the Beta Cephei Stars Gamma Pegasi and Delta Ceti", The Astrophysical Journal, 314, 1987, 261–265

[Peterson-1993] Peterson, I., "Netwon's Clock: Chaos in the Solar System", New York, W.H. Freeman and Company, 199–214

[Pettengill-1979] Pettengill, G.H., et al., "Venus: Preliminary Topographic and Surface Imaging Results from the Pioneer Orbiter", Science, 205, 1979, 90–93

[Pettengill-1980] Pettengill, G.H., Campbell, D.B., Masursky, H., "The Surface of Venus", Scientific American, February 1980, 54–65

[Pioneer-2004] Pioneer Project internet website

[Pletschacher-1976] Pletschacher, P., "Helios B", Flug Revue, March 1976, page unknown (in German)

[Pollack-1967] Pollack, J.B., Sagan, C., "An Analysis of the Mariner 2 Microwave Observations of Venus", The Astronomical Journal, 150, 1967, 327–344

[Pollack-1975] Pollack, J.B., "Mars", Scientific American, September 1975, 106–117

[Pollack-1978] Pollack, J.B., "Multicolor Observations of Phobos with the Viking Lander Cameras: Evidence for a Carbonaceous Condritic Composition", Science, 199, 1978, 66–68

[Pollack-1981] Pollack, J.B., Cuzzi, J.N., "Rings in the Solar System", Scientific American, November 1981, 104–129

[Poqérusse-1978] Poquérusse, M., Steinberg, J.L., "First Results of the STEREO-5 Experiment: Evidence of Ionospheric Intensity Scintillation of Solar Radio Bursts at Decameter Wavelengths?", Astronomy and Astrophysics, 65, 1978, L23–L26

[Poqérusse-2004] Poquérusse, M., personal correspondence with the author, 16 June 2004

[Porsche-1968] Porsche, H., "Projekt Einer Deutsch-Amerikanischen Sonnensonde" (Project of a German-American Solar Probe), Mitteilungen der Astronomischen Gesellschaft, 25, 1968, 55–63

[Porsche-1980] Porsche, H., et al., "Proposal for an Interplanetary Mission to Sound the Outer Regions of the Solar Corona". In: Solar and interplanetary dynamics; Proceedings of the Symposium, Cambridge, Mass., August 27–31, 1979, Dordrecht, D. Reidel Publishing Co., 1980, 541–545

[Portree-2001] Portree D.S.F., "Humans to Mars: Fifty Years of Mission Planning 1950–2000", Washington, NASA, 2001, 1–4

[Powell-1984] Powell, J.W., "Thor–Able and Atlas–Able", Journal of the British Interplanetary Society, 37, 1984, 224

[Powell-2004] Powell, J.W., "State of Collapse", Spaceflight, September 2004, 361–365

[Powell-2005a] Powell, J.W., Personal communication with the author, 24 April 2005

[Powell-2005b] Powell, J.W., "The Forgotten Mission of Pioneer 5", Spaceflight, May 2005, 188–191

[Poynter-1984a] Poynter, M., Lane, A.L., "Voyager: The Story of a Space Mission", New York, Atheneum, 1984, 17–24

[Poynter-1984b] ibid., 32–35

[Poynter-1984c] ibid., 31

[Poynter-1984d] ibid., 62–63

[Prakash-1975] Prakash, A., Brice, N., "Magnetospheres of Earth and Jupiter after Pioneer 10", Space Science Reviews, 17, 1975, 823–835

[Pritchard-1974] Pritchard, E.B., Harrison, E.F., Moore, J.W., "Options for Mars Exploration", Astronautics and Aeronautics, February 1974, 46–56

[Quimby-1964] Quimby, F.H. (ed.), "Concepts for Detection of Extraterrestrial Life", Washington, 1964

[Ragent-1979] Ragent, B., Blamont, J., "Preliminary Results of the Pioneer Venus Nephelometer Experiment", Science, 203, 1979, 790–792

[Rausch-1967] Rausch, H., "Early Cutoff of Transmissions from Venus 4 Unexplained", Aviation Week & Space Technology, 6 November 1967, 17–18

[Reasenberg-1979] Reasenberg, R.D., et al., "Viking Relativity Experiment: Verification of Signal Retardation by Solar Gravity", The Astrophysical Journal. 234, 1979, L219–L221

[Reed-1978] Reed, D.R., "High-Flying Mini-Sniffer RPV: Mars Bound?", Astronautics and Aeronautics, June 1978, 26–39

[Reeves-2003] Reeves, R., posting to the FPSpace discussion group, 17 February 2003

[Richardson-2004] Richardson, J., Lorenz, R.D., McEwen, A., "Titan's Surface and Rotation: New Results from Voyager 1 Images", Icarus, 170, 2004, 113–124

[Robinson-1997] Robinson. M.S., Lucey, P.G., "Recalibrated Mariner 10 Color Mosaics: Implications for Mercurian Volcanism", Science, 275, 1997, 197–200

[Rubashkin-1997] Rubashkin, D:, "Who Killed the Grand Tour? A Case Study in the Politics of Funding Expensive Space Science", Journal of the British Interplanetary Society, 50, 1997, 177–184

[Rudd-1996] Rudd, R.P., Hall, J.C., Spradlin, G.L., "The Voyager Search for the Heliopause and Interstellar Space", paper presented at the First IAA Symposium on Realistic Near-Term Advanced Scientific Space Missions, Aosta, 25–27 June 1996

[Rudd-1997] Rudd, R.P., Hall, J.C., Spradlin, G.L., "The Voyager Interstellar Mission", Acta Astronautica, 40, 1997, 383–396

[Russo-2000] Russo, A., "The Definition of ESA's Scientific Programme for the 1980s". In: Krige, J., Russo, A., Sebesta, L. (eds.) , "A History of the European Space Agency 1958–1987", Vol. 2, 138–179

[Russell-1984] Russell, C.T., et al., "Interplanetary Magnetic Field Enhancements and Their Association with the Asteroid 2201 Oljato", Science, 226, 1984, 43–45

[Russell-1991] Russell, C.T., "Venus Lightning", Space Science Reviews, 55, 1991, 317–356

[Ryne-1993] Ryne, M.S., et al., "Navigation of Pioneer 12 During Atmospheric Reentry at Venus", Paper AAS 93–712

[Sagan-1976] Sagan, C., Salpeter, E.E., "Particles, Environments, and Possible Ecologies in the Jovian Atmosphere", The Astrophysical Journal Supplement Series, 32, 1976, 737–755

[Sagan-1997] Sagan, C., "Pale Blue Dot: A Vision of the Human Future in Space", New York, Random House, 1994

[Sagdeev-1982] Sagdeev, R.Z., Moroz, V.I., "Venera 13 and Venera 14", Soviet Astronomy Letters, 8, 1982, 209–211

[Sagdeev-1994a] Sagdeev, R.Z., "The Making of a Soviet Scientist", New York, John Wiley & Sons, 1994, 232–243

[Sagdeev-1994b] ibid., 237

[SAL-1979] "The Venera 11 and Venera 12 Experiments: First Results", Soviet Astronomy Letters, 5, 1979, 1–3

[Sandel-1979] Sandel, B.R., et al., "Extreme Ultraviolet Observations from Voyager 2 Encounter with Jupiter", Science, 206, 1979, 962–966

[Sandel-1982] Sandel, B.R., et al., "Extreme Ultraviolet Observations from the Voyager 2 Encounter with Saturn", Science, 215, 1982, 548–553

[Santer-1985] Santer, R., et al., "Photopolarimetric Analysis of the Martian Atmosphere by the Soviet Mars-5 Orbiter: I. White Clouds and Dust Veils", Astronomy & Astrophysics, 150, 1985, 217–228

[Santer-1986] Santer, R., et al., "Photopolarimetry of Martian Aerosols: II. Limb and Terminator Measurements", Astronomy & Astrophysics, 158, 1986, 247–258

[Schaefer-2000] Schaefer, B.E., Schaefer, M.W., "Nereid has a Complex Large-Amplitude Photometric Variability", Arxiv astro–ph/0005050 preprint

[Schneiderman-1963] Schneiderman, D., "Mariner II – An Example of an Attitude-Stabilized Space Vehicle", paper presented at the XIV International Astronautical Congress, Paris, 1963

[Schmidt-1980] Schmidt, K.D., Grün, E., "Orbital Elements of Micrometeoroids Detected by the Helios 1 Space Probe in the Inner Solar System". In: Proceedings of the Symposium on Solid Particles in the Solar System, Ottawa, Canada, August 27–30, 1979, 321–324

[Schmidt-1981] Schmidt, R., Schwingenschuh, K., "Magnetic Field Measurements on Board Venera 13, Venera 14 and Venera-Halley", in "Proceedings of the Alpbach Summer School, 29 July–7 August 1981", Noordwjik, ESA, 245–247

[Schubert-1981] Schubert, G., Covey, G., "The Atmosphere of Venus", Scientific American, January 1981, 66–75

[Schuerman-1977] Schuerman, D.W., Weinberg, J.L., Beeson, D.E., "The Decrease in Zodiacal Light with Heliocentric Distance during the Passage of Pioneer 10 through the Asteroid Belt", Bulletin of the American Astronomical Society, 9, 1977, 313

[Schurmeier-1970] Schurmeier, H.M., "The 1969 Mariner View of Mars", paper presented at the XXI International Astronautical Congress, Constance, 1970

[Schwenn-1984] Schwenn, R., Rosenbauer, H., "10 Jahre Sonnenwind Experiment auf Helios 1 und 2". In: Porsche, H. (ed.) "10 Jahre Helios – 10 Years Helios", 66–79 (in German and English)

[Sebesta-1997] Sebesta, L., "The Good, the Bad, the Ugly: U.S.–European Relations and the Decision to Build a European Launch Vehicle". In: Butrica, A.J. (ed.), "Beyond The Ionosphere: Fifty Years of Satellite Communication", Washington, NASA, 1997, 145–147

[Sebesta-2003] Sebesta, L., "Alleati Competitivi: Origini e Sviluppo della Coopera-zione Spaziale tra Europa e Stati Uniti 1957–1973" (Competitive Allies: Origins and Development of the Space Cooperation Between Europe and the United States 1957–1973), Rome, Laterza, 2003, 207–213 (in Italian)

[Seiff-1979] Seiff, A., et al., "Structure of the Atmosphere of Venus up to 110 Kilometers: Preliminary Results from the Four Pioneer Venus Entry Probes", Science, 203, 23 February 1979, 787–790

[Selivanov-1982] Selivanov, A.S., et al., "Evolution of the Venera 13 Imagery", Soviet Astronomy Letters, 8, 1982, 235–236

[Semenov-1996a] Semenov, Yu.P. (ed.), "Rakyetno-Kosmiceskaya Korporaziya 'Energiya' Imieni S. P. Korolyova 1946–1996" (Space and Rocketry Corporation 'Energiya' named after S.P. Korolyov 1946–1996), Moscow, RKK Energhiya, 1996, 140–141 (in Russian)

[Semenov-1996b] ibid., 141–142

[Semenov-1996c] ibid., 142–143

[Semenov-1996d] ibid., 143

[Semenov-1996e] ibid., 144

[Senske-1987] Senske, D.A., Head, J.W., "Characterization of the Venus Equatorial Highlands Using Pioneer Venus Imaging Mode Data", paper presented at the XVII Lunar and Planetary Science Conference, Houston, 1987

[Senske-1989] Senske, D.A., Head, J.W., "Geology of the Venus Equatorial Region from Pioneer Venus Radar Imaging", in: Lunar and Planetary Institute Abstracts for the Venus Geoscience Tutorial and Venus Geologic Mapping Workshop, 1989, 43–44

[Senske-1990] Senske, D.A., "Geology of the Venus Equatorial Region from Pioneer Venus Radar Imaging", Earth, Moon, and Planets, 50/51, 1990, 305–327

[Shayler-2000] Shayler, D.J., "Disasters and Accidents in Manned Spaceflight", Chichester, Springer–Praxis, 2000, 169–199

[Sheehan-1996a] Sheehan, W., "The Planet Mars: A History of Observations and Discovery", Tucson, The University of Arizona Press, 1996, 166–167

[Sheehan-1996b] ibid., 176–177

[Sheehan-1996c] ibid., 180–181

[Sheehan-1998] Sheehan, W., O'Meara, S.J., "Phillip Sidney Coolidge: Harvard's Romantic Explorer of the Skies", Sky & Telescope, April 1998, 71–75

[Sheehan-1999] Sheehan, W., Dobbins, T.A., "Charles Boyer and the Clouds of Venus", Sky & Telescope, June 1999, 56–60

[Shorthill-1976a] Shorthill, R.W., et al., "Physical Properties of the Martian Surface from the Viking 1 Lander: Preliminary Results", Science, 193, 1976, 805–809

[Shorthill-1976b] Shorthill, R.W., et al., "The Environs of Viking 2 Lander", Science, 194, 1976, 1309–1318

[Showalter-1998] Showalter, M.R., "Detection of Centimeter-Sized Meteoroid Impact Events in Saturn's F Ring", Science, 282, 1998, 1099–1102

[Showalter-2006] Showalter, M.R., Lissauer, J.J., "The Second Ring-Moon System of Uranus: Discovery and Dynamics", Science, 311, 2006, 973–977

[Siddiqi-2000a] Siddiqi, A. A., "Challenge to Apollo", Washington, NASA, 2000, 256–260

[Siddiqi-2000b] ibid., 305–308

[Siddiqi-2000c] ibid., note on page 241

[Siddiqi-2002a] Siddiqi, A.A., "Deep Space Chronicle: A Chronology of Deep Space and Planetary Probes 1958–2000", Washington, NASA, 2002, 26–27

[Siddiqi-2002b] ibid., 29–31

[Siddiqi-2002c] ibid., 34–35

[Siddiqi-2002d] ibid., 36–37

[Siddiqi-2002e] ibid., 40

[Siddiqi-2002f] ibid., 42–43

[Siddiqi-2002g] ibid., 45

[Siddiqi-2002h] ibid., 50–51

[Siddiqi-2002i] ibid., 73–74

[Siddiqi-2002j] ibid., 75–77

[Siddiqi-2002k] ibid., 88–90

[Siddiqi-2002l] ibid., 86–88

[Siddiqi-2002m] ibid., 97–98

[Siddiqi-2002n] ibid., 103–105

[Siddiqi-2002o] ibid., 103–105

[Siddiqi-2002p] ibid., 108

[Siddiqi-2002q] ibid., 109–110

[Siddiqi-2002r] ibid., 125–127

[Siddiqi-2002s] ibid., 129–130

[Simmons-1977] Simmons, G.J., "Surface Penetrators – A Promising New Type of Planetary Lander", Journal of the British Interplanetary Society, 30, 1977, 243–256

[Simpson-1974] Simpson, J.A., et al., "Protons and Electrons in Jupiter's Magnetic Field: Results from the University of Chicago Experiment on Pioneer 10", Science, 183, 1974, 306–309

[Simpson-1975] Simpson, J.A., et al., "Jupiter Revisited: First Results from the University of Chicago Charged Particle Experiment on Pioneer 11", Science, 188, 1975, 456–459

[Simpson-1979] Simpson, R.A., et al., "Viking Bistatic Radar Observations of the Hellas Basin on Mars: Preliminary Results", Science, 203, 1979, 45–46

[Simpson-1980], Simpson, J.A., et al., "Saturnian Trapped Radiation and its Absorption by Satellites and Rings: The First Results from Pioneer 11", Science, 207, 1980, 411–415

[Sjorgren-1979] Sjorgren, W.L., "Mars Gravity: High Resolution Results from Viking Orbiter 2", Science, 203, 1979, 1006–1010

[Smith-1960] Smith, D.E., Smith, A.E., "Pioneer 5 and its Orbit", Flight, 1 April 1960, 437

[Smith-1974] Smith, E.J., at al., "Magnetic Field of Jupiter and its Interaction with the Solar Wind", Science, 183, 1974, 305–306

[Smith-1975] Smith, E.J., et al., "Jupiter's Magnetic Field, Magnetosphere, and Interaction with the Solar Wind: Pioneer 11", Science, 188, 1975, 451–455

[Smith-1977] Smith, B.A., et al., "Voyager Imaging Experiment", Space Science Reviews, 21, 1977, 103–127

[Smith-1979a] Smith, B.A., et al., "The Jupiter System Through the Eyes of Voyager 1", Science, 204, 1979, 951–972

[Smith-1979b] Smith, B.A., et al., "The Galilean Satellites and Jupiter: Voyager 2 Imaging Science Results", Science, 206, 1979, 927–950

[Smith-1980] Smith, E.J., et al., "Saturn's Magnetic Field and Magnetosphere", Science, 207, 1980, 407–410

[Smith-1981] Smith, B.A., et al., "Encounter with Saturn: Voyager 1 imaging Science Results", Science, 212, 1981, 163–191

[Smith-1982] Smith, B.A., et al., "A New Look at the Saturn System: The Voyager 2 Images", Science, 215, 1982, 504–537

[Smith-1983] Smith, B.A., "JPL Tries to Revive Link with Viking 1", Aviation Week & Space Technology, 4 April 1983, 16

[Smith-1986a] Smith, B.A., et al., "Voyager 2 in the Uranian System: Imaging Science Results", Science, 233, 1986, 43–64

[Smith-1986b] Smith, B.A., "Voyager 2's Uranus Flyby Provides Detailed Images of Moon System", Aviation Week & Space Technology, 3 February 1986, 66–67

[Smith-1986c] Smith, B.A., "Scientists Gain New Insights On Uranus From Voyager 2 Data", Aviation Week & Space Technology, 10 February 1986, 66–69

[Smith-1989a] Smith, B.A., et al., "Voyager 2 at Neptune: Imaging Science Results", Science, 246, 1989, 1422–1449

[Smith-1989b] Smith, B.A., "Neptune Rendezvous Will Mark Final Stage of Voyager 2's Mission", Aviation Week & Space Technology, 7 August 1989, 70–71

[Smith-1989c] Smith, B.A., "Voyager's Discoveries Mount on Final Rush to Neptune", Aviation Week & Space Technology, 28 August 1989, 16–20

[Smith-2006] Smith, B.A., personal correspondence with the author, 3 December 2006

[Smyth-1991] Smyth, W.H., Combi, M.R., Stewart, A.I.F., "Analysis of the Pioneer-Venus Lyman-Alpha Image of the Hydrogen Coma of Comet P/Halley", Science, 253, 1991, 1008–1010

[Snyder-1967] Snyder, C.W., "Mariner V Flight Past Venus", Science, 158, 1967, 1665–1669

[Soberman-1990] Soberman, R.K., Dubin, M., "Reexamination of Data from the Asteroid/Meteoroid Detector", NASA document CR–185875, 1990

[Soderblom-1980] Soderblom, L.A., "The Galilean Satellites of Jupiter", Scientific American, January 1980, 88–101

[Soderblom-1990] Soderblom, L.A., et al., "Triton's Geyser-Like Plumes: Discovery and Basic Characterization", Science, 250, 1990, 410–415

[Sonett-1963] Sonett, C.P., "A Summary Review of the Scientific Findings of the Mariner Venus Mission", Space Science Reviews, 2, 1963, 751–777

[Spaceflight-1976] Spaceflight, February 1976, 75

[Spaceflight-1977] "Space Research 'Down Under'", Spaceflight, June 1977, 205–206

[Spitzer-1980] Spitzer, C.R. (ed.), "Viking Orbiter Views of Mars", Washington, NASA, 1980

[S&T-1963] "Photographic Observations of the Mars Probe", Sky & Telescope, January 1963, page unknown. (Reprinted in: Page, T, Page, L.W., "Wanderers in the Sky", New York, Macmillan, 1965, 193)

[Standish-1993] Standish, E.M. Jr., "Planet X : No Dynamical Evidence in the Optical Observations", The Astronomical Journal, 105, 1993, 2000

[Stanley-2004] Stanley, S., Bloxham, J., "Convective-Region Geometry as the Cause of Uranus' and Neptune's Unusual Magnetic Fields", Nature, 428, 2004, 151–153

[Starrfield-1994] Starrfield, S., Shore, S.N., "Nova Cygni 1992: Nova of the Century", Sky & Telescope, February 1992, 20–25

[Steinberg-2001] Steinberg, J.L., "The Scientific Career of a Team Leader", Planetary and Space Science, 49, 2001, 511–522

[Stewart-1987] Stewart, A.I.F., "Pioneer Venus Measurements of H, O, and C Production in Comet P/Halley Near Perihelion", Astronomy and Astrophysics, 187, 1987, 369–374

[Stone-1961] Stone, I., "Mariner to Scan Venus' Surface on Flyby", Aviation Week, 12 June 1961, 52–57

[Stone-1963] Stone, I., "Mariner Design Modified for Mars Flyby", Aviation Week & Space Technology, 6 May 1963, 50–54

[Stone-1964] Stone, I., "Six-Month Lifetime Predicted for Pioneer", Aviation Week & Space Technology, 27 January 1964, 69–75

[Stone-1979a] Stone, E.C., Lane, A.L., "Voyager 1 Encounter with the Jovian System", Science, 204, 1979, 945–948

[Stone-1979b] Stone, E.C., Lane, A.L., "Voyager 2 Encounter with the Jovian System", Science, 206, 1979, 925–927

[Stone-1981a] Stone, E.C., Miner, E.D., "Voyager 1 Encounter with the Saturnian System", Science, 212, 1981, 159–163

[Stone-1986] Stone, E.C., Miner, E.D., "The Voyager 2 Encounter with the Uranian System", Science, 233, 1986, 39–43

[Stone-1989] Stone, E.C., Miner, E.D., "The Voyager 2 Encounter with the Neptunian System", Science, 246, 1989, 1417–1421

[Stone-2005] Stone, E.C., et al., "Voyager 1 Explores the Termination Shock Region and the Heliosheath Beyond", Science, 309, 2005, 2017–2020

[Stooke-1998] Stooke, P.J., "Locating the Viking 2 Landing Site", paper presented at the XXIX Lunar and Planetary Science Conference, Houston, 1998

[Stooke-1999] Stooke, P.J., "Revised Viking 1 Landing Site", paper presented at the XXX Lunar and Planetary Science Conference, Houston, 1999

[Strangeway-1993] Strangeway, R.J., "The Pioneer Venus Orbiter Entry Phase", Geophysical Research Letters, 20, 1993, 2715–2717

[Strom-1979] Strom, R.G., "Mercury: A Post-Mariner 10 Assessment", Space Science Reviews, 24, 1979, 3–70

[Strom-1987] Strom, R.G., "Mercury the Elusive Planet", Washington, Smithsonian Institution Press, 1987

[Stuart-1984] Stuart, J.E., "Interplanetary Navigation". In: "Mathematiques spatiales pour la preparation et la realisation de l'exploitation des satellites/ Space mathematics for the preparation and the development of satellites exploration", Toulouse, Cépaduès, 1984, 1015–1038

[Stuhlinger-1970] Stuhlinger, E., "Planetary Exploration with Electrically Propelled

Vehicles", paper presented at the Third Conference on Planetology and Space Mission Planning, New York, October 1970

[Sullivan-1965] Sullivan, W., "Mariner 4 Makes Flight Past Mars", The New York Times, 15 July 1965, 1

[Surkov-1977] Surkov, Yu. A., "Geochemical Studies of Venus by Venera 9 and 10 Automatic Interplanetary Stations". In: Proceedings of the VIII Lunar Science Conference", 1977, 2665–2689

[Surkov-1982a] Surkov, Yu.A., et al., "Venera 13 and Venera 14 Measurements of the Water Vapor Content in the Venus Atmosphere", Soviet Astronomy Letters, 8, 1982, 223–224

[Surkov-1982b] Surkov, Yu.A., et al., "Aerosols in the Clouds on Venus: Preliminary Venera 14 Data", Soviet Astronomy Letters, 8, 1982, 377–379

[Surkov-1982c] Surkov, Yu.A., et al., "Element Composition of Venus Rocks: Preliminary Results from Venera 13 and Venera 14", Soviet Astronomy Letters, 8, 1982, 237–240

[Surkov-1983] Surkov, Yu.A., et al., "New Data on the Composition, Structure and Properties of Venus Rocks Obtained by Venera-13 and Venera-14", LPI 1983

[Surkov-1997a] Surkov, Yu. A., "Exploration of Terrestrial Planets from Spacecraft", Chichester, Wiley–Praxis, 1997, 221–225

[Surkov-1997b] ibid., 356

[Surkov-1997c] ibid., 203–212

[Surkov-1997d] ibid., 345–349

[Surkov-1997e] ibid., 225–226

[Surkov-1997f] ibid., 229–234

[Surkov-1997g] ibid., 286–292

[Surkov-1997h] ibid., 349–352

[Surkov-1997i] ibid., 279–286

[Surkov-1997j] ibid., 252–276

[Swift-1997a] Swift, D.W., "Voyager Tales: Personal Views of the Grand Tour", Reston, AIAA, 1997, 61–74

[Swift-1997b] ibid., 75–82

[Swift-1997c] ibid., 103–112

[Swift-1997d] ibid., 232

[Swift-1997e] ibid., 149

[Swift-1997f] ibid, 96–97

[Swift-1997g] ibid, 225

[Swift-1997h] ibid, 150

[Swift-1997i] ibid, 228

[Swift-1997j] ibid, 155

[Swift-1997k] ibid, 158

[Swift-1997l] ibid, 153

[Swift-1997m] ibid, 277–279

[Swift-1997n] ibid, 267–268

[Swift-1997o] ibid, 266–271

[Swift-1997p] ibid, 186

[Swift-1997q] ibid, 194

[Swift-1997r] ibid, 305–306

[Swift-1997s] ibid, 323–324

[Swift-1997t] ibid, 269–270

[Swindell-1974] Swindell, W., Doose, L.R., Tomasko, M.G., "Spin-Scan Images of Jupiter from Pioneer 10", Bulletin of the American Astronomical Society, 6, 1974, 387

[Swindell-1975] Swindell, W., et al., "The Pioneer 11 Images of Jupiter", Bulletin of the American Astronomical Society, 7, 1975, 378

[Synnott-1980] Synnott, S.P., "1979J2: The Discovery of a Previously Unknown Jovian Satellite", Science, 210, 1980, 786–788

[Synnott-1979] Synnott, S.P., "1979J3: Discovery of a Previously Unknown Satellite of Jupiter", Science, 212, 1981, 1392

[Taylor-1979a] Taylor, F.W., et al., "Infrared Remote Soundings of the Middle Atmosphere of Venus from the Pioneer Orbiter", Science, 203, 1979, 779–781

[Taylor-1979b] Taylor, F.W., et al., "Temperature, Cloud Structure, and Dynamics of Venus Middle Atmosphere by Infrared Remote Sensing from Pioneer Orbiter", Science, 205, 1979, 65–67

[Taylor-1980] Taylor, J.W.R. (ed.), "Jane's All the World's Aircraft 1980–81", London, Jane's, 631–632

[Teegarden-1973] Teegarden, B.J., et al., "Pioneer-10 Measurements of the Differential and Integral Cosmic-Ray Gradient Between 1 and 3 Astronomical Units", The Astrophysical Journal, 185, 1973, L155–L159

[Terrile-1979] Terrile, R.J., et al., "Jupiter's Cloud Distribution Between the Voyager 1 and 2 Encounters: Results from 5-Micrometer Imaging", Science, 206, 1979, 995–996

[Teske-1993] Teske, R.G., "The Star that Blew a Hole in Space", Astronomy, December 1993, 30–37

[The Tech-670221] "MIT Satellite to Orbit the Sun", The Tech, 21 February 1967, 3

[Toller-1982] Toller, G.N., "A Study of Galactic Light Using Pioneer 10 Observations of Background Starlight", Bulletin of the American Astronomical Society, 1982, p.656

[Tolson-1978] Tolson, R.H., et al., "Viking First Encounter of Phobos: Preliminary Results", Science, 199, 1978, 61–64

[Tomasko-1979] Tomasko, M.G., et al., "Preliminary Results of the Solar Flux Radiometer Experiment Aboard the Pioneer Venus Multiprobe Mission", Science, 203, 1979, 795–797

[Tombaugh-1950] Tombaugh, C.W. The Astronomical Journal, 55, 1950, 184

[Trainor-1974] Trainor, J.H., et al., "Energetic Particle Population in the Jovian Magnetosphere: A Preliminary Note", Science, 183, 1974, 311–313

[Trainor-1980] Trainor, J.H., McDonald, F.B., Schardt, A.W., "Observations of Energetic Ions and Electrons in Saturn's Magnetosphere", Science, 207, 1980, 421–424

[Trainor-1984] Trainor, J.H., et al., "Results from the HELIOS Galactic and Solar

Cosmic Ray Experiment (E7)". In: Porsche, H. (ed.) "10 Jahre Helios – 10 Years Helios", 149–155 (in German and English)

[Travis-1979] Travis, L.D., et al., "Cloud Images from the Pioneer Venus Orbiter", Science, 205, 1979, 74–76

[Turner-2004] Turner, M.J.L, "Expedition Mars", Chichester, Springer–Praxis 2004, 33–57

[Turnill-1984a] Turnill, R. (ed.), "Jane's Spaceflight Directory 1984", London, Jane's Publishing, 84–86

[Turnill-1984b] ibid., 86

[Turnill-1984c] ibid., 88

[Tyler-1981a] Tyler, G.L., et al., "Radio Wave Scattering Observations of the Solar Corona: First-Order Measurements of Expansion Velocity and Turbulence Spectrum using Viking and Mariner 10 Spacecraft", The Astrophysical Journal, 249, 1981, 318–332

[Tyler-1981b] Tyler, G.L., et al., "Radio Science Investigations of the Saturn System with Voyager 1: Preliminary Results", Science, 212, 1981, 201–206

[Tyler-1982] Tyler, G.L., et al., "Radio Science with Voyager 2 at Saturn: Atmosphere and Ionosphere and the Masses of Mimas, Tethys and Iapetus", Science, 215, 1982, 553–558

[Tyler-1986] Tyler, G.L., et al., "Voyager 2 Radio Science Observations of the Uranian System: Atmosphere, Rings, and Satellites", Science, 233, 1986, 79–84

[Tyler-1989] Tyler, G.L., et al., "Voyager Radio Science Observations of Neptune and Triton", Science, 246, 1989, 1466–1473

[Ulivi-2004a] Ulivi P., with Harland D.M., "Lunar Exploration", Chichester, Springer–Praxis, 2004, 1–32

[Ulivi-2004b] ibid., 58–60

[Ulivi-2006] Ulivi, P., "ESRO and the deep space: European Planetary Exploration Planning before ESA", Journal of the British Interplanetary Society, 59, 2006, 204–223

[Unz-1970] Unz. F., "Solar Probe 'Helios'" , paper presented at the XXI International Astronautical Congress, Constance, 1970

[Vaisberg-1976] Vaisberg, O.L., et al., "Scientific Objectives and Preliminary Results of the Venus Reconnaissance by the Venera 9 and 10 Orbiters: Cloud Layer, Upper Atmosphere and Solar-Wind Interaction", Soviet Astronomy Letters, 2, 1976, 1–3

[Van Allen-1968] Van Allen, J.A., Drake, J.F., Gibson, J., "Solar X-Ray Observations with Explorer 33, Explorer 35 and Mariner V", Astronomical Journal, 73, 1968, S81

[Van Allen-1972] Van Allen, J.A., "Observations of Galactic Cosmic-Ray Intensity at Heliocentric Radial Distances of from 1.0 to 2.0 Astronomical Units", The Astrophysical Journal, 177, 1972, L49–L52

[Van Allen-1974] Van Allen, J.A., et al., "Energetic Electrons in the Magnetosphere of Jupiter", Science, 183, 1974, 309–311

[Van Allen-1980] Van Allen, J.A., et al., "Saturn's Magnetosphere, Rings and Inner Satellites", Science, 207, 1980, 415–421

[Van Allen-1984] Van Allen, J.A., "Geometrical Relationships of Pioneer 11 to Uranus and Voyager 2 in 1985–86", University of Iowa Department of Physics and Astronomy, 1984

[van der Kruit-1986] van der Kruit, P.C., "Surface Photometry of Edge-on Spiral Galaxies V. The Distribution of Luminosity of the Disk of the Galaxy Derived from the Pioneer 10 Background Experiment", Astronomy and Astrophysics, 157, 1986, 230–245

[van der Linden-1994] van der Linden, P., "Expert C Programming", Englewood Cliffs, Prentice–Hall, 1994, page unknown

[Varfolomeyev-1993] Varfolomeyev, T., "The Soviet Mars Programme", Spaceflight, July 1993, 230–231

[Varfolomeyev-1998a] Varfolomeyev, T., "Soviet Rocketry that Conquered Space: Part 4", Spaceflight, January 1998, 28–30

[Varfolomeyev-1998b] Varfolomeyev, T., "Soviet Rocketry that Conquered Space: Part 5", Spaceflight, March 1998, 85–88

[Varfolomeyev-1998c] Varfolomeyev, T., "Soviet Rocketry that Conquered Space: Part 6", Spaceflight, May 1998, 181–184

[Vedrenne-1979] Vedrenne, G., et al., "Observations of the X-Ray Burster 0525.9-66.1", Soviet Astronomy Letters, 5, 1979, 314–317

[Vekshin-1999] Vekshin, B., "Pisma Zhitateley" (reader's letters), Novosti Kosmonavtiki, No. 5, 1999, 53

[Verigin-1999] Verigin, V., "9 Let Granata" (9 years of Granat), Novosti Kosmonavtki, No.2 1999, 38–40 (in Russian)

[Verschuur-1993] Verschuur, G.L., "Race to the Sun's Edge", Air & Space, April/May 1993, 24–30

[Vervack-1994] Vervack, R.J. Jr. et al., "Voyager 2 UVS Observations of Jupiter During the Comet Shoemaker–Levy 9 Impact Events", Poster presented at the 26th Annual Meeting of the Division for Planetary Sciences Bethesda, Maryland, October 31–November 4, 1994

[Vervack-2006] Vervack, R.J. Jr. personal correspondence with the author, 13 December 2006

[Veverka-1977] Veverka, J., "Phobos and Deimos", Scientific American, February 1977, 30–37

[Villante-1984] Villante, U., "Il Campo Magnetico Interplanetario" (The Interplanetary Magnetic Field), Le Scienze, December 1984, 28–36 (in Italian)

[Vinogradov-1970] Vinogradov, A. P., Surkov, Yu. A., Marov, M. Ya., "Investigation of the Venus Atmosphere by Venera 4, Venera 5 and Venera 6 Probes", paper presented at the XXI International Astronautical Congress, Constance, 1970, 211–224

[Vinogradov-1976] Vinogradov, A.P., et al., "The First Panoramas of Venus – Preliminary Analysis", Soviet Astronomy Letters, 2, 1976, 26–8

[Vladimirov-1999] Vladimirov, A., "Kapustin Yar – Stranitsy Istoriy Kosmosa" (Kapustin Yar – Pages of Space History), Novosti Kosmonavtiki, No. 6, 1999, 9–11 (in Russian)

[VnIITransmash-1999] VnIITransmash, "Specimens of Space Technology, Earth

Based Demonstrators of Planetary Rovers, Running Mock-ups", Saint Petersburg, 1999

[VnIITransmash-2000] "Pages of history of VNIITransmash", Saint Petersburg, VnIITransmash, pages unknown (in Russian)

[Vogt-1982] Vogt, R.E., et al., "Energetic Charged Particles in Saturn's Magnetosphere: Voyager 2 Results", Science, 215, 1982, 577–582

[Volland-1984] Volland, H., et al., "Das Faraday-Rotations-Experiment". In: Porsche, H. (ed.) "10 Jahre Helios – 10 Years Helios", 118–121 (in German and English)

[von Zhan-1979] von Zahn, U., et al., "Venus Thermosphere: In Situ Composition Measurements, the Temperature Profile, and the Homopause Altitude", Science, 203, 1979, 768–770

[Warwick-1981] Warwick, J.W., et al., "Planetary Radio Astronomy Observations from Voyager 1 Near Saturn", Science, 212, 1981, 239–243

[Warwick-1982] Warwick, J.W., et al., "Planetary Radio Astronomy Observations from Voyager 2 Near Saturn", Science, 215, 1982, 582–587

[Warwick-1989] Warwick, J.W., et al., "Voyager Planetary Radio Astronomy at Neptune", Science, 246, 1989, 1498–1501

[Webber-1975] Webber, W.R., "Pioneer 10 Measurements of the Charge and Energy Spectrum of Solar Cosmic Rays During 1972 August", The Astrophysical Journal, 199, 1975, 482–493

[Weber-1977] Weber, R.R., et al., "Interplanetary Baseline Observations of Type III Solar Radio Bursts", Solar Physics, 54, 1977, 431–439

[Westphal-1965] Westphal, A.J., Wildey, R.D., Murray, B.C., "The 8–14 Micron Appearance of Venus Before the 1964 Conjunction", Astrophysical Journal, 142, 1965, 799–802

[Wetmore-1965] Wetmore, W.C., "Comet Flyby Studied for Mariner Backup", Aviation Week & Space Technology, 19 November 1965, 45–60

[Wiegert-2007] Wiegert, P., personal correspondence with the author, 10 January 2007

[Wilson-1966] Wilson, J.N., "Mechanical Design Evolution of the Mariner Spacecraft", paper presented at the XVII International Astronautical Congress, Madrid, 1966

[Wilson-1979] Wilson, A., "Scout – NASA's Small Satellite Launcher", Spaceflight, November 1979, 446–459

[Wilson-1982a] Wilson, A., "The Eagle has Wings: The Story of the American Space Exploration 1945–1975", London, The British Interplanetary Society, 1982, 33–34

[Wilson-1987a] Wilson, A., "Solar System Log", London, Jane's Publishing, 1987, 21

[Wilson-1987b] ibid., 22–23

[Wilson-1987c] ibid., 27–28

[Wilson-1987d] ibid., 47–48

[Wilson-1987e] ibid., 49–50

[Wilson-1987f] ibid., 55–56

[Wilson-1987g] ibid., 56–58

[Wilson-1987h] ibid., 59–60

[Wilson-1987i] ibid., 65–67

[Wilson-1987j] ibid., 67–69

[Wilson-1987k] ibid., 72–74

[Wilson-1987l] ibid., 76–77

[Wilson-1987m] ibid., 74–75

[Wilson-1987n] ibid., 78–80

[Wilson-1987o] ibid., 80–83

[Wilson-1987p] ibid., 84–87

[Wilson-1987q] ibid., 88–91

[Wilson-1987r] ibid., 100–106

[Wilson-1987s] ibid., 107–109

[Wilson-1987t] ibid., 109–112

[Wilson-1987u] ibid., 94–100

[Winkler-1976] Winkler, W., "Helios Assessment and Mission Results", Acta Astronautica, 3, 1976, 435–447

[Winkler-1983] Wikler, W., "Material Performance under Combined Stresses in the Hard Space Environment of the Sunprobe Helios-A", Acta Astronautica, 4, 1983, 189–205

[Woiceseyn-1974] Woiceseyn, P.M., Kliore, A.J., Sesplaukis, T.T., "Dynamics of Jupiter's Lower Atmosphere from Pioneer 10 S-Band Radio Occultation Measures", Bulletin of the American Astronomical Society, 6, 1974, 339

[Wolfe-1974] Wolfe, J.H., et al., "Preliminary Pioneer 10 Encounter Results from the Ames Research Center Plasma Analyzer Results", Science, 183, 1974, 303–305

[Wolfe-1980] Wolfe, J.H., et al., "Preliminary Results on the Plasma Environment of Saturn from Pioneer 11 Plasma Analyzer Experiment", Science, 207, 1980, 403–407

[Wolverton-2000] Wolverton, M., "Pathfinding the Rings: The Pioneer Saturn Trajectory Decision", Quest, 7, No. 4, 2000, 5–11

[Wolverton-2004a] Wolverton, M., "The Depths of Space: the Pioneer Planetary Probes", Washington, Joseph Henry Press, 2004, 7–39

[Wolverton-2004b] ibid., 46–51

[Wolverton-2004c] ibid., 195–198

[Wolverton-2004d] ibid., 203–209

[Wolverton-2004e] ibid., 171

[Woo-1978] Woo, R., "Radial Dependence of Solar Wind Properties Deduced from Helios 1/2 and Pioneer 10/11 Radio Scattering Observations", The Astrophysical Journal, 219, 1978, 727–739

[Wotzlaw-1998] Wotzlaw, S., Käsmann, F.C.W., Nagel, M., "Proton – Development of a Russian Launch Vehicle", Journal of the British Interplanetary Society, 51, 1998, 3–18

[Wu-1978] Wu, F.-M., Judge, D.L., Carlson, R.W., "Europa: Ultraviolet Emissions and the Possibility of Atomic Oxygen and Hydrogen Clouds", The Astrophysical Journal, 225, 1978, 325–334

[Wu-1988] Wu, F.M., et al., "The Hydrogen Density of the Local Interstellar Medium and an Upper Limit to the Galactic Glow Determined from Pioneer 10 Ultraviolet Photometer Observations", The Astrophysical Journal, 331, 1988, 1004–1012

[Zak-2004] "Planetary: Projects and Concepts", Anatoly Zak website

[Zellner-1972] Zellner, B.H., "Minor Planets and Related Objects. VIII. Deimos", The Astronomical Journal, 77, 1972, 183–185

[Zheleznyakov-2001] Zheleznyakov, A., Rozenblyum, L., "Yaderniye Vzryvyi v Kosmose" (Nuclear Explosions in Space), Novosti Kosmonavtiki, November 2001, page unknown (in Russian)

[Zwickl-1977] Zwickl, R.D., Webber, W.R., "Solar Particles Propagation from 1 to 5 AU", Solar Physics, 54, 1977, 457–504

Further reading

BOOKS

Briggs, G., Taylor, F., "The Cambridge Photographic Atlas of the Planets", Cambridge University Press, 1982

Burrows, W.E., "This New Ocean: The Story of the First Space Age", New York, The Modern Library, 1999

Godwin, R., (editor), "Deep Space: The NASA Mission Reports", Burlington, Apogee, 2005

Godwin, R., (editor), "Mars: The NASA Mission Reports", Burlington, Apogee, 2000

Isakowitz, S.J., Hopkins, J.P. Jr., Hopkins, J.B., "International Reference Guide to Space Launch Systems", 3rd edition, Reston, AIAA, 1999

Kelly Beatty, J., Collins Petersen, C., Chaikin, A. (editors), "The New Solar System", 4th ed., Cambridge University Press, 1999

Siddiqi, A. A., "Challenge to Apollo", Washington, NASA, 2000

Shirley, J.H., Fairbridge, R.W., "Encyclopedia of Planetary Sciences", Dordrecht, Kluwer Academic Publishers, 1997

Surkov, Yu.A., "Exploration of Terrestrial Planets from Spacecraft", Chichester, Wiley–Praxis, 1994

MAGAZINES

Aerospace America

l'Astronomia (in Italian)

Aviation Week & Space Technology

Espace Magazine (in French)

Flight International

Novosti Kosmonavtiki (in Russian)

Science

Scientific American

Sky & Telescope

Spaceflight

INTERNET SITES

Don P. Mitchell's "The Soviet Exploration of Venus" (www.mentallandscape.com/V_Venus.htm)

Encyclopedia Astronautica (www.astronautix.com)

Interplanetary Probes of the Soviet Union (sovams.narod.ru)

Jonathan's Space Home Page (planet4589.org/space/space.html)

JPL (www.jpl.nasa.gov)

NASA NSSDC (nssdc.gsfc.nasa.gov)

Novosti Kosmonavtiki (www.novosti-kosmonavtiki.ru)

NPO Imeni S.A. Lavochkina (www.laspace.ru)

Pioneer Project (www.nasa.gov/centers/ames/missions/archive/pioneer.html)

Space Daily (www.spacedaily.com)

Spaceflight Now (www.spaceflightnow.com)

Sven's Space Place (www.svengrahn.se)

The Planetary Society (planetary.org)

Voyager Project (voyager.jpl.nasa.gov)

Index

1M 1960 Mars probe, 12–15, 26
1M 1965 Mars probe, 48
1V 1960 Venus probe, 12, 15
1V 1967 Venus probe (see also Venera 4,
  Kosmos 167), 52–56, 59, 70
1VA Venus probe (see also Venera 1), 15–17,
  26
2M 1969 Mars probe: see M-69
2MV Mars and Venus probe, 26–28, 45
2V Venus probe (see also Venera 5, 6), 70–72
3M Mars probe: see M-71, M-71P, M-71S,
  M-73, M-73P, M-73S, Mars 2, 3, 4, 5, 6, 7
3MV Mars and Venus probe, 31–33, 45–48,
  52, 59
3V Venus probe (see also Venera 7, 8,
  Kosmos 359, 482), 97–97, 156
4M Mars probe and rover, 16
4NM Mars probe and rover, 167
4V-1 Venus probe (see also Venera 9, 10, 11,
  12), 209–212, 270–272, 290, 293
4V-1M Venus probe (see also Venera 13, 14),
  284–285
4V-2 Venus radar orbiter, 293
5M Mars probe, 168–170
5NM Mars probe, 167–168
8K71 missile, 1, 2, 12, 28
8K72 launcher, 2, 12
8K78 and 8K78M Molniya launcher, 12, 14,
  15, 28, 31, 32, 48, 72, 98, 157
8K82K Proton launcher, 52, 74, 78, 81, 100,
  112, 160, 168, 169, 209, 270, 273, 285

Abastumani astronomical observatory, 162
Adams, J.C., xlviii, 428

Adams, W.S., xxxi
Adrastea (Jupiter Satellite), 330, 350
Advanced Cooperation Project, 196
Aelita (novel and movie), xxxv
Aerobee missile, 5
Airy, G.B., xlviii–xlix, l
Amalthea (Jupiter satellite), xli, 137,
  328–330, 354
Ames Research Center, 48, 51, 69, 125, 146,
  154, 155, 259, 262, 305, 320, 397
Andrianov, A.M., 33
Antoniadi, E., xxvi–xxvii, xxxv, xli, xlvi
Apianus, P., xxxiv
Apollo 11, 81, 400
Apollo 16, 125
Apollo lunar manned program, 33, 68, 69,
  126, 208, 217, 218, 242, 307, 443
Apollo Lunar Module, 68, 220
Arecibo radiotelescope, 154
Ariane launcher, 291
Ariel (Uranus satellite), xlvi, 401, 406, 408,
  413–414, 415
Aristotle, liii
Arrhenius, S., xxxi
artificial meteors, 2, 5
Asteroid (1) Ceres, lii
Asteroid (2) Pallas, lii
Asteroid (3) Juno, lii
Asteroid (4) Vesta, lii
Asteroid (307) Nike, 133
Asteroid (433) Eros, lii–liii
Asteroid (588) Achilles, liii
Asteroid (944) Hidalgo, liii
Asteroid (1862) Apollo, liii

Asteroid (2060) Chiron, liii
Asteroid (2201) Oljato, 281
Asteroid (3352) McAuliffe, 319
Asteroid (69230) Hermes, liii
Asteroids, knowledge of, lii–liii
Asteroids, missions to, 97, 256
Astronomical Unit, determination, xxix, 8–9, 58
Astroplane: see Mars airplane
Atlas (Saturn moon), 366, 380, 397
Atlas missile and launcher, 9, 34, 305
Atlas-Agena launcher, 9, 19–22, 34, 38, 57, 59, 78
Atlas-Centaur launcher (see also Centaur stage), 33, 78, 89, 112, 132, 140, 156, 172, 176, 200, 218, 264
Atlas-Vega launcher, 9
ATS-E satellite, 78
Automated Biological Laboratory, 67, 223

Baade, W., 53
Babakin, G.N., 48, 52, 59, 73, 100, 114, 167, 290
Baldwin, R., 43
Barmin, I.V., 168–169, 272
Barnard, E.E., xxxv, xli, 42
Bartoli, D., xxxiii
Beagle 1964 Mars probe, 67–68
Beer, J.H., xxxiii
Belinda (Uranus satellite), 404, 407
Berlin observatory, xlviii
Bianca (Uranus satellite), 408
Biemann, K., 242
Blamont, J., 289
Bode, J.E., xlvi, xlviii
Bode's law, xlviii, xlix, lii
Bond, G.P., xl
Bond, W.C., xlv
Boyer, C., xxxii
Brenner, L., xxx–xxxi
Burgess, E., 4, 44, 131
Burke, B.F., xli
Burney, V., li
Burroughs, E.R., xxxv
Bush, G.W., 450

Cain, D.L., 38
Callisto (Jupiter satellite), xxxix, xliii, 136, 243, 327, 338–339, 348–350, 351, 353

Calypso (Saturn satellite), 387
Camichel, H., xxxii
Campbell, W.W., xxxv
Carter, J.E., 315
Cassini Division, xliv, 366, 372, 380, 395
Cassini Saturn orbiter, 256, 397
Cassini, G.D., xxx, xxxiii, xxxix–xl, xliv
Centaur stage, 9, 18, 78, 112, 141, 181, 200, 202, 216, 226, 228, 272, 304, 318, 320, 322
Cerulli, V., xxxv
Challis, J.C., xlix
Charon (Pluto satellite), 434
Chelomei, V.N., 30–31, 52
CNES (Center National d'Etudes Spatiales), 162, 229, 291
CNIE (Comision Nacional de Investigaciones Espaciales), 289
CNR (Centro Nazionale Ricerche), 198
Colombo, G., 172
Comas Solá, J., xlv
Comet 1P/Halley, liv, 50, 153, 281–283, 422
Comet 1P/Halley, missions to, 50, 207, 256, 400, 423
Comet 2P/Encke, lv, 57, 206, 207, 208, 281
Comet 6P/d'Arrest, 206
Comet 7P/Pons-Winnecke, 57
Comet 10P/Tempel 2, 57
Comet 16P/Brooks 2, 4
Comet 21P/Giacobini-Zinner, 207, 281
Comet 26P/Grigg-Skjellerup, 206
Comet 29P/Schwassmann-Wachmann 1, 153
Comet 31P/Schwassmann-Wachmann 2, 277
Comet 55P/Tempel-Tuttle, liv
Comet 109P/Swift-Tuttle, liv
Comet C/1969T1 Tago-Sato-Kosaka, 206
Comet C/1969Y1 Bennett, 206
Comet C/1973E1 Kohoutek, 49, 179–180, 206
Comet C/1975V1 West, 203
Comet C/1978H1 Meier, 203
Comet C/1979Y1 Bradfield, 203, 277
Comet C/1986P1 Wilson, 283
Comet C/1987B1 Nishikawa-Takamizawa-Tago, 283
Comet C/1987U3 McNaught, 283
Comet D/1770L1 Lexell, 4
Comet D/1895Q1 Swift, 44
Comet D/1993F2 Shoemaker-Levy 9, lv, 351, 449

Cometary Explorer, 206–208
Comets, knowledge of, liii–lv, 206
Comets, missions to, 11, 57–58, 89, 97, 151, 206–208
CONSCAN (Conical Scan), 128
Cook, J., xxix
Copernicus, N., xxv, xxxix
Cordelia (Uranus satellite), 405, 407
Cosmoids, 134
Cressida (Uranus satellite), 404
Crocco, G.A., 4, 301, 303
Crommelin, A.C.D., li
Cross, C.A., 4
Cutting, E., 301

d'Arrest, H.L., xlviii
Davidson, K., 44
Dawes, W.R., xxxiii, xli, xlv
Day Probe: See Pioneer Venus Multiprobe
Deimos (Mars satellite), xxxiii, 29, 38, 83, 112, 116, 121, 140, 228, 239, 250, 251, 330
Delta launcher, 262, 264
DeMarcus, W., xl
Denning, W.F., xxvi–xxvii
Desdemona (Uranus satellite), 404
Despina (Neptune satellite), 426, 428, 431, 435
Dione (Saturn satellite), xliv, 147, 364, 368, 369, 372, 375, 387, 397
Dollfus, A., xxxi, xxxvii, xxxix, xliii, 40, 42, 162
Drake, F., 131
Dryden Flight Research Center, 258
DSN (Deep Space Network) and the Canberra, Goldstone, Johannesburg, Madrid antennas, 40, 58, 81, 110, 125, 147, 150, 153, 174, 188, 189, 196, 200, 205, 253, 280, 311, 322–323, 326, 346, 354, 363, 382, 383, 398, 400, 404, 417, 422–423, 440–441, 447
Dunham, T., xxxi, xlv
DZhVS (Dolgozhivushaya Veneryanskaya Stanziya), 291–293

E-6 lunar probe, 102
E-8 lunar probe, 73, 74, 167
Edgeworth, K.E., lvi
Effelsberg radiotelescope, 200
Ehricke, K., 302
Einstein, A., lvi

ELDO (European Launcher Development Organization), 291
Enceladus (Saturn satellite), xliv, 147, 372, 373, 376, 383, 387, 389, 392, 397, 414, 417
Encke Division, xlv, 380, 389, 392, 395, 397
Encke, J.F., xlv
Eole satellite, 289
Eos (Eole-Venus) probe, 289–291
Epimetheus (Saturn satellite), 147, 150, 371, 373–374, 382
ESA (European Space Agency), 258, 291, 400
ESRO (European Space Research Organization), 38–39, 172, 200, 206, 208, 256, 262, 266, 291
ESTEC (European Space Technology Center), 200
Europa (Jupiter satellite), xxxix, xliii, 136, 143, 326, 328, 336, 346, 350, 352–354, 372, 387, 389, 413, 414
Explorer 1, 1, 9
Explorer 6, 6
Explorer 34, 63
Explorer 35, 133

Farquhar, R.W., 206, 207
Flammarion, C., xxxiv
Flandro, G.A., 301–303
Fontana, F., xxx, xxxii
Fracastorius, G., liv
Franklin, K.L., xli

Galactic Jupiter Probe, 125–126
Galatea (Neptune satellite), 426, 428, 437
Galilean satellites: see Io, Europa, Ganymede, Callisto
Galilei, G., xxv, xxxii, xxxix, xliii, xliv, xlix
Galileo Jupiter orbiter, 156, 315, 320, 449
Galle, J.G., xlviii, 430
gamma-ray bursts, 201–202, 204, 271, 277–278, 279, 284, 289
Ganymede (Jupiter satellite), xxxix, xli, xliii, 136, 143, 256, 305, 327–328, 331, 337–338, 339, 346, 350–351, 352, 353, 370, 414
Gauss, K.F., lii
GCMS (Gas Chromatograph Mass Spectrometer), 223, 228, 233, 234, 239, 242–243
GE (Gas Exchange) experiment, 225, 240–241, 243

Geminga neutron star, 446
General Relativity, 87, 153, 248–250
Giotto probe, 400
Goddard Space Flight Center, 5, 34, 76, 125–126, 198, 206, 262
Goddard, R.H., 2
Goldstone radiotelescope: see Deep Space Network
Grand Tour, 126, 134, 146, 200, 302–308, 309, 310, 315
Gravity assist, 4, 172, 302–303, 309–310, 331, 340, 355, 364, 382, 398, 402, 420, 441, 467–468
Green, N.E., xxxiv
Grimaldi, F., xxxiii
GSOC (German Space Operation Center), 202
Gulliver experiment, 67, 223
Guthnick, P., xli

Hades: see Sinope
Hall, A., xxxiii, 116, 250
Halley, E., xxix, liv
Harding, K.L., lii
'Heavy Sputnik' Venus probe, 16
Helene (Saturn moon), 387
Heliopause, 151, 302, 443–446, 449, 450–451
Helios 1, 189, 200–201, 202, 203, 204, 205–206, 226
Helios 2, 201–202, 203, 204, 205, 206, 228
Helios C, 206, 208
Helios solar probes, 196–206, 208, 262
Hencke, K.L., lii
Herschel, J., xlvi
Herschel, W., xxxiii, xliv, xlv–xlvi, lii, 371, 400
Hirayama, K., lii
Hohmann, W., 301, 465
Hooke, R., xxxix
Horowitz, N.H., 243
Horrock, J., xxix
Hoyle, F., xxxi
Hubble Space Telescope, 43, 307, 397, 407, 426
Huggins, W., xxxv
Humason, M.L., xv
Hussey, T.J., xlviii
Huygens Titan lander, 256
Huygens, C., xxxiii, xliii

Hyperion (Saturn satellite), xlv, 147, 368, 376, 384–386, 430

Iapetus (Saturn satellite), xliv, 147, 151, 366, 376, 382, 383–384, 386, 393, 397, 424
IBEX (Interstellar Boundary Explorer), 451
IKI (Institut Kosmicheskikh Isledovanii), 162, 163, 285
IMP (Interplanetary Monitoring Platform), 133, 204
Io (Jupiter satellite), xxxix, xli, xliii, 131, 136, 137–138, 143, 305, 310, 315, 326, 327, 328–329, 330, 331–334, 336, 346, 352, 363, 364, 372, 416, 417, 428
Io volcanoes, 340–343, 346, 347–348, 355–356, 361, 433
IRAS (InfraRed Astronomy Satellite), 422
Isayev, A.M., 26–28
ISEE 3 satellite, 204
IUE (International Ultraviolet Explorer), 422

Janssen, P.J.C., xxxvi
Janus (Saturn satellite), 147, 150, 370, 373–374, 382
Jeffreys, H., xl
Jet Propulsion Laboratory (JPL), 9–11, 18, 34, 38, 42, 48, 57, 58, 65, 67, 69, 76, 108, 110, 117, 126, 128, 134, 141, 151, 171, 172, 200, 206, 207, 208, 217, 218, 219, 220, 223, 253, 256, 258, 266, 301, 304, 306, 307, 308, 309, 310, 313, 315, 318, 320, 322, 323, 326, 340, 364, 366, 389, 391, 398, 399, 404, 419, 443, 450
Jodrell Bank radiotelescope, 7, 17, 33
JPL: see Jet Propulsion Laboratory
J-S-P trajectory, 302, 307, 310
J-S-U-N trajectory, 302, 310, 311, 318, 346
Judica-Cordiglia A. and G.B. , 16
Juliet (Uranus satellite), 404
J-U-N trajectory, 302, 307, 310
Jupiter Orbiter with Probe: see Galileo Jupiter orbiter
Jupiter, Great Red Spot, xxxix, xliii, 139–140, 143, 324, 325, 326, 343, 347, 361, 426
Jupiter, knowledge of, xxxix–xliii, 134–135, 138–140, 143–145, 343–345, 361
Jupiter, missions to, 11, 31, 88, 91, 97, 125–145, 152, 156, 196, 254, 256, 301–311, 315, 318, 320, 323–363

Jupiter, ring, 143, 330, 337, 350, 359–361, 401

Karkoschka, E., 407
Keeler Gap, 364, 366
Keeler, J.E., xliv, 366
Kepler, J., xxix, xxxii
Khrushchev, N.S., 15, 31
Kirkwood, D., xl, lii
Klein, H.P., 243
Kohlhase, C.E., 382
Kohoutek, L., 176
Korolyov, S.P., 12, 15, 26, 30, 31, 48, 55, 114, 167
Kosmoplan, 30–31
Kosmos 21, 31
Kosmos 27, 31
Kosmos 96, 45
Kosmos 167, 61
Kosmos 359, 98
Kosmos 429, 112
Kosmos 482, 157
Kosmos 1124, 150
Kotelnikov, V.A., 186–187
Kovtunenko, V.M., 170
Kowal, C.T., liii
Kozyrev, N.A., xxxi
Krynov, Y.L., xxxvii
Kryukov, S.S., 114, 170
Ksanfomaliti, L.V., 162
Kuiper belt, lvi, 153, 441
Kuiper, G.P., xxxi, xxxvii, xliii, xlv, xlvi, l, li, lii, lvi, 186, 194, 406
Kutyreva, A.P., xxxvii

Lagrange, J.-L. de, liii, 387
Lagrangian points, liii, lvi, 134, 387, 397
Lalande, J.J., xlvi
Langley Research Center, 69, 217–218, 220
Laplace, P.S. de, xliv, xlvi
Large Probe: See Pioneer Venus Multiprobe
Large Space Telescope: see Hubble Space Telescope
Larissa (Neptune satellite), 426, 430–431
Lassell, W., xlvi, xlix
Late Heavy Bombardment, 191
Lavochkin design bureau, 48, 55, 78, 97, 100, 114, 150, 160, 167, 170, 293
Leonardo da Vinci, xxix
Leonids meteor shower, liv, 45

LeVerrier, U.J.J., xlviii–xlix, l, lv, 428, 430
Levin, G., 242, 243
Lewis Research Center, 208, 218
Lexell, A.J., xlvi
Liais, E., xxxiv
Lick Observatory, 364
Local Bubble, 446
Lomonosov, M.V., xxix–xxx
Lowell, P., xxvi, xxx, xxxiv–xxxv, l, li, 40, 245
LR (Labeled Release) experiment, 225, 241, 242
Lucas, G., 371
Luna 1, 2, 5
Luna 2, 2
Luna 3, 2, 30, 45
Luna 13, 108, 211
Lunar Prospector, 51, 154
Lunokhod, 98, 101, 105, 108
Lyot, B., xxxi, xliii

M-69 Mars missions, 73–74, 78–81, 99, 169
M-71 Mars missions (see also Mars 2, 3), 100–108, 160, 169, 209
M-71P Mars missions (see also Mars 2, 3), 100–108, 112
M-71S Mars mission, 100, 102, 108, 112, 113, 168
M-73 Mars missions (see also Mars 4, 5, 6, 7), 160–163, 167, 209, 211
M-73P Mars missions (see also Mars 6, 7), 160, 162
M-73S Mars missions (see also Mars 4, 5), 160–162
M-79 Mars mission: see 5M
M-100 Sounding Rocket, 108
Mab (Uranus satellite), 407
Mädler, J.H., xxxiii
Magellan Venus orbiter, 256, 283, 287, 441
Maraldi, G., xxxiii
Mariner 1, 19–22
Mariner 2, 22–25, 26, 30, 34, 55, 58, 63, 182, 441, 443
Mariner 3, 38, 57
Mariner 4, 38–45, 48, 57, 58, 65, 66, 68, 74, 76, 78, 83, 87, 122, 124, 304
Mariner 5, 58–59, 63–65, 262, 304
Mariner 6, 76–78, 81–87, 108
Mariner 7, 76–78, 81–87, 108, 110, 122
Mariner 8, 108, 112

Mariner 9, 108–112, 114–123, 124–125, 160,
162, 174, 182, 208, 218, 220, 226, 229, 243,
248, 250, 251
Mariner 10, 172–196, 198, 200, 279, 410, 468
Mariner 11 and Mariner 12: see Voyager 1
and Voyager 2
Mariner A, 9–11, 18, 19, 25, 216, 266
Mariner B, 9, 11, 18, 34, 38, 40, 216
Mariner C (see also Mariner 3, 4, 5), 34–38,
45, 57, 58–59
Mariner H (see also Mariner 8), 108–110
Mariner I (see also Mariner 9), 110
Mariner Jupiter-Saturn: see Voyager
Mariner Jupiter-Uranus, 310
Mariner R (see also Mariner 1, 2), 18–19, 25,
35
Mariner Venus Mercury: see Mariner 10
Mars 1, 28–30, 32, 33
Mars 2, 112–114, 117, 123–124, 165
Mars 3, 112–114, 117, 123–124, 165
Mars 4, 163–164
Mars 5, 163, 164–165
Mars 6, 163, 164, 165–167, 226, 230
Mars 7, 163, 165
Mars Airplane, 258–259
Mars Ball, 257–258
Mars Pathfinder, 254
Mars penetrator, 259–260
Mars rovers (see also Marsokhod, PrOP-M),
256–257
Mars Sample Return mission (see also 5M,
5NM), 70, 234, 260
Mars, 1960 Soviet probes: see 1M
Mars, 1962 Soviet probes, 28–30
Mars, knowledge of, xxxii–xxxix, 40–44,
83–87, 118–124, 164–167, 230, 233–234,
239, 243–248, 251–253
Mars, missions to, 2–4, 9–11, 12–15, 26,
28–31, 32–33, 34–45, 52, 66–70, 73–87, 88,
97, 89–125, 160–170, 216–261, 308, 398
Mars, Olympus Mons, 82, 116, 118–120, 248,
252
Mars, search for life, 14, 15, 28, 44, 67,
223–225, 239–243
Mars, Valles Marineris, 116, 120, 243–245,
252
Marshall Space Flight Center, 11, 69,
304–305
Marsokhod (see also PrOP-M), 168

Massachusetts Institute of Technology
(MIT), 19, 52, 242
Masursky, H., 285
Maunder, E.W., xxxv
Maxwell, J.C., xliv
McLaughlin, D.B., xxxix
Menzel, D.H., xxxi
Mercury, 'moon accident', 187
Mercury, Caloris basin, 186–187, 193
Mercury, knowledge of, xxv–xxvii, xxix,
lv–lvi, 171–172, 186–187, 190–196
Mercury, missions to, 11, 89, 97, 127–196
Mercury, US manned program, 9
MESO (Mercury Sonde), 89, 172, 262
Metis (Jupiter satellite), 330, 350
Meudon observatory, xxvi, xxvii, xxxv, 88,
162
Meyer, S., xxxix
Miller, S.L., 335
Mimas (Saturn satellite), xliv, 147, 371, 372,
380, 387, 389, 397, 408–409
Mini Sniffer, 258
Minivator, 67
Minovitch, M.A., 172, 302
Miranda (Uranus satellite), xlvi, 401, 402,
404, 406, 407, 412, 414–416, 417
Molniya: see 8K78
Montgolfier J.M. and J.E., 291
Morabito, L., 340
Mount Palomar observatory, 5
Mount Wilson observatory, xxxi, xxxvii
Multivator, 67
Murray, B.C., 366
Mutch, T.A., 254–256

N-1 launcher, 78, 167, 168
Naiad (Neptune satellite), 428
Nançay Radioastronomy Station, 101
National Academy of Sciences, 38, 97, 126,
206, 225, 259, 307
Navigator program, 11, 31, 126, 301
'Nedelin catastrophe', 15
Neptune, knowledge of, xlviii–l, li, lv, 423,
426, 428, 437–439
Neptune, missions to, 91, 151, 302–303,
305–308, 310–311, 313, 322, 355, 382, 402,
404, 420, 422–441
Neptune, rings, 424–425, 426–430, 431, 432,
435–437

Nereid (Neptune satellite), l, 423–424, 426, 430, 437
NERVA (Nuclear Engine for Rocket Vehicle Applications), 308
New Horizons, 441
Nicholson, S.B., xxxv
Night Probe: See Pioneer Venus Multiprobe
Nix Olympica: see Mars, Olympus Mons
Nixon, R.M., 307
North Probe: See Pioneer Venus Multiprobe
Nova Cygni 1992, 449

OAO 2 satellite, 206
Oberon (Uranus satellite), xlvi, 401, 405, 406, 408, 412, 415
Olbers, H.W.M., lii
Oort cloud, lvi
Oort, J.H., lvi
Ophelia (Uranus satellite), 405, 407
Opik, E.J., 43
Orbital Quarantine Facility, 260
Out-of-Ecliptic missions, 155–156, 307
Oyama, V.I., 243

PAET (Planetary Atmosphere Experiment Test), 262
Pallene (Saturn satellite), 397
Pan (Saturn satellite), 397
Pandora (Saturn satellite), 366
Parker, E., 443
Parkes Radiotelescope, 400, 422
PEPP (Planetary Entry Parachute Program), 69, 226
Perdita (Uranus satellite), 407
Perseids meteor shower, liv
Pettit, E., xxxv
Phobos (Mars satellite), xxxiii, 29, 38, 82–83, 112, 116, 118, 121, 239, 250–251, 330
Phobos, Stickney crater, 116, 251
Phoebe (Saturn satellite), xlv, l, 147, 376, 383–384, 393–394, 397
Piazzi, G., lii
Pic du Midi observatory, xxvii, xxxii, xlii, l
Pickering, W.H. (19th century astronomer), xxxiv, xli, xlv, 42
Pickering, W.H. (JPL director), 42
Pimentel, G., xxxvii, 78
Pioneer 4, 2, 9
Pioneer 5, 5–9

Pioneer 6, 49–52
Pioneer 7, 49–51
Pioneer 8, 49–51
Pioneer 9, 49–51, 133
Pioneer 10, 131, 132–140, 141, 142, 143, 145, 151–154, 155, 204, 309, 350, 446, 449
Pioneer 11, 131, 140–153, 154, 155, 204, 280, 366, 368, 374, 376, 379, 382, 389, 451, 468
Pioneer 12: see Pioneer Venus Orbiter
Pioneer 13: see Pioneer Venus Multiprobe
Pioneer Anomaly, 153, 154
Pioneer E, 49
Pioneer F: see Pioneer 10
Pioneer G: see Pioneer 11
Pioneer H, 155–156
Pioneer Jupiter probes, 125–155, 207, 221, 265, 305, 309, 310, 313, 323, 325, 326, 330, 336, 343, 345, 347, 359
Pioneer Mars Orbiter, 259–260
Pioneer Outer Planets Orbiter, 156
Pioneer solar probes, 48–52, 57, 125, 128, 154, 198
Pioneer Venus 1: see Pioneer Venus Orbiter
Pioneer Venus 2: see Pioneer Venus Multiprobe
Pioneer Venus Multiprobe, 266–268, 272–273, 274–276, 277
Pioneer Venus Orbiter, 260, 264–266, 271, 272, 273–274, 279–284, 285, 290
Pioneer Venus program, 259, 262–268, 320, 326
Planetary Society, 154
Planet-X, lvi, 437
Pluto, knowledge of, l–li, lvi, 425, 434, 441
Pluto, missions to, 91, 302, 305–308, 310, 444
Portia (Uranus satellite), 404
PR (Pyrolytic Release) experiment, 225, 240, 241, 243
Proctor, R.A., xxxiii, xl
Prognoz 7 satellite, 277
Prometheus (Saturn satellite), 366, 397
PrOP-M 'Marsokhodik', 105–108, 168
Protazanov, Ya., xxxv
Proteus (Neptune satellite), 426, 430437
Proton launcher: see 8K82K
Puck (Uranus satellite), 404, 407, 413
'Purple Pigeons', 256

Quill spy satellite, 266

R-11A-MV sounding rocket, 14
R-16 missile, 15
R-7: see 8K71
Radio-occultation technique, 38, 40, 58,
  64–65, 78, 81, 83, 116, 122, 124, 137–138,
  144, 148, 164, 165, 174–176, 180, 181, 186,
  215, 307, 310, 331, 334–337, 356–359, 364,
  368–369, 372, 389–391, 399, 409–410, 423,
  425, 426, 431, 434
RAE 2 lunar orbiter, 204, 376–377
Ramsey, W.R., xl
Ranger 7, 18
Ranger US lunar probes, 18, 19, 35, 126, 186,
  223–225
Redstone missile, 1
Reed, D.R., 258
Reese, E.J., xli
Reinmuth, K.W., liii
Rhea (Saturn satellite), xliv, 147, 151, 364,
  368, 372, 374–376, 382, 383, 397, 399
Riccioli, G., xxix
Richardson, R.S., xxxi
Ristenpart, F., xliii
Roemer, O., xl
Rosalind (Uranus satellite), 404, 407
Ross, F.E., xxxi
RTG (Radioisotope Thermal Generator), 66,
  68, 69, 125, 126, 128, 132, 140, 141, 153,
  154, 156, 220, 221, 231, 257, 259, 293, 301,
  305, 313, 446, 451

Sagan, C., 19, 131, 315, 443
Sagdeev, R.Z., 163
Saheki, T., xxxix
Salzman, L., 131
Sandia Laboratories, 259
Saturn 1B launcher, 66
Saturn launcher, 11
Saturn V – Intermediate 20 launcher,
  304–305
Saturn V launcher, 68, 69, 76, 218, 301, 305
Saturn, knowledge of, xliii–xlv, 147–151, 372,
  376–379, 394–398
Saturn, missions to, 91, 141, 145–151, 156,
  256, 302–303, 305–311, 313, 331, 345, 355,
  363–398
Saturn, rings (see also Cassini Division,
  Encke Division, Keeler Gap), xliii–xlv,
  146, 147, 159, 150, 307, 310, 364–366, 372,

373, 374, 376, 377, 379–382, 383, 387,
  388–390, 391, 392, 395–397, 409, 431
Schaer, E., 150
Schiaparelli, G.V., xxvi, xxvii, xxx,
  xxxiii–xxxiv, xxxv, xlvi, liv, 82, 245
Schouette, C.H., lvi
Schröter, J., xxx
Scout C launcher, 52
Secchi, A., xxxiv, xlvi
Selivanov, A.S., 214
Semyorka: see 8K71, 8K72, 8K78
Seneca, L.A., liii–liv
SERT 2 satellite, 208
SETI (Search for Extraterrestrial
  Intelligence), 154
Shergottites-Nakhlites-Chassignites (SNC)
  meteorites, 233–234
Shklovsky, I.S., xli
Showalter, M.R., 397
Sinope (Jupiter satellite), xli, 134, 143, 326
Sinton , W.M., xxxvii
Sinton absorption bands, xxxvii, 14, 28, 78
SIRIO satellite, 263
Sisyphus asteroid-meteoroid detector,
  130–131, 133–134, 137, 145
Skylab, 206
Slipher, E.C., xxxv
Slipher, V.M., xxxv, xlv
Soderblom, L.A., 415
Soffen, G.A., 256
Solar Monitor, 91
Solar probes, lv, 11, 48–52, 91, 97, 125, 196–
  206
Solar wind, discovery of, 2, 22, 443
Solwind satellite (P78-1), 204, 205
SOREL (Solar Orbiting Relativity
  Experiment), 206
Soyuz 4, 72
Soyuz 5, 72
Soyuz Soviet manned program, 168, 169
Space Science Board of the National
  Academy of Sciences, 38, 225, 259, 307,
  309
Space Shuttle, 226, 258, 260, 285, 305, 306,
  307–308, 310, 322, 419, 440, 450
Space Shuttle flight STS-51L, 420, 441
Spacelab, 285
Spinrad, H., xxxvii
Sputnik satellites, 1, 2, 4, 5, 12

SR-71 spyplane, 238
STAR (Self-Test And Repair) computer, 307, 315
Starlite, 91
STEREO experiment, 101, 113, 123, 162, 165
Stewart, H.J., 303
Stickney, A., 116
Stone, E.C., 443
Strughold, H., xxxvii
Sunblazer, 52
Surveyor US lunar landers, 132, 220

Taurid meteor shower, 30–31
Telesto (Saturn satellite), 387
Tethys (Saturn satellite), xliv, 147, 368, 371, 372, 376, 387–388, 389, 391, 392, 397
TETR satellite, 49
Thalassa (Neptune satellite), 428
Thebe (Jupiter Satellite), 330, 350
Thor-Able launcher, 2, 7
Thor-Delta launcher, 49
Tikhov, G.A., xxxvii, 14
Titan (Saturn satellite), xliv, xlv, 146, 149–150, 151, 256, 310, 311, 363, 364, 366, 368–370, 372, 382, 383, 384, 385, 387, 389, 398, 424
Titan III launcher (see also Titan IIIE-Centaur), 66, 172, 218, 304, 305, 310
Titan IIIE-Centaur, 155, 200, 219, 226, 227, 228, 318, 320
Titania (Uranus satellite), xlvi, 401, 405, 408, 412–413, 414
Titius, J.D., xlviii
Tolstoy, A.N., xxxv
Tombaugh, C.W., l–li, lvi, 42
TOPS (Thermoelectric Outer Planet Spacecraft), 306–308, 309, 315
Tournesol satellite (D2A), 162
Triton (Neptune satellite), xlix–l, li, 423–424, 425, 426, 427, 431, 432–424, 437
Tsander, F.A., xxxv
TsNIIMash, 172
Tycho Brahe, xxxii, liv

Umbriel (Uranus satellite), xliv, 401, 406, 407, 408, 413, 415
UMVL probe, 170
Uranus, knowledge of, xlv–xlviii, l, lv, 400–402, 403–404, 405, 410–412

Uranus, missions to, 146, 151, 156, 302–303, 306–308, 310–311, 313, 315, 322, 355, 382, 389, 393, 398–420
Uranus, rings, 330, 399, 401, 403, 407–408, 409–410, 417–418, 431
Urey, H.C., xxxvii, 334
Usuda Radiotelescope, 423

V-2 missile, 2, 5
V-67 Venus mission: see Venera 4, Kosmos 167
V-69 Venus mission: see Venera 5, 6
V-70 Venus mission: see Venera 7, Kosmos 359
V-72 Venus mission: see Venera 8, Kosmos 482
Valles Marineris: see Mars, Valles Marineris
Vanguard satellite and launcher, 1
Vega, Soviet Venus-Halley probe, 289, 293
Vega, US launcher: see Atlas-Vega
Vega, US probe, 9
Vela satellites, 202
Venera 1, 16–17, 18, 46
Venera 2, 45–46, 61, 181
Venera 3, 45–46
Venera 4, 59, 61–63, 64, 65, 70, 72, 73, 262, 289, 290, 292
Venera 5, 72–73
Venera 6, 72–73
Venera 7, 98, 156, 159
Venera 8, 157–160, 211, 214
Venera 9, 212–216, 270, 271, 272, 291
Venera 10, 212–213, 214–216, 270, 271, 272, 291, 293
Venera 11, 273, 276–278
Venera 12, 273, 276–278, 288
Venera 13, 285–289
Venera 14, 285–289
Venus Orbiter, 262, 266
Venus Radar Mapper: see Magellan
Venus rover, 293
Venus, 1962 Soviet probes, 26–28
Venus, knowledge of, xxvii–xxxii, 17–18, 24–25, 58–59, 61–65, 73, 98, 124, 157–160, 181–182, 214–216, 266, 275–277, 279–281, 284
Venus, Maxwell Montes, 266, 281
Venus, missions to, 2, 5–6, 9–11, 12–14, 15–17, 18–28, 30–31, 31–32, 45–46, 52–56,

58–65, 66, 70–73, 88, 97–98, 156–160, 170, 172–174, 180–182, 209–216, 256, 262–293, 368

Venus, radar observations, 17–18, 58–59, 215–216, 266, 280–281, 287

Vetchinkin, V.P., 465

Viking 1, 228–234, 235, 238, 239, 241, 242, 243, 250–251, 254, 256

Viking 2, 202, 228, 229, 234–239, 240, 241, 243, 250, 251–253, 254, 256

'Viking 3', 256

Viking A: see Viking 2

Viking B: see Viking 1

Viking Dynamic Simulator satellite, 226

Viking Lander, 220–226

Viking Lander 1 see Viking 1

Viking Lander 2 see Viking 2

Viking Mars mission, 118, 125, 160, 167, 170, 200, 202, 211, 218–256, 258, 259, 260, 307, 309, 399

Viking Orbiter, 219–220

Viking Orbiter 1: see Viking 1

Viking Orbiter 2: see Viking 2

Viking site selection, 125, 226–227, 229, 234–235

Vinogradov, A.P., 168

Vishniac, W.V., 67

Vision for Space Exploration, 450

VLA (Very Large Array), 422–423

VNII Transmash, 105, 108, 168, 272, 293

Vogel, H.C., xxxv

von Braun, W., 2–4, 11

Voyager 1, 204, 322, 347, 348, 356, 359, 383, 385, 387, 389, 390, 395, 398, 399, 400, 422, 430, 441, 443, 446, 447, 449, 450–451

Voyager 1 Jupiter flyby, 323–346

Voyager 1 launch, 320

Voyager 1 Saturn flyby, 363–382

Voyager 2, 204, 311, 326, 328, 330, 368, 372, 376, 380, 443, 446, 447, 449, 450, 451, 468

Voyager 2 Jupiter flyby, 346–363

Voyager 2 launch, 318–320

Voyager 2 Neptune flyby, 422–441

Voyager 2 receiver problem, 322–323, 347, 354, 393–394, 400, 426, 447

Voyager 2 Saturn flyby, 382–398

Voyager 2 Uranus flyby, 151, 398–420, 422

Voyager, Mars and Venus probes, 11, 34, 65–70, 76, 108, 125, 216, 217, 225, 289, 301, 305

Voyager, outer solar system probes, 141, 143, 146, 151, 153, 188, 254, 309–311

Voyager Interstellar mission, 441–451

Voyager scientific instruments, 315–318

Voyager spacecraft, 311–315

VPM-73 (Visual Polarimeter-Mars), 162, 164, 210

Vsekhsvyatskii, S.K., 330

Vulcan (hypothetical planet), lv–lvi

Webb, J.E., 28

Wells, H.G., xxxv

Whipple. F.L., xxxi, lv, lvi

Wildt, R., xl, xlv, xlvi, 139

Witt, K.G., liii

Wolf Trap, 67

Wolf, M., liii

Yangel, M.K., 15

Yevpatoria deep space communication center, 14, 16, 30, 46

Young, C.A., xxxv

Zond 1, 31–32, 46

Zond 2, 32–33, 38, 45

Zond 3, 45, 46

Zond: see 3MV

Zupi, G.B., xxv

Zwicky, F., 5

Printing: Mercedes-Druck, Berlin
Binding: Stein+Lehmann, Berlin